Multimedia Communications and Video Coding

Multimedia Communications and Video Coding

Edited by

Yao Wang, Shivendra Panwar, Seung-Pil Kim, and Henry L. Bertoni

Polytechnic University
Brooklyn, New York

Springer Science+Business Media, LLC

Library of Congress Cataloging-in-Publication Data

Multimedia communications and video coding / edited by Yao Wang ...
[et al.].
        p.   cm.
    "Proceedings of a symposium ... held October 11-13, 1995, at
Polytechnic University, Brooklyn, New York"--T.p. verso.
    Includes bibliographical references and index.
    ISBN 978-1-4613-8036-8   ISBN 978-1-4613-0403-6 (eBook)
    DOI 10.1007/ 978-1-4613-0403-6

    1. Multimedia systems--Congresses. 2. Digital video--Congresses.
3. Coding theory--Congresses.   I. Wang, Yao, 1962-   .
QA76.575.M78   1996
006.6--dc20   96-21634
                                                         CIP

Proceedings of a symposium on Multimedia Communications and Video Coding,
held October 11 – 13, 1995, at Polytechnic University, Brooklyn, New York

ISBN 978-1-4613-8036-8

© 1996 Springer Science+Business Media New York
Originally published by Plenum Press, New York in 1996
Softcover reprint of the hardcover 1st edition 1996

PREFACE

This book constitutes the proceedings of the *International Symposium on Multimedia Communications and Video Coding* (ISMCVC95) held October 11 – 13, 1995, at the Polytechnic University in Brooklyn, New York. This Symposium was organized under the auspices of the New York State funded Center for Advanced Technology in Telecommunications (CATT), in cooperation with the Communications Society and the Signal Processing Society of the Institute of Electrical and Electronic Engineers (IEEE).

In preparing this book, we have summarized the topics presented in various sessions of the Symposium, including the keynote addresses, the Service Provider and Vendor Session, the Panel Discussion, as well as the twelve Technical Sessions. This summary is presented in the Introduction. Full papers submitted by the presenters are organized into eleven chapters, divided into three parts. Part I focuses on systems issues in multimedia communications. Part II concentrates on video coding algorithms. Part III discusses the interplay between video coding and network control for video delivery over various channels.

The ISMCVC95 is the fifth in a series of recent symposiums organized by the Electrical Engineering faculty of Polytechnic University to highlight topical areas of current research interest. Previously, four symposiums have been organized under the auspices of the Weber Research Institute, and dealt with:

> Directions in Electromagnetic Wave Modeling; October, 1990
> Ultra Broadband, Short Pulse Electromagnetics, I; October 1992
> Ultra Broadband, Short Pulse Electromagnetics, II; May 1994
> Integrated Optics: Devices and Modeling; October 1994

Two telecommunication workshops were organized under the auspices of CATT and in collaboration with NYNEX. They were titled:

> Network Management and Control Workshop; September, 1989
> Second Network Management and Control Workshop; September, 1993

Proceedings of the above conferences and workshops were published in a format similar to this volume by Plenum Press.

The organizers would like to thank the authors and invited speakers in the technical and special sessions, as well as all the reviewers, for making the Symposium an exciting look into the future of multimedia communication. We would also like to thank Ameena Mustafa and Carletta Lino for their professional assistance in the conference administration.

Yao Wang
Shivendra Panwar
Brooklyn, New York   Seung-Pil Kim
Henry L. Bertoni

CONTENTS

INTRODUCTION
— SCANNING THE SYMPOSIUM

The Editors

In his revolutionary experiments of 1895, Marconi achieved transmission of coded radio signals over distances up to 2 km. Six years later, he succeeded in the first transmission of telegraphic signals across the Atlantic from Cornwall to Newfoundland. The century that followed has seen the introduction of many new electronic communications services. Broadcast radio and television have dramatically changed our perception and understanding of the world we live in. As we enter the second century, another exciting communication service known as multimedia is emerging, which involves integrated transmission of text, graphics, voice, image, video, and computer data, between individuals, as well as in the broadcast mode.

The International Symposium on Multimedia Communications and Video Coding (ISM-CVC95) was organized to celebrate the 100th anniversary of Marconi's remarkable achievement, and to foster the exchange of recent results and new ideas among researchers in various fields of multimedia technology, especially telecommunications and video coding. It was also intended to review the status of commercial developments in multimedia communications systems, and to identify technological issues that are key to the future development of the field. The conference consisted of an Opening Address, two Keynote Addresses, a Service Provider and Vendor Session, a Panel Discussion on Environment for New Media Services, and twelve Technical Sessions.

The Symposium opened with a review of the life and achievements of Guglielmo Marconi by his daughter Gioia Marconi Braga, who is the Founder and Chairperson of the Marconi International Fellowship Council. On this 100th anniversary of Marconi's first successful transmission, it is important for us to note how his pioneering work in radio has led to the era of instant global communications and how our lives depend on the fruits of his invention. Starting with the sinking of the Titanic, where reception of radio distress signals saved 750 lives, electromagnetic communications has become a bridge between the individual and the larger world; be it via radio, television or fiber optics.

While his earliest work was done in Italy, it was in England that Marconi received the funding needed to bring his vision of communication across the Atlantic to fruition. It was on December 12, 1901 that he succeeded in transmitting Morse Code from Poldhu in Cornwall to St. John's, Newfoundland. As a businessman he presided over the Marconi Company, and built a network of wireless telegraph stations to provide worldwide communications. Prior to his death he was working on an early form of radar. He died in 1937, and in tribute to his

Multimedia Communications and Video Coding
Edited by Y. Wang *et al.*, Plenum Press, New York, 1996

memory, radio stations all over the world observed two minutes of silence. As we go forward into the second century of telecommunications, it seems inconceivable that there might ever again be period of worldwide radio silence.

Following these remarks, Robert W. Lucky of Bellcore, a recipient of the Marconi International Fellowship Award, delivered the Keynote Address entitled "The Prospect for Infinite Bandwidth." Fiber optic technology has made it possible to provide subscribers with essentially unlimited bandwidth for communication. How this technology be deployed will depend on potential applications, and on the willingness of users to pay. While video to the home can justify high capacity in one direction, use of the Internet may be the driver for high capacity connections both to and from the subscriber. Currently subscribers are constructing Web pages for others to browse. At one level, people establish pages out of the need or desire to gain recognition. On another level, new ways of conducting business are developing based on high capacity Internet connections. This approach may have distinct advantages for small companies, since larger and more established companies may have no distinction in this medium. As the joke goes, on the Internet no one knows you are a dog.

The economics of providing subscribers with high capacity two way connections remains in question. The public will pay up front for the purchase of a computer, so that its manufacture recoups its investment in a short time. However, the public expects the infrastructure cost of connecting the computer to the telecommunication network will be recovered over a period of many years. Even the willingness of subscribers to pay their share of the infrastructure cost up front may not be adequate from the point of view of the telecommunication industry.

Curtis R. Carlson of David Sarnoff Research Center delivered the second keynote address of the symposium, entitled "21st Century Communications: Vividness and Control". He made a strong case for why vividness and control are the keys to any successful information service. By vividness he means information matched to the human senses, and control implies portability, choice and interactivity. He gave several examples where advances in vividness or control or both were important in the successful introduction of a new product or service. These include the increase in vividness in broadcast services, starting with radio to black and white television to color television, but which have limited control. At the other extreme are services such as the World Wide Web, which offers some vividness, but complete personal control. Other examples of current and near future products that he predicted will be successes include Hughes' DirecTV, head-mounted displays, High Definition TV (HDTV), personalized advertising and video servers. Dr. Carlson closed by strongly urging that industry quickly converge on standards to make some of these promising new products commercially successful.

The *Service Provider and Vendor Session* was chaired by Ivan Frisch from Polytechnic University. The first speaker was John Sie, and his talk's title was, "Video on Demand: Killer App or Siren's Song ?" He made the case for why switched video service from telephone companies made no economic sense. His analysis is based on a revenue projection derived from the market demand. He shows that the revenue from switched video services can only cover a fraction of the expected price of the bandwidth and associated costs needed to deliver these services. Thus telephone companies that went ahead and introduced video–on–demand services would either have to subsidize this service using profits from other services, or would be driven to bankruptcy! On the other hand, he believes that near–video–on–demand can be profitable if delivered by cable or over the air. The details of his analysis are presented in the first paper in Chapter 1.

Kingsley Nelson of NYNEX spoke on the subject of "The Multimedia Challenge." He outlined the transition his company is facing as it moves from traditional telephone service to all kinds of information services. He echoed Curtis Carlson's theme of the customers' need for control over the services he or she gets. Services would have to be tailored to the needs of customers in the future. At home, there will be a convergence of TVs and PCs into a single appliance. Access to information services could be by wireless, cable TV or twisted-pair wire, though NYNEX was investing heavily on one approach: Multichannel Multipoint Distribution Service (MMDS). This wireless technology can deliver 100–150 video channels, and would be introduced by NYNEX during 1996. In this new era, companies like NYNEX were forming alliances with other content, software, hardware and service providers. He outlined joint ventures with Pacific Telesis, Bell Atlantic, Viacom and CAI. Cable companies could prove tough competitors to NYNEX in offering these services, but were burdened by debt.

Professor B. Gopinath of Rutgers University presented his views on "Operations and Business Support System Issues in Multimedia Services." He believes that the time and cost in developing operations and business support systems, such as network management, call processing and billing, was the main road block in the introduction of multimedia services. Currently the cost of a single line of code is $4 per month. This was in contrast to the recent progress and cost performance improvements made in other areas such as video coding, hardware, network optimization and algorithms. At the same time, services are getting more specialized and complex, and thus economies of scale are no longer possible. In his opinion, the challenges in network management have moved from network-centered algorithms to meeting the needs of customers. He believes that new approaches and research are needed in this area.

Richard Baker of PictureTel made the point of his presentation, "Being Everywhere, Now: Technology Trends in Desktop Visual Communications," in the way he presented it. He used a PictureTel desktop videoconferencing system, projected on the auditorium screen, to make a remote presentation from an office in the Boston area. A 128-Kbps ISDN line was used to make the connection. He explained how desktop video-conferencing differed from room-based video-conferencing, in terms of its significantly lower cost, use of standard PCs, ISDN and LAN/WAN connections. Besides the technology, he also described the cultural impact of desktop video-conferencing. He believed that it was a complement to, and not a replacement of, face-to-face meetings. It made many interactions possible which were not possible earlier. Application sharing on PCs was one such new possibility. Making the analogy to the introduction of copying and fax machines, he believes that many of the uses that this technology would be put to would be unexpected. His experience over the year that he has been using desktop conferencing was that it liberated him from many time and space constraints in his work.

The *Panel Discussion on The Environment for New Media Services* was moderated by Walter Johnston, NYNEX Science and Technology. Each of the panelists made a short presentation followed by questions from the audience. Eli Noam, from Columbia University, made a case for his views on two controversial issues. The first issue he discussed was the mergers in the media industry. Examples of these were the merger of Westinghouse with CBS, Time-Warner with TBS, and ABC with Disney. He claimed that these would be failures, since they made little business sense. The breakup of AT&T was the correct model, since the trend was in favor of smaller companies that specialized in various technologies and services. System integrators could then efficiently offer services using the best available technology. His other comments were on the spectrum auction being conducted by the FCC. He thought it represented a dis-investment by the Federal Government, which uses the spectrum as a cash cow. The public would pay for it in the future in the form of higher prices, and small companies would be prevented from entering the market because of the high initial expense. Prof. Noam then proposed an alternative approach with a usage-based pricing structure.

Lou Schulyer, from McGraw Hill Publishing, outlined the future strategy of content providers in the multimedia business. For ordinary consumers, due to the lack of affordable access to on-line services, the traditional approach using books, magazines, etc., will continue to dominate. For business users, who can afford the access fees, the most important considerations are issues like security and reliability. The content providers are currently only dipping their toes in the multimedia business, and wrestling with how to present content in different media. Barry Haskell, of AT&T Bell Laboratories, reviewed some future issues in multimedia. These include a need for tools for multimedia specification and creation. In the area of protecting intellectual property rights, he saw the need for one-time run software and copy protection. Scott King, from Hewlett Packard, stressed the importance of giving control and interactivity to users in any future service offering. Hewlett Packard believes that the market would grow in an evolutionary manner, with the formation of consortia, and with the packaging and branding of products and services becoming increasingly important. Ifay Chang, from PRIDE at Polytechnic University, focused on one particular multimedia application: Cyberspace Assisted Responsive Education (CARE). Technology would lead to a paradigm shift in how education is delivered to students. Thus distance and schedules would no longer be barriers to learning in distributed teaching. He went on to discuss the infrastructure and technology needed to make this vision possible.

One of the questions asked of the panelists was what would be the "killer app" that would drive the development and deployment of multimedia technologies. The usual suspects were mentioned: video-on-demand, Internet access, don't know, but the most interesting answer was from Scott King. Citing the success of the Home Shopping Network, he said that whatever the "killer app" would be, it would not be anything the members of the audience could think of as being a potential success. It was a wry comment on how far engineers are from mainstream culture.

Papers presented in twelve technical sessions are reorganized into eleven chapters, separated into three parts. Part One focuses on systems issues in multimedia communications, Part Two concentrates on video coding algorithms, and Part Three discusses the interplay between video coding and network control for video delivery over various channels. The topics presented in these papers are summarized below.

Part I includes four chapters. Chapter 1 includes nine papers discussing multimedia network architectures, implementations and standards. The first paper, by Sie, is based on his talk in the *Service Provider and Vendor Session*, as discussed previously. The next paper, by Raychaudhuri, provides a useful overview of the issues in ATM based multimedia networking. He describes an architecture to support a range of distributed multimedia networking scenarios. Asthana *et al.* describe Kaleido, an integrated multimedia system designed for multimedia desktop and multimedia server applications. Unlike present computers, the hardware and software in Kaleido are specifically designed to process and control real-time multimedia streams. Nishimura *et al.* have developed a Personal Multimedia–Multipoint Tele-Conference (PMTC) system. It is an integrated Groupware system using current Narrowband ISDN technology. Gradinariu and Prevot describe an experimental setup for distance learning. A TCP/IP network using protocols such as the Real time Transport Protocol (RTP), the Internet Multicast Backbone (Mbone) and the Reliable Multicast Protocol (RMP) was used for this purpose. Bugos and Deng study the transmission of MPEG-1 video streams over a 10 Mb/s Ethernet LAN. They show that up to six MPEG-1 video streams with light background traffic can be supported by an Ethernet. Liu describes an approach to video packetization, packet loss detection and error concealment, again for MPEG-1 video streams in a LAN environment. The Telecommunication Information Networking Architecture (TINA) is an open architecture that unifies the Telecommunication Management Network (TMN) and the Intelligent Network (IN) architectures. In his paper, Verma discusses the suitability of this architecture for video dial-tone and other interactive services. In the last paper in this section, Liu and Zhang describe the implementation of a multiparty communication application called the Multiway Talk Protocol (MTP). It is intended for use in a Motif/X window environment, with the UNIX operating system and the TCP/IP protocol suite.

Chapter 2 concentrates on system level performance issues in multimedia networks. The first four papers deal with the flow control of multimedia traffic streams. The paper by Knightly and Rossaro studies the tradeoff between the delay induced by smoothing VBR video sources close to the source and the resulting reduction in delay within the network. Thompson *et al.* propose using the smoothed Wigner-Ville distribution to detect bursty traffic, and then control it to improve multiplexed traffic performance. Jung and Meditch have devised a dynamic flow control scheme based on a new traffic descriptor based on the characterization of source underload and overload states. Performance gains over a static scheme are demonstrated. Campbell *et al.* propose a dynamic Quality of Service (QoS) management scheme for multi-layered coded flows. The paper also describes it's implementation on an experimental ATM network. The next paper, by Li and Crowther, describes a dynamic routing protocol, based on the ST-II Protocol, to support multimedia traffic in TCP/IP networks. Experimental results from a laboratory test environment are also presented. The paper by Ryoo and Panwar presents a video-on-demand network design problem and a numerical solution technique. A tree topology such as those used in hybrid fiber-coaxial cable networks is assumed. The final paper in this section, by Apteker *et al.* argues that satisfying the end uset receptivity should be the ultimate aim of a multimedia system, rather than technical QoS measures. A framework for using both end-user receptivity and technical QoS measures for system design is proposed.

Chapter 3 includes three papers that deal with the problem of synchronizing real-time multimedia traffic. The first paper, by Zuniga and Feig, describes the implementation of a frame dropping strategy in workstations for a software based MPEG decoder. By avoiding dropping I and P frames, and by using other enhancements, they could improve on existing software decoders. Liu *et al.* describe an adaptive synchronization method to deal with network changes. Results from an experimental network are presented. Cai and Pantziou introduce a synchronization model that admits both strict and weak synchronization models. Thus a two level model can be defined for different multimedia objects, where, for example, strict synchronization may be needed between images and text, but only week synchronization is required between images and audio.

Chapter 4 discusses issues related to video server architecture and hardware implementation. Video server research involves a variety of issues such as video storage, scheduling, communication protocol, hardware architecture and user demand, etc. Eleftheriadis introduced the efforts of the Digital Audio-Visual Council (DAVIC), established in June, 1994. The applications targeted for support in the current DAVIC specification include movie-on-demand, teleshopping, broadcast, near-video-on-demand, delayed broadcast, games, telework, and karaoke-on-demand. Chen *et al.* presented a video server based on the push-and-pull (PP) model, which is suitable for business video applications. Tetzlaff and Flynn discussed the problem of video server block allocation and scheduling. Dubey *et al.* considered the movie scheduling problem in a VOD system. They proposed a two layered scheme consisting of master and local servers for increased performance through local caching, which improved availability of a movie at the local server site. Bhat provided an analysis of the AT&T multimedia communication (MMC) server for server-centric and peer-to-peer scenarios based on a queuing network model. Finally, Gonzales and Linzer discussed the VLSI (Very Large Scale Integrated Circuits) implementation of video coders, which is important for efficient server implementation. They analyzed architectural trade-offs for implementing video encoders based on two different concepts — a performance scalable architecture (PSA) and a functional scalable architecture (FSA).

Part II includes four chapters. Chapter 5 present recent works in model-based and object-oriented video coding. The first paper by Huang and Tang reviewed major components in a model based coder for talking head sequences, including 3D face modeling, facial motion analysis, and facial expression synthesis, and identified several remaining challenging issues. The next five papers dealt with object-oriented coding, also known as content-based coding. Katsaggelos and Ozcelik proposed to perform spatial and temporal prediction of the motion and segmentation field, and only code unpredictable parts. Gu *et al.* described algorithms for moving object segmentation and tracking, using morphological operations in both spatial and temporal domains. Ostermann studied the problem of detecting moving objects in front of a static background. The algorithm makes use of temporal variations in edge and contrast, in addition to luminance, and is able to separate moving shadows from moving objects. The last two papers both use a 2D deformable mesh for object description. Altunbasak and Tekalp focused on mesh generation. They described an algorithm for occlusion detection and compared two methods for occlusion-adaptive mesh design. They also considered the mesh design problem when the object boundary is given. Wang *et al.* presented an object-oriented coder, which describes the shape and motion of each object by the initial positions and inter-frame displacements of the nodes in the object region, respectively.

Chapter 6 discusses the interaction between audio and visual signals in human communications and how to make use of this interaction in designing efficient communication systems. The first two papers dealt with speech-reading (more commonly known as lip-reading). Petajan and Graf reviewed the past research related to automatic lip reading and described recent work in facial image analysis required for mouth movement parameterization. Benoit presented results of speech reading tests accompanied by natural and synthetic faces. It was found that speech is better understood when presented with a natural face and that lip movement is an important visual cue. The next paper by Jacquin described the results of a study evaluating the perceptual qualities of the H.261 coder and a model-assisted H.261 coder, which uses a lower quantization parameter for the facial region. The latter was rated higher, mainly because it produced better eye contact and lip synchronization. The last two papers considered the use of speech information to improve video coding efficiency. Lavagetto described a way of using speech to improve motion compensation accuracy in the mouth region. Rao and Chen proposed a coder which uses the speech signal to predict the mouth shape, and compared two alternatives for designing the speech classifier (clustering audio first vs. classifying video first). They also described an algorithm for analyzing mouth shape parameters.

Chapter 7 includes five papers on motion estimation and other issues in video coding. Leonardi and Iocco presented a scheme which tracks motion using a constant acceleration model rather than a conventional constant velocity model. Chen and Wilson proposed an adaptive block-matching algorithm for estimating large displacements. It reduces the search complexity to $O(\log n)$ by utilizing motion correlation in the video sequence. Hsu and Van Dyck presented a pruned tree-structured Vector Quantization (VQ) technique for compression of high subbands in subband coding, which is suitable for video transmission over ATM networks. Chen et $al.$ described a neural network approach for adaptive quantizer design in video coders. Dugelay et $al.$ considered the application of fractal image coding techniques to video and compared two different approaches — an interframe coding method and a cubic approach which extends the 2D block used for images into 3D cubes.

Part II ends with Chapter 8, which discusses new video coding standards and other relatively new problems in video coding. In the first paper, Puri described the MPEG-4 standardization effort with details of MPEG-4 requirements, test methods, and chronological development plans. Due to the increased functionality requirements of MPEG-4, a flexible syntax structure is required. As a result, MPEG-4 Syntax Description Language (MSDL) layer has been introduced between the channel coding and presentation layers. Wiegand et $al.$ introduced an approach for rate control in H.263 coder, which is based on coding mode selection for group-of-blocks, instead of simply varying quantization step sizes. The optimization is based on the use of a Lagrangian function and a trellis decision tree. Stereoscopic video coding is becoming an important issue as witnessed by the inclusion of the multiview functionality in the MPEG-4 standard. Chiang and Zhang presented an approach which makes use of both motion compensation and disparity compensation, which predicts one channel (e.g., a left channel) from the other using an affine model. Content-based indexing and retrieval in a video database is a new and challenging research area. In the last paper of this chapter, Chang presented efficient ways for performing indexing, retrieval, and editing, directly over compressed bitstreams obtained by transform or subband coders.

Part III includes three chapters. Chapter 9 focuses on the transport of variable bit rate (VBR) video and image streams over ATM networks. The first four papers investigate the delivery of MPEG1 coded video. Heyman et $al.$ considered the statistical modeling of MPEG1 coded video. They compared the accuracies of several models when applied to a 10 minute sequence and evaluated the multiplexing gain over a bandwidth allocation scheme based on the peak rate. Mata and Sallent analyzed the relation between MPEG1 coding parameters, including the length of Group of Pictures (GOP), distance between P-pictures and quantization parameter (QP). They also described the statistical behavior of the generated bit rates under different parameter settings. Hamdi *et al.* developed a relation between the QP and the GOP bit rate, which is then used to maintain the VBR output conformity to ATM traffic contract. Hsu and Ortega introduced the concept of effective buffer size which establishes a link between the channel rate and the encoder/decoder buffer size required to prevent data loss. They described an algorithm that jointly adjusts the encoder and channel rates, to minimize the quantization distortion, subject to a given end-to-end delay constraint. The next two papers dealt with MPEG2 video which permit the use of scalable or layered coding. Chandra and Reibman considered the statistical modeling of MPEG2 video coded using SNR scalability. Specifically, they modeled the bit rate of the enhancement layer while assuming the base layer is constant bit rate (CBR), and examined the statistical multiplexing gain over one layer coding. Luo and El Zarki considered the interplay between leaky bucket control and two layered coding schemes: data partition and SNR scalability. They developed an adaptive layered source coding method which outperforms both the static and CBR layered source coding in terms of video quality and network resource utilization. Finally, the paper by Choi consider the transmission of JPEG coded image signals and described a way of jointly minimizing the rate and distortion fluctuations.

Chapter 10 discusses issues related to wireless communication of video signals. Tzou gave an overview of characteristics of the digital satellite channel and its potential applications, and described several challenging issues. Liu and El Zarki presented schemes for classification (layering) of video data and channel error protection when using H.263 coder over wireless channels. The study of video delivery over wireless channels requires an efficient software tool for performance evaluation. Wong *et al.* presented a software platform for simulating end-to-end wireless visual communication systems.

Chapter 11 focuses on image/video coding and post-processing techniques under lossy channel environments. Vetterli *et al.* described a successive approximation type of source coding technique, which aims to achieve an optimal trade-off between source and channel coding. Girod *et al.* described a scalable coding structure with unequal error protection. Wee *et al.* described a reconstruction-oriented layered source coding scheme incorporating VQ or LOT (Lapped Orthogonal Transform). In their approach, each decoded layer is first enhanced, before it is used to generate the next layer. Hemami proposed to embed in the coded streams optimal interpolation coefficients for recovering damaged image blocks due to channel errors, and compared the reconstruction performances with embedded coefficients vs. a fixed set of coefficients. The last paper in this chapter, by Kwak and Haddad, presented a decoding method for standard DCT coded images, which can reduce blocking artifacts and enhance directional features.

VIDEO ON DEMAND: KILLER APP OR SIREN'S SONG?

John J. Sie

President & CEO
Encore Media Corporation
5445 DTC Parkway, Suite 600
Englewood, CO 80111

Video On Demand (VOD), delivered through a switched digital video, or SDV architecture, is a killer application that the Telcos are touting that would transform the local residential telephone business into a full-service video provider. This paper will attempt to point out some very serious and fundamental incongruencies or flaws of using VOD as a driver to finance the SDV infrastructure as a public network, by using a boundary-value analysis of consumer demand, bandwidth utilization and realistic revenue and cost projections. Once the interrelationship between the pyramids of "activities" and "economic scope" in the food chain of any product development is understood, the importance of boundary-value analysis becomes evident.

FOOD CHAIN OF VIDEO CODING/TELECOMMUNICATIONS

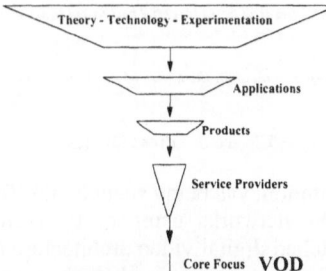

Theory - Technology - Experimentation

Applications

Products

Service Providers

Core Focus VOD

Figure 1. Activities

Figure 1 depicts the activities of the food chain of product development of VOD. First there are hundreds and thousands of activities: theories, technologies and experimentations conducted by universities and industrial laboratories from around the world. These activities are narrowed down to some specific applications like high-speed switching and Gigahertz transmissions, etc. These applications are further narrowed to come up with a system or product like the digital encoder or video file server. Ultimately, a handful of customers or service providers will buy the system, with all of them sharing a core focus -- VOD. It is an inverted pyramid in terms of activities leading to a specific application. However, the economic impact or scope of such a food chain is just the opposite (Figure 2).

FOOD CHAIN OF VIDEO CODING/TELECOMMUNICATIONS

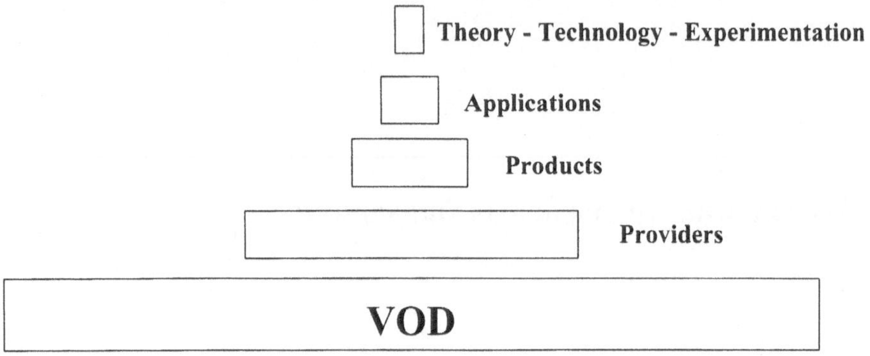

Theory - Technology - Experimentation

Applications

Products

Providers

VOD

Figure 2. Economic Scope

It may seem that collectively a lot of money is being spent doing research, but it is really very small when compared to the billions and billions of dollars of consumer spending needed to justify the VOD infrastructure deployment. In this case, the "economic scope" of the product development food chain is inversely proportional to the "activities" leading to the product. Boundary-value analysis will indicate whether the pyramid of activities is favorably inverted, because if there is no broad-based economic support for the VOD or SDV architecture, the result could be disastrous.

■ **TELCOs are racing ahead in the Global Competition to build the Information Superhighway**

- *"Ameritech increased its budget by $4.4 Billion to $29 Billion . . . to construct 750 MHz broadband networks." '94*

- *"[Ameritech] networks will employ ATM high-speed data switching (SDV)." 2/94*

- *"Milpitas is among the first cities targeted by PacTel in its seven-year, $16 Billion commitment to build a broadband network."*

- *Bell Atlantic Corp. put meat on the bones of its plan . . . for a project worth $11 Billion." 6/94*   Source Multichannel News

Figure 3. Press Quotes

Figure 3 indicates that a lot of money is being spent by the Telcos on SDV deployment. Most Telcos are rebuilding their public networks using costly (somewhere between $1,000.00 to $2,000.00 per line) fiber-rich switched digital video architecture that offers consumer voice, data, video and other multi-media services. But why are they busy deploying the so-called Broadband Networks when most of the interactive services that are envisioned and talked about can be adequately delivered by ISDN and twisted copper pair wire to the home?

The Telcos, through their technologist spokespersons, all point to the Killer Application for SDV . . . Video on Demand (VOD). As different from other more nebulous interactive applications, they would generally all state: 1) Consumer demand for VOD is proven from home video rental and Near Video on Demand provided by DBS services like DirecTV; (2) Other interactive services have no proven demand, nor cost or pricing data; and (3) Competing technologies like cable television, DBS, MMDS, LMDS cannot be transformed into SDV architecture like the Telcos can -- hence the Killer Application--a business that can generate enormous amounts of revenue and profit that its competitors cannot. *Is VOD the Killer App or a Siren's Song?*

Just as it's important with video coding to know what the source material is, to understand the VOD business, you have to know the source -- and the source is movies -- in particular, Hollywood. Therefore, an analysis of the business of movie theaters, home video and pay per view will be both necessary and helpful.

8 Major Studios Release   200 Movies/Year
Total Theatrical Production 450 Movies/Year

Movie Theaters

Total Box Office Revenue	$5.2 Billion
Average Ticket Price	$4.14
Tickets Purchased per Month	105 MM
Total Audience Potential	190 MM
Total Theatrical Buy Rate Per Week	14%
> $100 MM Blockbuster Buy Rate (8-10 per Year)	≥ 10%

Figure 4. Theatrical Movie Business

Figure 4 gives a summary of the theatrical movie business. There are eight (8) major studios in Hollywood that account for the bulk of the box office receipts with a release schedule of about 200 movies per year. Total movie production including independents, art films, etc. is about 450 movies. The total annual box office is 5.2 billion dollars (1994) with an average ticket price of $4.14. Of the total potential audience, about 190 million in America, the weekly buy rate, or movie going rate, is about 14% in total or 26 MM tickets sold per week. A buy rate of only 10% will boost a film to blockbuster status (box office receipts of 100 million dollars or more). The public perception of a blockbuster is that everyone has seen the it, however, 90% of the theater-going audience has not seen the film! Moreover, only about 2% of the movies produced each year actually become blockbusters.

Home Video Rentals

Total Rental Revenue	$10 Billion
Total New Release Revenue	$ 6 Billion
Average New Release Rental Price	$2.70
New Release Rentals per Month	185 MM
Total VCR HH	83 MM
New Release Buy Rate per Month	220%

Figure 5. Home Video Business

Figure 5 gives a summary of the home video business. Today, the video rental business is about $10 billion annually, with about $6 billion of that from new releases. The average price for renting a new release is $2.70. The new releases rented per month averages 185 million, out of 83 million VCR households. Therefore, the new release rental rate, or buy rate, is 220%, which means that the average VCR household rents 2.2 movies a month.

Armed with the data of consumer demand of movies in theaters and video stores, a projection can be made of what a VOD business might look like. Given the movie theater "buy rate" of about 55% per month at $4.14 a ticket and home video "buy rate" of about 220% at $2.70 per rental, the projected VOD "buy rate" at $3.00 per title is 210%. It is made up of the following categories: new releases at 110%, catalog or library at 40%, adult at 40% and others at 20%. The 210% projection is about 25% higher than the Near Video on Demand buy rate that DBS is getting. This tracks quite well with the VOD testing that TCI, US West and AT&T conducted about 2 years ago conducted in Colorado. Using projected pricing and buy rates of VOD, a macro economic model can be deduced without getting into the details of the internal specifics of SDV architecture.

■ ASSUMPTION: TELCOM Industry can Develop/Deploy a
Network Infrastructure ("Blob" Network) that can
Economically Support VOD Business

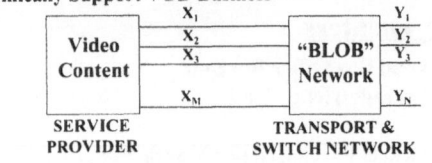

Digital Video Path Established:
Any X Title to Any Y Home with "VCR-like" Functions

Figure 6. "BLOB Network"

Figure 6 depicts the functional diagram of a "Blob Network." The assumption is that the Telecom industry can develop and deploy a network infrastructure, the "Blob Network," that can economically support a VOD business. The left side of the network is connected to the video content -- from X_1 to X_M number of movie titles, all stored in the service provider's digital video servers. The right side of the network is connected to individual homes, from Y_1 to Y_N. The "Blob Network" provides the transport and switching functions connecting, for example, 10,000 "X" titles to 100,000 "Y" homes. The digital video path is established from any "X" title to any "Y" home with VCR-like functions, such as pause, rewind and play, etc.

Monthly VOD Gross Revenues

210% Buy Rate	@ $3.00	$6.30
Hollywood	@ 45%	<$2.80>
Service Provider	@ 40%	<$2.50>
"BLOB" Network Revenue @ 15%		$1.00

Average Movie Leased Time = 2 Hours
Video Coding per NTSC Video = 6 Mbps

"BLOB" Network Lease Rate
@ 4.2 Hour/Month @ 6 Mbps = $1.00

Figure 7. "BLOB Network" Financial Pro Forma

Figure 7 derives a financial pro forma for the "Blob Network." With a VOD buy rate of 210% at $3.00 per title, that's $6.30 of revenue per household per month. Hollywood will take almost half of that, about $2.80. The service provider, which does the marketing, promotion, billing, collections and installs the video file servers would take at least 40%. This leaves a "Blob Network" revenue for transport and switching of about $1.00 per home per month. *One dollar per month.* The average viewing time for each movie is about 2 hours, or 4.2 hours a month using a 210% buy rate. The video coding for an NTSC video picture and sound is about 6 Mbps. The conclusion: If Telcos can deploy an SDV network that can profitably support a VOD business as a driver then they should be able to profit by leasing the same network to VOD service providers at $1.00 per VOD home per month.

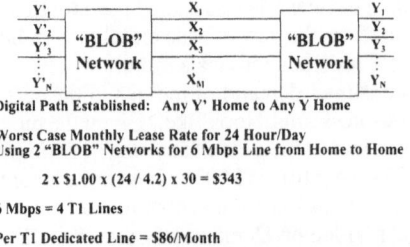

Digital Path Established: Any Y' Home to Any Y Home

Worst Case Monthly Lease Rate for 24 Hour/Day
Using 2 "BLOB" Networks for 6 Mbps Line from Home to Home

2 x $1.00 x (24 / 4.2) x 30 = $343

6 Mbps = 4 T1 Lines

Per T1 Dedicated Line = $86/Month

Figure 8. Economics of a True VDT Network (2 "BLOB Networks")

Although many characterize the "Blob Network" as the structure necessary for Video Dial Tone applications, it really is only one-half of a true VDT because the left and the right side of the network are asymmetrical, X_M vs. Y_N. However, by placing two "Blob Networks" back to back and if N to M is 10 to 1, it will work pretty well as a true end-to-end VDT network. Figure 8 indicates that a digital video path from any Y' home on the left can be connected to any Y home on the right. Using the financial pro forma derived in Figure 7, the worst case monthly lease rate for a full time 6 Mbps data line between any two homes in a community using two "Blob Networks" back-to-back comes out to be $343.00. Since a 6 Mbps data path can accommodate 4 T1 lines, a dedicated T1 line should only cost about $86.00 per month. Presently T1 lines are about $1,500.00 per line per month!

In a digital communications network, all information is transmitted as a modulated form of binary information, streams of 1's and 0's, thus the network does not know the nature of the source, whether it's a movie from VOD or a voice conversation from POTS (Plain Old Telephone Service). A VDT lease rate of $2.00 per month (2 "Blob Networks" back-to-back) for 6 Mbps at 4.2 hours per month, is equivalent to 50.4 Mbps-minutes per day. Ordinary POTS or Audio Dial Tone uses about 32 Kbps of bandwidth, 20 minutes per day or 0.64 Mbps-minutes per day. Therefore, completing the simple arithmetic provides a POTS network lease rate under such an economic structure that should be 1/78 of the VOD lease rate or $0.03 per month! Allocating about $2.00 a month for billing and administrative costs, results in residential local phone rates that should be only about $2.03 per month! *$2.03 per month?* Now there seems to be something wrong with that. The conclusion that there is about $1.00 per month per home to pay for the switching and transport of VOD seems logical, therefore if VOD is the *Killer App* that is going to drive the deployment of SDV architecture, then the $2.03 POTS rate is inevitable.

CONCLUSIONS

Possible conclusions to the boundary-value analysis of the "Blob Network":
1. If the Telcos can deploy a public switched digital video network that is viable for VOD applications, then a back-to-back dual network construction should be able to provide residential telephone service at $2.03 per month. Furthermore, under the proposed Telecommunications Bill, the Telcos are likely to be required to offer equal and unbundled access to the public switch network to unaffiliated service providers. Therefore, even for those states where the PUCs have already made agreements with the RBOCs for long-term fixed local phone rates, resellers could tap into the public switched network and compete for local telephone business at much below the Telco's current average rate of $18.00 per month.
2. Residential rate payers are totally subsidizing the Telco's entry into the VOD market with artificially low VOD tariffs.
3. The T1 tariff should be reduced to less than $100 per month instead of $1,500. Again, resellers can directly get into the T1 business by leasing VDT ports.
4. The SDV infrastructure is prohibitively expensive to deploy, certainly for VOD. And, the Telcos risk financial ruin, leaving the rate payers, the public, holding the bag. This nation cannot afford to be without telephones -- so if the Telcos believe that VOD, using SDV, is going to be the economic engine that drives it and commits the funds to deploy it, there will likely *have to be* a rate payer funded "bail out" to ensure the continuation of telephone service.

In reality, the truth is probably somewhere amongst all of these conclusions. Hopefully, this very simple but rigorous boundary-value analysis has demonstrated that VOD is not a Killer Application and will never work as a driver, no matter what kind of scenario, to deploy SDV architecture.

ATM-BASED SYSTEM ARCHITECTURE FOR MULTIMEDIA NETWORKING

D. Raychaudhuri

NEC USA, C&C Research Laboratories
4, Independence Way
Princeton, NJ 08540
ray@ccrl.nj.nec.com

ABSTRACT:

An ATM-based system architecture for multimedia networking is introduced in this paper. The design objective is to provide a seamless computing and communication environment with quality-of-service (QoS) control over a range of distributed multimedia computing scenarios. The major technical challenges facing the designer of such a system are identified as: (1) *speed*; (2) *quality-of-service*; (3) *scalability* (4) *flexibility*; and (5) *mobility*. These design goals can be substantially realized with an architecture centered around a QoS-based multiservices wireline + wireless ATM network that supports both fixed and portable multimedia terminals with suitable hardware and software. Selected network, terminal and software design considerations related to achieving QoS control and mobility are discussed, using illustrative experimental/simulation results from ongoing research where applicable. The presentation concludes with a brief view of the "WATMnet" and "multimedia C&C testbed" prototypes under development at NEC Princeton.

1. INTRODUCTION:

Distributed multimedia systems based on high-speed networks represent a historic convergence of computing and communication technologies. Emerging network-based multimedia applications include video conferencing, collaborative groupware/games, video-on-demand, multimedia information retrieval and distributed multimedia processing. Multimedia products of this type have recently become viable due to significant advances in core technologies during the early 1990's, such as ATM switching, MPEG video, high-speed CPU's, GB RAM and multimedia software. Early multimedia systems (e.g. CD-ROM video, MPEG TV, etc.) tend to use only a subset of these new technologies to achieve relatively specific functionalities. More generalized distributed multimedia applications will require a truly flexible and integrated computing and communications architecture, which is now the subject of various R&D efforts worldwide. It is our view that significant qualitative and quantitative advancements are necessary to provide network-based multimedia products that compete effectively with conventional alternatives such as books, newspapers, magazines, radio, TV, telephone, etc. A combination of many advanced technologies will be required to approach the user-level capabilities of conventional media as outlined in the table below:

User-level functionality	Applicable technologies
Low application latency	Fast CPU, broadband network, real-time OS
High-quality media display	HD display, broadband network, media compression
Ubiquitous usage	Seamless network, low-cost terminals, software..
Portability	Mobile terminal, wireless network,..
Interoperability	Network/media/API standards..
Simple user interface	I/O alternatives, GUI software, ...

Only after the basic capabilities of speed, display quality, portability, ubiquity, interoperability, etc. are realized will users begin to appreciate the "value-added" features of network-based digital multimedia (such as remote information access, hypertext links, context-based retrieval, editing and storage capabilities, user interface options, etc.).

In this paper, we propose a multimedia system architecture which approaches the above objectives using a multiservices wireline + wireless ATM network to support both fixed and portable multimedia terminals with suitably designed hardware and software. Selected design topics including quality-of-service (QoS) control and mobility support in ATM are discussed. The presentation concludes with a brief summary of related prototyping activities at NEC USA's C&C Research Laboratories in Princeton.

2. MULTIMEDIA SYSTEM ARCHITECTURE:

We consider a fairly general multimedia computing and communication ("C&C ") scenario, as illustrated in Fig. 1 below. In this environment, a variety of multimedia computing devices, both fixed and portable) communicate with each other and/or access information from remote media servers. Observe that the scenario shown is characterized by the coexistence of many applications, media types and service platforms within a single multiservices broadband networking framework. It is desirable that the same applications for multimedia information retrieval, video-on-demand, groupware, etc. will run on static PC/workstation/set-top box, as well as on portable devices such as notebook PC's, PDA's (personal digital assistant), or PIA's (personal information appliance).

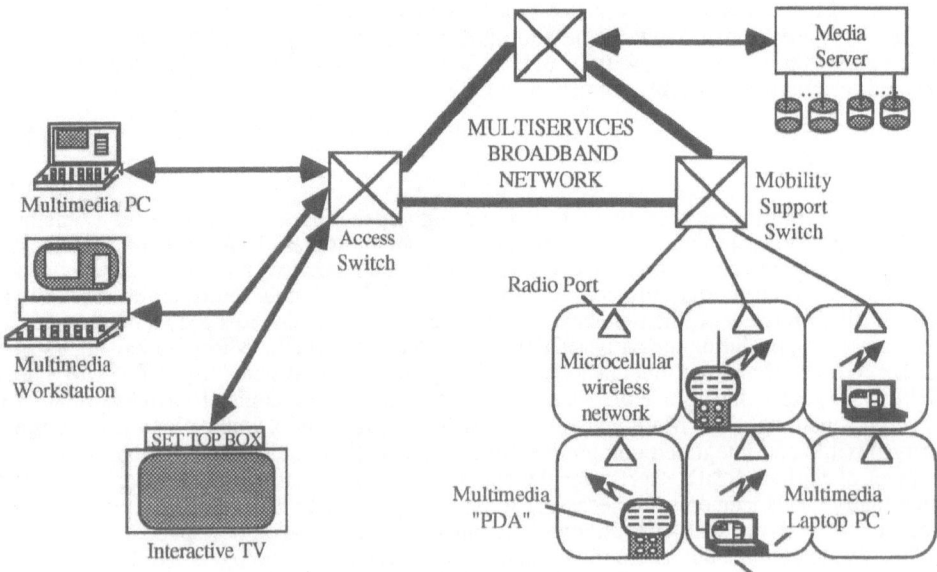

Fig. 1. Typical multimedia "C&C" scenario

In the system under consideration, network-based multimedia services are central to the application, and are equally accessible to both static and mobile users. It is recognized that in order to maximize utility and convenience, it is essential to design a seamless networking and software architecture which incorporates both wired and wireless portions of the system. Of course, there may be quantitative differences in the computing, media processing/display or communications capabilities of different multimedia terminals in the system, depending on available hardware speed or network bandwidth. The objective is to have a qualitatively uniform system architecture that applies across different platform and network types.

Based on the above considerations, high-level system design goals for the multimedia networking architecture under consideration include:

(1) *Speed*, in terms of high useful throughput in network, servers and terminals
(2) *Quality-of-service (QoS) control* at network and user levels
(3) *Scalability*, in terms of applications which adapt to network/platform capabilities
(4) *Flexibility* for easy application development, interoperability, etc.
(5) *Mobility* support for seamless networking of portable/wireless devices

ATM network technology provides a useful basis for such a system since it was designed for high-speed, service integration and explicit quality-of-service support. It is possible to build a flexible multimedia system based on ATM network services, along with suitably designed multimedia terminal hardware and software. A conceptual view of the multimedia system architecture under consideration is shown in Fig. 2 (note: shaded areas typically represent hardware). Observe that the proposed architecture is centered around an ATM network with unified wired + wireless services via mobility-enhanced ATM network and signaling/control protocols [1,2].

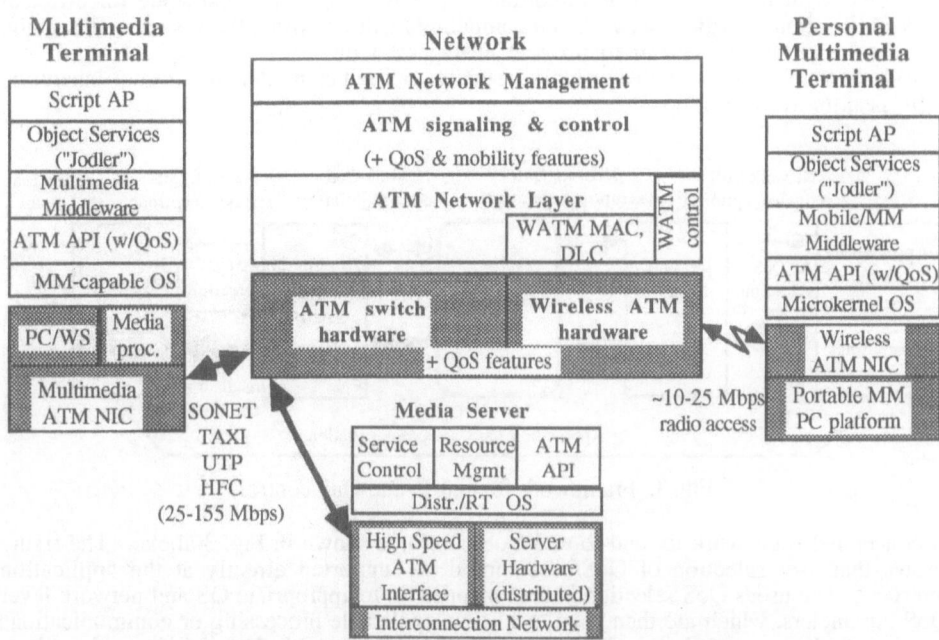

Fig. 2. Conceptual view of proposed multimedia system architecture

This generalized ATM network provides end-to-end broadband transport capabilities with resource sharing efficiencies and quality-of-service (QoS) control [3,4], in both wireline and wireless access scenarios. It also permits application developers to use a uniform QoS-based ATM application programming interface (API) for software developed on both fixed and portable platforms. As shown in the figure, a typical multimedia scenario will involve three types of terminal/computer equipment connected to the network: media server, (fixed) multimedia terminal and (portable) personal multimedia terminal. The portable terminal is connected to the network via a wireless ATM interface, and should be capable of running the same application programs as the fixed multimedia terminal, although there may be quantitative limits imposed by radio channel or CPU processing constraints. A software stack consisting of a multimedia-capable OS, ATM application programming interface (API) with QoS control, multimedia/mobile middleware and interpretive object script-based application programs [5,6] is proposed for use on both fixed and portable terminals in this system. This software approach supports scalability, QoS control and platform independence, while providing a high-level of abstraction for network and media related resources to the application developer.

In the following sections, we discuss technical approaches needed to realize two of the five high-level system design goals listed earlier, namely QoS control and mobility. Coverage of the other three major system design issues (i.e., speed, flexibility and scalability) is precluded by space limitations; the interested reader is referred to related work listed in the references.

2. QUALITY-OF-SERVICE CONTROL IN ATM:

Quality-of-Service (QoS) control in distributed multimedia is important because real-time multimedia applications containing audio/video tend to degrade rapidly below a threshold of impairment. A predictable performance level is mandatory for any widely used technology, so that it is essential for end-to-end application quality to be maintained at least on a statistical basis. In addition, given the widely varying and relatively high resource requirements of multimedia, it is important to provide a QoS vs. cost scale for efficient shared utilization of critical network and computing facilities within the system. The option of over-designing the system to provide near-perfect quality to all applications is not considered viable even with increasingly fast networks and computers because of inherently high "peak-to-average" ratios.

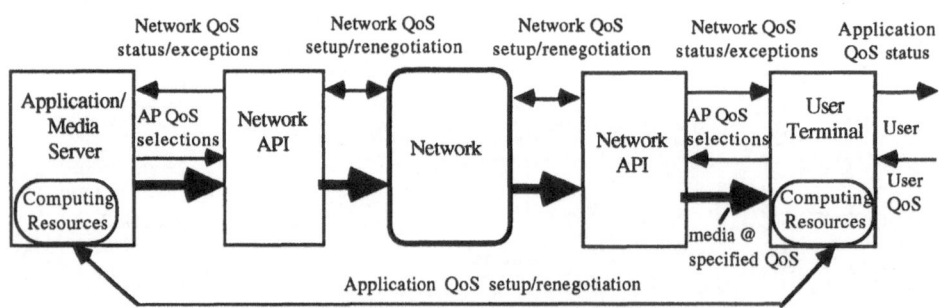

Fig. 3. Framework for end-to-end QoS control

A conceptual framework for end-to-end QoS control is shown in Fig. 3 above. The figure shows that user selection of QoS/cost should be supported directly at the application interface. The user's QoS selection is then mapped on to appropriate OS and network layer QoS parameters, which are then used to reserve applicable processing or communications resources. It is noted that each QoS interface is shown with 2-way information flow, corresponding to dynamic renegotiation of QoS parameters depending on current conditions. The notion of terminal <-> network negotiation is central to efficient and robust system operation in the presence of applications (such as video) with time-varying bandwidth and processing requirements [7].

An ATM-based framework for QoS control involves two major elements. The first is an ATM network designed to provide a range of transport services with QoS controls [8], while the second is a corresponding software QoS API at servers and terminals [9]. Current ATM network design is typically based on the notion of service classes such as available bit-rate (ABR), variable bit-rate (VBR) and constant bit-rate (CBR), each with its own service parameter list and QoS specification options. In general, ABR is viewed as a packet data service without explicit QoS indication, while CBR is viewed as a circuit-switched service with a prespecified high QoS requirement. A more continuous form of QoS control is envisioned for the VBR class via specification of a source traffic profile in the form of usage parameter control (UPC) parameters, typically implemented as dual leaky buckets with parameters: peak rate (R_p), sustained rate (R_s) and burst length (B). Statistical QoS guarantees on cell loss rate (CLR) and cell delay variance (CDV) are then provided by the network, which used an appropriate call admission control process (CAC) to admit new calls.

While this VBR mode is viewed as useful and potentially more efficient than static CBR, practical difficulties have been experienced in achieving significant statistical multiplexing gain for video or multimedia calls with conventional fixed UPC parameters, e.g. [10]. Accordingly, we are investigating the feasibility of a new "VBR+" service which supports dynamic bandwidth renegotiation during the course of a call [11,12]. With the VBR+ service, the ATM interface for a video connection can continually monitor source statistics to estimate required UPC values. When the required UPC values are significantly different from the current UPC setting, a renegotiation process is initiated. This approach has the benefit of letting applications operate at their "natural" bit-rate, while facilitating a relatively high degree of statistical multiplexing within the network.

Fig. 4 shows a typical experimentally obtained trace of the dynamic UPC used by a VBR+ NIC for a composite SIF MPEG-1 video sequence with bit-rate in the region of 2-5 Mbps. It is observed that the VBR+ UPC controller tracks the video source statistics quite well, and thus can provide better video quality than a static UPC averaged over the whole sequence. Corresponding results for peak SNR show significant improvements in video quality during stressful regions of the sequence [13], thus validating the basic concept.

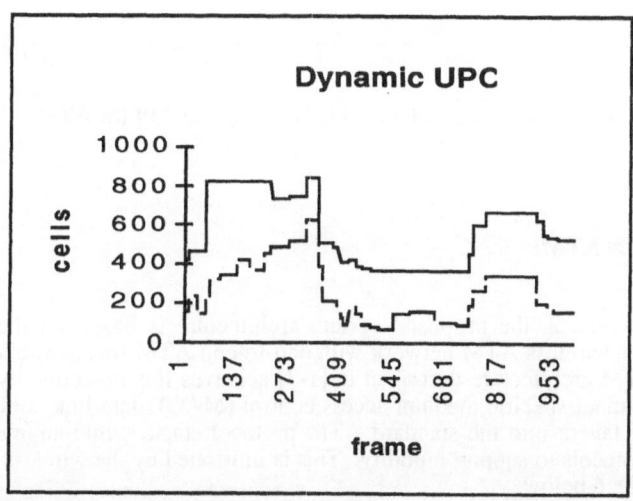

Fig. 4: Peak (upper trace) & sustained (lower trace) rates for MPEG video w/ dynamic UPC

In order to properly utilize ATM network QoS control features such as those discussed above, the application programmer needs access to an appropriate QoS-capable network API. A high-level API concept currently under investigation (called "ATM service manager" [9]) provides automatic mapping of application segments to a range of transport layer protocols, AAL's and ATM transport services (with associated UPC and QoS values). A schematic diagram of this high-level ATM API is shown in Fig. 5 below. The proposed software layer supports renegotiation of QoS between the application/ and network during the course of a multimedia session, with the objective of optimizing network cost/quality to meet high-level user needs. An ATM API of this type also supports the concept of "QoS control knobs" at the user interface level to permit the user to specify cost/quality trade-offs needed for efficient use of a multiservices broadband network.

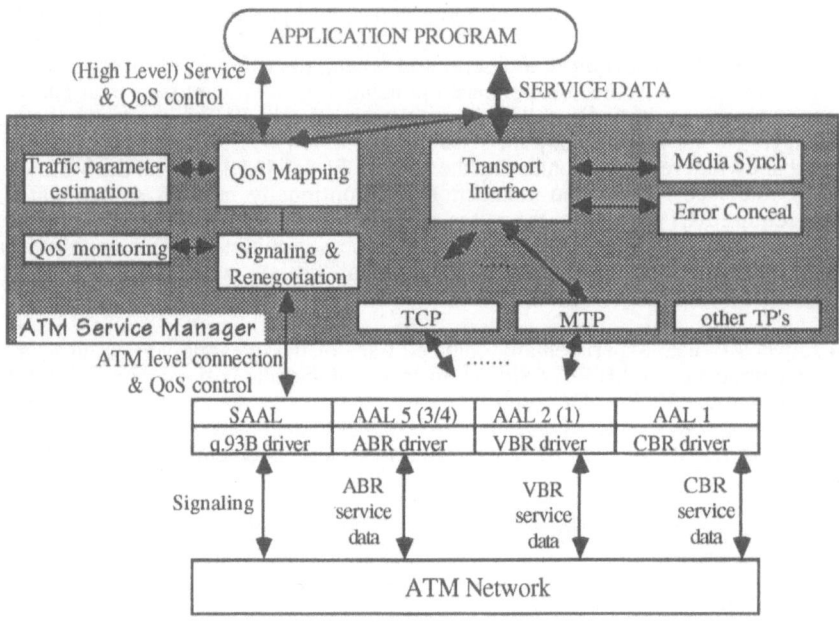

Fig. 5: Schematic of adaptive QoS-capable API for ATM

4. MOBILITY IN ATM:

As discussed in Sec. 2, the proposed system architecture is based on the concept of a seamless wired + wireless ATM network with end-to-end ATM transport and QoS control. The wireless ATM architecture described in [1-3] achieves this objective by incorporating new wireless channel specific medium access control (MAC), data link control (DLC) and wireless control layers into the standard ATM protocol stack, while augmenting existing ATM control protocols to support mobility. This is illustrated by the wireless ATM protocol stack given in Fig. 6 below.

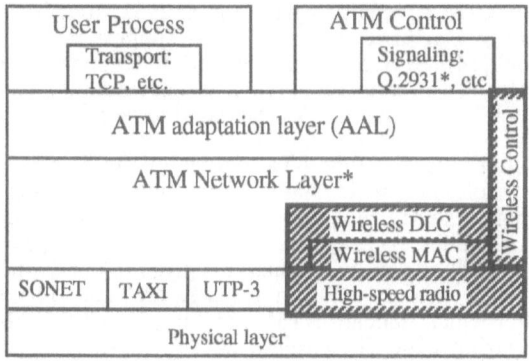

* Includes mobility extensions

Fig. 6. Wireless ATM protocol stack

The MAC protocol for wireless ATM access is required to support a mix of ABR, VBR and CBR services with QoS control. One approach is to use a dynamic TDMA/TDD protocol with frame-by-frame allocation of ABR and VBR slots and period assignment of isochronous CBR slots. A protocol of this type has been shown to provide reasonable performance at throughput levels up to about 0.6-0.7 after accounting for control and physical layer overheads. The reader is referred to [2,3] for further details. The DLC protocol [13] just above the MAC sublayer is designed to improve the effective error rate on the wireless access link before cells are transferred to/from the fixed ATM network. DLC protocols are applied not only to packet-mode ABR services, but also to stream-mode CBR and VBR services. For ABR, DLC operation follows traditional SREJ ARQ procedures on a burst-by-burst basis, without time limits for completion. For CBR and VBR, the DLC operates within a timeout interval that is specified by the application at call set-up time. In this case since CBR or VBR allocation is periodic, additional ABR allocations are made at the MAC layer to support retransmitted cells. Emulation results for DLC/MAC operation indicate that the proposed DLC procedures are effective over a range of typical channel fading models [13].

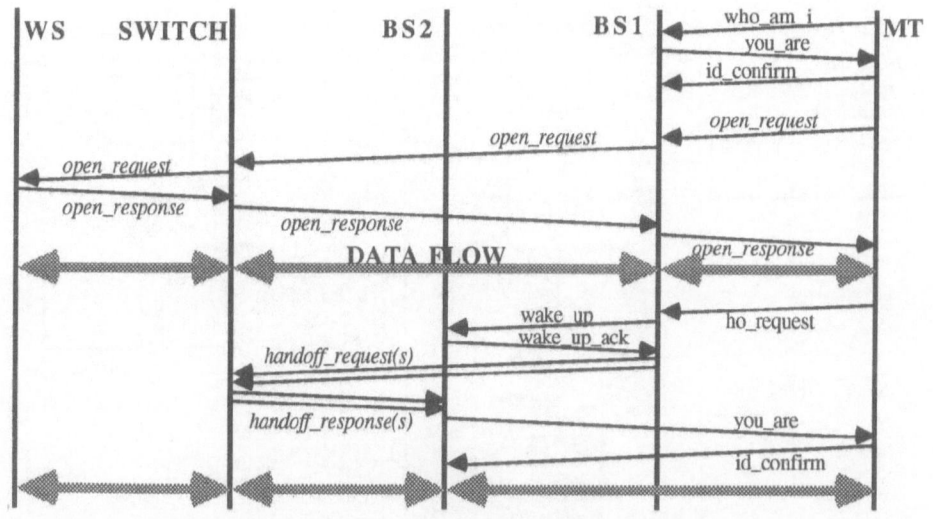

Fig. 7: Example handoff control procedure in wireless ATM

Selected extensions to ATM signaling syntax (e.g. Q.2931) together with a new wireless control ("meta-signaling") sublayer are needed to support mobile ATM users at the connection control level [14]. Specific mobile network functions requiring signaling/control support include address registration for mobile users, wireless network QoS parameter specification/renegotiation and handoff. A basic issue to be addressed is that of designing wireless control and ATM signaling to support relatively transparent handover operations. In general, this will require re-routing of the ATM connection from one base station to another, while also moving wireless network state information (e.g. DLC buffers and MAC permissions) to smoothly resume communication with a minimum of cell loss. An example of signaling and meta-signaling exchanges between fixed terminal (WS), base stations (BS) and mobile terminal (MT) during an ATM handoff is given in Fig. 7. It can be observed that wireless control messages such as "who_am_i", "you_are" and "id_confirm" are used for initial registration, after which standard ATM signaling is used for call control all the way from the mobile. During handoff, wireless control messages such as "ho_request", "wake_up" and "wake-up-ACK" are used between mobile and base stations, while new mobility-related signaling messages ("handoff-request/response") are used to communicate VC re-routing requirements to the switch. This general approach for ATM handoff has been validated experimentally, and results will be reported in a forthcoming paper.

5. PROTOTYPING ACTIVITIES:

We have developed an initial version of a wireless ATM prototype ("WATMnet") incorporating an 8 Mbps (2.4 Ghz ISM band) radio, (simplified) TDD MAC, DLC and mobility signaling enhancements for handoff [15]. The WATMnet prototype became operational in July 1995, and has been used to demonstrate wireless ATM services (initially TCP/IP using ABR/AAL5) at a wireless laptop PC. The prototype wireless network is integrated into our ATM-based "multimedia C&C testbed", which demonstrates adaptive QoS-based multimedia applications on both workstations and mobile personal terminals. A summary of major hardware and software (in italics) components being implemented on this experimental testbed is given in Fig. 8 below. A full description of the prototype system is beyond the scope of this paper.

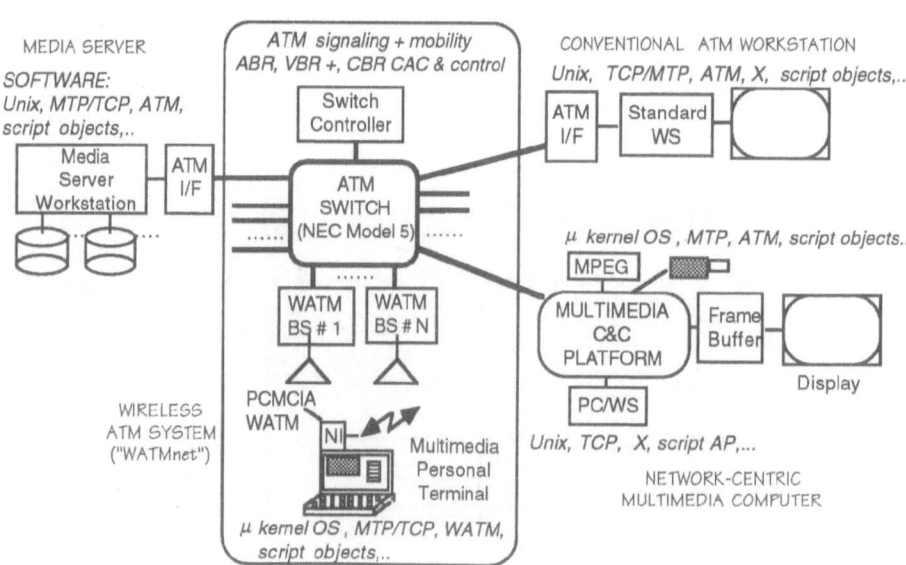

Fig. 8: NEC Princeton's Multimedia C&C Testbed (under development)

6. CONCLUDING REMARKS:

An ATM-based system architecture for multimedia networking has been presented in this paper. The proposed system design aims to provide high speed, quality-of-service control, flexibility, scalability and mobility within a unified ATM framework. We have presented a QoS control approach based on a dynamically renegotiated "VBR+" ATM service, together with an adaptive software API which supports QoS specification and renegotiation between application and network. A wireless ATM architecture capable of providing standard ATM services to mobile terminals has also been briefly outlined. Proof-of-concept prototyping activities related to the proposed multimedia networking architecture have been summarized. Early evaluation and prototyping results tend to confirm the viability of the QoS-based wireline + wireless ATM networking approach, which we believe offers important performance and functionality advantages over available alternatives.

7. REFERENCES:

[1]   D. Raychaudhuri, "ATM Based Transport Architecture for Multiservices Wireless Personal Communication Networks", *Proc. ICC'94*, pp. 317.1.1-7.

[2]   D. Raychaudhuri and N. Wilson, "ATM Based Transport Architecture for Multiservices Wireless Personal Communication Network", *IEEE J. Selected Areas in Comm.*, Oct. 1994, pp. 1401-1414.

[3]   D. Raychaudhuri, "Wireless ATM: An Enabling Technology for Personal Multimedia Communications", *Proc. Mobile Multimedia Comm. Workshop*, April 1995, Bristol, UK.

[4]   D. Raychaudhuri, "Multimedia Networking Technologies: A Systems Perspective", *CTR Multimedia Networking Symposium*, Oct 28, 1994, Columbia University, NY.

[5]   M. Ott, V. Bansal, R.J. Siracusa, J.P. Hearn, R. Dighe, N. Mori, A. Iwata, H. Suzuki and D. Raychaudhuri, "Prototype ATM LAN System Multimedia On Demand Applications", *NEC R&D Journal*, Oct. 1994, pp. 366-377.

[6]   M. Ott,   "Jodler: A Distributed Scripting Language", NEC USA, C&C Research Laboratories Technical Report, 1994.

[7]   H. Kanakia, et. al., An Adaptive Congestion Control Scheme for Real-Time Packet Video Transport, in *Proc. SIGCOMM'93*, Ithaca, N.Y., September 1993.

[8]   G. Ramamurthy and R. Dighe, "A Multidimensional Framework for Congestion Control in BISDN", *IEEE J. Selected Areas in Comm.*, Dec. 1991.

[9]   V. Bansal, et. al., "Adaptive QoS-based API for ATM Networking", Proc. 5th International Workshop on Network & Operating System Support for Digital Audio and Video (*NOSSDAV'95*), Durham New Hampshire, April 1995.

[10] H. Harasaki and M. Yano, "A Study on VBR Coder Control Under Usage Parameter Control", in *Proc. Fifth International Packet Video Workshop*, Berlin, Germany, March 1993.

[11] D. Reininger, G. Ramamurthy and D. Raychaudhuri, "VBR+, an enhanced service class for multimedia traffic", ATM Forum, Contribution 94-0353, Munich, Germany, May 1994.

[12] D. Reininger, G. Ramamurthy and D. Raychaudhuri, "VBR MPEG Video Coding with Dynamic Bandwidth Renegotiation", *Proc. ICC'95*, Seattle, WA, June 1995, pp. 1773-1777.

[13] H. Xie, P. Narasimhan, R. Yuan and D. Raychaudhuri, "Data Link Control Protocols for Wireless ATM Access Channels", *Proc. ICUPC'95*, Tokyo, Japan, Nov. 1995.

[14] R. Yuan, S.K. Biswas and D. Raychaudhuri, "An Architecture for Mobility Support in an ATM Network", *Proc. 5th WINLAB Workshop on Third Generation Wireless Networks*, April 1995.

[15] L.J. French and D. Raychaudhuri, "The WATMnet System: Rationale, Architecture and Implementation", *IEEE Computer Communications Workshop*, Eastsound, WA, Sept. 15-17, 1995.

KALEIDO: A SYSTEM FOR DYNAMIC COMPOSITION AND PROCESSING OF MULTIMEDIA FLOWS

Abhaya Asthana, Mani Srivastava, Venkatesh Krishnaswamy

AT&T Bell Laboratories
Murray Hill, New Jersey, 07974

INTRODUCTION

Emerging multimedia technologies have the potential to revolutionize the way humans organize, communicate and consume information [1,3,4,5,6,7,8]. However, the full benefits of this technology have yet to reach the vast majority of computer users. One reason for this is that high-bandwidth networks are not widely deployed or accessible. But beyond the bandwidth barrier, there also exists a *multimedia computing* barrier. The architectures of present day computers are not significantly different from their earlier generation counterparts: fundamentally oriented towards *data processing* -- they are simply much faster and better at it. Consequently, these systems (hardware and system software alike) are ill-equipped to cater to the special requirements imposed by multimedia. Most commonly used approach is to attach 'multimedia' devices such as cameras, sound-cards, CD-ROM drives as I/O peripherals to an existing computer system, and hope that a 'fast enough' CPU will run a few targeted applications reasonably well. It is possible, using this approach, to build *some* highly optimized (possibly through the use of custom hardware peripherals) stand alone multimedia applications that perform tolerably. However, such systems do not provide a general-purpose infra-structure to support the integration of multimedia capability into an *arbitrary* user application. To enable this, it is necessary to incorporate support for multimedia into the hardware and software fabric of the system, not just as 'add-ons'. Kaleido is an experimental approach to designing such an integrated multimedia system. It is an on-going project and in this paper we present the current snap shot of our architecture and implementation.

SYSTEM ARCHITECTURE

Our goal is to create a high performance, flexible system that provides architectural support for user defined composition and control of media flows in real-time, collaborative multimedia applications. The scope of our target applications includes both multimedia desk-top as well as multimedia server environments. Specifically, the following requirements have guided our design:

- Time Management: Continuous media information has an inherent time component that distinguishes it from other types of computer data. This time component is characterized by such properties as stream rates, delivery deadlines for individual media units, synchronization requirements within and across media and response time constraints. A multimedia system needs to incorporate support for timing requirements at the system infrastructure level.

- I/O Support: Multimedia computers have to accommodate 'new' I/O devices such as microphones, cameras, speakers and multimedia storage and playback devices as well as interfaces to high-bandwidth communication networks. Further, one needs to be able to explicitly allocate and schedule I/O resources such as interconnect bus bandwidth.

- Computing Support: Processing of high-bandwidth video and audio information requires high-speed computing support. Examples of such computing requirements are compression/decompression, video composition and scaling, audio processing and mixing, and content-based processing.

- Media Flow Management: Multiple media streams may simultaneously flow through a multimedia system. The system has to allow these flow topologies involving the various I/O and computing modules to be flexibly configured. This calls for an interconnection network within the system that not only provides adequate aggregate bandwidth but also provides support for the coexistence of multiple simultaneous streams as well as pipelining of individual streams through various system components.

- Application development and run-time environment: In order to support the goal of being able to incorporate multimedia into any arbitrary application, the system needs to provide tools for application composition and execution. This includes the appropriate abstractions for device control, topology management, specification of timing properties and storage management.

Our design approach may be characterized as *disassembly followed by reassembly*. We break the multimedia processing system down into its constituent components and then provide the methodology and tools to reassemble those components into building blocks and services.

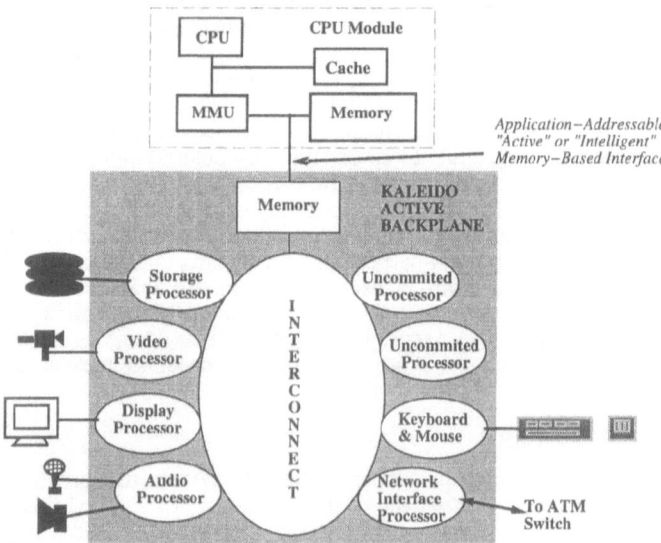

Figure 1. The Active Backplane

The key ideas in Kaleido are:

- Repartitioning of the hardware and software system with an *I/O Centric View* in which the general purpose computing subsystem is separated from the real-time communication and media flow processing subsystem.

- At the very elemental level in the system architecture, allowing autonomous components to interact over an *active backplane*. The active backplane provides programmable building blocks for pipelined media flows, in-band processing of media streams, self managing communication resources for direct data movement between network, disk, monitor, audio, camera, etc., and coordinates interaction between these components.

- Allowing direct application level access to resources in the active backplane. For example, application driven scheduling of interconnect and I/O resources.

- Providing tools for composition of intra and inter service media flows.

I/O Centric Approach

In our model, a multimedia computer system consists of autonomous components that are connected by a high-speed, high bandwidth *interconnect*. Each of these components encapsulates a set of I/O or processing capabilities together with the intelligence necessary to manage these capabilities. These components interact directly with each other using mechanisms provided by a *services kernel*. The kernel provides communication primitives, service registries, connection and communication resources management, and service interaction management.

Together, the interconnect and the services kernel constitute an *active backplane*. The active backplane provides not only the physical connectivity between the sub-systems and their components but also the logical data and control channels for their interaction. Such an active backplane-based architecture represents a communication and I/O-oriented repartitioning of a conventional computer architecture as shown in Figure 1. The interconnect facilitates efficient and flexible data transfer between autonomous components. The components include devices such as disk, camera, video/graphics display monitor, speaker, microphone, camera, and pen; computing resources such as processors for encryption, decryption, compression, and decompression and interfaces to external networks such as ATM, ISDN and Ethernet adaptors. The host CPU+cache+main memory that are the centerpiece of the traditional computer systems are now adjunct to the active backplane, and in general neither the CPU or the CPU-memory bus need to be involved in the transfer of information between the various components.

Active Buffers and Dynamic Composition of Media Flows

Multimedia information flows through the Kaleido system in the form of streams. We define a stream as a sequence of data items that obey some timing properties. Each stream has a source which generates the stream and a sink which terminates it. The entity that terminates such a stream may, in turn, transform the information and send it along to another entity; it then becomes a source for this new stream derived from the initial stream. Pipelines may be constructed in the above manner to contain streams emanating from multiple sources and which interact with each other in a number of ways before they are terminated at one or more sinks.

The basic data-structuring abstraction provided in Kaleido for the specification of the connectivity, data transformations and the resource and timing requirements of such pipelined streams is the *active buffer*. An active buffer is defined by a set of input ports, a set of transformation functions that may be applied to the data that is presented at the input ports, a set of output ports to which the transformed data is routed and a control port which allows for the selection of the transformation function to be applied and to set parameters. A set of timing specifications that the input and output data must meet are explicitly specified for the active buffer. The notion of active buffer in Kaleido is a generalization of the notion of process or task in conventional kernels, to also explicitly encapsulate timing and communication information.

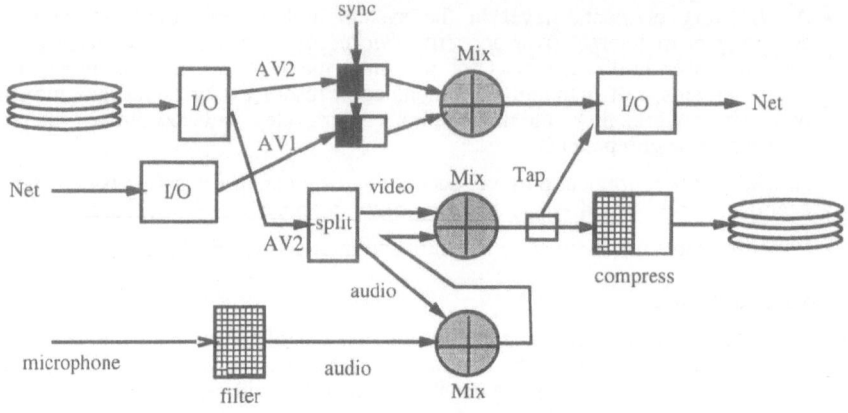

Figure 2. Dynamic Composition of Services Using Active Buffers

Pipelined multimedia streams are constructed by connecting the sources and the sinks via a network of active buffers. To illustrate the idea of constructing pipelined streams out of active buffers, consider the example shown in Figure 2. A compressed audio/video stream, AV1, enters the system via the network interface. The stream is synchronized and mixed with another audio/video stream, AV2, retrieved from a local disk, and transmitted back to the network. Concurrently, the stream AV2 is split into its audio and video components, the audio mixed with live speech from the microphone, the resultant stream combined with the original video, compressed and stored back onto the disk as well as sent over the network. The figure illustrates the active buffer network that implements this application.

The above example illustrates the powerful support provided by the active buffer abstraction in specifying multimedia applications. One, the stream topology of the application can be completely specified by specifying a network of active buffers. Two, stream processing and transformation requirements can be specified by chosing the appropriate active buffers that meet these functional specifications to be interconnected. Finally, the active buffers chosen to build the application are well-characterized in terms of their timing requirements. Hence, the abstraction serves to specify the real-time performance requirements of the application.

The Kaleido system provides a 'kit-set' of active buffers that can be used to construct multimedia applications. The system derives the computing and network resource requirements of these from their functional and timing specifications. For each active buffer, the system uses the resource requirements to dynamically construct a mapping on to the active backplane which guarantees that its timing and functional specifications are met.

The system provided active buffer set may be extended by adding user-specified active buffers. As in the case of system-provided active buffers, these must also be fully characterized in terms of their functional specifications, timing characteristics and resource requirements. The Kaleido run-time system will provide the same mapping functions and guarantees to these active buffers as for system active buffers. Hence, the Kaleido system provides the framework for functional scalability and support for a wide variety of real-time stream-oriented applications.

ACTIVE BACKPLANE DESIGN

The active backplane enables the real-time transfer and processing of multimedia streams through built in support for: 1) multithreaded execution that allows multiple contexts to be active concurrently, 2) fast context switching and event handling, 3) high bandwidth, low latency communication between tasks executing on various processors.

The goal is to be able to separate real-time sensitive media flow communication, composition, and processing from conventional computing tasks. In doing this, the active backplane mechanism enables application-driven requirements to drive the scheduling of I/O interconnect medium bandwidth and of processing resources embedded within the autonomous components.

Applications are mapped to the Kaleido hardware as a set of threads, some of which run on the main CPU and the rest on the various processing units within the active backplane. The two sets of threads communicate across the CPU-subsystem boundary through shared data structures maintained in memory pages that are mapped to the application address space on the CPU [2]. Typically the application threads running in the active backplane will establish the data transfer pathways between the various I/O peripheral and network interface devices using the appropriate active buffers. The application threads running on the CPU will, in general, be outside the data paths of the streams within the active backplane, but will participate in their set-up and control. The role of the active backplane is to allocate the active buffers, map them on to the threads and the interconnect and to manage their run-time operation to meet the real-time specifications of the application.

Each processing node in the active backplane is identified with an address space, and the basic service provided by the interconnect module to a processing node is that of reliable read or write of packets from or to another address space. The interface of a processing node to the interconnect module is via a transfer controller. The transfer controller is on one side a bus master on the local bus of the processing node, and connects to the interconnect fabric on the other side.

A distinguishing characteristic of the interconnect module is that its bandwidth is explicitly reservable and schedulable using a credit-based mechanism. This is in contrast to the lack of any direct scheduling of interconnect resources in conventional workstation interconnect busses. In Kaleido's interconnect scheduling mechanism, a cyclic scheduling is used where each application specifies a certain amount of bandwidth credits between pairs of processing nodes in a time period T (This specification is, in fact, derived from the active buffers allocated by the application). A small pool of application ids is used to distinguish between different applications, with the ids playing a role somewhat analogous to VCIs in ATM. The Kaleido service kernel does admission control based on these credit requests, and passes this information down to a central scheduler module as well as to the transfer controllers. The central scheduler and the transfer controllers realize a distributed and hierarchical policing and scheduling algorithm whereby an application is guaranteed to receive bandwidth credits allocated to it, and unused bandwidth credits are allocated to a best-effort category of applications. A key role in this scheme is played by the transfer controllers which can preempt the transfer of one packet and switch to another one. Kaleido's interconnect architecture retains the simplicity of a memory-like interface, reliable communication, sender as well as receiver initiated transfers, and efficient handling of small-sized transfers that are characteristic of traditional end-system interconnects, while at the same time allowing explicit allocation and scheduling of interconnect resources.

SOFTWARE INFRASTRUCTURE AND SERVICE COMPOSITION

The services kernel forms the backbone of a decentralized software infrastructure wherein components belonging to the autonomous sub-systems interact with each other to provide services. *Composition* forms the basis of service creation. In addition to the physical I/O resources available to the user there are logical level building blocks available such as mixers, splitters, transformers, bridges, filters, selectors and taps. These fundamental operations can be applied to any media: voice, video or character streams. Using these physical and logical building blocks, the composition tools in Kaleido enable the creation of powerful multimedia services and applications. These applications may create pipelines of operators and other building blocks through which streams of data may be processed.

The services architecture is a hierarchical one with complex services being constructed out of simpler ones. At the bottom layer are *basic services* which are the most primitive resources available to the user of the system. Typically, a device may be packaged in the form of a basic service which provides an interface for accessing the device. Basic services are, however, not restricted to manipulating hardware devices. Software resources such as queues, data filters and so on are also examples of basic services. A basic service can be declared to be an active buffer. In this case, the kernel incorporates the knowledge of its computational and timing requirements and manages the system resources required to implement it and guarantee that its specifications are met.

Aggregate services are composite services which typically require the involvement of other basic and aggregate services. This collection is typically be tied together using a *control thread* that co-ordinates the invocation of other services. An aggregate service can be invoked on its own by the user or linked to other services using well-defined interfaces to form larger aggregates. Activation of an aggregate service causes activation of its component services as well as the connections between them.

Programmers use an interactive tool to assemble and configure the constituent services of their multimedia applications.

SUMMARY AND PROTOTYPE STATUS

The Kaleido system is intended to address the limitations of current computer systems in their ability to support multimedia applications. Specifically, we address the issues of real-time performance, media flow configurability and functional scalability of a computer system. We have proposed an architecture based on intelligent multimedia devices and processors interacting over an active backplane under the direct control of applications. We have defined a powerful abstraction, the active buffer, for the specification of media flow topologies, stream transformations and their real-time requirements. We have outlined an architecture that implements the active backplane with the necessary resource scheduling support.

Our initial focus is on implementing an architectural simulator for the active backplane, and the tools for service composition and application program development on standard hardware/software platforms. Currently we are using SUN/SPARC stations equipped with Parallax X-video boards running Solaris as the Kaleido desktops, and a Pentium/PCI system with multiple DSP arrays as the platform for Kaleido server. An ATM switch from Fore System provides the connectivity between the desktops and the server. Within a desktop system, the active backplane is emulated using a Solaris process with LWPs/threads modeling active buffers and basic services.

REFERENCE

[1]   D. Tennenhouse, "Telemedia, Networks and Systems," Annual Report, MIT, Lab. for Computer Science, July 1992.

[2]   Asthana, A. et.al, "A Memory Participative Architecture for High-Performance Communications Systems," IEEE INFOCOM 94, Toronoto, Canada, June 12-16, 1994.

[3]   D. Hindus and C. Schmandt, "Ubiquitous Audio, Capturing Spontaneous Collaboration," Proc. CSCW 92, pp. 210-217, Nov. 1992.

[4]   P. Resnick, "Hypervoice: A Phone-Based CSCW Platform," Proc. CSCW 92, pp. 218-225, Nov. 1992.

[5]   C. Schmandt and S. Casner, "Phone Tool: Integrating Telephones and Workstations," Proc. IEEE Global Telecomm Conf. Nov 27-30, 1989.

[6]   R. Kamel, et.al., "PX: Supporting Voice in Workstations," IEEE Computer Mag., pp. 73-80, August 1990.

[7]   J.R. Ensor, et.al. "Control Issues in Multimedia Conferencing," Proc. TriComm 91, Chapel Hill, NC, April 18-19, 1991.

[8]   A. Hopper, "Improving Communications at the Desktop," Rep. Royal Soc. Discussion Mtg. on Communications After 2000 AD, Mar. 18-19, 1992.

A SYNCHRONOUS MULTIMEDIA COMMUNICATION SYSTEM OVER N-ISDN

Design and Implementation of Networked Collaborative Environment

Takashi Nishimura[1], Shigeki Masaki[2], and Kazunori Shimamura[1]

[1]NTT Human Interface Laboratories
[2]NTT System Service Department
1-2356 Take, Yokosuka, 238-03, JAPAN
nishi@nttvdt.hil.ntt.jp

1. Introduction

The advent of digitized network system and high performance / low cost WS/PCs has made it possible to perform office tasks more and more efficiently. It enabled us to develop a Computer Supported Cooperative Work (CSCW) system as a working environment.

To utilize conventional video conference systems or groupware systems, we need to enlarge multimedia communication capabilities and multipoint connections. We developed a Personal Multimedia-multipoint TeleConference system (PMTC)[1] which makes it possible to use a WS/PC at own desk connected to an ATM based B-ISDN by integrating both computer processing and telecommunication functions. PMTC/B-ISDN is a distributed synchronous groupware system that integrates various multimedia data such as audio, video, text, graphics, and drawings. Through internal testing, we evaluated its functions. Thanks to the high-speed network environment, the system made users feel as if they were actually in a face to face conference through its high quality audio and video communication. However, it will take time to before high-speed broadband networks are used in the office environment.

Groupware systems or enhanced tele-conference systems such as CRUISER[2], RAPPORT[3], and MERMAID[4] have been developed and utilized in actual remote collaboration.

We have developed PMTC which is a new practical groupware system for use with the current N-ISDN. It has been developed as a tool of group work or conference based on the experimental evaluation of PMTC/B-ISDN functions such as virtual conference environment, conference management, and synchronized multimedia handling etc. .

This paper describes the design concepts, hardware architecture and its implementation from the view point of multimedia data handling over multipoint connected N-ISDN, and conference service management issues. The experimental evaluation is also discussed.

2. PMTC Design Concept

The design concept of PMTC/N-ISDN basically follows the concept of PMTC/B-ISDN. PMTC focuses on the possibility of integrating different types of work in the office. Considering collaborative environments such as conferencing, it is essential for the participants to be able to simultaneously share information. Regarding face-to-face meetings, groupware systems should support three kinds of virtual conference spaces; common space (sharing the same information with all participants), closed space (sharing information with selected participants), and local space for personal workspace as shown in Figure 1.

Figure 1 The Concept of Virtual Conference Space

The system utilizes multimedia data to ensure an easy to use user interface. For example, the system displays images of all conference participants and data image simultaneously in multiple windows. It also provides the same display and operation format as graphic windows used for displaying documents on general usage of WSs or PCs.

Moreover, we are aiming for this system to be used as a PC/WS based desktop conferencing system with a shared computing environment. Therefore, we are considering the following functions:

(1) Connecting up to five locations for one conference by utilizing audio and video for conversation, and shared data such as text, graphics, and pointing data.

(2) Applying commercial application software with minor modification to run on the shared environment

(3) Conference process management such as holding sub-conferences, convener changes, conference participants withdrawal or addition during a conference.

(4) Sophisticated multi-user interfaces that not only incorporate display management but also audio management

(5) Introduce available technologies and network capabilities.

3. PMTC/N-ISDN   Hardware Architecture and its Implementation

The hardware architecture of PMTC/N-ISDN is shown in Figure 2. The system is configured from a general purpose Unix workstation by adding auxiliary hardware. The workstation controls the auxiliary hardware through the SCSI interface. The auxiliary hardware consists of a main controller (CTL), which controls each module, an N-ISDN basic rate interface module (NIF) with B-channel switchers, a multimedia multiplexer/demultiplexer module (MUX), audio and video codecs (CODECs), and an input/output audio-video control module (AVC). The NIF has four N-ISDN lines to process incoming and outgoing calls. It also has a B channel control switch to originate and terminate calls on the four lines via a pilot number function, and can connect any B channel and coding unit. The video and audio codec modules consist of an ITU-T Standard H.261 video codec, H.221 multimedia MUX, H.242 communication protocol processor and G.711,722,728 audio codec units. Specifications of PMTC/N-ISDN are shown in Table1.

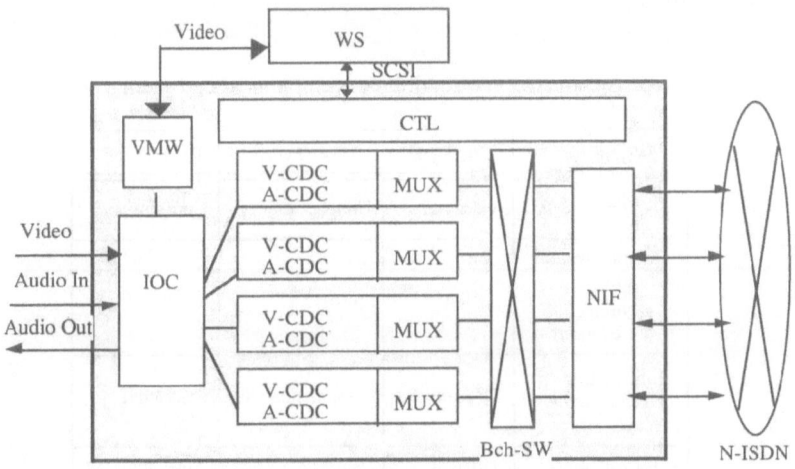

Figure 2 PMTC/N-ISDN Hardware Architecture

Table 1 PMTC/N-ISDN Hardware Specification

Items		Specifications
Network Interface		ISDN Basic Rate Interface, 2B x 4 lines
AV Communication	Communications Protocol	ITU-T H.242, Max 5 connected remote conference
	Multiplexing Method	ITU-T H.221 HSD LSD based data communication available
	Audio Codec	ITU-T G.711 (ulaw), G.722, G.728
	Video Codec	ITU-T H.261
Data Communication		HDLC
Audio Input/Output Function		Echo Canceler at every SL Position
		Sound Localization Function Max 5 Positions on the Display

3.1 Multipoint Network Configuration and Call Setup

PMTC/N-ISDN was designed to be connected to the ISDN by the basic rate interface in up to five locations. For multipoint communications over N-ISDN, we selected a distributed mesh type connection. This is because the centralized system, or star type connection, needs a Multipoint Connection Unit (MCU), and is more cost effective than a mesh type system. It has many limitations in handling multimedia such as audio windowing, and sub-conference capabilities. A comparison of multipoint communication systems is shown in Table 2.

PMTC/N-ISDN requires four lines of 2 B-channels; for the user's convenience, we assigned only one telephone number for the four lines. This method makes call setup procedures, easier at the multipoint connection. However, the two B-channels with synchronous communication to the particular site may not be guaranteed with this method because channel selection is determined on a first come first serve basis. We propose that the B-channel switcher mechanism enable synchronous connection between two arbitrary lines at the NIF to the MUX.

Table 2 A comparison of multipoint communication systems

System	Star Type	Mesh Type	Ring Type
MCU Node	Necessary	Unnecessary	Unnecessary
Communication Circuits (n:Number of Location)	n	n(n-1)/2	n
Information Exchange Ways	Restrictive	Sophisticated	Restrictive
Sub-Conference	Difficult	Easy	Difficult
Latency of Information Transmission	Medium	Low	High

3.2 Audio and Video Data Handling

Audio signals are mixed with those all other participant's audio data in the AVC. A sound localization technique as shown in Figure 3, is implemented so that users can discriminate participants easily. When a speaker is placed on either side of the monitor, the audio signals are concurrently distributed between the right and left speaker. As shown in the figure, the voice of person A is evenly distributed between the right and left speakers. The voice of person B is stronger through the left speaker while the voice of speaker C is stronger through the right speaker. To the listener, it seems that the voices of speakers match their respective locations. This makes it easy to identify the participants and to enhance the "feeling of being there."

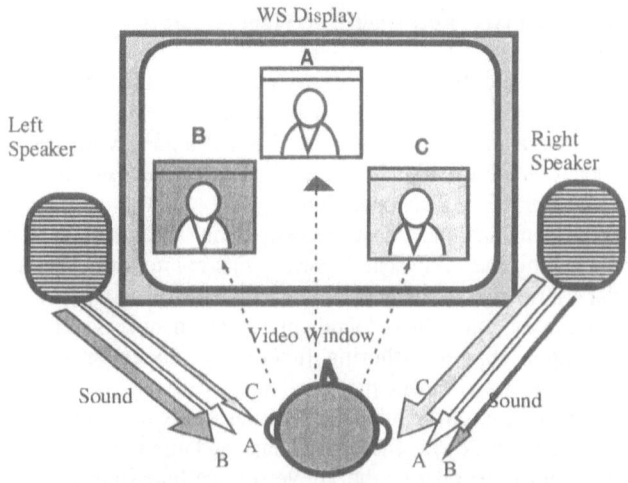

Figure 3 Audio window control corresponding to a video window position

However, as long as a speaker system in the PMTC is used, echo may result. We designed a multipoint communication echo canceling system which works in conjunction with a sound localization position on the display as shown in Figure 4. The echo canceling uses the same number of channels as the sound localization function.

A lot of visual data also needs to be displayed, such as the conference participants facial images, topic videos, and documents. This information needs to be organized effectively within the limited display area of the workstation. PMTC adopts the multiwindow system introduced by PMTC/B-ISDN, which works in conjunction with the video multiwindow controller and the video enhanced X Window System.

Topic videos or multimedia presentation scenarios are very useful as aids for easy-to-understand and vivid presentation. Multimedia presentations composed of audio, video, texts, graphics, and pointer data can be edited and stored on the workstation, and transmitted to other conference members and played back during the conference. Audio data and video data are encoded and decoded separately, however, and their synchronization can be adjusted by using a multimedia authoring tool which we have developed.

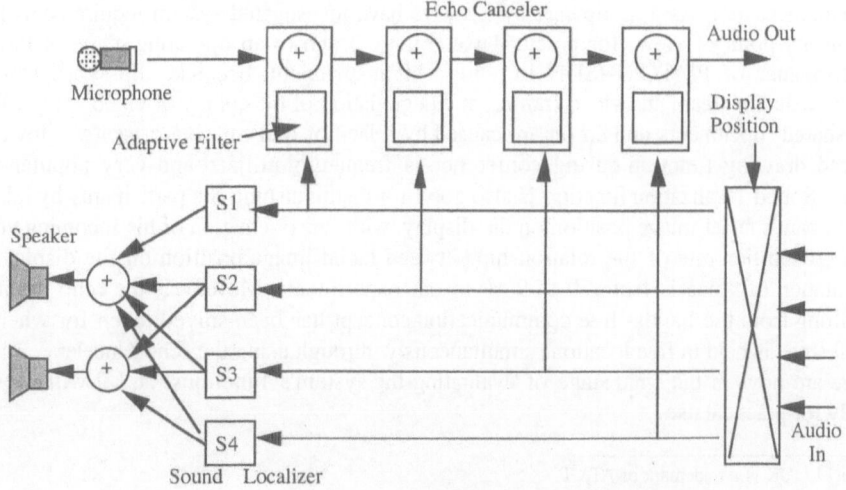

Figure 4 Configuration of multipoint echo canceler system

35

4. PMTC/N-ISDN Service and User Interface Design

The PMTC provides the following services to assist the operation of conference management on the desktop.

(1) PMTC/N-ISDN as a practical system has been designed with up to five locations connected by remote conference. The design is based on our investigation of office activities and our evaluation of PMTC/B-ISDN operation.

(2) Conference participants include a host or chairperson, and members. The host is given full power for managing the conference. Participants communicate with each other and operate multimedia information on shared virtual conference spaces.

(3) The system provides conference document management services such as document authoring, including a shared authoring function, a document registration function, multimedia authoring, and its automatic presentation.

(4) The system also provides some conference assistance services such as automatic delivery of conference announcements, conference documents, and minutes.

The PMTC allows users to operate the above service functions during the conference. Service functions described here require a relatively complicated interface as an operation. Therefore, easy-to-use and user friendly interface design is crucial. For easy-to-use multimedia operation such as media selection, audio and video control, shared drawing, and file handling over the virtual conference spaces, the following points have been discussed for user interface design.

(1) The system provides three kinds of conference spaces, common space, closed space ,and local space on the desktop. The spaces are set up as a conference desktop, a sub-conference desktop, and an ordinary desktop, respectively, as shown in Figure 6.

(2) Media operations over each desktop environment are designed with a look and feel concept based on the OPEN LOOK^{TM} GUI.

(3) PMTC is used on the networked environment, so that its security functions such as document filing, file transmission, authentication of conference participants are implemented as well.

5. System Evaluation

PMTC/N-ISDN has already been introduced in some locations as a distributed software development tool. As a groupware system, we have investigated system requirements from the user's point of view for practical use. Figure 5 shows an operating scene of PMTC. Performance of PMTC/N-ISDN is a little bit insufficient, because lip-sync problems between audio stream and video frames, and degradation of the quality of video while editing the shared documents and so on are caused by a lack of transmission capacity. However, shared drawing function during conference is frequently utilized and very popular with users. Sound localization function is also good for distinguishing the participants by relating participant's facial image position on the display with the orientation of his incoming voice. The recognition rate of the relationship between facial image position on the display and orientation of voice is from 70 to 90% in our experiments. Moreover, the echo problems resulting from the hands- free communication concept has been solved, even for when the conference is held in five locations simultaneously, through using the echo canceler.

We are now in the final stage of evaluating the system's functions, and it will soon be ready for practical use.

OPEN LOOK is a trademark of AT&T

Figure 5 An operating scene of PMTC

6. Conclusion

PMTC/N-ISDN is a sophisticated desktop multimedia multipoint communication system connected to the N-ISDN, that supports multimedia tele-conferences using texts, graphics and pointing data, high quality audio and video data based on the ITU-T H.320 terminal requirements, and incorporates document preparation, and database management functions. PMTC/N-ISDN provides convener change, conference participants withdrawal or addition during conference as part of conference management services, document registration and delivery before the conference as document management services, and multimedia authoring and its automatic presentation as conference assistance services.

For these multiple purposes, the system has been designed with an easy-to-use and friendly user interface. Graphical user interface with a look and feel concept, audio windowing, user customization functions are all adopted in our system. Based on user's evaluations, PMTC/N-ISDN effectively integrates various tasks of office work through a friendly user interface.

Acknowledgments

The authors would like to thank their colleagues for their helpful discussion.

REFERENCES

1. K.Shimamura and S. Masaki, A promising application for broadband ISDN: Hypermedia Teleconference, Symposium on Integrated Services Digital Networks for Telecommunications, Hawaii, 1989

2. E. J. Addeo, A. D. Gelman and V. F. Massa, Multimedia multipoint communications service capability for broadband networks, Proc. IEEE ISS, 1987

3. A. R. Ahuja, J. R. Ensor and D. N. Horn, The RAPPORT multimedia conferencing system, Proc. COIS , 1988

4. S. Sakata and T. Ueda, Real-time desktop conference system based on integrated group communication protocols, Proc. Int'l. Phoenix Conf. on Com. and Comm., 1989

DISTANT AND LOCAL ASPECTS OF EXCHANGING REAL-TIME MULTIMEDIA INFORMATION WITHIN DISTRIBUTED CLASSROOM COMMUNICATION SUBSYSTEM

Lucia GRADINARIU[1], Patrick PREVOT[2]

[1]CISM - Claude Bernard University
[2]GRACIMP - INSA de Lyon
Lyon FRANCE

INTRODUCTION

Applications based on distributed computing and multimedia information have changed performance requirements of network technologies and communication protocols. People have thought for a long-time that computer networks could offer only asynchronous communication services so that real-time transmissions wouldn't be possible. Since 1992, experiences with audio and video conference applications across the Internet have proven that there are strategies to integrate real-time services across the network chiefly when the "real-time" is cleverly traded between what an application really needs and what the network can offer.

The explosion of a new generation of applications across computer networks is the natural consequence of finding better methods to manage information in digital format and of integrating these methods in computer environments. Distributed Computing Environments are classes of objects (e.g.: information servers, communication services, networks, graphical interfaces, media drivers) and associated methods (e.g. methods to access, to transmit or to present information). Some of these objects are instantiated to create a particular environment where users can develop a particular activity. Users must "emerge" in this environment in order to pursuit their activity. Activity is mostly human. Session is a better term to express the temporary existence of a particular environment. A session might be started by users or by applications. Objects may be shared between different sessions when they have appropriate interface for each session. They belong to a session only when the session "needs" them otherwise they rest in a "potential" state. A session may associate states to objects that are involved in . A common or a distributed manager (under human control or following preconceived rules for that type of activity) may use information about these states in order to take decisions about session evolution. When session ends, all states are cleared and "active" objects returned to their "potential" existence.

The new generation of applications aims at providing users with such computer mediated environments to accomplish their activities better than in other environments. Computer Mediated Communication is a class of environments that enable transmission and reception of messages using computers as input, storage, output and routing devices[1]. Such

environments may support session oriented activities that use multimedia, multi-party communication:
- synchronized distance teaching (from teaching to collaborative learning);
- cooperative work in telepresence (from tele-meeting to real-time co-design);
- multimedia information recover (from media on demand to intelligent query multimedia databases);
- tele-work in virtual reality (from data collection to local intervention in remote fabrication process);

Designing computer mediated communication environments is a complex task. This is the reason why we've started with an experimental platform for distant teaching and training. This platform is flexible enough to let us analyze different aspects of exchanging real-time multimedia information in distributed environments.

FROM COMPUTER ASSISTED TO COMPUTER MEDIATED LEARNING SYSTEMS

An efficient learning system must be a robust and flexible tool serving teachers and students on achieving high quality of taught information. Computer assisted learning systems were designed to facilitate knowledge organization and representation independently of teacher's presence or student's location. Most of them are based on built-in learning scenarios and an associated knowledge database dedicated to the set of concepts which students have to learn. Multimedia authoring tools offer a fascinating framework for new architectures of computer assisted learning applications. This new perceptual quality of learning incites students to interactivity and enforces their need of being integrated to the learning environment. The only way to fully achieve these goals is to add communication means to this environment and let users collaborate and share learning resources independently of distance. Hence we'll be able to set up distributed learning systems that offer real-time interactivity between students and a real or a virtual teacher. These systems must be reusable in various domains and must allow users to dynamically choose a learning scenario. In this context, asynchronous individual learning and synchronous real-time collaborative learning are no further antagonist but they complement each other to finely complete education.

Recent evolution of network technology and widespread development of multimedia applications enable the transition from the classroom social space to a "cyberspace" of computer supported distributed classrooms at least for experiments. We need experiments as there is no other way to capture feed-back from user behavior during a computer mediated collaborative learning session. We designed our platform to be flexible that is to allow dynamic composition of object instances within a session (like setting-up or tearing down an audio channel between a group of participants, coherent presentation of information issued from distributed servers, temporarily sharing the same work-space between distant users). The result is that students and teachers may tailor the learning environment to their specific needs on the spur of the moment. This is the consequence of multimedia integration in distributed applications and of communication subsystems ability to accommodate transmission requirements of these new applications.

GENERAL PRINCIPLES IN DESIGNING DISTRIBUTED CLASSROOMS

We define distributed classrooms environment as a "world class" with access points (multimedia terminals or workstations) spread across a wide area network. Networks make distance transparent to end users and support distributed access to information and services or applications. Teachers may teach to this class while standing in front of their friendly desktop workstation. The teaching support could be the "hypercours"[2] which is a temporarily association of knowledge representations as multimedia information on distributed servers. Students may start training applications to achieve understanding by practice whilst interacting with the teacher or with other active group of students. We think this approach for distributed classroom environments is able to support real integration of learning, training and communicating thus meeting the need "to hear, to see and to do by myself" in order to maximize understanding.

We chose to implement a distributed class-rooms infrastructure (communication support and applications) as this is the emerging one for our campus main activity: teaching and training. Moreover such a platform, which we consider the simplest both as mechanism and policy of communication subsystem, offers us experience for future implementations of groupware environments that behave more complex group communication and management aspects[3]. We looked for an architecture which allows widespread development and exploitation of multi-user, distributed, multimedia applications whilst isolating applications and users from the complexity of multimedia, multi-party, communications. The horizontal decomposition of this architecture emphasizes on two main tasks for the communication subsystem:

- to transport multimedia information across heterogeneous networks;
- to retrieve and to present multimedia information on user's access point to the learning environment.

There is also a vertical decomposition which mainly reveals management and control tasks of a basic mechanism for establishing communication channels and exchanging multimedia information in a heterogeneous and dynamically changeable environment.

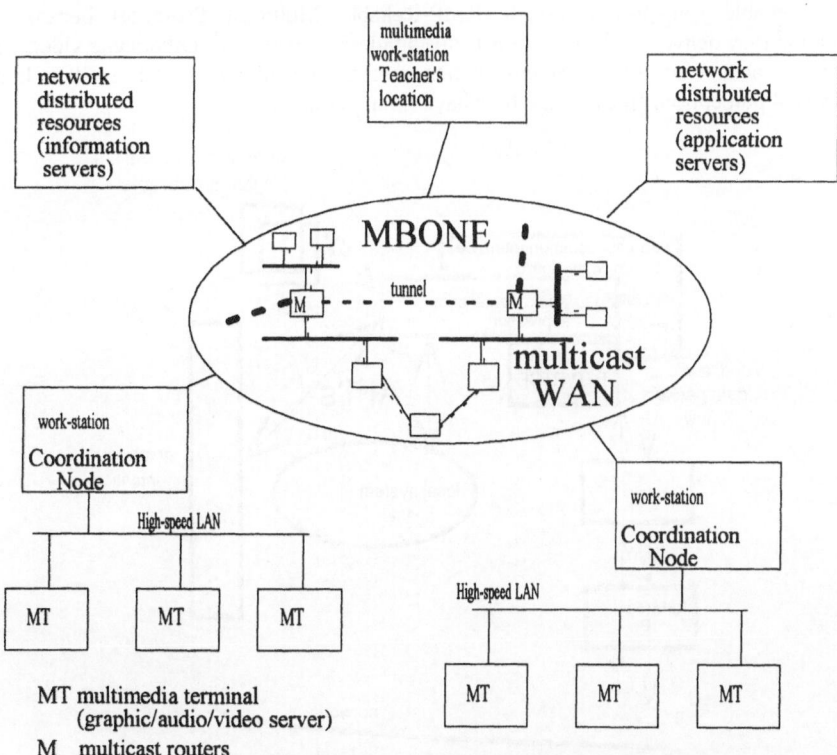

Figure 1. Communication infrastructure for distributed classrooms

ARCHITECTURE OF AN EXPERIMENTAL PLATFORM

Our platform reuses good, existing "bricks" like multicast transmission on computer networks and conferencing applications developed on its principles. We've extended interactivity at terminal level while multiplexing user access points at node station level.
Figure 1 identifies a "backbone" of nodes which are UNIX workstations charged to support optimal transfer of multimedia information flows across long distances. These nodes accommodate services charged to do this transfer actually by experimental means (IP[4] multicast, resource reservation, transport protocols, adaptative applications for real-time audio/video communication across the Internet). Due to real-time transmission constraints to respect while using an unreliable network as support of communication, these services are loosely coupled instances of the same application (audio/video conference, shared whiteboard) running on each workstation which is member of the distributed. The transport protocol underlying these communication applications (e.g. RTP, the Real time Transport Protocol) has to encapsulate information about the continuous media flows (video, audio, images) it delivers: timing, source of synchronization, encoding format, etc. It might also provide information which is useful for session management coordination of application instances across the multicast platform like source identifier or packet sequence number.

This is the distant aspect of exchanging multimedia information in distributed classrooms. The platform can scale to very large geographical areas as it uses Mbone, Internet Multicast Backbone[5], for multicast distribution to distant sites. Workstations which participate in the group play the role of switching nodes relaying student terminals to the session. Mbone conferencing applications are examples of available tools providing computer mediated communication services. These applications use specific transport protocols which enable point-to-multipoint exchange of multimedia information (in a reliable or unreliable manner) across wide area networks. Figure 2 presents local instances of such applications. Some of them (the multicast web browser and the shared whiteboard) need a reliable transport protocol (RMP-Reliable Multicast Protocol) because the information they deliver to the final user is meaningless if corrupted. Others like video/audio conferences accept some loss of information as this information is addressed to human audio/visual perception. In exchange they have timing constraints.

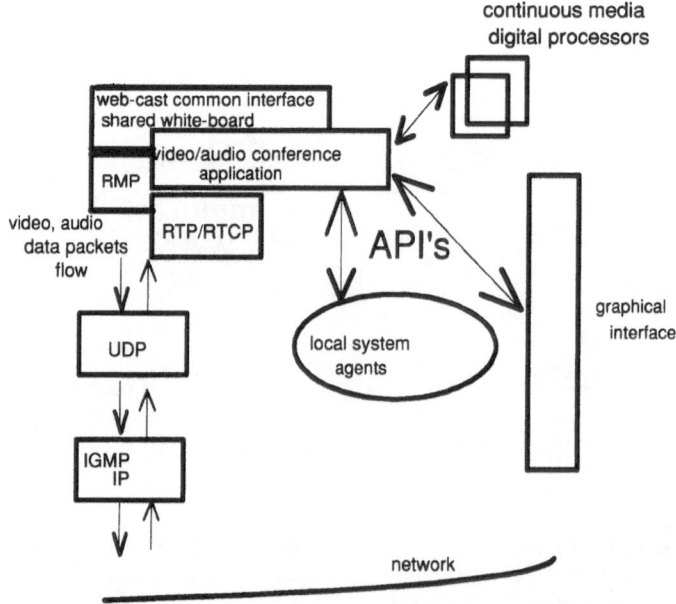

Figure 2. Local instances of multimedia multi-party communication applications.

The real-time quality is acquired by dynamically adapting application output rate and receiving buffer size to network variations. These methods are not (yet) enough to assure a good presentation quality as perceived by humans. Network resource reservation policy and specifications for real-time services integration should improve this aspect. An important role to correctly use these network services is to characterize distributed application traffic flows. Our platform contains traffic analyzers to capture and to compute by statistical methods models for traffic flows engaged by distributed classrooms specific applications.

Another important aspect deals with information exchange between nodes and user access points in order to offer to them a consistent classroom environment: view of the lecture, teacher video/audio presence, awareness of other colleges, group interactivity, private working space, etc. A student should enter in this environment by means as simple as possible (like a logging command) which enable him to access to all communication and presentation services, in a natural way, eventually moderated by the session policy (floor control, access authorization, communicating only with a specified group or person, etc.). We consider user access point as being a multimedia terminal doing "input" and "output" of multiple types of information accordingly to the semantic of the ongoing session. The goal is assuring synchronous, real-time, computer mediated communication as services that might be called-up by users at terminal level. The communication applications running on the node stations must be started with parameters that can assure the service quality requested by users (or a coordination entity) at "connection" set-up time. Users can decide when and how they use communications services. This approach was chosen because:
- this is the natural way people work and/or communicate that is each user can choose how, when and whom he wants to talk to;
- it's easier to characterize and thus to coordinate and assure a certain quality to each specific service;
- service provider (communication application) can deal much easier with applications/platforms heterogeneity/dependency;
- service implementation can be modified or upgraded to improve quality in a transparent way for users/applications.
Meanwhile there are aspects of using such services which are very dependent of session environment (distant teaching, collaborative working, entertainment, etc.):
- specifying the quality of service and, for a set-up connection, dynamically changing QoS parameters following application needs;
- making a decision when a certain service or a certain QoS is refused or temporarily inaccessible;
- managing user access to services in terms of authorization, limited number, aggregation, redundancy, explicit or implicit on transmission or on reception, etc.

MULTIMEDIA TERMINALS AS USER ACCESS POINTS

The multi-party communication applications we've analyzed before offer direct user access on multimedia workstation console. This speeds-up presentation of multimedia information to user (addressing local media drivers, using shared memory for image displaying). We can't afford to offer a multimedia workstation to each student but we dispose of terminals (X-terminals or PCs) with soft servers and associated libraries of functions for different media drivers. These terminals allow presentation of multimedia information received by the node station. They can also send information from the user to the node station and thus to other users connected to the same environment. The multi-party communication applications which are instantiated during a session on the node station have to be associated with some "gateway" applications. These gateways fulfill tasks related to media flows management: encoding format conversion, timing and synchronization of different streams, mixing several streams of the same type on a single output channel, etc. A user control agent manages user access and permissions from the moment when he connects to the environment until the end of the session. This agent has only a local existence as it controls the access points (terminals) connected to a node station. The node session control agent knows about user agent activity and manages session fulfillment at the node station in coordination with session control agents from other node stations connected to the environment. Figure 3 is a description of a remote presentation system configuration.

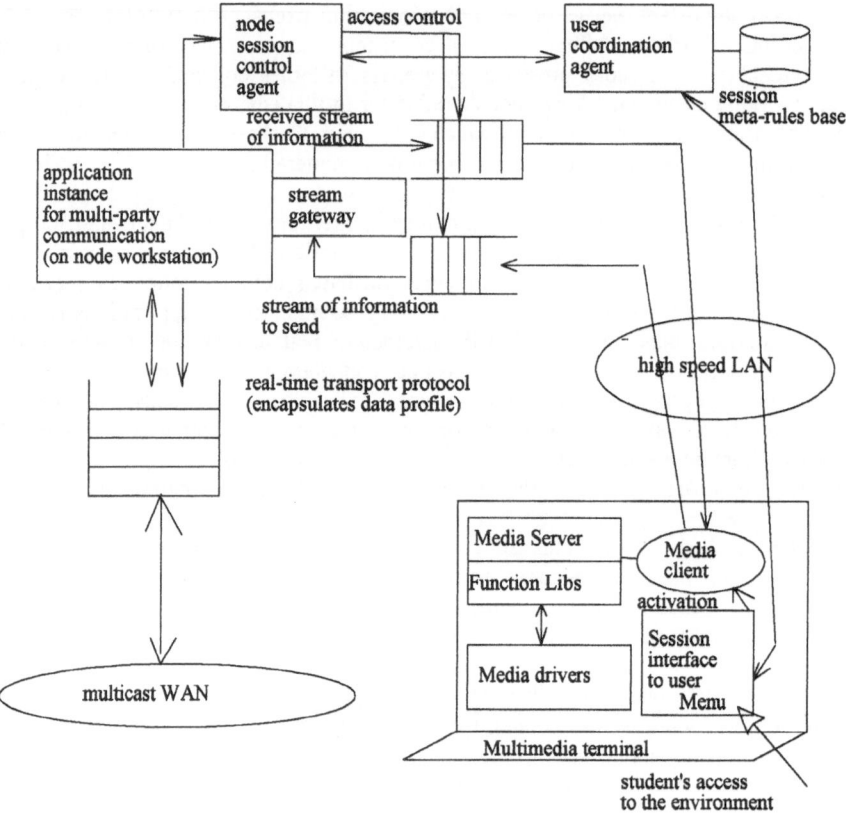

Figure 3. Remote presentation on multimedia terminal.

Using terminals makes it hard to acquire the same performances on presentation speed as if we had used only node station graphical console. This is another goal of our experiments: to optimize QoS parameters values for long distance transmission with respect of local presentation needs in terms of delay, error recovery, rate of information flows. We're studying some issues for local exchange of multimedia information in order to achieve real-time perception of end-to-end communication service at least at the level of conformance a user connected to a multimedia workstation console may perceive it. Some factors as the processing charge of media presentation deported to terminal CPU and use local multicasting and high speed connection on LAN have largely ameliorated time constraints for local information broadcast.

PERFORMANCE ASPECTS AND CONCLUSION

We identified some important parameters that influence the degree of performance a multimedia communication environment might offer to its users:

- quality of media presentation necessary to meet user perceptual needs;
- duration of media utilization, distribution of duration during a session
- number of communication channels (per used medium) and transmission direction;
- distance and/or distribution of media presentation end-points.

All these parameters are very dependent of session context. The platform we've set up is in service for experimental distance lecturing to students in various disciplines in order to discover more detailed aspects of this dependence related to course contents, people interactions and media utilization for knowledge representation. The most important thing during classroom activity is that students must acquire and understand concepts they are taught. The time they need for this has to be minimal and is correlated with communication system performance and with media utilization to represent elementary knowledge. These two factors determine the accessibility of lecturer's presentation and the level of interactivity during a session. Hence we may conclude that the efficiency of a computer mediated learning system is founded both upon teacher's ability to judiciously blend different media to present information to his students and upon communication applications performance to offer all functions user needs to fulfill his activity during a session.

REFERENCES

1. M.F. Paulsen *The Online Report on Pedagogical Techniques for Computer-Mediated Communication*, NKI, Oslo, Norway (1995).
2. N. Le Than, Ph. Roussel, A. Cavarero, ARCHE: atelier de rédaction de cours HypErmédia, *Hypertextes et Hypermedia, Realisations, Outils et Méthodes*, HERMES Editions, Paris, France, (1995).
3. L. Gradinariu, G. Beuchot, Conferencing Framework: Applications and Communication Aspects, *Proceedings of the Second Conference of Concurrent Engineering CE'95*, Washington, DC (1995).
4. D. E. Comer. *Internetworking with TCP/IP*, vol. 1, Prentice Hall, Engelwood Cliffs, New Jersey (1991).
5. M. R Macedonia, D. P. Burtzman, MBone provides audio and video across the Internet, *IEEE Computer*, (April 1994)

ANALYSIS OF MPEG-1 VIDEO TRANSMISSION OVER A SHARED ETHERNET LAN

Alan R. Bugos and Shuang Deng

GTE Laboratories, Incorporated
Advanced Transport Techniques Department
40 Sylvan Road
Waltham, MA 02254

abugos@gte.com and sdeng@gte.com

ABSTRACT

The transmission of real-time full-motion digital MPEG video at 30 frames per second over local area networks will be an important service for educational networks, business networks, and possibly in the home on residential LANs. Many LANs have been designed with shared Ethernet technology which provides robust low-cost data communications networking supporting 10 Mbps of shared bandwidth between interconnected computer users and data servers. Digital video standards such as MPEG-1 and MPEG-2 have been developed to provide full-motion video compression for transmission at approximately 1.5 Mbps and 3 to 6 Mbps respectively. The study described in this paper details an analysis of MPEG-1 video and audio transmission over a shared Ethernet network. The network performance is characterized for various MPEG-1 streams with light bursty background traffic loads. The paper presents results for up to 6 MPEG-1 video streams in the presence of background traffic where cumulative data rate, packet lengths, packet size distribution, packet inter-arrival time and packet jitter are described for MPEG-1 video clips. In addition, the network traffic measurements are compared to network simulations. Hence it is shown that a shared Ethernet network environment may be suitable for limited MPEG-1 video transmission.

INTRODUCTION

The use of real-time digital video for multimedia applications over local- and wide-area networks is on the rise. Many computers are multimedia capable and have the ability to decompress digital video in real-time as the video is delivered over a LAN or WAN. The growth in data communications networks has increased rapidly over the last few years. Many schools, universities, businesses, and industries use Ethernet or token ring local area networks to link computers and computing resources for users on enterprise networks. It is likely that residential LANs in the near future may consist of low-cost shared (or switched) Ethernet between a residential router connected to a service provider via asymmetrical digital subscriber loops (ADSL) or hybrid-fiber coax (HFC) networks. Central office video servers may someday provide MPEG-1 or MPEG-2 video transmission to the local users.

Multimedia Communications and Video Coding
Edited by Y. Wang *et al.*, Plenum Press, New York, 1996

Additionally, schools, universities, information kiosks, and businesses may use low-cost shared Ethernet networks to transport real-time MPEG-1 video for informational, educational or training purposes. Many applications for networked digital video are being developed and include video-on-demand (VOD), video mail, interactive video-based training and education, sales videos, information kiosks, interactive video games and simulations, video informational databases, and digital video archives for video editing, TV commercials, and advertisements.

The research described in this papers details an analysis of MPEG-1 video transmission over a shared Ethernet local area network conducted within the Advanced LAN Testbed at GTE Laboratories. The experimental configuration consists of a video server with MPEG-1 video content that is distributed through a shared 10BaseT Ethernet network to client computers with MPEG-1 decoder cards. All video traffic is captured over the network and post-processed and analyzed in detail. The network performance is characterized for several MPEG-1 video (and audio) streams with light bursty background traffic loads. Each video stream was analyzed for packet length or size, inter-arrival time, cumulative data rate, and packet size histogram. Additionally, network traffic measurements are compared to network simulations which were included as part of this study. The results of this study may be useful in designing local and wide area networks which utilize MPEG-1 digital video applications.

MPEG-1 VIDEO

MPEG compression standards are set by the Moving Pictures Expert Group (MPEG) of the International Standards Organization and the International Telecommunications Union (ITU). MPEG-1 is designed for image sequence compression and unlike still image compression, takes advantage of the correlation in the time dimension. The development of the standard is segmented into two phases. MPEG-1 is intended for image resolutions of approximately 360 pixels x 240 lines and data rates of 1.15 to 3.0 Mbps. This rate is notable because it is the data rate of uncompressed audio CD-ROM and digital audio tape (DAT). MPEG-2 is designed for broadcast video and telecommunications applications, including interlaced video and higher resolutions, and higher bit rates of 3 to 15 Mbps. [1,2]

MPEG-1 starts off with 352 x 240 pixels at 30 frames per second (NTSC). The color images are converted to YUV space, and the chrominance channels are decimated further to 176 x 120 pixels. As mentioned previously, less resolution is needed in those channels because the human eye is less sensitive to color than brightness. Motion prediction is done in the Y channel on 16 x 16 blocks: given the 16 x 16 block in the current frame, a close match is desired to that block in either a previous or future frame. Discrete cosine transforms (DCT) are done on 8 x 8 blocks to organize the spatial redundancy and the DCT coefficients are quantized. The results of all of this, which include the DCT coefficients, the motion vectors, and the quantization parameters, are Huffman coded using fixed tables. [2,3]

In coding an image sequence, the individual pictures are classified into three types: I or intra frames, P or predicted frames, and B or bi-directional frames. I frames are coded without referring to other frames and are therefore the functional equivalent of a JPEG image. P frames are coded relative to a motion-compensated prediction from a previous I or P frame. B frames are coded relative to a previous and/or a future frame, and the prediction is motion-compensated. The prediction of the B frame can be forward or backward relative to a P or I picture or to a simple motion-compensated average of both preceding and following P or I pictures.[3] The sequence of decoded frames usually is similar to the sequence 'IBBPBBPBBPBBIBBPBB ...', where there are 12 frames from I to I frame. For the decoder to work, the first P frame must be sent before the first two B frames.[2,3] An MPEG-1 stream consists of variable length audio and video packets multiplexed together and are known as Elementary Stream (ES) packets.

Each ES packet has a header and a payload where the payload contains contiguous bytes of Access Units (AUs) which are coded representations of the audio/video frames. The principal function of an MPEG decoder is to decode AUs from the bit stream. MPEG-1 is obviously a very asymmetrical algorithm, and encoders producing high quality video are very somewhat costly. Decoders offering MPEG playback are becoming much more affordable, however, and the image quality of MPEG video is by far superior to any of the other compressed digital video formats.

NETWORKED VIDEO SERVER AND VIDEO PROTOCOLS

Many multimedia servers are now available on the commercial market and these servers are designed to efficiently transport multiple continuous video and audio data streams from a server to a client computer. Additionally, protocols for continuous data streams to be transported over local- and wide-area networks have been developed and include protocols such as ST-II, Real-time Transport Protocol (RTP), Xpress Transfer Protocol (XTP), and Versatile Message Transfer Protocol (VMTP), to name a few. [4, 5]   In this study, a video server and video server software designed by StarLight Networks of Mountain View, CA was used on the network. StarWorks™ is a digital video networking software designed to support multiple clients simultaneously.   It consists of server software, which converts a high-end PC into a streaming video server, and client software, which allows multiple remote clients to transmit and receive video stream data over the network. The server software handles storage, session, and stream management. The storage system uses an array of magnetic disks and a disk access algorithm developed by StarLight known as "Streaming RAID" (RAID - Redundant Array of Independent Disks). Multiple SCSI drives are used in order to achieve a larger total throughput, and magnetic disks are used because of their random access capability, wide availability, and low cost. [6]

The transport layer protocol used is Media Transport Protocol (MTP), a proprietary rate-controlled protocol that controls the flow of video on the network. It allows more video traffic than could be obtained with other protocols not designed for video while still remaining compatible with existing data network protocols such as TCP/IP and IPX, although it is non-routable.   MTP was developed by StarLight Networks and is loosely based on XTP (Xpress Transfer Protocol), an efficient real-time transfer protocol designed to run on high-speed networks. [5]   MTP supports streaming data better than other protocols such as IP or IPX because it maintains contiguous data transfers while providing a way for control messages to get through to the server operating system. MTP allows large amounts of data to be sent without an acknowledgment, making it very well suited to video applications. The server and client computers are "connected" or virtually mapped to each other through Ethernet MAC layer addresses which will allow video to be transported on any bridged network with the proper bandwidth.

EXPERIMENTAL NETWORK CONFIGURATION

Experiments were conducted within the Advanced LAN Testbed at GTE Laboratories, Inc. to demonstrate the feasibility that several MPEG-1 video/audio streams can be supported over a shared Ethernet network. A shared 10BaseT Ethernet LAN configured in a star topology is used in the experimentation primarily because of its simple design and low cost. Since switched Ethernet offers much better performance than shared Ethernet, the use of a shared Ethernet network enables us to investigate the worst case scenario.

During the MPEG video experiments, several MPEG-1 video/audio streams were captured over a test network as shown in Figure 5. The StarLight video server has the capability to deliver 50-60 simultaneous MPEG-1 video streams to client computers over the Ethernet LAN (aggregate data rate up to 90 Mbps). Several network analyzers and data generators are connected to generate light background traffic (at 2% to 10% using TCP/IP or UDP) and to capture Ethernet data packets on the network. Each computer is connected to the shared Ethernet hub via a 10BaseT Ethernet connection (CAT-5 wiring).

Several different shared Ethernet hubs were used in the study and all of these hubs were designed around a Multiport Ethernet repeater integrated circuit (Advanced Micro Devices AM79C981 or AT&T T7201AMC). SNMP agents built into the hubs were used to observed collisions, frame alignment errors, FCS errors, and so forth. Each computer contains either a Sigma Design's or an Opti-Base MPEG-1 decoder card. The MPEG-1 video streams are ISO 11172-1 compliant video clips with audio and were encoded with either a Compression Labs Incorporated (CLI), Opti-Base, or Sigma Designs MPEG-1 encoder. One minute video sessions were chosen to simplify the post-processing analysis. All raw data packets on the Ethernet network were captured with *snoop*, a Unix packet capture routine or *EtherPeek*, a data capture program developed by the AG Group, Inc. and post-processed using network analysis and standard graphing tools. The time resolution of EtherPeek is 40 μsec.

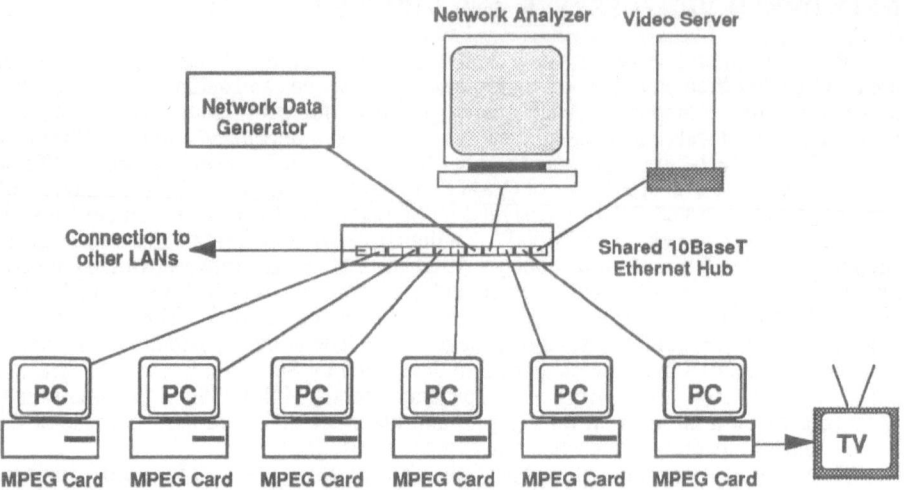

Figure 1. Experimental setup for MPEG-1 traffic analysis over a shared Ethernet network.

Typically the cumulative data rate (or bandwidth), packet length as a function of time, inter-arrival time, and packet size distributions (or packet size histograms) are shown. The cumulative data rate (in kbps) is defined here as the summation of the packet size (in bits) as a function of time divided by session time (in seconds) and is shown as

$$\text{Cumulative Data Rate}_i (kbps) = \frac{\sum_{i=0}^{n} PS(i)}{t_i}$$

where PS is the packet size in bits and t_i is the session time. The inter-arrival time is defined as the delta time between arriving data packets at either the client and/or server and are usually given in milliseconds.

DISCUSSION OF RESULTS

The results of this study are divided into measured LAN traffic measurements and network simulations using a computer workstation and are summarized in the following two sections. Note the overall data accumulated in this study of MPEG-1 traffic measurements is vast and only a small part of the results are presented in this brief paper.

Measured Ethernet LAN Traffic Observations and Analysis

The packet length versus session time (one second window) for a single MPEG-1 stream (upstream and downstream packets) is shown in Figure 2 and characterizes the temporal nature of the video stream in terms of packet size. Downstream packets sizes from the server to the client are either 1508 or 1200 bytes, whereas the 100 byte packets are both upstream and downstream acknowledgment or stream control packets. It can be seen that the server delivers twenty-two 1508 byte packets and one 1200 byte packet before it sends a 100 byte control packet to the client. The average time spacing between stream control packets is approximately 65 milliseconds. The client uses a buffer to hold the MPEG data before playout, and requests the next frame when the buffer is below a certain threshold. The packet size histogram for the same MPEG-1 stream is presented in Figure 3 which shows the total number of packet occurrences over the Ethernet network as a function of packet length. Clearly the majority of the packets transported over the network are 1508 bytes long, which is nearly the maximum size on an Ethernet packet. The 1200 byte packets are typically the remnants of an MPEG-1 stream frame after a series of the maximum length Ethernet packets and the 100 byte packets are stream control packets.

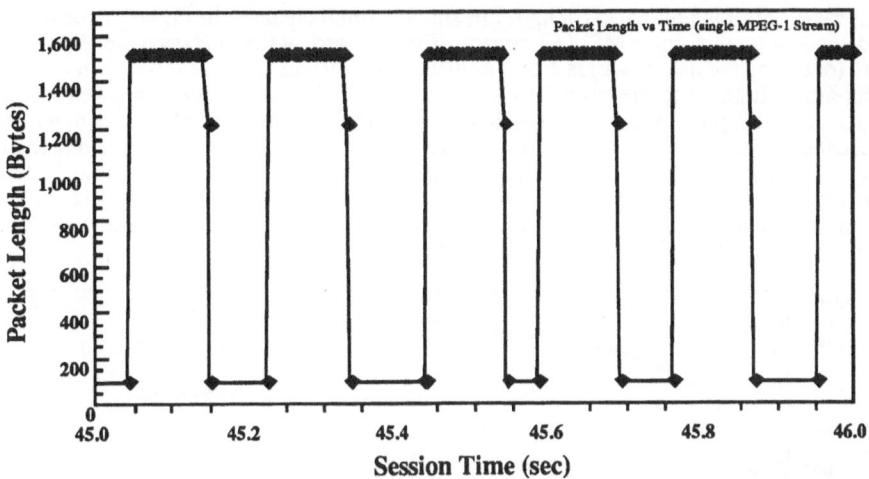

Figure 2. Packet length versus session time for a single MPEG-1 stream.

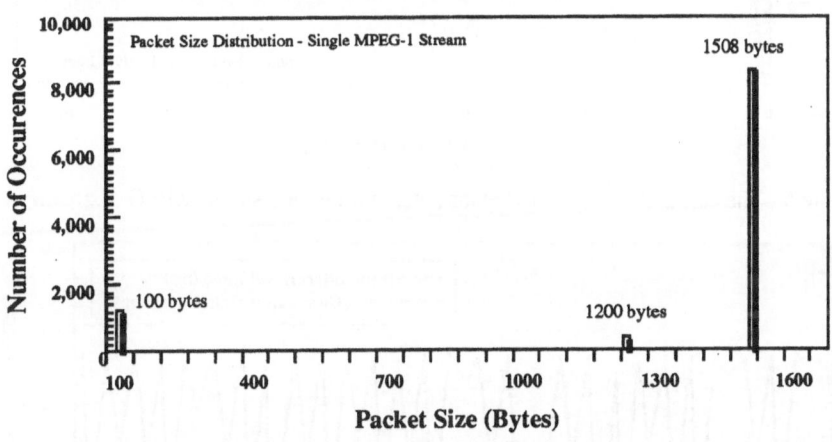

Figure 3. Packet size histogram for a single MPEG-1 stream.

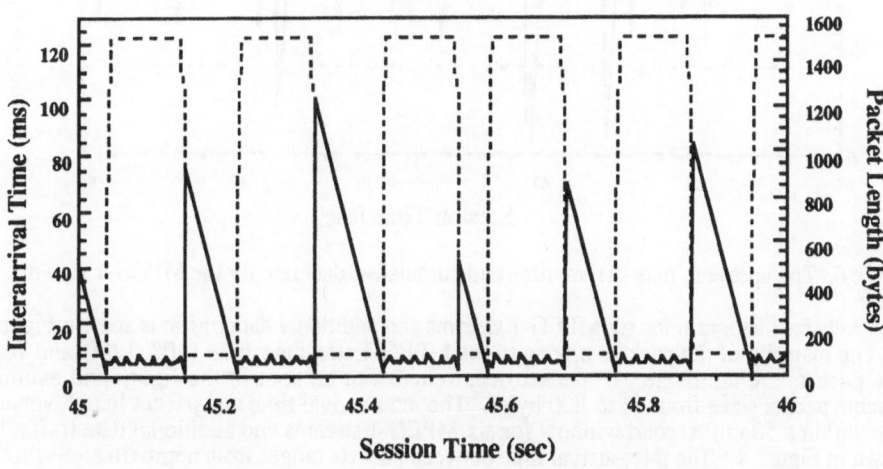

Figure 4. The inter-arrival time and packet length for a single MPEG-1 stream.

The inter-arrival time and packet length are shown simultaneously in Figure 4 for a one second window for this single MPEG-1 stream. The inter-arrival time between the 1508 byte packets (or during the data block) is 1 or 2 milliseconds whereas between the stream control the inter-arrival time ranges between 40 to 95 milliseconds with an average of approximately 65 milliseconds. Figure 5 shows the cumulative data rate for a single MPEG-1 stream which is 67 seconds in length. The data rate for this particular video clip has a sustain average of roughly 1.51 Mbps over the session time. The upstream inter-arrival time and upstream cumulative data rate for a single MPEG-1 stream are shown in Figure 6. The upstream data rate from the client is relatively low and averages to 10.4 kbps and the average upstream inter-arrival time is approximately 70 milliseconds.

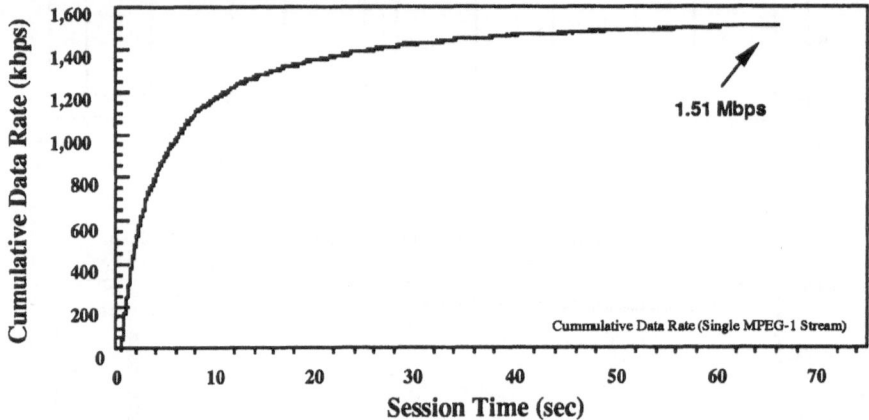

Figure 5. The cumulative data rate versus session time for a single MPEG-1 stream.

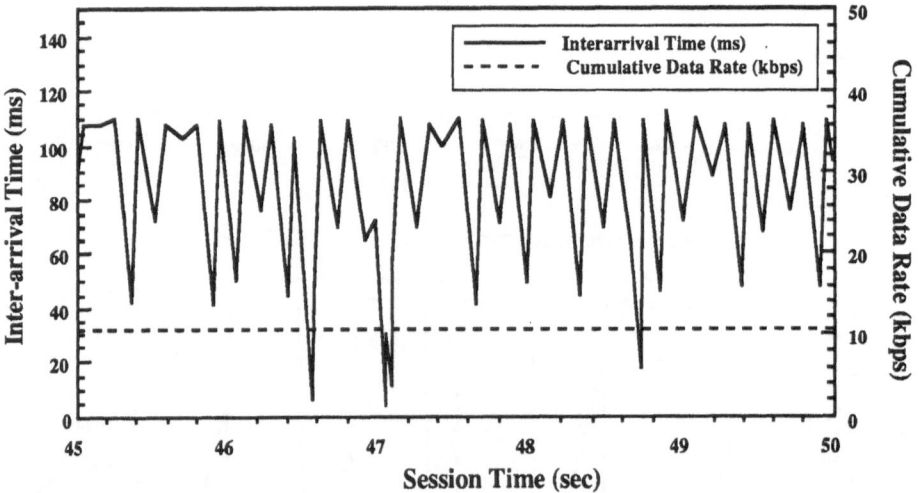

Figure 6. The upstream inter-arrival time and cumulative data rate for the MPEG-1 stream.

A packet size histogram for six MPEG-1 streams and additional data traffic is seen in Figure 7. The majority of the packets belong to the MPEG-1 streams where 1508, 1200, and 100 byte packets are dominate. IP packet occurrences can be seen in the figure and exhibit variable packet sizes from 64 to 800 bytes. The inter-arrival time and packet length versus time within a 50 millisecond window for six MPEG-1 streams and additional data traffic is shown in Figure 8. The inter-arrival time between packets ranges from approximately 1 to 3 milliseconds on the network during peak usage. The total and upstream cumulative bandwidth for six MPEG-1 streams and additional data traffic are shown in Figure 9. The combined six MPEG-1 streams reaches a cumulative data rate of roughly 9.251 Mbps. The upstream cumulative data rate is approximately 50 kbps. Since the video streams are not data coordinated, and have different rates, the aggregate traffic appears to be more bursty than single streams.

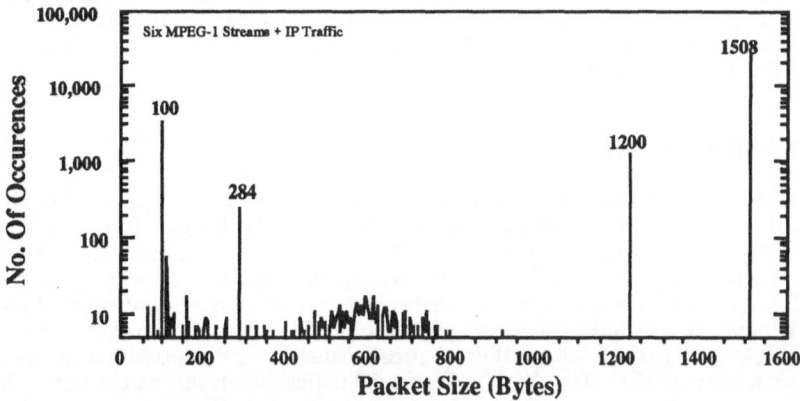

Figure 7.   Packet size histogram for six MPEG-1 streams and additional data traffic.

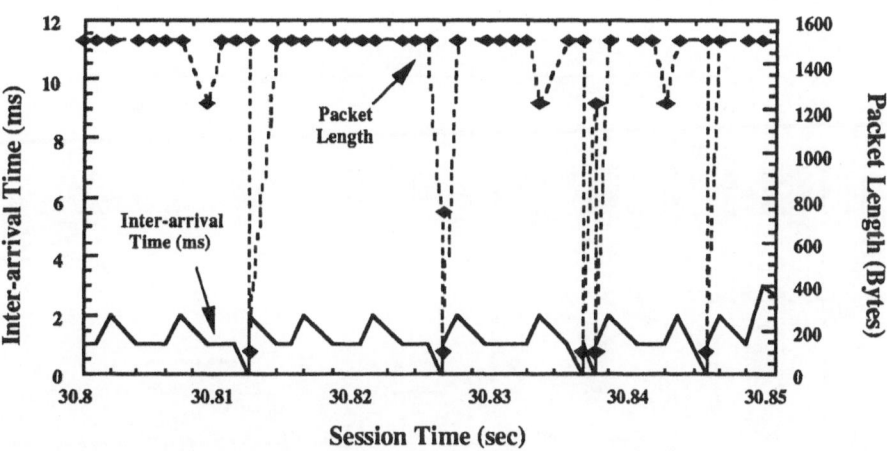

Figure 8.   The inter-arrival time and packet length versus time within a 50 millisecond window for six MPEG-1 streams and additional data traffic.

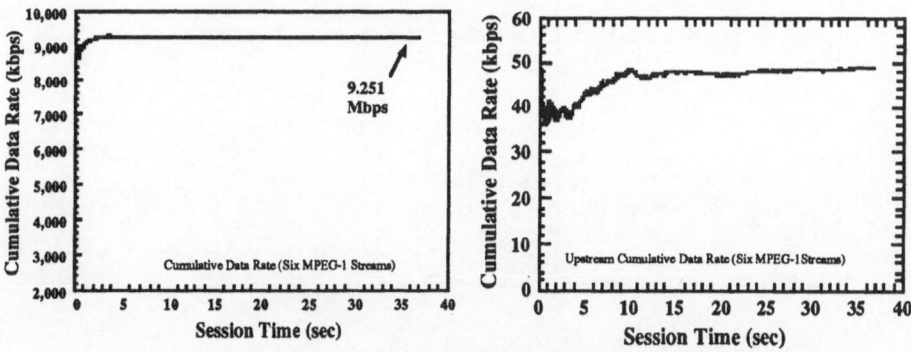

Figure 9.   The total and upstream cumulative bandwidth for six MPEG-1 streams and additional data traffic.

The jitter caused by the collision and deferment on the Ethernet network appears to be well within the range of tolerance of the MPEG decoder card and application. Total collisions and errors seen on the network were less than two percent and had no major effect on the video performance. The overall subjective quality of all six MPEG-1 video pictures was very good and no noticeable visual or audible degradation were detected. Hence, six MPEG-1 video streams with light bursty background traffic were transported over the Ethernet network with little or no contention. The use of playout buffers as a means to improve any jitter will be investigated in future research.

Network Simulation Results

A simulation model using OpNet simulation software and a Sun Sparcstation 20 was created to study the video packet delay characteristics and establish the performance baseline for current MPEG-1 video transmission within the testbed. Data packets were assumed to arrive according to a two-state MMPP. The average data rate was set to very from 0.75 Mbps to 2 Mbps. Video packets can experience delay, due to collisions with the background data traffic including the upstream control packets of MTP. Figure 10 shows the simulation results of packet delay of a 6 Mbps MPEG-1 stream caused by collision with different level of background data traffic. The delay in Figure 10 noticeably occurs as clusters. This is caused by two possible factors. First, the data and video traffic may arrive in bursts. Secondly, Ethernet itself has a "capture effect" that leads to a station, or a group of stations, fail to obtain Ethernet access for a prolonged period of time [7,8]. The overall delay effect, however, appears to be limited.

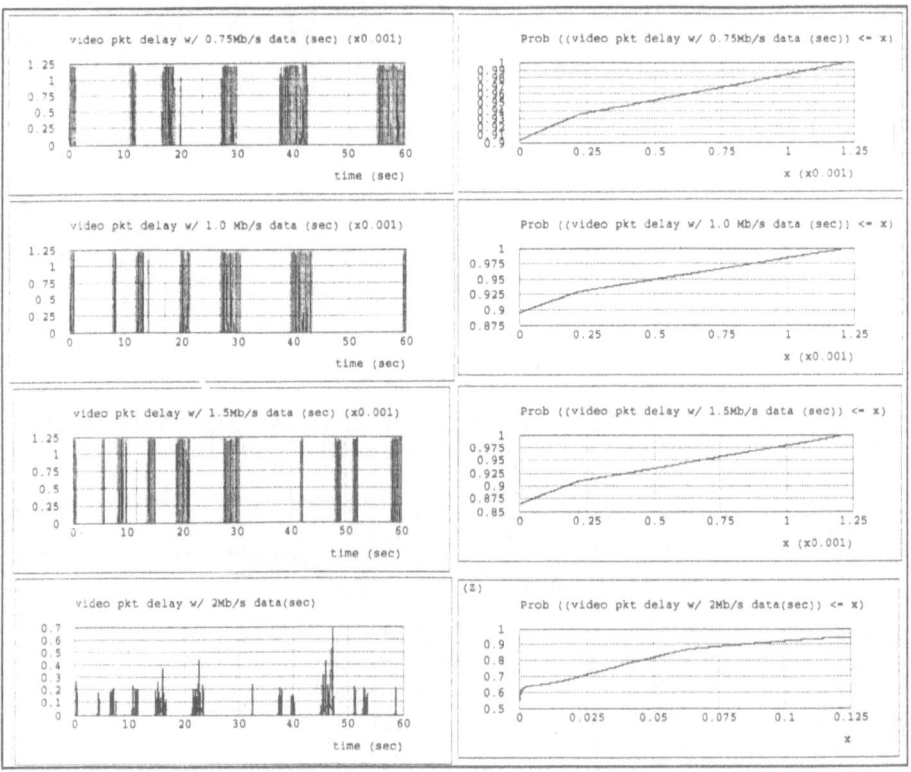

Figure 10. Packet delay for a 6 Mbps MPEG-1 stream.

The results suggest that MPEG-1 packet delays are bounded at below 1.25 ms when the background traffic is at 1.5 Mbps or less. Video packet delays vary in a wide range, up to 125 ms, with 2 Mbps background data traffic. It is believed that this is mainly due the high asymmetry ratio of video traffic in Ethernet. There are far fewer collisions with video data than, say, with a typical data applications LAN. These simulation results also correlate to our observation of satisfactory performance at six video streams, and noticeable degradation at six streams, with network load of about 7 Mbps and 8 Mbps respectively. In addition, one can infer from the results that the current application appears to be capable of tolerating a few millisecond of delay jitter.

SUMMARY AND CONCLUSION

The study described in this paper details an analysis of MPEG-1 video and audio transmission over a shared Ethernet network with light bursty background traffic loads. The results show that up to 6 MPEG-1 video streams in the presence of background traffic were sustained at an aggregate data rate of approximately 9.251 Mbps with no noticeable subjective degradation of video or audio quality. Packet length as a function of time, inter-arrival time, and packet size histograms were shown for a single MPEG-1 stream and for six MPEG-1 streams with background traffic. In addition, the network traffic measurements were compared to network simulations. Hence, it is shown that a shared Ethernet network environment may be suitable for limited MPEG-1 video transmission.

ACKNOWLEDGMENTS

The authors wishes to express words of thanks to their colleagues, Robert Olshansky and Paul Hill at GTE Laboratories for many useful discussions on digital video transport over LAN and WAN data networks.

REFERENCES

1. William Pennebaker and Joan Mitchell, *JPEG Still Image Data Compression Standard,* New York, Van Nostrand Reinhold (1993).

2. Le Gall, D., "MPEG: A Video Standard for Multimedia Applications," Communications of the ACM, April 1991, Vol. 34, No. 4, pp. 47 - 58.

3. Mark Adler, "The MPEG FAQ," http://eitech.com/techinfo/mpeg/mpeg.html, August 94.

4. Scütt, et al., "XTP and Multimedia," Proceedings of GLOBECOM '93, Vol. 2, pp. 877 - 882, 1993. (IEEE Document 07803-0917-0/93)

5. Weaver, A., "XTP: A Communications Protocol for Real-Time Distributed Systems," Proceedings of IECON '93, Vol. 1, pp. 502-508, 1993. (IEEE Doc. 0-7803-0891-3/93)

6. Tobagi, Fouad, et al., "Streaming RAID - A Disk Array Management System for Video Files," StarLight Networks, Inc., Mountain View, CA.

7. M. L. Molle, "A New Binary Logarithmic Arbitration Method for Ethernet", Technical Report CSRI-298, University of Toronto, 1994.

8. S. Deng, "Capture Effects of Residential Ethernet LAN", to be presented at IEEE ICC '95, Singapore, November 1995.

MPEG VIDEO FOR LAN-BASED
VIDEO CONFERENCING

Lurng-Kuo Liu

IBM T. J. Watson Research Center
Yorktown Heights, New York 10598

ABSTRACT

This paper describes the research done on a prototype LAN-based desktop video conferencing system using MPEG-1 video. The system is based on a PC with a codec engine for video/audio compression, and a network controller for LAN network communications. It is clear that network jitter and packet loss are likely to occur in a packet-based LAN environment. The design issues, problems encountered, and possible solutions are described. We seperate the tasks of packet loss detection and error concealment, and assign them to communications subsystem and video decoder subsystem, respectively. The mechanisms for handling MPEG-1 video packetization and error concealment are implemented in the system to minimize the impact of network jitter and packet loss on the performance of the video conferencing.

INTRODUCTION

It is evident that desktop video conferencing is no more just a dream today and has potential for widespread use in the near future, due to the advances in low-cost VLSI technology, networking, and compression standards [1,2,3,4]. In an integrated collaborative environment, video conferencing allows people to collaborate, exchange, and share on multimedia documents, while maintaining natural communications (video and audio) in a real-time manner. "A picture is worth a thousand words." By doig so, video conferencing systems help people to convey their ideas more accurately and perform their works more effectively.

With the fact that the vast majority of desktop computers are interconnected via packet-based Local Area Networks (LAN), it is desired to have a LAN-based video conferencing so that the users can take the advantage of installed LANs. However, the packet-based LAN environment may not provide a guaranteed Quality of Service (QoS). Therefore, network jitter and packet loss are likely to occur in a LAN environment. To minimize the impact of network jitter and packet loss on the performance of the video conferencing, care must be taken in the system and packet designs. This paper describes the research done on a prototype LAN-based desktop video conferencing system using MPEG-1 [1] video. The design issues, problems encountered, and possible solutions are described. Our discussion will be focusing on the design of sending MPEG-1 video over LANs, and the error concealment mechansim on the receiver site. The potential of extending to multipoint video conferencing is also discussed.

MPEG ALGORITHM OVERVIEW

To better understand how the network jitter and packet loss affect the performance of the video conferencing, the MPEG video bitstream syntax and decoder behavior are described briefly below. We are especially interested in the decoder behavior when errors occur in the MPEG bitstream. A more tutorial overview of MPEG is given by LeGall [5].

The video part of the MPEG standard defines the syntax and semantics of the compressed bitstream in layers. It is a hierarchical bitstream syntax, which includes sequence layer, Group of Pictures (GOP) layer, picture layer, slice layer, macroblock layer, and block layer. The highest syntactic structure of the compressed bitstream is the video sequence. It is divided into a set of disjoint groups of pictures to assist random access into the sequence. One out of three picture types (I, P, and B-pictures) can be chose to encode a picture in a GOP. I-picture is an intra-frame coded picture using information only from itself. P-picture is an unidirectionally predicted, inter-frame coded pictures with reference to earlier picture; and B-picture is a bidirectionally predicted, inter-frame coded picture with references to both earlier and later pictures. The picture in a GOP is further divided into a set of nonoverlapped slice which is a series of an arbitrary number of macroblocks. A slice is the smallest resynchronization unit when data corruption occurs in the bitstream. The decoder should be able to resynchronize with bitstream by looking for the next slice start code.

A macroblock consists of four 16×16 luminance blocks and two 8×8 chrominance blocks. When coded, a macroblock can be coded using either intra or non-intra mode. As in picture coding, an intra macroblock is coded using information only from itself while a non-intra macroblock is predictive coded using information from reference frame(s). Once the mode for the macroblock is decided, a two dimensional 8×8 discrete cosine transform (DCT) is operated on each coded block, the DCT coefficients are then quantized and Huffman encoded to produce MPEG bitstream.

In MPEG syntax, start codes are used for several purposes, such as providing extension data and identifying the layers in the coding syntax. They are specific bit patterns embedded in the compressed bitstream that can not be emulated anywhere else. These start codes can be used as synchronization patterns in the decoding process.

THE SYSTEM DESIGN OVERVIEW

The system is based on a PC with a codec engine for video/audio compression, and a network controller for LAN network communications. The block diagram of the desktop video conferencing system is shown in Figure 1. It consists of video compression

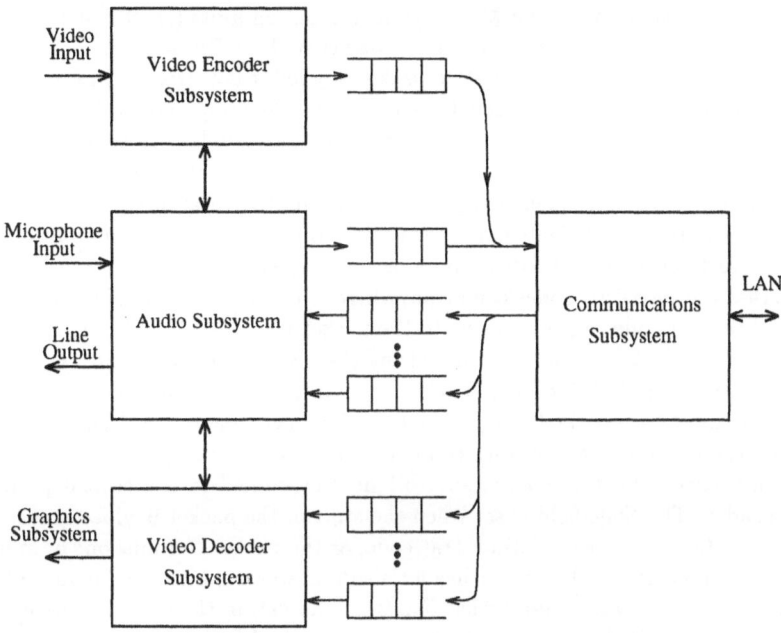

Figure 1: Block diagram of video conferencing system.

subsystem, video decompression subsystem, audio compression/decompression subsystem, and communication subsystem. The video compression subsystem receives digitalized pictures from camera via video decoder and then compresses them into MPEG-1 video stream. On the other hand, the video decompression subsystem receives compressed MPEG-1 video stream, decompresses it, and sends it to the computer graphics subsystem for display. The compression/decompression of audio signals from/to microphone/speaker are handled by audio subsystem. The G.722 algorithm is used for audio compression in our system. The communication subsystem handles the streams packetization/depacketization and the interface to the physical links. In this paper, our discussion will be focusing on the design of sending MPEG-1 video over LANs, and the error concealment mechansim on the receiver site.

Due to network jitter and packet loss within the LAN environment, the video and audio streams are encapsulated within packets separately. The video and audio streams are packetized at each end-station and sent across over LAN. The packetization scheme shall provide appropriate information for transport of video and audio streams over LAN networks. There are several packetization schemes have been proposed different compression algorithms [6,7]. In our approach, we seperate the tasks of packet loss detection and error concealment, and assign them to communications subsystem and video decoder subsystem, respectively. The first part of the packet header we proposed contains an one-bit *Ptype* field for packet type to distinguish between video and audio packets, an one-byte *ChID* field for channel ID to identify the source channels, a two-byte *Seq* field for sequence number to detect packet loss, and a 32-bit *TS* for timestamp to reflect the sampling time of the first data in the packet for the purpose of synchronization. The channel ID information will be used for multipoint video conferencing applications.

The second part of the packet header is the header for MPEG-1 video. In order to minimize the impact of packet loss on the performance of the video presentation, we associate packetization with the MPEG synchronization units (start codes at different layers). The packetization is done in such a way that the MPEG-1 sequence start code, GOP start code, picture start code are packet-aligned if not preceded immediately by a higher layer start code. This can be achieved by inserting zero-stuffing bits before the strat codes to make them packet-aligned. Slice start codes are treated differently. In our case, we define each slice as a macroblock row. Since the compressed bitstream of a picture in a video sequence is generally distributed in several packets, it is not likely that the packet size will match exactly with the size of one or multiple slices. The slice start codes are not forced to be packet-aligned due to coding efficiency. This implies that the slice start codes can be anywhere in the packet. Therefore, the spurious MPEG slice start code may occur due to the packet loss, even the MPEG start code is an unique string that can not be emulated anywhere else in the bitstream. Our research has shown that the design of sync-packet, in addition to the MPEG start code, will help the receiver to recover from packet loss. The sync-packet is designed in such a way that the synchronization start codes received by decoder are not spurious ones. The sync-packet indicator *Sync* is an 1-bit field contained in the second part of the packet header. The *Sync* field is set when the start of the packet payload is a sequence start code, GOP start code, picture strat code, or the packet contains one or more slice start codes. In addition, the packet header also contains a 2-bit field *Pic* for indicating current picture type, and a 9-bit filed *ScOfs* fro indicating the offset of the first start code in the packet if the packet is a sync-packet. The MPEG video packet format is shown in Figure 2.

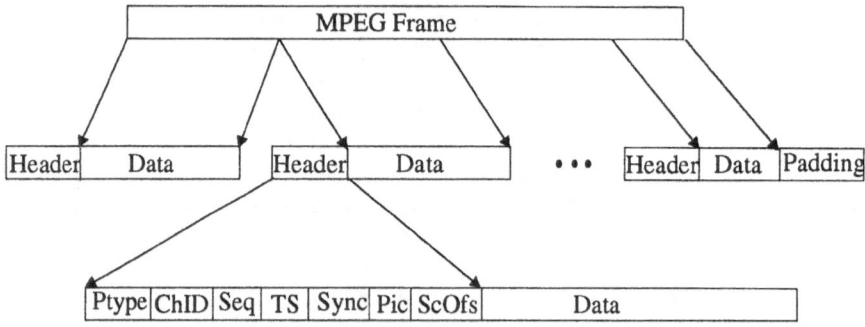

Figure 2: MPEG video packet format

Since the slice start code is the smallest resynchronization unit, the MPEG decoder will search for the next slice start code when data corruption occurs in the bitstream. When resynchronization is required, the communications subsystem passes the sync-packet to the decoder subsystem with the first slice start code leading by zeros.

ERROR CONCEALMENT

Packet loss is likely to happen in LAN environment and may cause decoding problems in the receiver site. For MPEG video, the errors in the bitstream generally produce colorful blocks, macroblocks mispatching, frame freezing, or even worse decoder hanging if care is not taken in the decoder design. Different approaches have been proposed to deal with the problems. Among these approaches, error concealment approach [8,9] is intended to prevent perceivable video quality degradation by using information available in the decoded frames. The areas for error comcealment is identified first. Based on the concept of the sync-packet design described in previous section, we ensure the synchronization start codes received by decoder are not spurious ones. By looking at the embeded information in slice start codes, we classify the areas into two categories: error concealment area within a frame and cross frame error concealment area. These two categories are treated differently.

For the case which error concealment area is within a frame, it is mainly due to the loss of some slices. Motion field replication, if available, is used in this case. The motion vectors for the preceding macroblock row of the same frame are used to perform prediction for the error concealment area. This acts as that the preceding macroblock row has macroblock size of 16×32, if the error concealment area has one macroblock height. When the motion fields replication can not be done, the area is replaced by the corresponding macroblocks in the previous frame. For the cross frame error concealment area, it is mainly due to the loss of picture start code. The first part of error concealment area, which is in the current frame, will be processed as in the previous case. The second part of the error concealment area, which is in the new frame, will be replaced by the corresponding macroblocks in its previous frame if the picture type of the new frame is Intra-coded. Otherwise, the decoder should skip the frame and search for the next sync-packet with picture start code.

CONCLUSIONS

In this paper, a LAN-based video conferencing system is presented and the packetization and error concealment scheme for MPEG-1 video is discussed. We seperate the tasks of packet loss detection and error concealment, and assign them to communications subsystem and video decoder subsystem, respectively. We associate packetization with the MPEG synchronization unit. Our research has shown that the design of sync-packet, in addition to the MPEG start code, will help the receiver to recover from packet loss. It is because that the spurious MPEG start code may occur due to the packet loss, even the MPEG start code is an unique string that can not be emulated anywhere else in the bitstream.

The system and mechanism described above have been tested by having video conferencing using two prototype LAN-based desktop video conferencing systems located 20 miles away from each other. It provides a very smooth real time video conferencing with MPEG-1 video quality.

REFERENCES

1. ISO/IEC JTC1/SC29/WG11, ISO/IEC CD 11172: Information technology, *MPEG-1 Committee Draft*, (December 1991).

2. International Telecommunication Union, Video codec for audiovisual services at $p \times 64$ kbits, *ITU-T Recommendation H.261*, (March 1993).

3. International Telecommunication Union, Video coding for low bitrate communication, *ITU-T Draft H.263*, (July 1995).

4. ISO/IEC JTC1/SC29/WG11, ISO/IEC CD 13818: Information technology, *MPEG-2 Committee Draft*, (December 1993).

5. D. LeGall, MPEG: A video compression standard for multimedia applications, *Communications of the ACM*, vol. 34, pp. 46–58 (1991).

6. T. Turletti and C. Huitema, Packetization of H.261 video streams, *Internet Draft*, (September 1994).

7. D. Hoffman, G. Fernando, and V. Goyal, RTP payload format for MPEG1/MPEG2 video, *Internet Draft*, (June 1995).

8. D. Raychaudhuri, H. Sun, and R. Girons, ATM transport and cell-loss concealment techniques for MPEG video, in *IEEE International Conference on Acoustics, Speech and Signal Processing*, vol. 1, pp. I–117–I–120 (1993).

9. L. Chia, D. Parish, and J. Griffiths, On the treatment of video cell loss in the transmission of motion-JPEG and JPEG images, *Computers & Graphics*, vol. 18, no. 1, pp. 11–19 (1994).

SUITABILITY OF TINA FOR VDT AND INTERACTIVE SERVICES

S. Devendra K. Verma

NYNEX Science & Technology, Inc.
Research and Development
Broadband/Video Systems Engineering Laboratory
500 Westchester Avenue
White Plains, N. Y. 10604
Tel. (914) 644-2436
Fax (914) 644-2301
E-mail: sdkv@nynexst.com

ABSTRACT

The Broadband and Multimedia are changing our life style. It provides the opportunity for scanning 500 channels, Video-On-Demand (VOD) in real-time, interactive games, home shopping, distance learning, telecommuting, etc. The national information infrastructure, regional information highways, and other broadband initiatives have created the need for new service architecture. The Broadband development can be revolutionized by a flexible operations and network architecture.

The Telecommunication Information Networking Architecture (TINA) is an open architecture unifying the Telecommunication Management Network (TMN) and the Intelligent Network (IN) architectures together. It provides more flexibility in application interoperability, and facilitates a faster application development. The Information Superhighway is considered as highway of Broadband Communications Networks with ramps connecting different service/shopping centers of Information Providers. The Video Dial Tone (VDT) and Interactive services are based on Integrated Broadband Communications Network. The VDT and Interactive Services have both Service and Management oriented features and functionalities. The TINA having both capabilities TMN (to support Management aspect) and IN (to support Service aspect) is most suitable for the VDT and Interactive services.

INTRODUCTION

The Subscribers/Customers wish to have complete control over what they want to receive and when they want to receive; and as a result the Video-On-Demand (VOD) and Interactive Services are included in the services featuring Monthly subscription, A la carte subscription, Pay per view, and Near VOD. The migration of technology and application is driving the residential broadband market (i.e., NVOD & VOD movies/events, PPV/IPPV/EPPV, ITV/IMTV, Videogames, Interactive marketing/home shopping, Distance learning/education, ...) supported by Analog broadcast, Digital broadcast, and Digital switched services.

The architecture delivering Video Dial Tone (VDT) services comprises of the Video Services Platform (VSP), VDT Service Provider (VDT_SP), Video Information Provider (VIP), Video Information User (VIU), and Settop Box (STB)/Digital Entertainment Terminal (DET). The VIP provides televised broadcast or libraries of video, image, graphics, textual, and audio information to be accessed by the subscribers/customers interactively. The VSP connects various VIPs and the Subscribers/Customers. Interactive video services (i.e., VOD, Home shopping, ...) require on-demand and real-time delivery of video information to the Subscriber/Customers.

A combination of open management system and intelligent network seems to be the ideal solution for meeting the customers' future needs. Several standards are evolving and being adopted such as Common Management Information Protocol/Common Management Information Services (CMIP/CMIS), Distributed Management Environment/Distributed Computing Environment (DME/DCE), Systems Network Architecture (SNA), Simple Network Management Protocol (SNMP), etc.

The current network is viewed as inefficient and costly in managing services and technologies. The Advanced Intelligent Network (AIN) and Telecommunications Management Network (TMN) technologies address some of these problems but they do not provide a common network platform. The TINA provides an unified platform, supported by ATM, SONET, and ISDN. TINA also facilitates a multivendor environment.

DESCRIPTION OF TINA

The Telecommunication Information Network Architecture Consortium (TINA-C), a project group of about 40 companies, led by Bellcore is set up to establish specifications for an open platform for multivendor, broadband networks.

The TINA architecture consists of three layers:
a) Computing Layer: It provides a distributed-processing environment.
b) Network Resource Layer: It provides the bandwidth and hardware devices in the network for service provisioning, and
c) Service Architecture Layer: It provides the software components to provide services.

TINA is a combination of IN (a generic, service-oriented architecture) and TMN (a generic, management-oriented architecture). The TINA architecture is based on the Broadband communication and Distributed computing technologies for information networks to transport multimedia information and manage multimedia communication. It is viewed as an Open architecture for telecommunication services in the emerging Broadband, Multimedia, and Information Superhighway technologies.

The Distributed Processing and Object Oriented Software are the salient features of the TINA application to enable new services. All the network elements are structured as distributed/independent processing units connected over networks. The data and logic are defined as objects. The software building blocks facilitates its reusability in enhancing/developing new services.

The TMN principles recommend the use of independent management networks to manage telecommunications networks based on standard interfaces. TMN is envisioned as solution provider for complex problem of telecommunications networks and services pertaining to Operation, Administration, Maintenance & Provisioning (OAM&P) in open and multivendor environment. The TMN uses the concept of Functional Architecture based on Function Blocks and Reference Points. The Function Blocks are logical entities. The Reference Points represent the exchange of information between two function blocks. The TMN nodes are identified as Operation System (OS), Mediation Device (MD), Q-Adaptor (QA), Work Station (WS), and Network Element (NE), supported by function blocks OSF, MF, QAF, WSF, and NEF respectively.

The IN functional model comprises of three levels - a) Service Switching Level, b) Service Control Level, and c) Management and Service Creation Level. A simplified IN management functional model is specified with layers - i) Network Element (NE), ii) Network Element Management (NEM), iii) Network Management (NM), and iv) Service Management. Besides these layers, a Business Management Layer is included to support the TMN management functionality. The current IN architecture is evolving to an Information Network Architecture (INA), moving the physical network to the software system with distributed processing capabilities. The Distributed Processing Environment is supported by Service and Delivery segment building blocks.

The software to support the functionalities of the VDT and Interactive Services can be decomposed in Building blocks (Object modules) and grouped across the three layers of TINA to provide and enhance the services efficiently and economically.

CONCLUSION

Services are being personalized to individual needs with a user-friendly presentation and simplicity. TINA being more general and unified architecture, allows more flexibility and application operability. It will facilitate end-to-end network management with more intelligent, fully automated and integrated features to support growing Broadband market, having VDT and Interactive services. Turning the TINA vision into reality is not an easy task; but, this is the smart and efficient way to support our customers' growing needs in a timely fashion.

REFERENCES

1. CCITT Rec. M.3010, Principles for a TMN; CCITT, February 1992.
2. CCITT Rec. M.3020, TMN Interface, Specification Mrthodology; CCITT, February 1992.
3. CCITT Rec. (Draft) X.cnma Architecture for Customer Network Management Service for Public Data Networks; CCITT, February 1992.
4. TINA Consortium, The TINA-C Architecture as Related to IN and TMN; Bellcore, Red Bank, New Jersey, 1994.
5. TINA Consortium, Application of TINA-C Architecture to Management Services; Bellcore, Red Bank, New Jersey, 1994.
6. TINA Consortium, Service Specification Concepts in TINA-C; Bellcore, Red Bank, New Jersey, 1994.
7. TINA Consortium, Practical Issues Involved in Architectural Evolution from IN to TINA; Bellcore, Red Bank, New Jersey, 1994.
8. TINA Consortium, Towards Networking Telecommunications Information Services; Bellcore, Red Bank, New Jersey, 1994.
9. TINA Consortium, Service Architecture, Service Session ans Service Federation in TINA-C; Bellcore, Red Bank, New Jersey, 1994.
10. Bellcore, Presentation Control for Information Networking Services, TM-ARH-022909; Bellcore, Red Bank, New Jersey, April 1993.
11. Bellcore, Making the TINA Distribution Support Environment Fault Tolerant, TM-ARH-020823; Bellcore, Red Bank, New Jersey, January 1992.
12. Robin Smith, and E. H. Mamdani, The Management of Telecommunications Networks; Ellis Horwood Ltd., New York, 1992.

13. Daniel Minoli, Video Dialtone Technology: Digital Video over ADSL, HFC, FTTC, and ATM; McGraw-Hill, Inc., New York, 1995.
14. Jan van Duuren, Peter kastelein, and Frits C. Schoute, Telecommunication Network and Services; Addison-Wesley Publishing Co., New York, 1992.
15. Robert K. heldman, T. F. Madison, and T. A. Bystrzycki, Future Telecommunications: Information Applications, Services, & Infrastructure; McGraw-Hill, Inc., New York, 1993.
16. Mark J. Bunzel and Sandra K. Morris, Multimedia Applications Development: Using Indeo Video and DVI Technology, Second Edition; McGraw-Hill, Inc., New York, 1994.
17. Arch C. Luther, Authoring Interactive Multimedia; A. P. Professional, New York, 1992.
18. Edward Cooper, Broadband Network Technology: An Overview for the Data and Telecommunications Industries, Prentice-Hall, Inc., Englewood Cliffs, New Jersey 1994.
19. Byeong Gi Lee, Minho Kang, and Jonghee Lee, Broadband Telecommunications Technology; Artech House, Inc., Boston, MA, 1993.
20. Bellcore, A Guide to New Technologies and Services, Issue 7; Bellcore, 1993.
21. Proceedings, International Conference on Multimedia Computing and Systems; IEEE Computer Society, Washington, D.C., 1995.
22. ICC'95 Proceedings, Communications - Gateway to Globalization; IEEE Communications Society and the IEEE Seattle Section, Washington, 1995.

A MULTIWAY TALK PROTOCOL[*]

Hong Liu[1] and Zhiqiang Zhang[2]

[1]Department of Computer and Information Science
University of Massachusetts Dartmouth
285 Old Westport Road, North Dartmouth, MA 02747-2300
(liu@cis.umassd.edu)
[2]Peritus Software Services, Inc.
304 Concord Road, Billerica, MA 01821-3485
(zzhang@peritus.com)

INTRODUCTION

As the Internet turns the world into a global village, numerous communication applications bloom. It is changing the way we live and work. A protocol for multi-party communications independent to applications, functionality, and supporting techniques is under urgent demand. It is to cater to ever-changing networking techniques, to offer various system functions, and to meet unforeseen computer network applications.

This paper proposes a protocol to provide a framework for upper layer control of multi-party communications, called Multiway Talk Protocol (MTP). We study popular multi-party communication applications at levels of both underlying network architecture and user application requirements. Such a study enables us to abstract generic communication applications with three MTP control classes that are independent to network architecture. This innovative idea in the MTP design makes MTP flexible and versatile to implement computer communication applications on any supporting network architecture. The idea comes from imperative high-level programming languages, which contains three basic control structures for generic programming on any computer architecture. We demonstrate our assertion by applying the MTP concept to implement teleconferencing applications. We point out the further work in the conclusion of this paper.

A SURVEY ON MULTI-PARTY COMMUNICATION APPLICATIONS

Our survey on multi-party communication applications concentrates on those software packages running over the Internet. These applications demonstrate the unique advantage of digital computer networks like the Internet for multimedia communications. The digital forms of multimedia components can be individually manipulated to accomplish computer-supported cooperative work (CSCW) or to deliver interactive entertainment. However, how to implement efficient and reliable multi-party communications on top of the Internet's unreliable multicast protocol is a major technical challenge. We study the techniques from

---

[*]This work is partially supported by a professional development grant for Distance Learning and Instructional Technology by the President's Office of the University of Massachusetts and the Information Technology Council.

the two ends of software development: one from the bottom is the supporting network architecture; the other from the top is the application requirements. Merging of the two ends reveals the necessity of a framework independent to the network architecture and generic to communication applications. Such a process resembles the emergence of high-level programming languages, which are independent to computer architecture and offer generic control to programming.

Based on underlying network architecture, current communication control frameworks can be classified into two: centralized model or distributed model. Centralized abstraction collects information common to all users into a single place. A central process handles all user input and display output events to the shared information. Thus, the information is consistent among all the users. However, the system performance could be intolerable when the number of users explodes and the failure modes arise from spreading over wide areas, which is called scalability problem. Examples of centralized models are MMConf, Etherphone, and Touring Machine. MMConf (the MultiMedia Conference Control program) by Crowley et al.[1] is designed to control a multimedia conferencing system and it is also a single conferencing application that enforces a predetermined set of control policy. Xerox PARC's work on Etherphone[2] a system for audio, with a later extension to video[3], is a desktop conferencing application over Ethernets and it allows access negotiation among peers. Touring Machine[4,5] at Bellcore Information Networking Research Laboratory provides a system infrastructure on which applications requiring complex multimedia communications can be developed, independent of the actual network fabric used to provide transport by isolating application development from communication technical details.

Distributed model collectively maintains common information to all users. Any change to the shared information must be broadcast to all users. Thus, the system is scalable and the accessing policy can be easily tailored to a user's preferences. However, it must permit certain degree of data inconsistency as discussed by Bentley et al.[6] Fortunately, the observations[7] show that humans can cope with a degree of inconsistency that arises from partitioned networks and lost messages, as long as the distributed state will tend to converge in time. Examples of distributed models are VAT, WB, and CCCP. VAT[8] (the Visual Audio Tool) and WB[9] (WhiteBoard) are both developed by Jacobson's team at the University of California at Berkeley and are both based on the lightweight session model[10]. The lightweight session model uses multicast so that data delivery is independent to the number of receivers and a sender does not have to keep track of receivers' status. CCCP (the Conference Control Channel Protocol) by Handley et al.[7] is a scheme to control conferences ranging from small, tightly coupled meetings, to extremely large loosely coupled seminars. Tightly coupled means the application cannot run unless all processes are available and contactable. Loosely coupled allows some of the sites become unavailable because the application has a number of cooperating instantiations.

For application requirement study, we concentrate on data sharing policy and user views. These applications can be divided into two categories: with or without a shared access window. Most chat programs fall into the category of no shared access window. Each user retains a local window where others' messages are displayed and the user's own message is echoed. Various examples differ in their access policies and user interfaces, which present a spectrum of protocol processing efficiency and user interface enhancement. *YTalk* by Yenne[11] is a multi-user chat program whose predecessor is Talk. As an extension of *Talk* -- a UNIX chat program, YTalk allows more than two users in a conversation. Currently, the protocol for YTalk is an iteration of Talk by communicating with a Talk daemon per user participating the conversation. Each user has one window divided into slots, one for every other user to display messages. *IRC* by Sandrof[12] is an Internet Relay Chat system. The protocol adopts client-server model. Each user delivers a message to a server who relays the message to all users participating the conversation including the speaker himself. Again, each user has one window. But the window is split into two parts. The large upper part displays messages from other users relayed by the server. The lower part, containing a single line, echoes the user's own keyboard input.

Most game programs need a shared access window. The shared access window can be updated by any eligible users and it keeps stringent consistency among all users' views.

XIGC by Coffin[13] is an X window system interface for the Internet Go Competition system. It allows a user to play or observe a game remotely. It also records the scores for each game as well as each player. The protocol is based on client-server model. A user contacts the Internet Go Server (IGS) to join in a game or to observe games. The IGS relays the users' moves to all participants' screens and maintains the game records for review. The user interface has three windows: a Go board window which is the shared access window, a menu window, and a window for communications between the user and the IGS. *ICS* is an Internet Chess System[14]. It is similar to XIGC. However, its user interface contains two windows instead of three: a chess board window, which is the shared access window, with a menu bar and a communications window.

THE MULTIWAY TALK PROTOCOL:
A FRAMEWORK TO CONTROL MULTI-PARTY COMMUNICATIONS

Our multiway talk protocol, *MTP*, is a framework to control multi-party communications over the Internet. It provides an interface between communication applications and supporting infrastructure, which resembles a high-level programming language separating computer applications from computer architecture. The protocol has three dimensions of control parameters: The first dimension is interaction control, e.g., one-to-one talk, one-to-many talk, or many-to-many talk. The second dimension is access control, stretching from all access rights so that all information is readable and writable by all participants to one privileged access right so that only one central process has the sole right to access the shared information. The last dimension is sequencing control, communication streams among various parties being separated or steamed according to their time sequences.

By choosing different options on these three dimensions of control parameters, we can stitch out any multi-party communication applications from this MTP protocol. For example, YTalk mentioned in the previous section would be many-to-many talk with all access rights and separated communication streams. Many-to-many talk with one privileged access right and full time sequencing would turn out to be IRC. Both XIGC and ICS are one-to-one talk with all-read-rights one-privileged-write-right and time sequencing. In the later section of application examples, we will demonstrate how the concept of the MTP protocol is applied to implement communication applications.

The format of MTP protocol messages (packets) is like any protocol data unit with two parts: packet header and packet body. A packet header includes control information and it has a predefined length in octets, named HEADER_LEN. A packet body contains the user message from the upper layer. The MTP protocol can be viewed as a middleware supporting networking applications or a lower sublayer of the application layer in the ISO/OSI 7 layer reference model. The length of a packet body varies from 0 to a predefined maximum number in octets, named MAX_BODY_LEN. Figure 1 shows the structure of a packet header.

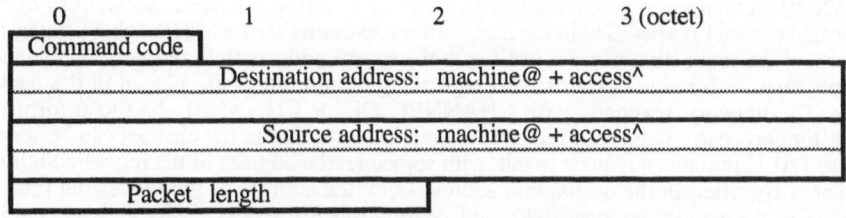

Figure 1. Structure of a Packet Header

Destination address and *source address* specify the identity of both the intended recipient(s) and the originator, respectively. Each address field is 8 octets: 6 for the machine address in either the IP address or the physical address, and 2 for the access point of the associated application process such as socket number. *Packet length* is a count of octets in the MTP protocol packet, including the packet header and the packet body. A user message from the upper layer is ended with two characters CR-LF. A user message could be segmented into several packets. Thus, the actual maximum body length of the last packet of a user message is MAX_BODY_LEN - 2. *Command code* indicates the control type. The table in Figure 2 lists the contents. The second column contains the names of the commands while the third column gives their descriptions.

Command Code	Command name	Description
0 - 5	-	Reserved
6	CHANNEL	Create or tune in a channel specified in the destination
7	CHANNEL_OK	Create or tune-in success
8	CHANNEL_NO	Create or tune-in fail, reason in the packet body
9	TALK	Request to talk someone specified in the destination
10	RESPONSE	Respond to a talk request, accept or decline
11	TALK_OK	Talk success
12	TALK_NO	Talk fail, reason in the packet body
13	QUIT	Quit from a channel or a talking group
14	STATUS	Status of a channel or someone
15	STATUS_OK	Status retrieved, contents in the packet body
16	STATUS_NO	Status retrieval fail, reason in the packet body
17 - 247	-	Unused
248	SEND_NO	Send fail, reason in the packet body
249	SEND	Send a packet to a channel or a talking group
250 - 255	-	Reserved

Figure 2. Command Code

The commands are categorized into three classes: 1) communication management; 2) status inquiry; and 3) message delivery. The class of communication management controls the way in which a communication is conducted by selecting a set of options from the three control dimensions discussed before. Channels model after public broadcast stations while talking groups mimic private meetings. Each can have various ways of interaction where access policy and sequencing requirement are determined by the upper layer applications. Application policy or requirement is different, and therefore, is separate from the protocol mechanism. For instance, how to advertise channels or how to look for talking partners is up to each application and is not addressed in the MTP. However, an access policy is supported by the protocol's mechanism. For example, a public talk show channel (where a host moderates the show, invited guests are allowed to talk, and everyone else can listen) can be implemented by manipulating access control parameters (giving one privilege of updating the access control to the public information, assigning some with rights to write the public information, and granting all with rights to read the public information). CHANNEL creates a new channel or tunes in an existing channel: the address of the requested channel is specified in the destination address field of a MTP packet header, the address of the requesting process itself is in the source address field of the packet header, and negotiation for access rights and sequencing requirements are placed in the packet body. The network responds with CHANNEL_OK or CHANNEL_NO to inform the requesting process whether the request for creating or tuning in the channel can be served or not. TALK initiates a request to talk with someone: the address of the requested talking partner is specified in the destination address field, the address of the requesting process itself is in the source address field, and negotiation for access rights and sequencing requirements are placed in the packet body. Upon receiving the TALK request, the requested talking partner replies with RESPONSE, indicating his/her willingness to talk or not. The network informs both talking partners with TALK_OK or TALK_NO. One can tune out a channel or leave a talking group with the QUIT command.

The class of status inquiry retrieves the information about channels or personnel. It supports applications for advertisement or browses and assists searching interesting channels or potential talking partners. STATUS issues a request to retrieve the information of a channel or someone specified in the destination address field of a MTP packet header. The network responds the requesting process with STATUS_OK or STATUS_NO for the information requested or the reason that the requested information is not available. Some special channel numbers can be set up for directories like TV Guide, Yellow Pages, or White Pages. Some special addresses can be set up, where someone can be fingered to find out his/her name, address, plan, and availability.

The class of message delivery contains a pair of commands to send messages with negative acknowledgment. SEND issues a packet to a channel or a talking group specified in the destination address field of the packet header, which carries the upper layer messages in the packet body. The network informs the message originator with SEND_NO if the network is unable to deliver the SEND packet and the reason can be found in the packet body of SEND_NO. Typical reasons are, for instance, the message originator does not have the right access rights to the channel at this moment, the talking group does not exist any more since everybody has quitted, or the network system is down. The MTP protocol is responsible to the message delivery only. The upper layer applications deal with the issues such as whether the message is received by all parties or how the recipients manipulate the message.

THE SUPPORTING ARCHITECTURE

The supporting architecture for MTP comprises hardware, software, and the network. The hardware contains the target machine and input/output devices. The software is subdivided into the operating system, the communication subsystem, and the user interface. The network refers to the physical communication infrastructure with associated transfer protocols and capacities. The minimum hardware requirement is a workstation with keyboard and mouse. The software works on UNIX operating system, TCP/IP protocol suite, and Motif/X window. MTP is decoupled from the underlying network architecture. Thus, MTP can be implemented with either centralized approaches such as client-server model or distributed approaches such as MBone concept.

The client-server model, as well understood,[14] consists of a set of server processes, each acting as a resource manager for a collection of resources of a given type, and a set of client processes, each performing a task that requires access to some shared resources. Client processes issue requests to servers whenever they need to access some resources. The server process performs the requested action and sends a reply back to the client process. Thus, each server process is viewed as a centralized provider of the type of resources that it manages.

MBone stands for the Internet Multicast backBone developed by Casner[15,16] in the Information Sciences Institute at the University of Southern California. MBone is a virtual network layered on top of portions of the physical Internet, providing a semi-permanent IP multicast facility. It uses a network of routers with multicasting ability, called *mrouters*. A host establishes or joins a common shared session by making an announcement. The associated mrouter forwards that announcement to the other mrouters in the network. Groups are disbanded when everyone leaves, freeing up the IP multicast address for reuse. The mrouters occasionally poll hosts to determine if any are still group members. If there is no reply by a host, the mrouter stops advertising that host's group membership to the other mrouters. The MBone topology and the scheduling of multicast sessions are handled cooperatively. No single entity is in charge of either local topology changes or event scheduling. Thus, it is a distributed model. We have implemented two versions of MTP with client-server model. The MBone concept is used in the third version, which will be presented in a sequel to this paper.

Under client-server model, there are two kinds of servers: one is asynchronous and is used in Version One of MTP implementation; the other is concurrent and is used in Version Two. Clients need not to know what kind of sever that is currently in use because they perform the same functionality. Concurrent servers are more efficient to use but are more complicated to implement than asynchronous ones. Figure 3 depicts the structure of MTP under client-server model. Initially, the server runs the master process with the master socket, a well-known port number (permanent), waiting for connection requests from clients. A client issues a connection request to the master socket for services. Upon receiving the request, the master process either selects an active slave socket if it is an asynchronous server or forks a slave process with a new socket if it is a concurrent server. Then each slave socket communicates with one client to provide the services directly. The slave socket is assigned by the operating system dynamically (temporary) and it is released for reuse when the client is disconnected from the server.

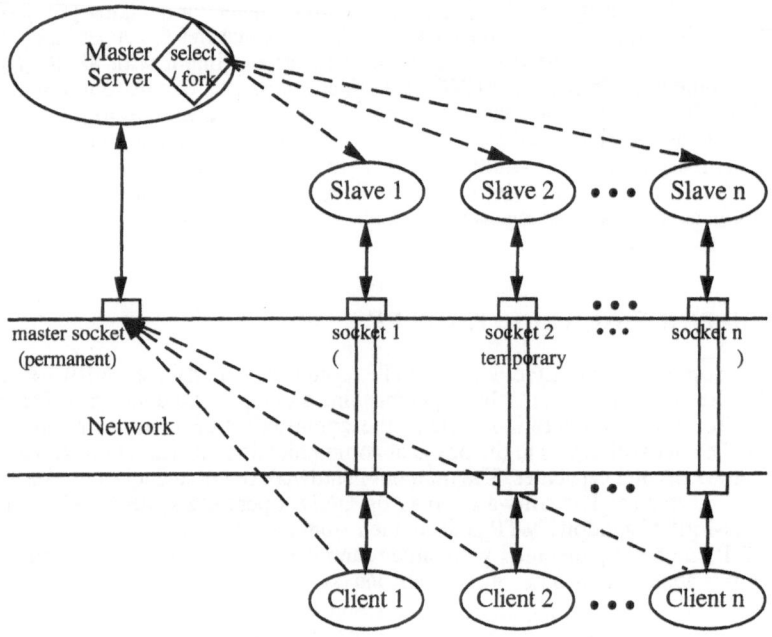

Figure 3. Structure of MTP under Client-Server Model

The master server maintains a central database for the status of all clients. Figure 4 shows an entry of the database. *Client's address* specifies the identity of the client. The format is the same as the address field of the MTP packet header shown in Figure 1. *Client's name* is a string of the user name. *Channel/Group #* identifies the address of a channel or a talking group that the client is participating. The format is also the same as the address field of the MTP packet header. *Access rights* specify the client's rights to the channel or the group, which is negotiated at the time when the client joins in or is updated later. A client can participate in several channels or talking groups simultaneously. The list of channels or the groups is ended with a *null* symbol.

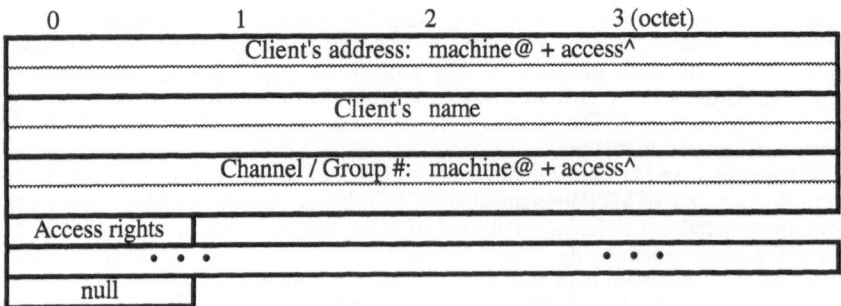

Figure 4. Entry of the Database

We choose the shared memory mechanism to support channels. Each channel has a shared memory in the master server for clients to access. A set of semaphores guide the shared memory to synchronize the clients and to assure their access rights. We choose the named pipe mechanism to support a talking group. Each member has a pipe to every other member in a talking group for message delivery. Signal mechanism is used for members to set up a talking group and to negotiate the access rights. These mechanisms are well understood and widely used.[17,18]

APPLICATION EXAMPLES

While we are prototyping MTP, we tested its feasibility and advantage by applying the concept on implementing teleconferencing applications. Using MTP Version One, a simple teleconferencing software called "Electronic Workshop" has been assigned as a semester-long project accompanying a graduate computer network course by Liu.[19] *Electronic Workshop* models a single track conference with features such as registering participants, moderating speakers, delivering messages to eligible members, and posting questions from participants for discussions. It is a combination of one-to-many and many-to-many talk with a central control on access rights and on time sequence. Students were given the code of MTP and then designed and implemented their Electronic Workshop based on this control framework. For each of the last two years, over two thirds of the students in the class completed the project and some students added more features or enhanced the user interface in addition to the project assignment. It demonstrated the feasibility of MTP supporting fast development of customized teleconferencing applications.

Using MTP Version Two, a sophisticated teleconferencing software called "Multi-way Talk" is under development as a Master's project by Zhang.[20] *Multi-way Talk* facilitates communications for geographically apart users through the Internet besides modeling multi-track multi-session conference. Its functionality includes advertising public channels, coordinating a channel arrangement such as selecting a moderator and soliciting speakers, registering participants, setting up a side conversation, and recording a session contents for re-play. It is a combination of one-to-one, one-to-many, and many-to-many talk with the choice of a central control or a cooperative control on access rights and on time sequence. This application example shows the flexibility of MTP supporting versatile networking applications.

CONCLUSION

Our experience has shown that the three classes of MTP control commands abstract well for multi-party communication applications. It is like imperative programming languages abstracting programming by the three basic control structures -- sequencing, selection, and repetition. MTP simplifies and speeds the development of multi-party communication applications. A unique value of MTP from other communication control frameworks is that it is independent to the supporting network architecture. Thus, it does not have the scalability problem inherent in centralized models and it controls the inconsistency problem encountered in distributed models.

The MTP project is still evolving. On one hand, MTP supports development of multi-party communication applications. On the other hand, multi-party communication applications challenge MTP's ability. An immediate extension is to sequence multimedia. Our next application project is to apply MTP on integrating an audio tool for packetized voice transmission and a shared digital whiteboard for distributed image access into a multimedia instructional system. This system will be used to support seminars and highly specialized courses over the Internet across the campuses of the University of Massachusetts.

REFERENCES

1. T. Crowley et al., MMConf: an infrastructure for building shared multimedia applications, in: Proceedings of Computer-Supported Cooperative Work, pp.329-342, ACM Press, New York (1990).
2. P.T. Zellweger, D.B. Terry, and D.C. Swinehart, An overview of the Etherphone system and its applications, in: Proceedings of the Second IEEE Conference on Computer Workstations, pp.160-168, IEEE Press, New York (1988).
3. H.V. Vin, P.T. Zellweger, D.C. Swinehart, and P.V. Rangan, Multimedia conferencing in the Etherphone environment, IEEE Computer, 24(10):69-79 (1991).
4. M. Arango et al, Touring machine: a software platform for distributed multimedia applications, in: IFIP Upper Layer Protocols, Architectures and Applications, Vancouver, Canada (1992).
5. Bellcore Information Networking Research Laboratory, The Touring machine system, Communications of the ACM, 36(1):68-77 (1993).
6. R. Bentley et al., Architectural support for cooperative multiuser interfaces, IEEE Computer, 27(5):37-46 (1994).
7. M. Handley, I. Wakeman, and J. Crowcroft, The conference control channel protocol (CCCP): a scalable base for building conference control applications, in: Proceedings of ACM SIGCOMM'95, pp.275-287, Boston, USA (1995).
8. V. Jacobson and S. McCanne, VAT -- UNIX Manual Pages, Lawrence Berkeley Laboratory, University of California at Berkeley, CA.
9. V. Jacobson and S. McCanne, Using the LBL network whiteboard, Lawrence Berkeley Laboratory, University of California at Berkeley, CA.
10. S. Floyd, V. Jacobson, S. McCanne, C.G. Liu, and L. Zhang, A reliable multicast framework for light-weight sessions and application level framing, in: Proceedings of ACM SIGCOMM'95, pp.342-356, Boston, USA (1995).
11. B. Yenne, YTalk -- a multi-user chat program, ytalk or yenne@austin.eds.com.
12. M. Sandrof, IRC -- interface to the Internet Relay Chat system, ms5n+@andrew.cmu.edu.
13. S. Coffin, XIGC -- X window system interface for the Internet Go Server, v3.6, November 1994, scoffin@advtech.uswest.com.
14. G. Coulouris, J. Dollimore, and T. Kindberg, Distributed Systems: Concepts and Design, 2nd Ed., Addison-Wesley, Massachusetts (1994).
15. S. Casner, Frequently asked questions (FAQ) on the multicast backbone," May 6, 1993, file://venera.isi.edu/mbone/faq.txt.
16. M.R. Macedonia and D.P. Brutzman, MBone provides audio and video acrosss the Internet, IEEE Computer, 27(4):30-36 (1994).
17. W.R. Stevens, UNIX Network Programming, Prentice-Hall, New Jersey (1990).
18. W.R. Stevens, Advanced Programming in the UNIX Environment, Addison-Wesley, Massachusetts (1992).
19. H. Liu, Courseware of CIS 577 -- Computer Networks at the University of Massachusetts Dartmouth, Springs of 1994 and 1995. Available upon request from liu@cis.umassd.edu.
20. Z. Zhang, Multi-way Talk, a Master's project to fulfill a Master's degree in Computer Science at the University of Massachusetts Dartmouth, to be completed in December 1995. Available upon request from liu@cis.umassd.edu.

EFFECTS OF SMOOTHING ON END-TO-END PERFORMANCE GUARANTEES FOR VBR VIDEO

Edward W. Knightly[†] and Paola Rossaro[‡]

The Tenet Group
[†]EECS Department, U.C. Berkeley
[‡]International Computer Science Institute, Berkeley, CA

Abstract

Integrated services networks have the framework needed to provide performance guarantees to multimedia traffic sources such as Variable Bit Rate (VBR) video. This paper investigates the effects of smoothing VBR traffic sources on their end-to-end delay bounds and on the achievable utilization inside the network. We provide a set of rules that determine if smoothing results in a net advantage to the source and use experiments with traces of MPEG-compressed video to quantify the effectiveness of smoothing in practical networking situations.

INTRODUCTION

Many distributed multimedia applications are delay- and loss-sensitive and have stringent requirements in terms of network performance. A resource reservation scheme together with priority service disciplines inside the network, as in (Ferrari and Verma, 1990), provide the means for giving network clients end-to-end Quality of Service (QoS) guarantees on delay, loss, and throughput. The guaranteed service can be *deterministic*, in which all packets of a connection are guaranteed to meet the promised QoS, or *statistical* in which probabilistic performance bounds are guaranteed. The two services can be viewed as trading off QoS for network utilization in that deterministic service provides the best QoS with its no-loss, no-delay-violation guarantee, but it cannot employ statistical multiplexing to achieve higher network utilization since that would result in a reduced (non-deterministic) QoS.

An important issue in providing end-to-end performance guarantees is the effect of *smoothing* traffic sources on QoS and network utilization. In this paper, we use analysis and traces of MPEG compressed video to explore the effects of smoothing VBR video sources on end-to-end deterministic performance guarantees. We consider the case of smoothing by a FIFO (a first-in first-out queue) so that before the packets of a VBR video source are transmitted to the network, they are sent through a FIFO at the source. This source-FIFO services packets at a smoothing rate R_S where R_S is less

Multimedia Communications and Video Coding
Edited by Y. Wang *et al.*, Plenum Press, New York, 1996

than the unsmoothed source's peak rate and greater than its long term average rate. The smoothing rate can be viewed within the context of the Deterministic Bounding Interval Dependent (D-BIND) traffic model (Knightly and Zhang, 1995) in which the bounding rate R_S represents an upper bound on the unsmoothed source's rate over every interval of length I_S. The D-BIND model provides a more accurate traffic specification via multiple rate-interval pairs which in turn allows a higher utilization than is possible with a peak-rate-allocation scheme.

The traffic shaping implemented by the FIFO results in a traffic stream that is "less bursty" than the original traffic stream, where "less bursty" is defined in terms of the D-BIND model in a manner similar to (Low and Varaiya, 1993). When the smoothed traffic sources are multiplexed at queues inside the network, the resulting bound in queueing delay will be reduced. Or in other words, for a given queueing delay bound, smoother traffic results in a higher achievable network utilization. However, the smoothing-FIFO at the source has introduced additional end-to-end delay by adding a smoothing delay.

Thus, we are considering the effect of smoothing VBR video sources on the total end-to-end delay bound where the end-to-end delay bound considers both smoothing and queueing delays at all hops along the path. Since smoothing decreases the bound on queueing delay, but increases the bound on smoothing delay, or the time that packets spend in the traffic shaper before entering the network, smoothing can be considered a tradeoff between buffering at the source and buffering inside the network.

In this paper, we propose a set of conditions for determining if traffic smoothing results in a net improvement in end-to-end QoS for a source. We then investigate these results in a practical environment by analyzing traces of MPEG compressed video. The results show that over a single congested hop with homogeneous sources, smoothing is an ineffective means for achieving higher network utilization. That is, the savings in queueing delay achieved by transmitting smoother traffic are outweighed by the smoothing delay incurred at the source. Alternatively, if a source traverses multiple congested nodes, smoothing can result in considerable improvements in achievable network utilization for a given QoS guarantee. Intuitively, the reason for this is that the smoothing delay is incurred only once at the source, whereas the queueing delay may be incurred at each hop along the path. Thus, in many cases, a higher network utilization or lower end-to-end delay can be achieved by smoothing. Moreover, the analysis indicates that the advantages of smoothing increase with the number of congested hops traversed by the connection.

This work differs from other approaches to transporting VBR video (e.g., (Kanakia, Mishra, and Reibman, 1993; Zhang and Knightly, 1995)) in that once a connection is established, the network provides constant quality video transmission without being affected by the current network congestion. Thus, we do not adjust a coder's "Q-factor" (a quality factor in many compression algorithms to reduce the bit-rate), nor do we drop packets. Other related works include (de Veciana and Walrand, 1992; Shroff and Schwartz, 1994), in which smoothing is considered in the realm of statistical performance guarantees. Our work considers a *lossless smoothing*, in which we provide an end-to-end deterministic QoS and introduce only delay at the source in order to better shape the traffic for the network.

In the remainder of this paper, we first describe the D-BIND model and the associated admission control test for a FCFS scheduler. Next, we provide a definition of smoothing and introduce a set of rules to determine when a source should smooth. Finally, we use traces of MPEG-compressed video to examine the rules in a realistic scenario and we introduce various performance metrics to further investigate the effectiveness of smoothing and the tradeoffs involved.

DETERMINISTIC QOS

Compared to statistical service, deterministic service provides better QoS in that it provides a no-loss, no-delay-violation service. While this does preclude statistical multiplexing, as shown below, it does not require a peak-rate-allocation scheme. For the network to deliver such a service, it needs a deterministic upper bound on all sources receiving the service. This approach has the added advantage that a source's traffic specification can be enforced. For example, if a source promises that its minimum packet inter-arrival time is $Xmin$, this may be easily verified and enforced by the network. Alternatively, statistical models of the source are inherently much more difficult to enforce.

The Deterministic-BIND Model

As shown in (Knightly and Zhang, 1995), previous deterministic traffic models such as the (σ, ρ) model (Cruz, 1991) and the $(Xmin, Xave, I, Smax)$ model (Ferrari and Verma, 1990) cannot capture the property that sources exhibit burstiness over a wide variety of interval lengths. The Deterministic Bounding Interval Dependent traffic model was introduced to address this issue. The key components of the D-BIND model are that it is *bounding*, required to provide deterministic QoS guarantees, and *interval-dependent*, needed to capture important burstiness properties of sources which in turn allows for a higher network utilization (see (Knightly and Zhang, 1995)).

Each deterministic traffic model uses parameters to define a traffic constraint function $b(t)$ which constrains or bounds the source over every interval of length t. Denoting $A[t_1, t_2]$ a connection's arrivals in the interval $[t_1, t_2]$, the traffic constraint function $b(t)$ requires that $A[s, s+t] \leq b(t)$, $\forall s, t > 0$. Note that $b(t)$ is a time-invariant deterministic bound since it constrains the traffic source over every interval of length t. For example, the (σ, ρ) is defined such that $A[s, s+t] \leq \sigma + \rho t$ for all t.

The D-BIND model is defined via rate-interval pairs $\{(R_k, I_k)|k = 1, 2, \cdots, P\}$ so that the constraint function is given by a piece-wise linear function

$$b(t) = \frac{R_k I_k - R_{k-1} I_{k-1}}{I_k - I_{k-1}}(t - I_k) + R_k I_k, \quad I_{k-1} \leq t \leq I_k \tag{1}$$

with $b(0) = 0$. Thus the rates R_k can be viewed as an upper bound on the rate over every interval of length I_k so that

$$A[t, t + I_k]/I_k \leq R_k \quad \forall t > 0, k = 1, 2, \cdots, P. \tag{2}$$

Figure 1(a) shows a plot of the D-BIND rate-interval pairs for a 30 minute trace of an MPEG-compressed action movie. Plotting the bounding rate R_k vs. interval length I_k, the figure shows that the model captures the source's burstiness over multiple interval lengths. For example, for small interval lengths, R_k approaches the source's peak rate (5.87 Mbps in this case) while for longer interval lengths it approaches the long term average rate (the total number of bits in the MPEG sequence divided by the length of the sequence, 583 kbps).

Figure 1(b) shows the movie's D-BIND constraint function $b(t)$ as described by Equation (1) using the rate-interval pairs of Figure 1(a). The figure shows maximum bits (in kbits) that the source transmits over any interval of length t (shown in seconds) and indicates that the D-BIND model is capturing the temporal properties of the MPEG video. For example, the peak rate shown in Figure 1(a) is caused by transmission of the largest I frame of the sequence. This can be seen in the portion of the constraint function with the largest slope (slopes indicating rates) between $t = 0$ and $t = 42$ msec (the frame rate is 24 frames per second). Importantly, even in the worst case, a large I frame is followed by two typically smaller B frames, which is captured by the constraint curve's slope decreasing in the interval $t = 42$ msec to $t = 125$ msec. Next, a P frame is transmitted (even in the worst case) and these tend to be smaller than I frames but larger than B frames. In (Knightly and Zhang, 1995) it was shown that the D-BIND model's ability to capture both micro- and macro-level burstiness of the video sequence, lead to considerably higher utilization than that achieved with previous models.

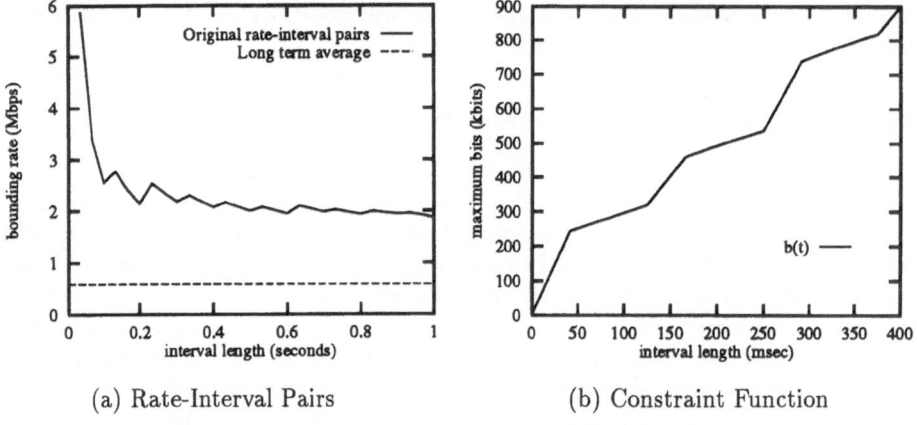

(a) Rate-Interval Pairs   (b) Constraint Function

Figure 1: D-BIND Characterization for Action Movie

Connection Admission Control

Deterministic admission control conditions rely on the delay analysis techniques of (Cruz, 1991; Zhang and Ferrari, 1994). These works show that if a FCFS scheduler with link speed l is serving N connections each bounded by its respective constraint functions $b_j(t)$, then an upper bound on delay for all connections is given by:

$$d = \frac{1}{l} \max_{t \geq 0} \{ \sum_{j=1}^{N} b_j(t) - lt + \bar{s} \}. \tag{3}$$

where $\bar{s}$ is the transmission time of the largest sized packet. The proof starts with an expression for the work in the system at time t, $W(t) = \max_{s \leq t} (\sum_j A_j[s, s+t] - l(t-s))$ and uses bounds on individual sources ($b_j(t) \geq A_j[s, s+t] \; \forall s, t > 0$) to bound the aggregate. Delay bounds for priority service disciplines that are more suitable for providing integrated services may be expressed in a similar manner (Knightly et al., 1995).

SMOOTHING OF VBR TRAFFIC

No-loss smoothing of VBR traffic in the case of the D-BIND model may be viewed from either of two equivalent view-points: the (R_k, I_k) rate-interval pairs or the $b(t)$ constraint function. In the first case, smoothing can be viewed as transforming a source with upper bounds $\{(R_k, I_k) \mid k = 1, 2, \cdots, P\}$ to $\{(\hat{R}_k, \hat{I}_k) \mid k = 1, 2, \cdots, P\}$, with $\hat{R}_k \leq R_k$ if $\hat{I}_k = I_k$. This transformation is realized with the smoothing or buffering at the traffic shaper.

A second view of traffic shaping may be seen from the smoothed source's new D-BIND constraint function $\hat{b}(t)$:

$$\hat{b}(t) = \frac{\hat{R}_k \hat{I}_k - \hat{R}_{k-1} \hat{I}_{k-1}}{\hat{I}_k - \hat{I}_{k-1}}(t - \hat{I}_k) + \hat{R}_k \hat{I}_k, \quad \hat{I}_{k-1} \leq t \leq \hat{I}_k. \tag{4}$$

With $\hat{R}_k \leq R_k$ and Equations (1) and (4), we have that $\hat{b}(t) \leq b(t) \; \forall t$. More formally, we define smoother traffic in the following manner (similar to the definition in (Low and Varaiya, 1993) for "burstiness curves"):

Definition 1 *If $lim_{t\to\infty}b(t)/t = lim_{t\to\infty}\hat{b}(t)/t$, then $\hat{A}(t)$ can be considered smoother or less bursty than $A(t)$ if $\hat{b}(t) \leq b(t) \; \forall t$.*

In this paper, we consider smoothing over S video frames to be defined in the following manner: a source smoothed over S frames is shaped by a FIFO queue with rate R_S (a bound on the source's rate over an interval of length I_S) and maximum buffer size B. B can be chosen with Equation (3) so that no packets are dropped by the traffic shaper. The buffer occupancy of the smoothing FIFO is then given by

$$q(t) = \sup_{0 \leq s \leq t} \{A[s,t] - R_S \cdot (t - s)\}. \tag{5}$$

The smoothed arrival process $\hat{A}$ is therefore

$$\hat{A}[t, t + \delta] = \min(R_S \cdot \delta, A[t, t + \delta] + q(t)). \tag{6}$$

Note that while the smoothed process will be less bursty than the original process (as in Definition 1), it will not in general be constant bit rate (see (Low and Varaiya, 1993) for a proof).

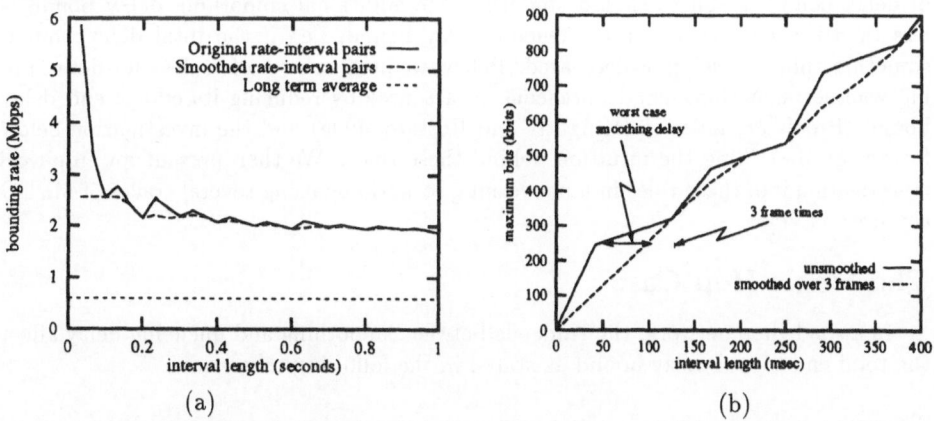

(a) (b)

Figure 2: Smoothed D-BIND Rate-Interval and Constraint Curves

An example of the transformation of smoothing on a source's rate-interval pairs is shown in Figure 2(a). The figure shows the original and smoothed D-BIND rate-interval curve using empirical data from the MPEG trace of an action movie. In this example, the source's original peak rate is $R_1 = 5.87Mbps$. From $I_1 = 1$ frame time, the bounding rate decreases towards the long-term average rate. After smoothing over $S = 3$ frame times (via a FIFO with rate $R_3 = 2.55Mbps$), for a given interval length, the bounding rate of the smoothed source $\hat{R}_j$ is less than or equal to the bounding rate of the unsmoothed source R_j. As well, for interval lengths of up to S frame times, $\hat{R}_j$ tends to reach the smoothing rate R_S.

Figure 2(b) shows the effect of smoothing on a source's constraint function. As shown, τ_S, the worst case smoothing delay when smoothing over S frames (using a FIFO with rate R_S) may be calculated as the maximum horizontal time distance between the two curves b and $\hat{b}$ (Cruz, 1991). That is,

$$\tau_S = \max_{t_2 > t_1}\{t_2 - t_1 | \hat{b}(t_2) = b(t_1)\}. \tag{7}$$

Tradeoffs

Intuitively, a less bursty source can better utilize network resources. This is stated in the following Lemma. We use d to represent the delay bound for an *unsmoothed* source $A(t)$ with constraint function $b(t)$, and $\hat{d}$ to represent the delay bound for the *smoothed* source with smoothed arrivals $\hat{A}(t)$ and smoothed constraint function $\hat{b}(t)$.

Lemma 1 *If source j is smoothed so that arrival process $\hat{A}(t)$ is less bursty than $A(t)$, then the queueing delay bound for a FCFS scheduler is reduced. Equivalently, smoother or less bursty sources require fewer network resources.*

Proof: The FCFS queueing delay bound is given by Equation (3) as $d = \frac{1}{l}\max_{t \geq 0}$ $\{\sum_{j=1}^{N} b_j(t) - lt + \bar{s}\}$. If source j is smoothed, then the jth term of the summation $b_j(t)$ is replaced with $\hat{b}_j(t)$. Since $\hat{b}(t) \leq b(t)$ for all t, it can only reduce the delay bound to $\hat{d} \leq d$. $\square$

However, while smoothing does reduce the queueing delay bound, it also introduces an additional delay τ_S, due to buffering at the FIFO queue. This delay contributes to the total end-to-end delay perceived by the source. Therefore, from the perspective of delay bound, a source should smooth if the additional smoothing delay bound is less than the reduction in the queueing delay bound, i.e., if the total delay bound, smoothing plus queueing, is decreased. Below, we present a set of "rules" for determining whether smoothing is advantageous to a source by reducing its end-to-end delay bound. Proofs are given in (Knightly and Rossaro, 1995) and the investigation below focuses on describing the intuition behind these rules. We then present an empirical investigation into these rules in a networking scenario by using several traces of MPEG compressed video.

The Single-Hop Case

In a single hop network, the tradeoffs between smoothing and queueing delay affect the total end-to-end delay bound as stated in the following theorem.

Theorem 1 *In the single-hop network with homogeneous sources and deterministic delay bounds, the reduction in queueing delay bound for a FCFS server introduced by smoothing is outweighed by the additional smoothing delay. That is, smoothing will not cause the total delay bound (smoothing plus queueing) to decrease, or $\tau_S > d - \hat{d}$.*

The proof is given in (Knightly and Rossaro, 1995). The theorem states that from a delay bound perspective, smoothing of homogeneous sources should never be performed in the single hop case since it will result in a net increase in total end-to-end delay bound. However, other factors may influence the decision to smooth such as the respective prices of buffers on an end system and inside the network. As well, the theorem is not always true for *heterogeneous* sources.

The Multi-Hop Case

Theorem 1 showed that over a single hop, smoothing can be an ineffective means for achieving higher network utilization. However, over multiple hops, the analysis has an important additional component: while the smoothing delay is incurred only once at the source's traffic shaper, in a congested network, queueing delays may be incurred at multiple nodes. Thus, a smoother source can reduce its queueing delay at each congested hop resulting in a considerable decrease in its end-to-end delay bound. Then, in many cases, over multiple congested hops this reduction in queueing delay outweighs the added delay caused by smoothing so that smoothing can often result in a net benefit to a network client.

The following proposition provides a rule for determining if smoothing provides a net advantage in terms of end-to-end delay bound for networks that use rate controlled service disciplines such as Rate Controlled Static Priority, Earliest Deadline First, or Hierarchical Round Robin (Zhang and Ferrari, 1994). The rate-controlling aspect of these service disciplines reshapes the traffic at each hop by using (for example) leaky buckets at each node rather than only at the entrance of the network.

Proposition 1 *Consider a network in which each node i has a rate-controlled service discipline that can provide an upper bound on queueing delay. If $\hat{d}_i$ is the queueing delay bound at hop i for the smoothed source and d_i is the original queueing delay bound at hop i, a source will obtain a net reduction in end-to-end delay bound due to smoothing if the following condition holds:*

$$\tau_S < \sum_{i=1}^{H}(d_i - \hat{d}_i) \tag{8}$$

where H is the number of hops between the source and destination.

Proof: The proposition states that if the bound on smoothing delay τ_S is less than the total reduction in queueing delay across multiple hops, then smoothing is advantageous to the source. Because of the decoupling effect of per-hop rate control, the end-to-end delay bound D may be calculated by summing delay bounds d_i at individual nodes (Ferrari and Verma, 1990). Thus, for a smoothed source, we can bound the end-to-end delay as $\hat{D} = \tau_S + \sum_{i=1}^{H} \hat{d}_i + \sum_{i=1}^{H} \pi_i$ where π_i is the propagation delay of the i'th hop. Without smoothing, the end-to-end delay is given by $D = \sum_{i=1}^{H} d_i + \sum_{i=1}^{H} \pi_i$. A source will have a reduction in end-to-end delay if $\hat{D} < D$. □

Equation 8 show the importance of the number of *congested* hops in determining smoothing conditions. The additional benefit of smoothing can only increase with H, since the summands in the right hand side of the inequality are always positive (see Lemma 1). If only a single hop is congested, then for all but one of the hops, $d_i \approx \hat{d}_i$, so that the situation will be similar to the single hop result of Theorem 1.

EXPERIMENTAL RESULTS

In this section, we use two 30-minute traces of MPEG compressed video to quantify the effects of smoothing on the total end-to-end delay bounds guaranteed to a source. One trace is of an action movie and the other is of a newscast. Both were digitized and compressed at 24 frames per second with frame pattern **IBBPBBPBBPBB**. We consider only a deterministic guarantee in which the network provides a no-loss, no-delay-violation guarantee. The smoothing is also deterministic in that the smoothing

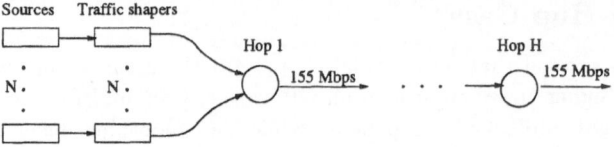

Figure 3: Network Topology

FIFO's parameters are selected such that no packets are dropped and the smoothing delay is bounded.

The experiments consider both the single and the multiple hop cases with a simple network topology shown in Figure 3. N connections are shaped by their respective FIFOs and then they are multiplexed at the network nodes. The connections then traverse H hops before reaching their destination.

The experiments are conducted as follows. First, we use the traces to calculate a source's D-BIND parameters.[1] We also calculate the D-BIND parameters under various smoothing intervals S. Next, for the given source characterization, we use the admission control test of Equation 3 to determine the maximum number of homogeneous admissible connections on the 155 Mbps link for a given delay bound. We then vary S, the number of frames smoothed over, H, the number of hops, and use the two different traces. To evaluate the effectiveness of smoothing, we use the following performance metrics: the utilization of the network which is directly proportional to the number of schedulable connections N; the net end-to-end savings in delay bound from smoothing connection j, $D_j - \hat{D}_j = \tau_{S,j} + \sum_{i=1}^{H}(d_{i,j} - \hat{d}_{i,j})$; and the total end-to-end delay bound $\hat{D}_j = \sum_{i=1}^{H} \hat{d}_{i,j} + \tau_{S,j}$.

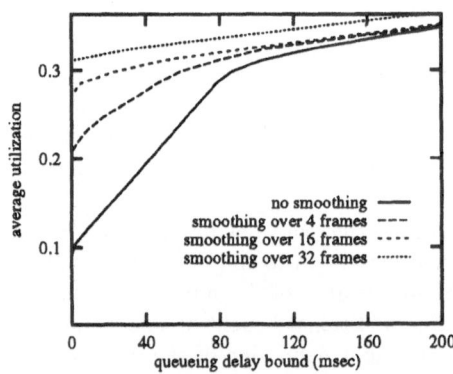

Figure 4: Utilization vs. Queueing Delay for Various Smoothing Policies

The first performance index used in the experiments is average utilization of the multiplexer. It is determined by calculating the maximum number of admissible connections and computing the resulting average network utilization for deterministic service as the sum of the admitted sources' long-term average rates divided by the link speed. Figure 4 shows the effect of smoothing on queueing delay only, that is, the delay expe-

---

[1]In the more practical case where the trace is not available in advance, algorithms in (Zhang and Knightly, 1995) can be used to obtain the D-BIND parameters for a "live" source.

rienced at a network node (ignoring for the moment smoothing delay). The horizontal axis is queueing delay bound and the vertical axis is the average utilization of the node. The general shape of the curves indicates that as the queueing delay bound increases, more connections are admissible so that a higher utilization is possible. As stated in Lemma 1, as the smoothing interval increases (in this case, from 0 to 4, 16, and 32 frames), or equivalently as the FIFO smoothing rate decreases (to R_4, R_{16}, and R_{32}), the traffic transmitted to the network becomes smoother so that the queueing delay bound is reduced. Equivalently, for a given queueing delay, higher utilizations are achievable inside the network.

However, when the total end-to-end delay bound is considered including both smoothing and queueing delay, the resulting utilization is quite different. As predicted by Theorem 1, smoothing never results in an increased network utilization for a given delay bound in the single-hop homogeneous case. This is shown in Figure 5(a) which depicts average utilization versus total delay bound for various smoothing rates in a single-hop network. For example, for an *unsmoothed* source and a delay bound of 80 msec, a 29% utilization is achievable. Alternatively, when the source is smoothed over $S = 16$ frames, a smoothing delay of 76 msec is introduced so that for a total delay bound of 80 msec, the achievable utilization for 16-frame smoothing is reduced to 27%.

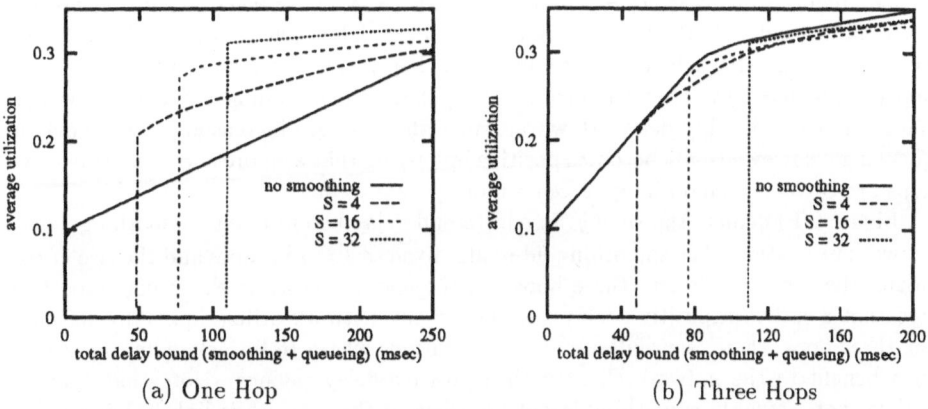

(a) One Hop   (b) Three Hops

Figure 5: Average Utilization vs. Total Delay Bound for Various Smoothing Rates

In the case of networks with *multiple* congested hops, the results can be considerably different from the above scenario. Figure 5(b) shows the same experiment described above except that the sources traverse three hops rather than one. In this case, smoothing results in a substantial reduction in a source's end-to-end delay bound. For example, for a total end-to-end delay bound of 100 msec, without smoothing, an average utilization of 17% is achieved. With smoothing over $S = 16$ frame times (with a FIFO of rate $R_{16} = 2.07$ Mbps), this utilization is improved to 29%. Thus, in this case, smoothing has resulted in a 70% increase in the number of admissible connections for the same end-to-end delay bound. This utilization improvement is explained by the rule presented in Proposition 1: while the smoothing delay occurs only once at the source, there is a queueing delay at each of the congested network nodes, thus favoring smoothing over multiple congested hops.

Figure 6(a) shows the effect of the number of congested *hops* (for the network topology of Figure 3) on the savings in end-to-end delay bound. That is, the figure shows the reduction in total delay $(D - \hat{D})$ versus the number of hops (H). The number of connections is fixed to 24 so that the average network utilization is 31% in all cases.

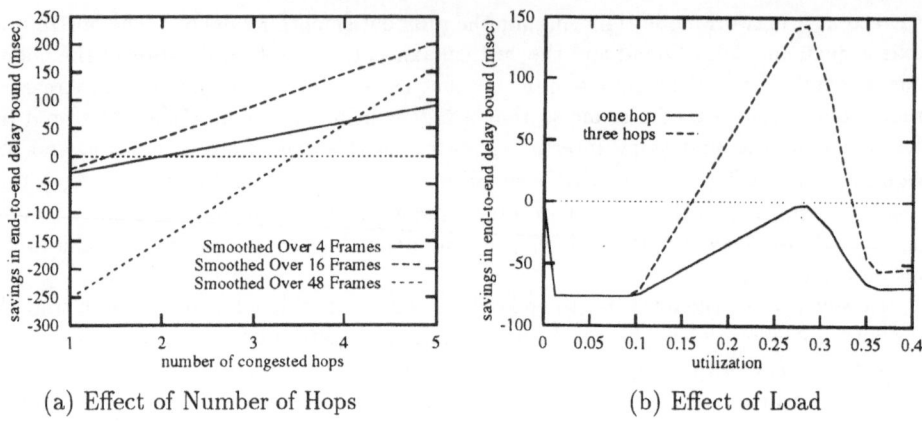

(a) Effect of Number of Hops　　　　　(b) Effect of Load

Figure 6: Effect of the Number of Congested Hops and Load on End-to-end Delay Savings

In the homogeneous case, $D - \hat{D}$ reduces to $H(d_j - \hat{d}_j) - \tau_S$ where d_j and $\hat{d}_j$ are the unsmoothed and smoothed delay bounds at node j. Thus, the savings in delay bound $D - \hat{D}$ increases linearly with the number of hops. The lines start negative at one hop (as predicted by Theorem 1) indicating that the smoothing delay τ_S outweighs the savings in queueing delay. However, since the lines are increasing, this savings in queueing delay eventually becomes positive indicating that as more and more congested hops are traversed, smoothing becomes more advantageous.

Figure 6(b) depicts the *savings* in end-to-end delay bound due to smoothing versus network utilization. The smoothing interval is fixed to $I = 16$ frames and the two curves depict the cases of one and three hops. In the one-hop case, as expected, smoothing does not result in a positive delay savings for any network utilization. Alternatively, for three hops, there is a wide range of utilizations where 16-frame smoothing results in a benefit to the network client with a positive delay savings. Note that even for a three hop network, smoothing is not beneficial if the network is lightly loaded (i.e., operating under low utilization). The reason for this is that in a lightly loaded switch $d \approx \hat{d}$ for any reasonable range of smoothing delays. This emphasizes that our results for a single hop are also an approximation for the case of a single *congested* hop. Also, for higher loads above 0.34, the savings in end-to-end delay again is less than zero even for the three-hop case. The reason for this is that the fixed smoothing interval $I = 16$ frames is inappropriate for higher utilizations and a larger smoothing interval would be required.

Figure 7 shows the effect of the smoothing interval (or equivalently, the smoothing rate or number of frames smoothed over) on savings in end-to-end delay $D - \hat{D}$. For the one-hop case, Figure 7(a) shows that more smoothing only worsens the end-to-end delay (decreasing the savings) with the two curves depicting utilizations of 33% and 19%. Figure 7(b) shows the same experiment for a network with three hops. At 33% utilization, there is a clear range of smoothing intervals from $S = 34$ and $S = 39$ where smoothing results in a significant savings in end-to-end delay of up to 320 msec. For the lower network utilization of 19%, smoothing is less effective as described above. The curves of Figure 7 indicate the importance of choosing the correct smoothing interval for a given network load according to the rules of Proposition 1 since improperly chosen smoothing policies can be far worse than doing no smoothing at all.

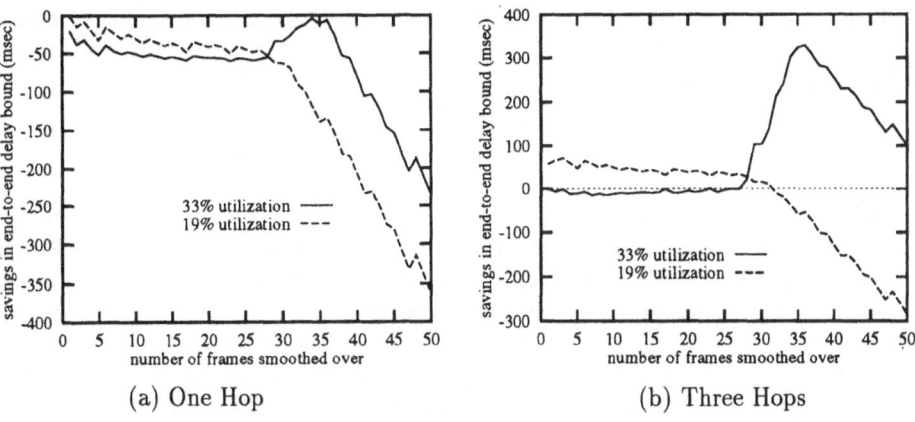

(a) One Hop   (b) Three Hops

Figure 7: Effect of Smoothing Interval on Savings in Delay Bound

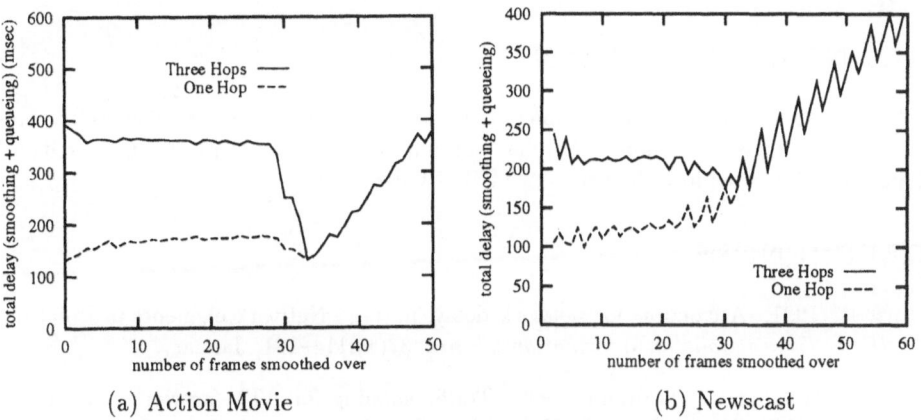

(a) Action Movie   (b) Newscast

Figure 8: Effect of Smoothing Rate on Total End-to-end Delay

A network client is ultimately interested in its achievable end-to-end delay bound and Figure 8 shows the total end-to-end delay bound including smoothing and queueing delay versus the number of frames smoothed over. Figure 8(a) shows the curves for the MPEG trace of the action movie (the trace used thus far) with 32% network utilization and Figure 8(b) shows the curves for a trace of a newscast with 28% node utilization. The figures indicate that the total delay bound achieved by the network client is quite sensitive to the smoothing interval. For example, if a 150 msec end-to-end delay bound is required for the action movie sequence traversing three congested hops, Figure 8(a) shows that if the network load is 32%, the only two admissible smoothing intervals are 33 and 34 frames. Any other smoothing interval would result in a rejection of connections by the admission control algorithm indicating that the requested 150 msec delay bound cannot be guaranteed at a utilization of 32% for the "improperly" shaped sources.

Finally, we note that the utilizations presented here are for deterministic service only. Unused network resources can always be utilized by sources receiving a statistical performance guarantee or a best effort service. We also note that most of the curves above did not tend to be as smooth as one may expect (i.e., they are jagged in many regions). The primary reason for this can be seen from the rate-interval pairs of Figure

1 which shows that for these traces of MPEG video, the bounding rate does not monotonically decrease with interval length, primarily because of the IBP frame structure of the MPEG algorithm.

CONCLUSIONS

In this paper we have investigated the effects of traffic smoothing on end-to-end deterministic performance guarantees to VBR video. With analysis and experimentation with traces of MPEG-compressed video, we have shown that over a single congested hop, smoothing of homogeneous sources never results in a net reduction in end-to-end delay bound, i.e., the reduction in queueing delay is outweighed by the addition of a smoothing delay. Alternatively, over multiple congested hops, we showed that smoothing can indeed result in a considerable reduction in end-to-end delay bound in many cases. We provided a set of rules for determining whether or not a source should smooth under a given network load.

ACKNOWLEDGEMENTS

The authors are grateful to Domenico Ferrari and Hui Zhang for their useful comments and suggestions and to Sandia National Laboratories and the University of Wuerzburg for providing the traces of MPEG video.

REFERENCES

Cruz, R. 1991. A calculus for network delay, part I : Network elements in isolation. *IEEE Transactions on Information Theory*, 37(1):114–121, January.

de Veciana, G. and J. Walrand. 1992. Traffic shaping for ATM networks. Technical Report UCB/ERL M92/135, University of California at Berkeley, Berkeley, CA, December.

Ferrari, D. and D. Verma. 1990. A scheme for real-time channel establishment in wide-area networks. *IEEE Journal on Selected Areas in Communications*, 8(3):368–379, April.

Kanakia, H., P. Mishra, and A. Reibman. 1993. An adaptive congestion control scheme for real-time packet video transport. In *Proceedings of ACM SIGCOMM'94*, pages 20–31, San Francisco, CA, September.

Knightly, E. and P. Rossaro. 1995. Smoothing and multiplexing tradeoffs for deterministic performance guarantees to VBR video. Technical Report TR-95-033, International Computer Science Institute, Berkeley, CA, July.

Knightly, E., D. Wrege, J. Liebeherr, and H. Zhang. 1995. Fundamental limits and tradeoffs for providing deterministic guarantees to VBR video traffic. In *Proceedings of ACM SIGMETRICS'95*, Ottowa, Ontario, May.

Knightly, E. and H. Zhang. 1995. Traffic characterization and switch utilization using deterministic bounding interval dependent traffic models. In *Proceedings of IEEE INFOCOM'95*, pages 1137–1145, Boston, MA, April.

Low, S. and P. Varaiya. 1993. Burstiness bounds for some burst reducing servers. In *Proceedings of IEEE INFOCOM'93*, pages 2–9, San Francisco, CA, March.

Shroff, N. and M. Schwartz. 1994. Video modeling within networks using deterministic smoothing at the source. In *Proceedings of IEEE INFOCOM'94*, pages 342–349, Toronto, Ontario, June.

Zhang, H. and D. Ferrari. 1994. Rate-controlled service disciplines. *Journal of High Speed Networks*, 3(4):389–412.

Zhang, H. and E. Knightly. 1995. RED-VBR: A new approach to support VBR video in packet-switching networks. In *Proceedings of IEEE Workshop on Network and Operating System Support for Digital Audio and Video (NOSSDAV'95)*, pages 275–286, Durham, NH, April.

DETECTION AND CONTROL OF
BURSTY VIDEO AND PACKET DATA TRAFFIC

C. Thompson, V. Harpanahalli, S. Mulpur and B. Jang

Center for Advanced Computation and Telecommunications
University of Massachusetts, Lowell, MA 01854

INTRODUCTION

Numerous investigations have focused on what statistical features can be used to characterize burst arrivals. In ISDN[1] and VBR[2] packet traffic the probability density functions (pdf) of the interarrival time and scene-length respectively asymptotically decay as $t^{-\lambda}$. The heavy tailed pdfs suggest high variance in the arrival process. High variance and long-range dependence can also be the result of clustered packet arrivals. Such clusters can have a significant impact on multiplexing and performance of high speed networks[3,4]. Consequently, methods for detecting and characterizing such events are needed. Recently it has been shown that the bursts can be adequately modeled using a chaotic model and self similar model for the packet run-length[5,6].

This paper examines the application of time-frequency analysis to the problem of burst arrival detection in packet data traffic. In particular the spectral features of the burst run-length time series will be used to detect the occurrence of bursts. Event time-series comprised of the number of packets generated during an event epoch will be used in our analysis. It is shown that the smoothed Wigner-Ville distribution holds promise in this regard. The influence of burst arrivals on multiplexing source channels is also examined.

SOURCE MODEL

ISDN Traffic Model

Recently ISDN traffic was found to be composed of basic structural units, referred to as events[5]. An event is a sequence of packet arrivals having short interarrival times followed by packet arrivals having long interarrival times. Each packet arrival is classified as either a burst or a jump arrival based on the packet interarrival time. A burst arrival has an interarrival time drawn from a probability density function (pdf) $f_b(t)$ which is a gamma distribution while a jump interarrival time is drawn from the pdf $f_j(t)$ which is asymptotic to $t^{-\lambda}$. These burst and jump arrivals are grouped to form an event, where each event is comprised of a sequence of burst arrivals followed by one or more jump arrivals.

The total run-length or number of arrivals x_s in an event is determined by the sum of the number of burst and jump arrivals namely n_b and n_j. The total run-length and number of burst arrivals are modeled by the pdf $r(x_s, n_b)$ which takes the form of a product of hyper-exponential functions.

Multimedia Communications and Video Coding
Edited by Y. Wang *et al.*, Plenum Press, New York, 1996

$$r\ (x_s, n_b) = C_O \sum_{m=1}^{12} \gamma^m e^{-\alpha^m} x_s \sum_{k=1}^{12} \gamma^k e^{-\alpha^k} (x_s - n_b - 1) \quad (1)$$

where C_O is a normalizing constant, $\gamma = 0.85$, and $\alpha = 0.825$. In the current analysis, x_s will be taken to range from 2 to 40 arrivals and n_b from 1 to x_{s-1}. The number of jump arrivals equals the difference $x_s - n_b$. The parameters of this model are determined from the analysis of 16 thousand packets of ISDN traffic data. The model and ISDN data agrees to within 2 packets in the mean and variance. Bursts are the result of long run-lengths of packets having short interarrival times occurring over a successive number of events. As a result of these intermittently occurring clusters in the data stream, the moments in the packet count do not exhibit asymptotic convergence with increase in the observation interval. These packet arrivals cannot be modeled as an iid process. One can consider that for a given event the run-length x is the sum of a stationary random variable and a noise term $x\ (e) = x_s\ (e) + x_c\ (e)$ where x_s is drawn from the pdf given by Eqn. 1 and x_c represents an additive noise term. The noise term x_c varies as a function of the event counter e. What differentiates x_c from x_s is that it takes on nonzero values intermittently. This intermittent activity gives rise to clusters in the run-length sequence. The spacing between clusters is not geometrically distributed. When this noise term is modeled by an intermittent chaotic difference equation it has been shown that the nonstationary features in the ISDN data traffic can be captured[5]. The additive noise variable is given by the solution of the difference equation

$$c\ (e) = \begin{array}{ll} \varepsilon + c\ (e-1) + L\ c\ (e-1)^2 & c\ (e-1) < D \\ (\ c\ (e-1) - D\ )\ /\ (1 - D\ ) & 1 > c\ (e-1) \geq D \end{array} \quad (2)$$

where $L = (1 - \varepsilon - D\ )\ /\ D^2$ and the scale transformation $x_c\ (e) = A\ c(e)$. The parameters $\varepsilon = 10^{-6}$, $A=15$ and $D= .09$. These values were found to yield the correct first and second moments for the run-lengths for intermittently occurring clusters. To generate the time-series, the number of burst arrivals n_b at a given trial is drawn from the marginal pdf $r\ (n_b\ )$. The run-length x_s is then sampled from the conditional pdf $r\ (x_s \mid n_b)$. Given x_s, n_b and x_c the time-series x can be calculated.

Video Traffic Model

Chandra et al. [8] examined H.261 and MPEG encoded video traffic generated in entertainment and teleconferencing applications. The VBR video traffic stream was comprised of intra(I) and predicted(P) coded frames. The transitions I to P, P to P and P to I frames were modeled using a Markov chain, with transition probabilities obtained from clips of video data. To better model the long holding times at some of the observed frame rates, the P-state was partitioned into K-classes. The allowable frame rate from each of these classes was described by a normal distribution. The number of classes needed to accurately model teleconferencing and entertainment videos was found to be 14 and 18 respectively. All I frames were modeled as a single class.

Variability in the traffic stream was modeled by considering the interframe correlation of the data rate within each of the classes. To this end a first order autoregressive model for each class was used. It was found that the linear increase in the index of dispersion in the rate was the result of the temporal variation in the correlation that occurs when one makes a transition between classes.

WIGNER-VILLE DISTRIBUTION

When the random variable x at each trial is iid, the resulting run-length sequence yields samples of a white-noise process. In such a case an N-point estimate of the autocorrelation will drop below a prescribed confidence bound for lags greater than one. In an observation interval where clusters having long run-lengths occur, the autocorrelation function decays at a slower rate. Therefore intermittent or bursty regions in x are manifest in low frequency characteristics in its power spectrum. For this reason spectral analysis based on short-time estimates of the autocorrelation allows one to identify clusters of burst events in the event series $x(e)$. To successfully segment the data, resolution in time and frequency must be considered. In this analysis time corresponds to the event index e.

The Wigner-Ville distribution (WVD) and its windowed and smoothed counterpart, the smoothed pseudo Wigner-Ville distribution (SPWVD) estimate the power-spectrum in an analysis window as a function of the observation time. Its quadratic dependence on the analyzed signal x allows the instantaneous energy and spectral density to be evaluated from the marginals of the WVD. Martin and Flandrin[7] demonstrated that the nonstationary second order statistics can be ascertained from averaging the adjacent short-time estimates of the Wigner-Ville distribution. This averaging, smoothed the spectrum while retaining the temporal resolution of the WVD. Hence time-frequency analysis based on the SPWVD provides short-time spectral estimates with better temporal resolution than those of the short-time Fourier transform. For this reason the SPWVD may be suitable for locating the burst clusters in data traffic.

In our analysis we will consider the sequence $x(e)$ where the independent variable e is the event index. The event sequence is first partitioned using a sliding rectangular window. The process of windowing localizes time segments by centering a window function of duration $2N$ points at successive locations eT, where T represents the magnitude of shift of the window between successive observations. The SPWVD provides a spectral estimate at each location eT in the segmented event sequence. The SPWVD is defined as

$$W_s(eT, k) = \frac{2}{P} \sum_{p=0}^{P-1} \sum_{m=0}^{N-1} x(eT + p + m) \, x^*(eT + p - m) \, e^{-j2\pi mk/N} \qquad (3)$$

The distribution W_s represents the windowed WVD averaged over P adjacent realizations. Since W_s has period $N/2$ in frequency, care must be taken to ensure that aliasing is avoided. The averaged expression is identical to evaluating the Fourier transform of the P-point correlation $R(m,-m)$. When the number of Wigner spectra P is equal to N and T is equal to one, the SPWVD reverts to $2N$ times the power spectrum of the signal.

BURST DETECTION AND CHARACTERIZATION

Time-frequency analysis of packet traffic data was performed using simulation data from ISDN and video traffic models. Packet arrivals in ISDN traffic can be split into events, where each event corresponds to a group of packet arrivals. The packet interarrival-time, time duration of an event and number of packet arrivals in each event is described by a random process. The smoothed pseudo Wigner-Ville distribution (SPWVD) will be used to segment and identify burst events in the traffic stream. The time-frequency feature afforded by the SPWVD allows one to identify clusters in the event time-series.

SPWVD Based Traffic Characterization

In this section we will examine the SPWVD derived characteristics of bursts arising in a burst run-length sequence. Burst events are generated using the ISDN traffic model. In this model the burst run-lengths are drawn from an iid hyper-exponential distribution and the result x_s is added to a chaotic noise signal x_c. The run-length sequence x is equal to $x_s + x_c$ and is shown in Fig. 1. The presence of chaotic noise gives rise to sporadically occurring bursts having long run-lengths. The SPWVD used in this section was computed using data analysis segments of $N = 128$ in duration. The result was then averaged over 5 adjacent events. This corresponds to a value of $P = 5$ in Eqn 5.. Each SPWVD estimate was evaluated on an interval of T equal to 32 events.

The SPWVD analysis of 2500 events is depicted in Fig. 2. At each observation point eT, $W_s(eT, k)$ is plotted as a function of discrete frequency index k. In the figure the horizontal axis represents the event index e and the vertical axis corresponds to the discrete frequency index k. During a burst period, the SPWVD exhibits dominant low frequency components. The ensemble average spectrum for these segments decays at a power law rate in frequency $k^{-\beta}$, where the exponent of this decay, β, is 0.909. A power law exponent of less than one is characteristic of chaotic noise.

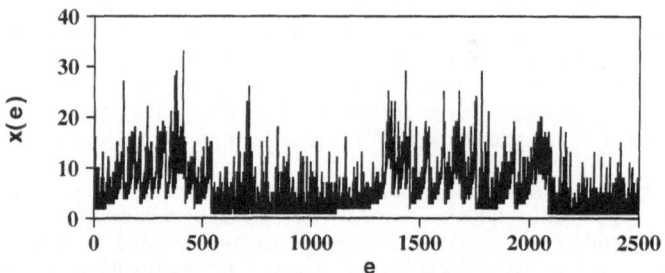

Figure 1. Sample path of the run-length $x(e)$

The ensemble averaged spectrum of the stationary segments is uniform in magnitude with the frequency index k. For bursty regions to be identified the disparent energy amplitudes between the stationary and chaotic signals at low-frequency can be exploited. To this end the magnitude of the energy in an effective bandwidth σ for each slice of the SPWVD estimate will be used to identify the occurrence of a burst event in the signal x. This parameter is calculated from the ensemble averaged spectrum. The centroid $|k_e|$ and the normalized second moment $|k_e^2|$ of W_s are evaluated for each observation location of the SPWVD. The effective bandwidth σ of the SPWVD estimates for each slice is evaluated by $\sigma^2 = |k_e^2| - |k_e|^2$. Averaging the centroid and bandwidth estimates over the slices of the SPWVD for the signal, yields the values $|k_e| = 7$ and $\sigma = 4$.

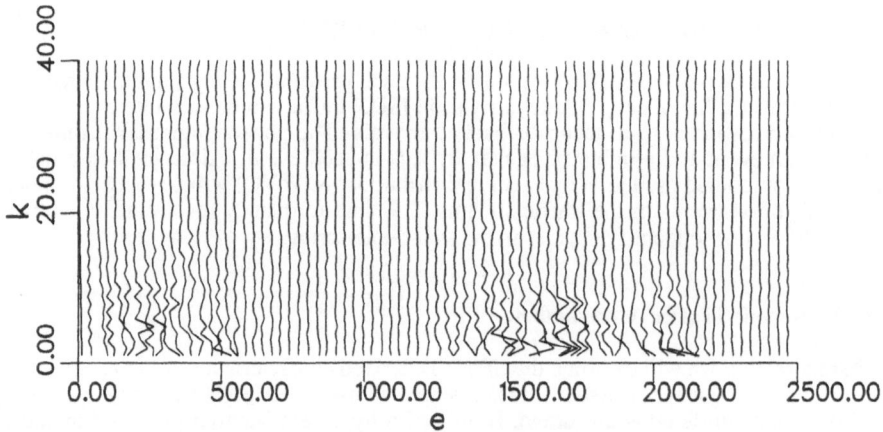

Figure 2. Smoothed Pseudo Wigner-Wille distribution of $x(e)$.

A burst event is detected by calculating the energy in the effective bandwidth σ evaluated from the spectral estimate obtained at an observation point. The energy in the passband for the chaotic signal ranges from 300 to 3200 in amplitude . For the stationary noise signal the maximum energy amplitude was found to be equal to 207.0. The difference in the energy amplitudes in the passband between the chaotic and stationary noise signals, allows one to detect chaotic burst events. The bursty run-lengths may be segmented from the stream using an amplitude threshold based on the subband energy of the SPWVD estimate. A burst event is tagged when the passband energy of the SPWVD is greater than 207.0. To be effective, the temporal location of the burst event must be faithfully detected in the presence of random noise.

To illustrate the detection performance, the event run-lengths tagged as burst events were removed from the traffic stream. The interarrival times and packet traffic are generated using this residual set of run-lengths. The mean interarrival time is 0.5 msec and the shortest interarrival time is 0.07 msec. The index of dispersion for the arrival counts which is the ratio of the expected number and variance in the arrivals over a time interval t, is shown in Fig. 3 for the packet stream generated from the run-lengths. The traffic stream obtained by removing the SPWVD tagged events exhibits asymptotic convergence in the index of dispersion in the count. The index of dispersion in the count for the original sequence increases with the observation interval at a power law rate. Therefore the SPWVD is useful in identifying clusters of bursts which result in the high traffic variability .

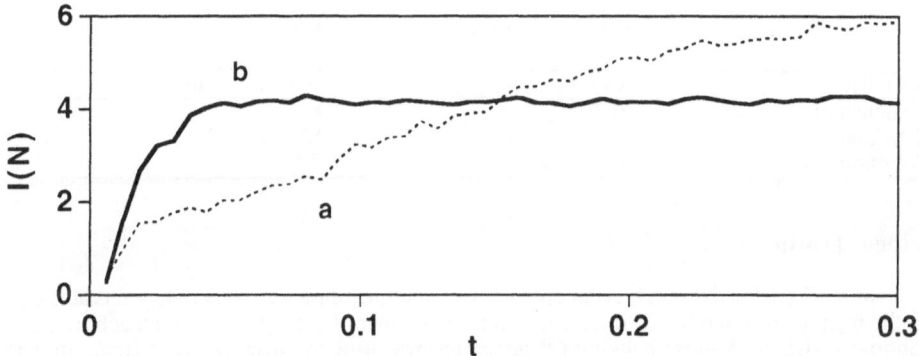

Figure 3. Index of dispersion in counts; (a) Original traffic sequence, (b)Traffic stream with bursts removed.

RESULTS

ISDN Traffic

Traffic streams were generated using the ISDN run-length model and merged into a single data stream using a time-division multiplexing scheme. Packet arrivals and event duration were generated using the model given in [5]. The mean and shortest interarrival time for the resulting stream of 11-byte data packets were chosen to be 0.5 msec and 0.07 msec respectively. As a result the traffic has a peak bit-rate of 1.3 Mbps (14000 packets/sec) and an average bit-rate of 0.18 Mbps (2000 packets/sec). A buffer size of 20 packets and a service rate of 1.5 Mbps will be used here. Given the deterministic service rate of μ packets/sec and the buffer size B, the time delay t_d is 1 ms. The multiplexed stream served as input to a G/D/1/K queue. Each stream was comprised of 20000 packets. Given a service rate of 17000 packets/sec and the average arrival rate of 2000 packets/sec, one estimates that approximately 8 streams can be multiplexed at the average arrival rate. Further if each stream multiplexed, generates packets at the peak rate of 14000 packets/sec then only one channel can be accommodated. The ratio of the number of sources that can be multiplexed at the average rate to the number of sources to that can be multiplexed at the peak rate is 8. These results were confirmed by simulation of 8 multiplexed streams having the aforementioned average rate. A time slot of one microsecond was chosen as the switch interval. Multiplexing 8 streams yielded a cell loss (CL) 0.1% out of 160000 cells. With tagged burst events removed using the SPWVD detection, 25 streams could be multiplexed with no loss.

Utilizing the location of bursts in the data stream, one can reduce cell loss. A simple traffic shaper for the tagged bursts was examined. For each tagged event all constituent packets are tagged. The protocol followed by the traffic shaper is as follows. When the multiplexer buffer is full, a time-delay is imposed on tagged packets from the selected channel. In other words, the traffic shaper forces the time slot for the selected channel to expire. As a result, as long as the buffer is full, a time delay is imposed on every channel generating a tagged packet. Therefore all the tagged packets from a particular burst event are not allowed into the buffer. This ensures that the packets arriving from a burst event will not be lost. The multiplexer moves on to the next channel and repeats the same procedure if a tagged burst appears. Packets from non-tagged events enter the buffer at any time and the tagged events enter the buffer as space becomes available. The longer interarrival times of the non-tagged packets ensure low cell loss. As channels are added the queue parameters, such as the service rate, buffer size and the input rate are held constant.

Closing a bursty channel when the buffer is full allows 8 streams to be transmitted with no loss. Further it is observed that, if the CL is increased to 0.1%, 30 streams can be multiplexed. The CL is due to the arrival of non-tagged packets when the buffer is full. The scheduling policy proposed results in a maximum time delay of 10 msec for the tagged packets.

Table 1. Multiplexed ISDN Traffic

Scheme	Cell Loss(%)	No. of Streams
no detection	0.1	8
detection	0	8
detection	0.1	30

Video Traffic

In VBR packet data each event epoch is of fixed time duration. This time interval is equal to the reciprocal of the frame-rate. However the number of cells generated in each frame is a random variable. A loose-packing [9] procedure was used to packetize each frame into 384 bit ATM cells. In an ATM cell, the cell header comprises 22 bits out of which 8 bits are used for the cell number, 5 bits for the frame number, and 9 bits for the absolute address. The remaining 362 bits are used for the video data. If N bits are coded at the e-th frame, the number of cells generated per frame is given by $x(e) = N/362$. Therefore the instantaneous cell arrival rate $R_c = 30x(e)$ cells/sec, given that the frame rate is 30 frames/sec. A sample path of the cell-count/frame for an encoded video data sequence is shown in Fig. 4. In this figure the cell-count/frame is plotted as a function of the frame index. The mean number of cells/frame is 148 and the peak number of cells/frame is 760 cells. The data is comprised of 5000 frames.

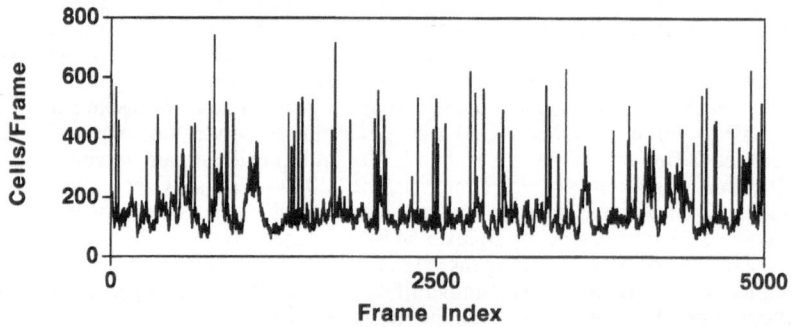

Figure 4. Sample path of the number of ATM cells per frame generated in a video sequence

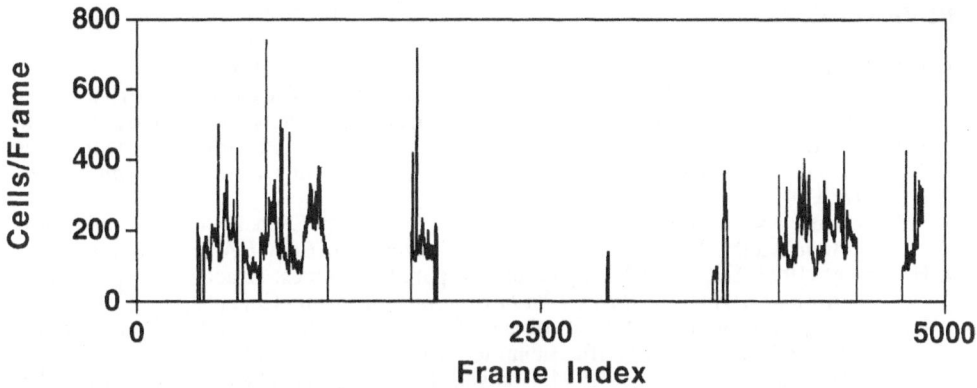

Figure 5. Frames tagged using the passband energy of the smoothed pseudo Wigner-Wille distribution.

Bursts in the cell-count/frame are tagged using a procedure similar to that described in the ISDN case. The energy threshold is set to the ensemble mean of the passband energy of the SPWVD and this yields satisfactory results as the index of dispersion in counts asymptotically converges for the stream with the detected bursts removed. The tagged bursts obtained from the SPWVD are shown in Fig. 5. For each tagged frame all the constituent cells are tagged as bursts. The average bit-rate is 1.8 Mbps and the peak bit-rate is 7.5 Mbps.

The service rate, μ, for transmission is chosen to be 15 Mbps. For this service rate, to achieve a one frame delay, the buffer size required is 148 cells. The one frame delay yields a time delay of 4 msec. For seven streams multiplexed, having the aforementioned average rate, the maximum CL is 0.1% where each source has 5000 frames. Closing a bursty channel based on tagged bursts allows 7 streams to be transmitted with no loss. From the sample path it can be seen that some of the non-tagged frames have more than 148 cells/frame which is greater than the buffer size. The high service rate, relative to the buffer size allows the data from the channels to be serviced without loss. Further it is observed that if the CL is increased to 0.1% 15 streams can be multiplexed. This cell loss is due to the arrival of non-tagged cells when the buffer is full. The cost of the scheduling policy is that the maximum time delay of a burst packet is of one frame interval.

Table 2. Multiplexed video streams

Scheme	Cell Loss(%)	No. of Streams
no detection	0.1	7
detection	0	7
detection	0.1	15

SUMMARY

The smoothed pseudo Wigner-Ville distribution has been used to detect the occurrence of bursts in packet data traffic. Categorizing the incoming data as a burst allows for modest traffic shaping with small transmission delays when data streams are multiplexed.

REFERENCES

1. K.S. Meier-Hellstern, P.E. Wirth, Y. Yan and D.A. Hoeflin, 1991, Traffic Models for ISDN data users: Office Automation Application, Teletraffic and data traffic, a period of change, Eds. A. Jensen and V. B. Iversen, Elsevier Science Publishers, 167-172.
2. M.R. Frater, J.F. Arnold and P. Tan, 1994, A New Statistical Model for Traffic Generated by VBR Coders for Television and BISDN, *IEEE Trans. on Cir., Sys. and Video Tech.*, 4; 521-526
3. R. Guerin, H. Ahmadi and M. Naghshineh, 1995, Equivalent Capacity and its Application to Bandwidth Allocation in High-Speed Networks, *IEEE Sel. Areas in Comm.*, 9;
4. H. Kroner, 1992, Statistical Multiplexing of Sporadic Sources: Exact and Approximate Performance Analysis, *Proc. 13th Intrr. Teletraffic Conf.*, Copenhagen.
5. K. Chandra, C. Thompson, P.E. Wirth and K. S. Meier-Hellstern, 1995, Source Model for ISDN Packet Data Traffic, submitted *IEEE Trans. on Networks*
6. M.W. Garrett and W. Willinger, 1994, Analysis, Modeling and Generation of Self-Similar VBR video traffic, *Proc. ACM Sigcomm*, London
7. W. Martin and P. Flandrin, 1985, Detection of Changes of Signal Structure by using the Wigner-Ville Spectrum, *Signal Processing*, 8; 215-233
8. K. Chandra and A. R. Reibman, 1995, Modeling traffic and statistical gains in multimedia applications, *Second Workshop on Community Networking*, Princeton, NJ.
9. M. Ghanbari and C.J. Hughes, 1993, Packing Coded Video Signals into ATM Cells, *IEEE/ACM Trans. on Networks*, 1; 505-509.

A TRAFFIC DESCRIPTOR-BASED FLOW CONTROL SCHEME FOR EFFICIENT VIDEO TRANSMISSION OVER ATM

Souhwan Jung and James S. Meditch

Department of Electrical Engineering
University of Washington
Seattle, WA 98195

ABSTRACT

Video transmission requires high transmission bandwidth and a sophisticated flow control scheme in order to guarantee the quality of service. In this paper, we design a traffic descriptor model for variable-bit-rate sources in ATM networks, and develop a rate-based flow control scheme to reduce queueing delay and possible buffer overflow. VBR traffic flow can be modeled as a two-state Markov Chain via an overload state and a underload state. Each state is characterized by its load index and duration. Based on the utilization factor of the queueing system, the transmission rate is determined from the traffic descriptor parameters. Since the traffic model gives a macroscopic view for traffic variations over time, our scheme is feasible for implementation in real-time flow control systems.

INTRODUCTION

With the spread of multimedia communications, efficient transmission of VBR video over asynchronous transfer mode (ATM) networks has been emerged as one of the important issues that is not yet solved. A number of studies have focused on the modeling of VBR video sources[1-3] in order to characterize source traffic variations over time. These studies have focused on generating source traffic rather than applying analytical results to investigate the effects of traffic variations on network congestion and queueing delay. Since even compressed video generates a large volume of synchronous traffic that must be transmitted within a given delay, video services generally require large transmission bandwidth and strict average and maximum delay constraints to guarantee quality of service[4-5].

Where video sources generate variable bit rate traffic as a result of compression, it is not easy to find an appropriate service rate for the source. A simple scheme is to allocate a fixed service rate during the entire connection. The service rate can be a value between the average and the peak rate of the source. If the peak rate is allocated, network utilization will be quite low due to wasted bandwidth. If the average rate is allocated, the buffer may overflow during long overload intervals, and may result in a large queueing delay. Hence, estimating an appropriate service rate that is typically larger than the average rate of a source and allocating the service rate to the source can significantly reduce queueing delay, while maintaining high efficiency of network utilization given a finite buffer size.

Multimedia Communications and Video Coding
Edited by Y. Wang *et al.*, Plenum Press, New York, 1996

In this paper, we design a traffic descriptor which characterizes traffic variations over time, and investigate the properties of the description parameters. Using these parameters, we estimate the average number of cell arrivals for each connection. The average load for a source is equivalent to the current service rate plus the queue change rate for a time frame. The new service rate based on the rate of queue size change is allocated to the source synchronously to reduce queueing delay and buffer overflow.

TRAFFIC DESCRIPTOR MODEL

Traffic description for variable bit rate ATM sources has been studied to understand the characteristics of the sources and to model traffic variations over time[6-7]. The purposes of a traffic descriptor are as follows: (1) to quantify traffic variations over time and (2) to give a macroscopic perspective for the effect of bursty traffic on network congestion and queueing delay. Since traffic variations over time must be expressed in terms of queueing delay, our traffic descriptor uses an approach to characterize the degree of overload during overload periods and find an appropriate service rate for each source.

Macroscopic Perspective for Traffic Flow

A VBR video source generates dynamically varying traffic that alternates between active and inactive periods. The duration of the active and the inactive periods depend on the video coding scheme, motion changes, and scene changes. If a service rate is allocated to a video source, the fluctuating traffic stream can be modeled as a sequence of bursts and interbursts periods[8]. Figure 1 illustrates the definition of burst and interburst periods for a given service rate. In the figure, given the service rate μ, a burst period is defined as the time interval in which the cell rate of traffic is greater than the service rate, and an interburst period is defined as the time interval in which the cell rate is less or equal to the service rate. For the same traffic pattern, the periods of burst and interburst depend on the service rate allocated to the source. The time unit for the definition of burst and interburst periods is a time interval in which the cell rate of traffic is measured. For example, the unit can be a slice or a frame for VBR video. The burst and interburst periods are measured at the slice level in our study. It is assumed that the slice is a horizontal set of macroblocks in a picture.

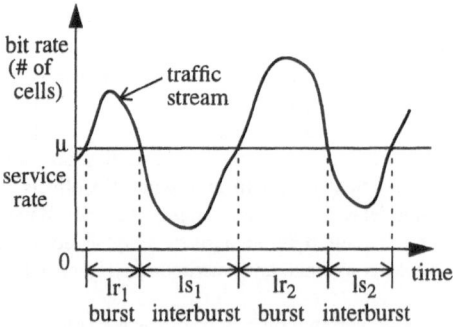

Fig. 1. Definition of burst and interburst

When the cell arrival rate is greater than the service rate, which means that more cells than can be served arrive, the network is overloaded during a burst period and the overloaded cells accumulate in a buffer. On the other hand, the network is underloaded during an interburst period and the cells in the buffer are drained. This gives us a macroscopic perspective for VBR traffic flow than the cell-level description with respect to the overload and the underload on the network server. A sophisticated traffic descriptor has been developed to effectively characterize traffic variations with respect to overload and underload.

Traffic Descriptor Parameters

In this section, a new traffic descriptor is defined using the concept of overload and underload. To evaluate how much overload (underload) a burst (an interburst) brings to the network, the total number of cells during a burst (an interburst) period is measured, and the ratio of the total number of cells during a burst (an interburst) period to the expected number of cells during the same time period can be parameterized to give the degree of the overload (underload). Since there will be a number of bursts and interbursts that alternate over time, the average value of the overload ratio and the underload ratio can be significant parameters to specify overall traffic variations. These parameters are called overload index and underload index. Hence, the definition for the overload index (r) and the underload index (s) are defined by

$$r = mean\left(\frac{B_k}{E_k}\right) \tag{1}$$

where B_k is the total number of cells that arrive during the k-th burst period and E_k is the expected load during the same period; and

$$s = mean\left(\frac{I_k}{E_k}\right) \tag{2}$$

where I_k is the total number of cells that arrive during the k-th interburst and E_k is the expected load during the same period.

Figure 2 illustrates the traffic model using overload and underload indexes. The overload index specifies the degree of overload during a time frame, while the underload index specifies the degree of underload during the same period. The overload index can also be used to measure the traffic burstiness[9]. The burst and interburst periods are important since the distribution of the parameters gives the dispersion of the intervals. The average burst period and interburst period are represented by l_r and l_s respectively.

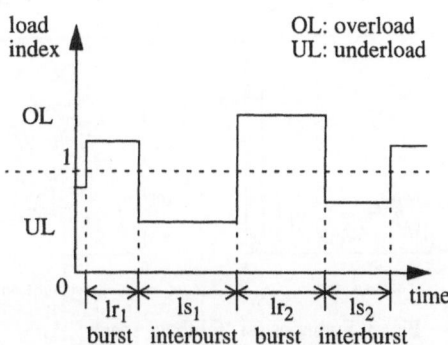

Fig. 2. Traffic flow model

So far, we have defined the traffic parameters called overload index, underload index, burst period, and interburst period. A traffic descriptor that is composed of the five parameters (μ, r, s, l_r, l_s) gives a complete description for traffic variations over time. For example, an average load factor, α, can be defined by

$$\alpha = \frac{rl_r + sl_s}{l_r + l_s} \tag{3}$$

since the term specifies a weighted average of two load indexes by their periods. The statistics of the traffic descriptor parameters were inspected in the following section.

Statistics of Traffic Descriptor Parameters

The properties of our traffic descriptor have been investigated using real MPEG video streams. A 50 second 'football' scene at the slice level has been generated from a software MPEG encoder developed at Stanford University. The resolution of a picture is 352x240 where a slice is assumed to be a horizontal strip of 16 row pixels which is a set of macroblocks. However, the slice can be any sequence of macroblocks in a frame. Hence, a picture consists of 15 slices in our simulation. It is explicit from the definition that the overload index decreases as the reference service rate increases, and the opposite holds for the underload index.

Figure 3-(a) shows the sample traffic stream of the 'football' scene and the distributions of traffic parameters. Since the traffic stream shows a characteristic that is fluctuating quite a bit, the allocation of the average rate to the source resulted in large queueing delay. The distributions of overload indexes, underload indexes, and average load factors for the whole stream are shown in Figure 3-(b), (c), and (d) respectively, given the service rate of the average source rate. It is seen that the overload index goes up to 5. This implies that five times more cells than expected have arrived during the overload period. The most probable value of the underload indices is about 0.5, which implies that about half of the expected cells have arrived during an interburst. Overall, the average burst period is longer than the average interburst period. The most probable value of the average load factor is about 1 as expected from the definition.

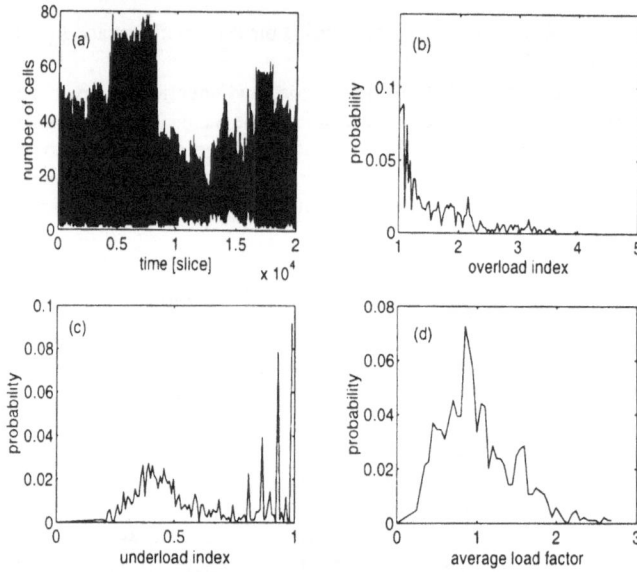

Fig. 3. Statistics for 'football' scene
(a) slice-level traffic   (b) overload index
(c) underload index   (d) average load factor

RATE-BASED FLOW CONTROL

The goals of flow control can be described as follows[10]: (1) control the service rate to prevent buffer overflow, (2) regulate the burstiness of traffic that may cause degradation of quality of service due to cell loss, and (3) retain high utilization of network resources. In this section, we are going to develop a flow control mechanism for a video source based on the queueing model of single server system. We demonstrate that a two-state, Markov chain model provides a viable approach for modeling time-varying traffic flow.

Two-State Markov Chain Model

In the previous sections, we have developed a traffic descriptor with five traffic parameters. The parameters are useful to model the traffic flow as a two-state, Markov chain. As shown in Figure 4, there are two states in the Markov chain, namely, the OL (overload state) and UL (underload state). The average cell arrivals in the overload state is $r\mu$, and that of the underload state is $s\mu$. This is similar to the Markov Modulated Poisson Process (MMPP) except that the cell arrival process is not Poisson.

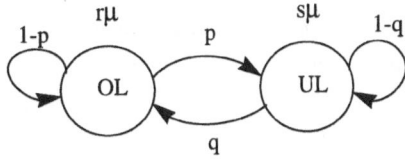

OL: overload   UL: underload

Fig. 4. Two-state Markov chain

The transition probabilities from one state to the other state, p and q, are determined from the statistics of the burst and interburst periods. Since the average duration of the OL state (UL state) is equivalent to the average burst (interburst) period, the transition probabilities are determined by

$$p = \frac{1}{l_r} \qquad q = \frac{1}{l_s} \tag{4}$$

The equilibrium probabilities P_1, P_2 of the two states are then given by

$$P_1 = \frac{q}{p+q} = \frac{l_r}{l_r+l_s} \tag{5}$$

$$P_2 = \frac{p}{p+q} = \frac{l_s}{l_r+l_s} \tag{6}$$

respectively. Further, the average cell arrival rate λ of the stationary, two-state, Markov chain is easily shown to be

$$\lambda = \frac{(r\mu)\cdot q + (s\mu)\cdot p}{p+q} = \frac{rq+sp}{p+q}\cdot\mu = \frac{rl_r+sl_s}{l_r+l_s}\cdot\mu \tag{7}$$

It then follows that utilization factor ρ is

$$\rho = \frac{\lambda}{\mu} = \frac{rl_r+sl_s}{l_r+l_s} \tag{8}$$

Since the last term in Equation (7) is a weighted average load factor, Equation (8) gives us an intuitive result wherein the utilization factor of a single server system is equivalent to the average load factor of the system. In the next section, we present a queueing analysis of the system.

Queueing Analysis

Using the traffic descriptor parameters designed in the previous section, the number of cells accumulated in a queue for a time frame δ is given by

$$\delta = \mu \cdot (r - 1) \cdot l_r - (\mu \cdot (1 - s) \cdot l_s) \tag{9}$$

While the first term $\mu \cdot (r - 1) \cdot l_r$ is the amount of pure queueing during overload periods, the second term $\mu \cdot (1 - s) \cdot l_s$ denotes the amount of pure draining during underload periods. Note that $s < 1$. Rearranging Equation (9) as follows,

$$\delta = \mu \cdot (rl_r + sl_s - l_r - l_s) = \mu \cdot (\rho - 1) \cdot (l_r + l_s) \tag{10}$$

we obtain

$$\rho = 1 + \frac{\delta}{\mu \cdot (l_r + l_s)} \tag{11}$$

Substituting ρ into Equation (7) results in

$$\lambda = \left(1 + \frac{\delta}{\mu \cdot (l_r + l_s)}\right) \cdot \mu = \mu + \frac{\delta}{l_r + l_s} \tag{12}$$

The last term in Equation (12) is the queue variation divided by the sum of the duration of a burst and an interburst. Denoting the queue variation rate by β, we have

$$\beta = \frac{\delta}{l_r + l_s} \tag{13}$$

from which the average cell arrival rate is expressed by

$$\lambda = \mu + \beta \tag{14}$$

From Equation (14), an important factor in estimating the average cell arrival rate is the average queue variation rate. In particular, the implication here is that if the queue is growing at a certain rate, then the service rate should be increased to the same rate in order to control the queue. On the other hand, the service rate should be decreased at the same rate once the queue is decreased. Given the current service rate for a VBR video source, the service rate should be updated via Equation (14).

Flow Control Mechanism

Our traffic descriptor approach makes it possible to model VBR video traffic flow as a two-state Markov chain. From the model, we can estimate the average cell arrival rate of the system. Since the difference between the cell arrival rate and the service rate may cause buffer overflows, the service rate should be greater than or equal to the cell arrival rate for system stability. Otherwise, the system buffer will overflow. Once the estimated cell arrival rate is greater than the service rate, the service rate should be updated to be greater than the arrival rate. The service rate can be updated periodically to control the queueing delay. From Equation (14), the service rate for the next time frame can be updated by

$$\mu(n + 1) \geq \lambda(n) = \mu(n) + \beta \tag{15}$$

where β is given by Equation (13), and n is the n-th time frame. If the service rate is updated to be greater than or equal to the arrival rate, the ratio of the arrival rate to the service rate f is given by

$$f = \frac{\lambda(n)}{\mu(n+1)} \leq 1 \qquad (16)$$

EXPERIMENTAL RESULTS

Experiments have been conducted to test the validity of our traffic model and the performance of the proposed flow control scheme. Some preliminary results are shown in the following figures. To inspect the validity of the two-state Markov chain model, the utilization factor of the queueing system is measured, and the results are compared with the values of utilization from Equation (8). In our simulation, the 'football' video scene was used, and the average source rate was given as the service rate. The utilization factor was measured for every 1000 slices. Figure 5 shows that the real values are quite near to the values from our model. The result is that our two-state Markov chain model with the states of overload and underload is effective for describing VBR video traffic.

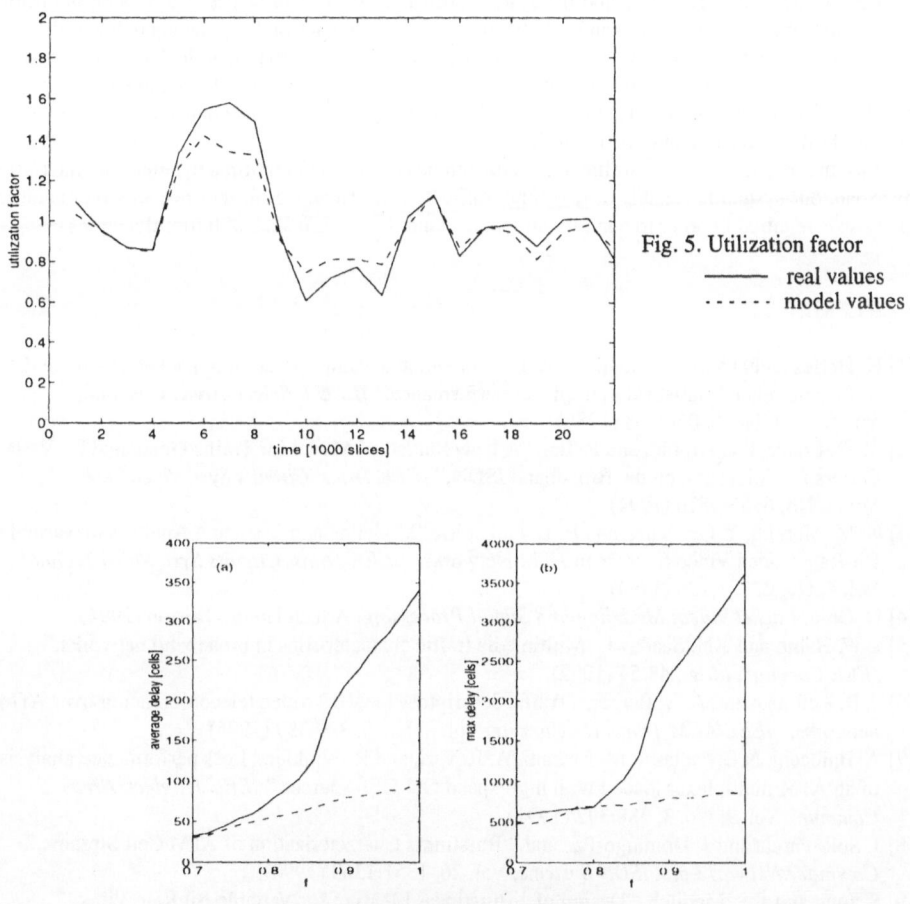

Fig. 5. Utilization factor
——— real values
· · · · · model values

Fig. 6. Performance of fixed service rate
vs. controlled service rate
——— no control, · · · · · flow control
(a) average delay
(b) maximum delay

Figure 6 shows the simulation results for flow control. The average and maximum queueing delay were investigated for different values of f in Equation (16). The solid line is fixed service rate and the dashed line is controlled service rate with the same average rate.

If the service rate for next time frame is updated as the arrival rate for the current time frame ($f = 1$), then the average number of cells in the queue with no control is about 350 cells, but the number decreases to about 100 with flow control. The maximum queue shows a similar result. In general, with high network utilization (high f values), our flow control scheme is quite effective in reducing queueing delay. From our experiments, the proposed flow control scheme looks promising for applications in dynamic bandwidth allocation for multiple virtual-circuit connections. Our system can be easily extended to multiplexed multimedia traffic streams.

CONCLUSIONS

A two-state Markov chain model with overload and underload states has been developed to characterize VBR traffic flow using traffic descriptor parameters. With the given service rate of a single-server system, VBR traffic flow can be modeled as a sequence of alternating overload and underload periods. The states of overload and underload respectively constitutes the two states for our Markov chain. This model provides us a macroscopic perspective for time-varying traffic flow.

The average cell arrival rate and the system utilization factor can be estimated using the traffic parameters, and the service rate can be updated periodically to control the queueing delay of the system buffer. The service rate is updated based on the rate of queue variation at the buffer. Our experiments have shown that both high utilization and small queueing delay can be obtained by applying our flow control scheme to real MPEG traffic. Queueing system with flow control has reduced delay by one-third relative to that of no control.

Since the parameters of our traffic descriptor can be computed in real-time to determine the traffic overload, our system is capable of real-time traffic control. However, further research on queueing analysis is required in order to quantify and understand the performance of buffer dynamics.

RERFERENCES

[1] H. Heffes and D.M. Lucantoni, "A Markov modulated characterization of packetized voice data traffic and related statistical multiplexer performance," *IEEE J. Select. Areas Commun.*, Vol. SAC-4, No. 6, 856:868 (1986).

[2] M. R. Frater, J. F. Arnold, and P. Tan, "A New Statistical Model for Traffic Generated by VBR Coders for Television on the Broadband ISDN," *IEEE Trans. Circuits Syst. Video Technol.*, Vol. 4, No. 6, 521:526 (1994).

[3] N. M. Marafih, Y. Q. Zhang, and R. L. Pickholtz, "Modeling and Queueing Analysis of Variable-Bit-Rate Coded Video Sources in ATM Networks," *IEEE Trans. Circuits Syst. Video Technol.*, Vol. 4, No. 2, 121:128 (1994).

[4] N. Ohta, *Packet Video: Modeling and Signal Processing*, Artech House, Boston (1994).

[5] I. W. Habib and T.N. Saadawi, "Multimedia traffic characteristics in broadband networks," *IEEE Commun. Mag.*, 48:54 (1992).

[6] A.R. Reibman and A.W. Berger, "Traffic descriptors for VBR video teleconferencing over ATM networks," *IEEE/ACM Trans. Networking*, Vol. 3, No. 3, 329:339 (1995).

[7] A. Baiocchi, N.B. Melazzi, M. Listanti, A. Roveri, and R. Winkler, "Loss performance analysis of an ATM multiplexer loaded with high-speed ON-OFF sources," *IEEE J. Select. Areas Commun.*, Vol. 9, No. 3, 388:392 (1991).

[8] J. Sole-Pareta and J. Domingo-Pascual, "Burstiness Characterization of ATM Cell Streams," *Computer Networks and ISDN Systems*, Vol. 26, 1351:1363 (1994).

[9] S. Jung and J. S. Meditch, "Design of a Burstiness Measure for Variable Bit Rate Video," *Proceedings of IEEE SICON/ICIE'95*, 483:487 (1995).

[10] T. Konstantopoulos and V. Anantharam, "Optimal flow control schemes that regulates the burstiness of traffic," *IEEE/ACM Trans. Networking*, Vol. 3, No. 4, 423:432 (1995).

END-TO-END QoS MANAGEMENT FOR ADAPTIVE VIDEO FLOWS

Andrew Campbell, Alexandros Eleftheriadis and Cristina Aurrecoechea

Center for Telecommunications Research
Columbia University
New York, NY 10027-6699
{campbell, elef, cris}@ctr.columbia.edu

INTRODUCTION

Distributed audio and video applications need to adapt to fluctuations in delivered *quality of service (QoS)*. By trading off temporal and spatial quality to available bandwidth, or manipulating the playout time of continuous media in response to variation in delay, audio and video flows can be made to adapt to fluctuating QoS with minimal perceptual distortion. In this paper we introduce *dynamic QoS management (DQM)* for the control and management of multi-layer coded flows operating in heterogeneous multimedia networking environments. Two key techniques are proposed: i) an end-to-end *dynamic rate shaping* scheme which adapts the rate of MPEG-coded [1] flows to the available network resources while minimising the distortion observed at the receiver; and ii) an *adaptive network service*, which offers "hard" guarantees to the base layer of multi-layer coded flows, and "fairness" guarantees to the enhancement layers based on a bandwidth allocation technique called *weighted fair sharing*. We also discuss a number of types of *media scaling object[1]* which are used to manage and control end-to-end QoS. These include *QoS filters* which manipulate multi-layer coded flows [2] as they progress through the communications system, *QoS adaptors* which scale flows at end-systems based on the flow's measured performance and user supplied QoS scaling policy, and *QoS groups* which provide baseline QoS for multicast flows.

MEDIA SCALING OBJECTS AND QoS-BASED API

In this section, we introduce a set of *media scaling objects* used to manipulate flows as they progress through the communications system. These comprise QoS adaptors, QoS filters and QoS groups:

Media Scaling Objects

QoS adaptors are used in conjunction with flow monitoring [8] function to ensure that the user and provider QoS specified in the service contract are actually maintained. In this role QoS adaptors are seen as quality of service arbiters between the user and network. QoS adaptors scale flows at the end-systems based on a user supplied QoS scaling policy (see QoS specification API section) and the measured performance of on-going flows.

---

[1] *Media scaling* is a general term, first proposed by [3], we use to refer to the dynamic manipulation of media flows as they pass through a communications channel.

QoS filters manipulate multi-layer coded flows [4] [5] [6] at the end-systems and as they progress through the network . We describe three distinct styles of QoS filters:

i)   *shaping filters*, which manipulate coded video and audio by exploiting the structural composition of flows to match network, end-system or application QoS capability; shaping filters are generally situated at the edge of the network at the source; they require non-trivial computational power; examples are the dynamic rate shaping (DRS) filter and the source bit rate (SBR) filter - see section on sender-oriented DQM;

ii) *selection filters*, which are used for sub-signal selection and media dropping (e.g. video frame dropping) are of low complexity and low computational intensity - selection filters are designed to operate in the network and are located at switches; they require only minimal computational power; examples are sub-signal filter, hierarch filter, hybrid filter - see section on network filtering;

iii) *temporal filters*, which manipulate the timing characteristics of media to meet delay bound QoS are also low in complexity and trivial computationally - temporal filters are generally placed at receivers or sinks of continuous media where jitter compensation or orchestration of multiple related media is required; examples are sync filter, orch filter - see section on receiver-oriented DQM).

Before potential senders and receivers can communicate they must first join a QoS group [7]. The concept of a QoS group is used to associate a baseline QoS capability to a particular flow. All sub-signals of a multi-layer stream can be mapped into a single flow and multicast to multiple receivers [1]. Then, each receiver can select to take either the complete signal advertised by the QoS group or a partial signal based on resource availability. Alternatively each sub-signal can be associated with a distinct QoS group. In this case, receivers "tune" into different QoS groups (using signal selection) to build up the overall signal. Both methods are supported in DQM. Receivers and senders interact with QoS groups to determine what the baseline service is, and tailor their capability to consume the signal by selecting filter styles and specifying the degree of adaptability sustainable; that is whether flows are discrete or continuous adaptive [3].

QoS-Based API

In [8] we designed a service contract based API which formalised the end-to-end QoS requirements of the user and the potential degree of service commitment of the provider. An important aspect of our API is that it shield the application from the complexity of QoS management and control [29]. In this section, we detail extensions to the *flow specification*, *QoS commitment* and *QoS scaling* clauses of the service contract required to accommodate adaptive multi-layer flows. The API presented here is not complete in that there are no primitives given for establishing and renegotiating connections or for manipulating QoS groups. Full details of these aspects are given in [8].

```
typedef enum {MPEG1, MPEG2, H261, JPEG}              mediaType;
typedef enum {besteffort, adaptive, deterministic} commit;
typedef enum {continuous, discrete}                  adaptMode;

typedef enum {
    DRS, SBR, sub_signal, hierarch, hybrid, sync, orch
} filterStyle;

typedef struct {                    typedef struct {
    adaptMode      adaptation;          gid              flow_id;
    filterStyle    filtering;           mediaType        media;
    events         adaptEvents;         commit           commitment;
    actions        newQoS;             subFlow          BL;
    signal         bandwidth;          subFlow          E1;
    signal         loss;               subFlow          E2;
    signal         delay;              int              delay;
    signal         jitter;             int              loss;
} QoSscalingPolicy;                    int              jitter;
                                       QoSscalingPolicy qospolicy;
                                   } flowSpec;
```

Multi-layered flows are characterised by three sub-signals in the *flowSpec* flow specification: a base layer (BL) and up to two enhancement layers (E1 and E2). Each layer is represented by a frame size and subjective or perceptive QoS as illustrated in [9]. Based on these characteristics, the MPEG-2 coder [9], [10] determines approximate bit rate for each sub-layer. In the case of MPEG-2's hybrid scalability [9], BL would represent the main profile bit rate requirement (e.g. 0.32 Mbps) for basic quality, E1 would represent the spatial scalability mode bit rate requirement (e.g. 0.83 Mbps) for enhancement, and E2 would represent the SNR scalability mode bit rate requirement (e.g. 1.85 Mbps) for further enhancement. The remaining flow specification performance parameters for *jitter*, *delay* and *loss* are assumed to be common across the all sub-signals (i.e. a single layer of a multi-layer video flow). The QoS commitment field has been extended to offer an *adaptive* network service that specifically caters for the needs of scalable audio and video flows in heterogeneous networking environments (see adaptive network service section).

The *QoSscalingPolicy* field of the *flowSpec* characterises the degree of adaptation that a flow can tolerate and still achieve meaningful QoS. The scaling policy consists of clauses that cover *adaptation modes*, *QoS filter styles*, and *event/ action* pairings for QoS management purposes. Two types of adaptation mode are supported: *continuous mode*, for applications that can exploit any availability of bandwidth above the base layer; and *discrete mode* for applications which can only accept discrete improvement in bandwidth based on a full enhancement (viz. E1, E2). The QoS scaling policy provides user-selectable QoS adaptation and QoS filtering. While receivers select filter styles to match their capability to consume media at the receiver (from the set of temporal filters), senders select filter styles to shape flows in response to the availability of network resources such as bandwidth and delay (from the set of shaping filters). Network oriented filters (i.e. selection filters) can be chosen by either senders or receivers. In addition, senders and receivers can both select periodic performance notifications including available bandwidth, measured delay, jitter and losses for on-going flows. The *signal* fields in the scaling policy allow the user to specify the interval over which a QoS parameter is to be monitored and the user informed. Multiple signals can be selected depending on application needs.

DQM ARCHITECTURAL COMPONENTS

Based on the receiver supplied QoS scaling policy, QoS adaptors take remedial action to scale flows, inform the user of a QoS indication and degradation, fine tune resources and initiate complete end-to-end QoS renegotiation based on a new *flowSpec* [8]. DQM consists of two sub-components: *QoS group management* maintains and advertises QoS groups created by senders for the benefit of potential receivers; *filter management* [11] instantiates and reconfigures filters in a flow at optimal points in the media path at flow establishment time, when new receivers join QoS groups or when a new *flowSpec* is given on a QoS renegotiation. In implementation, each of these architectural modules has well defined interfaces and methods defined in CORBA IDL. CORBA [12] runs on the end-systems and in the ATM switches, providing a seamless object oriented environment throughout the communication system base (see [7] for full details).

Illustrative Scenario

DQM can be viewed as operating in three distinct domains:

i) *sender-oriented DQM*, where senders select source filters and adaptation modes, and establish flow specifications. The sender-side transport protocol provides periodic bandwidth and delay assessments to the source filters (i.e. DRS or SBR filters) which regulate the source flow. Senders create QoS groups which announce the QoS of the flow to receivers via QoS group management;

ii) *receiver-oriented DQM*, where receivers join QoS groups and select the portion of the signal which matches their QoS capability. Receiver selected network based filters propagate through the network and perform source and signal selection. In addition, receiver-based QoS filters (i.e. sync-filter and orch-filter) are instantiated by default unless otherwise directed. These filters are used to smooth and synchronise multiple media. The receiver-side transport protocol provides bandwidth management and produces adaptation signals according to the QoS scaling policy;

iii) *network-oriented DQM*, which provides an adaptive network service (see later) to receivers and senders. Network level QoS filters (i.e. sub-signal, hierarch and hybrid-filters) are instantiated based on user selection, and propagated in the network under the control of filter management.

In Figure 1 a sender at end-system A creates a flow by instantiating a QoS group which announces the characteristics of the flow (viz. layer, frame size, subjective quality) and its adaptation mode. Receivers at end-systems B, C and D join the QoS group. In the example scenario shown the receivers each "tune" into different parts of the multi-layer signal: C takes BL, the main profile (which constitutes a bandwidth of 0.32 Mbps for VHS perceptual QoS), B takes BL and E1 (which constitutes an aggregate bandwidth of 1.15 Mbps for super VHS perceptual QoS), and D takes the complete signal BL+E1+E2 (which constitutes an aggregate bandwidth of 3 Mbps for laser disc perceptual QoS). In this example the complete signal is multiplexed onto a single flow, therefore, sub-signal selection filters are propagated by filter management. Receivers, senders, or any third party or filter management can select, instantiate and modify source, network and receiver-based QoS filters.

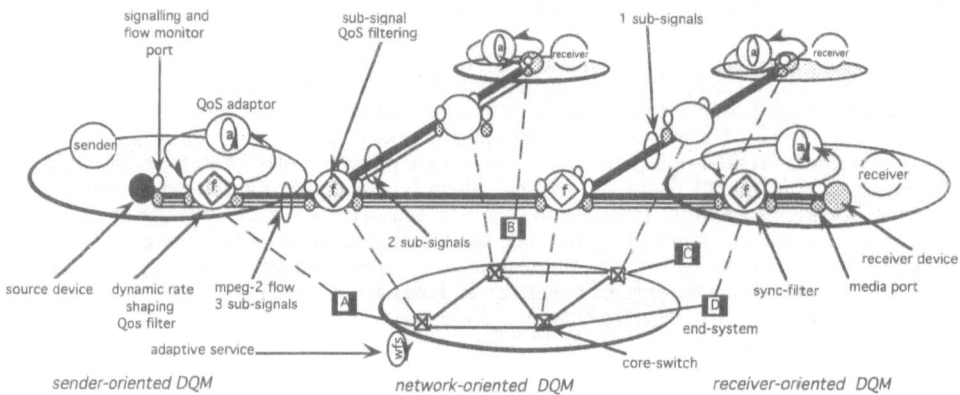

Figure 1: Dynamic QoS management of scalable flows

Sender-Oriented DQM. Figure 2 shows the functions of the sender-side transport protocol supporting dynamic QoS management, and the interface to a dynamic rate shaping filter. Currently, senders can select from two types of shaping filter at the source: dynamic rate shaping (DRS) and source bite rate (SBR) QoS filters. Both of these QoS filters manipulate the signal to meet the available bandwidth by keeping the signal meaningful at the receiver. The sender-side transport mechanisms includes a QoS adaptor, flow monitor and media scheduler. Bandwidth updates are synchronously received by the flow monitor mechanism from the network as part of the adaptive service (described a later section). The QoS adaptor is responsible for synchronously informing the source filter of the current bandwidth availability (B_{flow}) and measured delay (D_{flow}), and calculating new schedules and deadlines for transport service data units [13]. Media progresses from the source filter to the TSAP, and is scheduled by the media scheduler to the network at the NSAP based on the calculated deadlines.

The QoS adaptor is also responsible for informing the sending application of the on-going QoS based on options selected in the QoS scaling policy. Informing the application of the current state of the resources associated with a specific flow is key in implementing adaptive applications in end-systems. In this case the application manages the flow by receiving updates and interacting with the QoS adaptor to adjust the flow, e.g. change adaptation mode from continuous to discrete, request more bandwidth for BL, E1 and E2, or change the characteristics of the source filter, etc.

The DRS Filter. We define rate shaping as an operation which, given an input video bitstream and a set of rate constraints, produces a video bitstream that complies with these constraints. For our purposes, both bitstreams are assumed to meet the same syntax specification, and we also assume that a motion compensated block-based transform coding scheme is used. This includes both MPEG-1 and MPEG-2, as well as H.261 and so-called "motion" JPEG.

Although a number of techniques have been developed for the rate shaping of *live* sources [14] these cannot be used for the transmission of pre-compressed material (e.g. in VoD systems). The dynamic rate shaping filter is interposed between the encoder and the network and ensures that the encoder's output can be perfectly matched to the network's quality of service characteristics. The filter does not require interaction with the encoder and hence is fully applicable to both live and stored video applications.

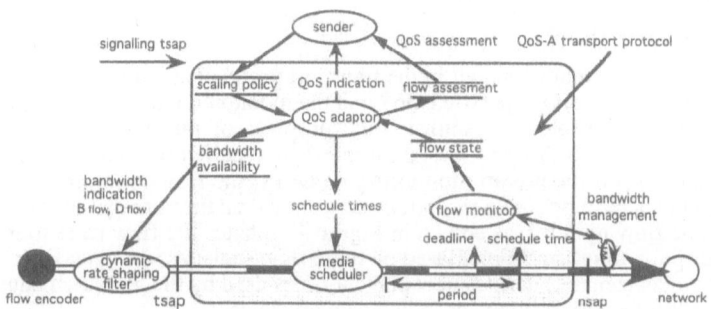

Figure 2: Sender-side transport QoS mechanisms

Because the encoder and the network are decoupled, universal interoperability can be achieved both between codecs and networks, and also among codecs with different specifications. An attractive aspect is the existence of low-complexity algorithms which allow software-based implementation in high-end computers. In order for rate shaping to be viable it has to be implementable with reasonable ease while yielding acceptable visual quality. With respect to complexity, the straightforward approach of decoding the video bitstream and recoding it at the target rate would be obviously unacceptable; the delay incurred would also be an important deterrent. Hence only algorithms of complexity less than that of a cascaded decoder and encoder are of practical interest. These algorithms operate directly in the compressed domain of the video signal, manipulating the bitstream so that rate reduction can be effected. In terms of quality, it should be noted that recoding does not necessarily yield optimal conversion; in fact, since an optimal encoder (in an operational rate-distortion sense) is impractical due to its complexity, recoding can only serve as an indicator of an acceptable quality range. In fact, regular recoding can be quite lacking in terms of quality, with dynamic rate shaping providing significantly superior results.

The rate shaping operation is depicted in Figure 3. Of particular interest is the source of the rate constraints $B_{flow}(t)$. In the simplest of cases, $B_{flow}(t)$ may be just a constant and known a priori (e.g. the bandwidth of a circuit-switched connection). It is also possible that $B_{flow}(t)$ has a well known statistical characterisation (e.g. a policing function). In our approach $B_{flow}(t)$ is generated by the adaptive network service.

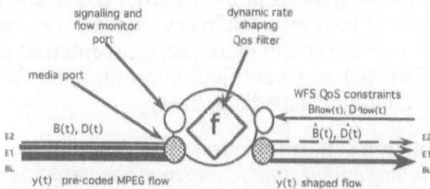

Figure 3: Dynamic rate shaping scheme

The objective of a rate shaping algorithm is to minimise the conversion distortion, i.e.:

$$\min_{B(t) < B_T(t)} \| y(t) - \hat{y}(t) \|$$

The attainable rate variation $(\hat{B}/B)$ is in practice limited, and depends primarily on the number of B pictures of the bitstream; no assumption is made on the rate properties of the input bitstream, which can indeed be arbitrary. There are two fundamental ways to reduce the rate: i) by modifying the quantised transform coefficients by employing coarser quantisation, and ii) by eliminating transform coefficients. In general, both schemes could be used to perform rate shaping; requantisation, however, leads to recoding-like algorithms which are not amenable to fast implementation and do not perform as well as selective-transmission ones. A selective transmission approach gives rise to a family of different algorithms, that perform optimally under different constraints; for full details see [15].

Receiver-Oriented DQM

QoS adaptors, which are resident in the transport protocol at both senders and receivers, arbitrate between the receiver specified QoS and the monitored QoS of the on-going flow. In essence the transport protocol "controls" the progress of the media while the receiver "monitors and adapts" to the flow based on the flow specification and the scaling policy. When the transport protocol is in monitoring mode [8] the flow monitor uses an absolute timing method to determine frame receptions times based on timestamps/sample-stamps [16], [17], [18]. The flow monitor, as shown in Figure 4, updates the flow state to include these measured reception times statistics. Based on these flow statistics, the sync-filter (see section on delay jitter management) derives new playout times used by the media scheduler to adjust the playout point of the flows to the decoding delivery device.

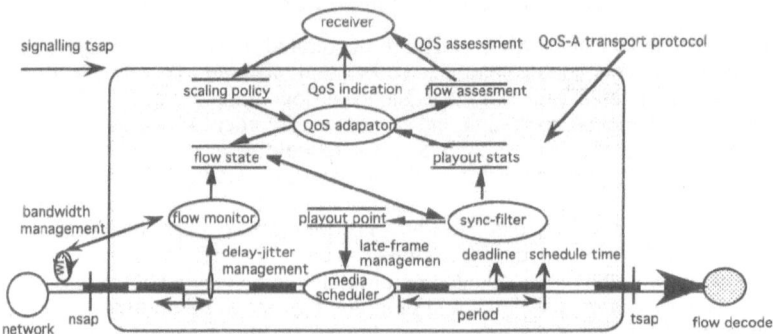

Figure 4: Receiver-side transport QoS mechanisms

QoS mechanisms that intrinsically support such adaptive approaches were first recognised in the late 70's by Cohen [19] as part of research in carrying voice over packet-switched networks. More recently, adaptive QoS mechanisms have been introduced at part of the Internet suite of application-level multimedia tools (e.g., *vat* [17], *ivs* [20], and *vic* [21]). Vat which is used for voice conferencing, recreates the timing characteristics of voice flows by having the sender timestamp on-going voice samples. The receiver then uses these timestamps as a basis to reconstruct initial flow, removing any network induced jitter prior to playout. These multimedia tools are widely used in the Internet today and have proved moderately successful - given the nature of best effort delivery systems, i.e., no resource reservation is made. In the near future, however, an integrated services Internet [22] will offer support for flow reservation (e.g., RSVP [6] and new QoS commitments (e.g., predictive QoS [22]) which are more suitable for continuous media delivery.

Receiver-oriented adaptation can be broken down into a number of receiver-side transport functions, i.e. *bandwidth management*, *late frame management* and *delay jitter management*, which are described in the following sub-sections (please refer to figure 4). We argue in this paper that these adaptive QoS mechanisms are inherently part of the transport protocol and not, as in the case of vat, ivs and vic, part of the application domain itself.

Bandwidth Management. Bandwidth management receives bandwidth indications in the control message portion of the TSDU (or in separate control messages) and adapts the receiver appropriately. The adaptive service, built on the notion of weighted fair share resource allocation, (see section on weighted fair share resource partitioning) periodically informs the receiver that more bandwidth is available or announces that the flow is being throttled back. Bandwidth management only covers the enhancement signals of multi-resolution flows. The base layer is not included since resources are guaranteed to the base layer. The announcement of available bandwidth on a flow allows the receiver to take either a full or partial enhancement layer. The choice depends on whether the flow is in continuous or discrete adaptation mode.

Late Frame Management. Late frame management monitors late arrivals in relation to the loss metric and the current playout times and takes appropriate action to trade off timeliness and loss. Packets that arrive after their expected playout points are discarded by the media scheduler and the late-packet metrics in the playout statistics are updated. The media scheduler is based on a split-level scheduler architecture [13] which provides hard deadline guarantees to base layer flows via admission control, and best effort deadlines to enhancements layers. Some remedial action may be taken by the QoS adaptor should the loss metric exceed the loss parameter in the flow specification. If the QoS adaptor determines that too many packet losses have occurred over an era, it pushes out the playout time to counteract the late state of packets from the network. Similarly, if loss remains well within the prescribed ranges then the QoS adaptor will automatically and incrementally "pull in" the playout time until loss is detected.

Delay Jitter Management. Our transport protocol utilises *sync-filters* for delay-jitter management by calculating the playout times of flows based on the user supplied jitter parameter in the flow specification[1]. Sync filters calculate the mean and variation in the end-to-end delay based on reception times measured by the flow monitor. Sync filters take the absolute, mean and variation in delay into account when calculating the playout estimate. A smoothing factor based on a linear recursive filtering mechanism characterised by a smoothing constant is used to dampen the movement of the playout adjustment. Intuitively, the playout time needs to be set "far enough" beyond the delay estimate so that only a small fraction of the arriving packets are lost due to late packets. The QoS adaptor trades off late packets versus timeliness based on the delay and loss parameters in the flow specification. The objective of delay jitter management is to pull in the playout offset, while the objective of late-packet management is not to exceed the loss characterised in the service contract. The QoS adaptation manager moderates between timeliness and loss. Based on these metrics the adaptation policy can adjust the damping factor, and acceptable ranges over which the playout point can operate.

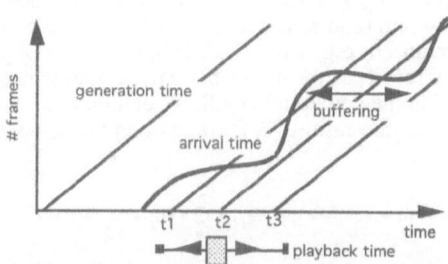

Figure 5: Sync-filter: timeliness and packet loss regulation

Figure 5 [23] shows packets arriving at the receiving end-system. Each packet includes a timestamp used in calculating the flow statistics for delay-jitter management. Selection of the playout point is important: an aggressive playout time which favours timeliness (such as t_1) will results in a large number of late-packets. In contrast, a conservative playout point (such as t_3) will be less responsive and timely but will result in no identifiable packet loss. In the DQM scheme late packets are the same as lost packets, and therefore the loss parameter in the flow specification moderates. An optimum playout schedule is represented by t_2 in the

---

[1] Temporal filters can also operate on multiple related audio and video flows to provide low-level orchestration management (in conjunction with the orch-filter). These filter types, however, are not discussed further in this paper.

diagram; here, continuous media delivery benefits from timely delivery with the exception of some packet loss - which may be deemed acceptable to the receiver in media perception terms.

ADAPTIVE NETWORK SERVICE

The adaptive network service provides "hard" guarantees to the base layer (BL) of a multi-layer flow and "weighted fair share" (WFS) guarantees to each of the enhancement layers (E1 and E2). To achieve this, the base layer undergoes a full end-to-end admission control test [13]. On the other hand, enhancement layers are admitted without any such test but must compete for residual bandwidth among all other adaptive flows. Enhancement layers are rate controlled based on explicit feed back about the current state of the on-going flow and the availability of residual bandwidth.

Weighted Fair Share Resource Partitioning

Both end-system and network communication resources are partitioned between the deterministic and adaptive service commitment classes. This is achieved by creating and maintaining "firewall" capacity regions for each class. Resources reserved for each class, but not currently in use can be "borrowed" by the best effort service class on condition of pre-emption [13]. The adaptive service capacity region (called the available capacity region and denoted by B_{avail}) is further sub-divided into two regions: i) guaranteed capacity region (B_{guar}), which is used to guarantee all base rate layer flow requirements; and ii) residual capacity region (B_{resid}), which is used to accommodate all enhancement rates where competing flows share the residual bandwidth.

Three goals motivate our adaptive service design. The first goal is to admit as many base layer (BL) sub-signals as possible. As more base layers are admitted the guaranteed capacity region B_{guar} grows to meet the hard guarantees for all base signals. In contrast, the residual capacity region B_{resid} shrinks as enhancement layers compete for diminishing residual bandwidth resources. The following invariants must be maintained at each end system and switch:

$$B_{avail} = B_{guar} + B_{resid} \text{, and } \sum_{i=1}^{N} BL_{(i)} \leq B_{avail}$$

Our second goal is to share [24], [25] the residual capacity B_{resid} among competing enhancement sub-signals based on a flow specific *weighting factor, W*, which allocates residual bandwidth in proportion to the range of bandwidth requested that in turn is related to the range of perceptual QoS acceptable to the user. In DQM, residual resources are allocated based on the range of bandwidth requirements specified by the users (i.e. BL.. BL+E1+E2 is the range of bandwidth required, e.g. from 0.32 Mbps to 3 Mbps for the hybrid scalable MPEG-2 flow in [9]). As a result, as resources become available each flow experiences the same "percentage increase" in the perceptible QoS, we call this *weighted fair share* (WFS). W is calculated for each flow as the ratio of a flow's perceptual QoS range to the sum of all perceptual QoS ranges.

$$W_{(i)} = (BL_i + E1_i + E2_i) / \sum_{j=1}^{N} (BL_j + E1_j + E2_j)$$

All residual resources B_{resid} are allocated in proportion to the W metric. Using this factor we calculate the proportion of residual bandwidth allocated to a flow to be $B_{wfs}(i)=W(i).B_{resid}$ and the proportion of the available bandwidth allocated to be $B_{flow}(i)= B_{wfs}(i)+ BL(i)$.

Our third and final goal is to adapt flows both discretely and continuously, based on the adaptation mode. In the discrete mode no residual bandwidth is allocated by the WFS mechanism unless a complete enhancement can be accommodated (i.e., $B_{wfs}(i) = E1(i)$ | E1(i)+E2(i), e.g. 0.83 Mbps or 2.68 Mbps from [9]). In continuous mode any increment of residual bandwidth $B_{wfs}(i)$ can be utilised (i.e. $0 < B_{wfs}(i) \leq E1(i) + E2(i)$, e.g. from 0 to 2.68 Mbps from [9]).

Rate Control Scheme

We build on the rate-based scheme described in [8] where the transport protocol at the receiver measures the bandwidth, delay, jitter and loss over an interval which we call an "era". An era is simply defined as the reciprocal of the frame rate in the flow specification (e.g. for a frame rate of 24 frames per second as shown in [9] the interval era is approximately 42 ms). The receiver-side transport protocol periodically informs the sender-side about the currently available bandwidth, and the measured delay, loss and jitter. This information is used by the source or *virtual source*[1] to calculate the rate to use over the next interval. The reported rate is temporally correlated with the on-going flow. An important result in [14] shows that variable rate encoders can track QoS variations as long as feedback is available within four frame times or less. This feedback is used by the dynamic rate shaping filter and network based filters to control the data generation rate of the video or the selection of the signal respectively. In the case of dynamic rate shaping, the rate is adjusted while keeping the perceptual quality of the video flow meaningful to the user.

Based on the concept of eras, control messages are forwarded from the receiver-side transport protocol to either virtual source or the source-side transport protocol using reverse path forwarding. A core-switch [26] where flows are filtered is always considered to be a virtual source for one or more receivers; for full details see [7]. The WFS mechanism updates the advertised rate as the control messages traverse the switches on the reserve path to the source or virtual source. Therefore any switch can adjust the flow's advertised rate before the source or virtual source receives the rate based control message. The source-side transport protocol hands the measured delay and aggregate bandwidth off (B_{flow}) to the dynamic rate shaping filter.

DQM maintains flow state at each end-system and switch that a flow traverses. Flow state is updated by the WFS algorithm and the rate-based flow control mechanism and comprises:

i) *capacity* (viz. B_{avail}, B_{guar}, B_{resid});

ii) *policy* (viz. filterStyle, adaptMode);

iii) *flowSpec* (viz. BL, E1, E2) ;

iv) *WFS share* (viz., B_{flow}, B_{wfs}, W).

The end-systems hold an expanded share tuple for measured delay, loss and jitter metrics. An admission control test is conducted at each end-system and switch on route to the core for the base layer signal. This test simply determines whether there is sufficient bandwidth available to guarantee the base layer BL given the current network load:

$$\sum_{j=1}^{N} BL_{(j)} \leq B_{avail}$$

If the admission control test is successful, WFS determines the additional percentage of the residual bandwidth made available (B_{wfs}) to meet any enhancement requirements in the flowSpec:

$$B_{wfs(i)} = W_{fact(i)}.(B_{avail} - \sum_{j=1}^{N} BL_{(j)})$$

The WFS rate computation mechanism causes new B_{wfs} rates to be computed for all adaptive enhancement signals that traverse the output link of a switch; switches are typically non-blocking which means the critical resources are the output links, however, our scheme can be generalised to other switch architectures [13].

Network Filtering

Currently our scheme supports two types of selection filters in the network. These are low complexity and computationally simple filters for selecting sub-signals. Selection filters do not transform the structure of the internal stream, i.e. they have no knowledge of the format of the encoded flow above differentiating between BL, E1 and E2 sub-signals. The two basic types of section filter used are:

---

[1]   We use the term *virtual source* to represent a network switch that modifies the source flow via filtering.

i) *sub-signal filters*: these manipulate base and enhancement layers of multi-layer video multiplexed on a single flow. The definition of sub-signals is kept general here; a flow may be comprised of an anchor and scalable extensions or the I and P pictures of MPEG-2's simple profile, or the individual hybrid scalable profile. Sub-signal filters are installed in switches when a receiver joins an on-going flow;

ii) *hierarchical filters*: these manipulate base and enhancement layers which are transmitted and received on independent flows in a non multiplexed fashion. In functional terms sub-signal and hierarchical filters can be considered to be equivalent in some cases. In sub-signal filtering one flow characterises the complete signal and in hierarchical-filtering a set of flows characterise the complete signal.

In addition, *hybrid filters* combine the characteristics of sub-signal and hierarchical filtering techniques to meet the needs of complex sub-signal selection. For example hierarchical filters allow the BL, E1 and E2 to be carried over distinct flows, and the user can accordingly tune into each sub-signal as required. As an example, the base and enhancement layers of the hybrid scalable MPEG-2 flow are each in turn made up of I, P and B pictures at each layer i.e. BL (I,P,B), E1 (I,P,B) and E2 (I,P,B). Using hybrid filters, the receiver can join the BL QoS group for the main profile and the E1 QoS group for the spatial enhancement and then select sub-signals within each profile as required (e.g. the I and P pictures of the BL).

CONCLUSION

At Lancaster University we are investigating heterogeneity issues present in applications, communications systems and networks. Resolving heterogeneous QoS demands in networked multimedia systems is a particularly acute problem that we are addressing within the framework of our Quality of Service Architecture (QoS-A) [8]. As part of that work we have described a scheme for the dynamic management of multi-layer flows in heterogeneous multimedia and multicast networking environments. Dynamic QoS management manipulates and adapts hierarchically coded flows at the end-systems and in the network using a set of scaling objects. The approach is based on three basic concepts: the scalable profiles of the MPEG-2 standard that can provide discrete adaptation, dynamic rate shaping algorithms for compressed digital video that provide continuous adaptation, and the weighted fair share service for adaptive flows. At the present time DQM is being implemented at Lancaster University. The experimental infrastructure at Lancaster is based on 80486 machines running a multimedia enhanced Chorus micro-kernel [13] and connected by programmable Olivetti Research Limited 4x4 ATM switches. This work is being carried out collaboratively with Columbia University in the US. At Columbia we are currently using CORBA [27] to propagate selection filters in the network using ASX200 switches. In addition to the implementation we are conducting an extensive simulation study into the feasibility of the adaptive network service for large scale use, and investigating the feasibility of extending the adaptive service concept into our enhance Chorus micro-kernel itself [13]. In other QoS related research at Lancaster we are developing a networked scalable multimedia storage system [28] based on the dynamic allocation of compressed file components within a hierarchy of heterogeneous servers, and specialised filter servers [11] which implement transcoder QoS filters.

REFERENCES

1. H.262, (1994), "Information Technology - Generic Coding of Moving Pictures and Associated Audio", Committee Draft, ISO/IEC 13818-2, International Standards Organisation, UK, March 1994.

2. Shacham, N, (1992) "Multipoint Communication by Hierarchically Encoded Data", Proc. IEEE INFOCOM'92, Florence, Italy, Vol.3, pp. 2107-2114.

3. Delgrossi, L., Halstrinck, C., Henhmann, D.B, Herrtwich R.G, Krone, J., Sandvoss, C., and C. Vogt, (1993), "Media Scaling for Audio-visual Communication with the Heidelberg Transport System", Proc ACM Multimedia'93 Anaheim, USA.

4. Pasquale, G., Polyzos, E., Anderson, E., and V. Kompella, (1993), "Fitter Propagation in Dissemination Trees: Trading Off Bandwidth and Processing in Continuos Media Networks", Proc. Forth International Workshop on Network and Operating System Support for Digital Audio and Video, Lancaster, UK.

5. Hoffman, D., Speer, M. and G. Fernando, (1993), "Network Support for Dynamically Scaled Multimedia Data Streams", Fourth International Workshop on Network and Operating System Support for Digital Audio and Video, Lancaster, UK.

6. Zhang, L., et. al., (1995), "RSVP Functional Specification", Working Draft, draft-ietf-rsvp-spec-07.ps.

7. Aurrecoechea, C., Campbell, A., Hauw, L. and Hisaya Hadama, (1995), "A Model for Multicast for the Binding Architecture". Technical Report, Center for Telecommunications Research, Columbia University, USA.

8. Campbell, A., Coulson, G. and Hutchison, D., (1994), "A Quality of Service Architecture", ACM Computer Communications Review.

9. Paek, S., Bocheck, P., and Chang S.-F., (1995), "Scalable MPEG-2 Video Servers with Heterogeneous QoS on Parallel Disk Arrays", Fifth International Workshop on Network and Operating System Support for Digital Audio and Video, Durham, New Hampshire, USA.

10. Eleftheriadis, A., and D. Anastassiou, (1995), "Meeting Arbitrary QoS Constraints Using Dynamic Rate Shaping of Code Digital Video", Fifth International Workshop on Network and Operating System Support for Digital Audio and Video, Durham, New Hampshire, USA.

11. Yeadon, N., Garcia, F., Campbell, A and D. Hutchison, (1994), "QoS Adaptation and Flow Filtering in ATM Networks", 2nd International Workshop on Advanced Teleservices and High Speed Communication Architectures, Heidelberg, Germany.

12. OMG, (1993), "The Common Object Request Broker: Architecture & Specification, Rev 1.3., December 1993.

13. Coulson, G., Campbell, A and P. Robin, (1995), "Design of a QoS Controlled ATM Based Communication System in Chorus", IEEE Journal of Selected Areas in Communications (JSAC), Special Issue on ATM LANs: Implementation and Experiences with Emerging Technology.

14. Kanakia, H., Mishra, P., and A. Reibman, (1993), "An Adaptive Congestion Control Scheme for Real Time Packet Video Transport", Proc. ACM SIGCOMM '93, San Francisco, USA, October 1993.

15. Eleftheriadis, A., (1995b), "Dynamic Rate Shaping of Compressed Digital Video", Ph.D. Thesis, Columbia University, USA.

16. Jeffay K., Stone, D.L., Talley, T. and F.D. Smith, (1992), "Adaptive, Best Effort Delivery of Digital Audio and Video Across Packet-Switched Networks", Proc. Third International Workshop on Network and Operating System Support for Digital Audio and Video, San Diego, USA.

17. Jacobson, V., (1993) "VAT: Visual Audio Tool", vat manual pages.

18. Shenker, S., Clark, D., and L. Zhang, (1993), "A Scheduling Service Model and a Scheduling Architecture for an Integrated Service Packet Network", Working Draft available via anonymous ftp from parcftp.xerox.com: /transient/service-model.ps.Z.

19. Cohen, D., (1977), "Issues in Transit Packetized Voice Communication", Proc. Fifth Data Communications Symposium, Snowbrid, USA.

20. Turletti, T, (1993), "A H.261 Software Codec for Video-conferencing over the Internet", INRIA Technical Report 1834, France.

21. McCanne, S., Jacobson, V., (1994), "VIC: Video Conference" U.C. Berkeley and Lawrence Berkeley Laboratory. Software available via ftp://ftp.ee.lbl.gov/conferencing/vic

22. Braden R., Clark, D., and S. Shenker, (1994), "Integrated Services in the Internet Architecture: an Overview", Request for Comments, RFC-1633.

23. Zhang, L., (1994), Symposium on Multimedia Networking, Columbia University, USA.

24. Steenstrup, M., (1992), "Fair Share for Resource Allocation", pre-print.

25. Tokuda, H., Tobe, Y., Chou, S.T.C. and Moura, J.M.F., (1992), "Continuous Media Communication with Dynamic QoS Control Using ARTS with an FDDI Network", Proc. ACM SIGCOMM '92, Baltimore, USA.

26. Ballardie, T., Francis, P. and Jon Crowcroft, (1993), "Core Based Tree (CBT) An Architecture for Scalable Inter-Domain Multicast Routing", Proc. ACM SIGCOMM '93, San Francisco, USA.

27. Lazar, A. A., Bhonsle S., Lim, K.S., (1994) "A Binding Architecture for Multimedia Networks", Proceedings of COST-237 Conference on Multimedia Transport and Teleservices, Vienna, Austria.

28. Pegler, D., Hutchison, D., Lougher, P. and D Shepherd, (1995), "A Scalable Multimedia Storage Hierarchy", Technical Report, MPG-01-95, Lancaster University, England.

29. Bansal, V., Siracusa, R.J, Hearn, J. P., Ramamurthy and D. Raychaudhuri, "Adaptive QoS-based API for Networking, Fifth International Workshop on Network and Operating System Support for Digital Audio and Video, Durham, New Hampshire, April, 1995

A DYNAMIC ROUTING SCHEME FOR MULTIMEDIA TRAFFIC

David Li and Nancy Crowther

BBN Systems and Technologies
10 Moulton Street
Cambridge, MA 02138
Email: dli@bbn.com, nsc@bbn.com

INTRODUCTION

We have developed a large TCP/IP internet supporting both multimedia applications in the form of telephony, intercom, and fax, and traditional bulk data transfer and messaging applications. Since the applications supported include conference calls and intercom, multicast must be provided by the underlying protocols. The multimedia and traditional data applications operate together in an integrated environment.

Multimedia traffic is supported by using the ST-II protocol. ST-II (see RFC 1190, 1990) was developed by BBN to support efficient delivery of streams of packets to either single or multiple destinations. It is designed for applications requiring guaranteed data rates and controlled delay characteristics. It is thus ideally suited to multimedia applications such as telephony, intercom, and fax.

Starting with an existing internet and ST-II router, the T/20, developed by BBN, (see Elliott and Lynn, 1994) we modified the routing method in this software to improve network resource utilization in support of both voice and IP data traffic. The result is a dynamic routing scheme which adjusts to changing traffic loads and uses the network as efficiently as possible.

The paper is organized as follows. First we give an overview of the ST-II protocol and how Routing supports it. Second is a brief explanation of how standard SPF routing works. Third, we explain the modification we made to Routing which allows it to spread ST-II calls over multiple paths through the network, thus improving network utilization. Finally we give some experimental results from a laboratory test environment.

ST-II PROTOCOL

ST-II is an internet protocol at the same layer as IP, but unlike IP it organizes data flow into virtual circuits, called "streams". That is, once a stream is set up from origin to one or more targets, all packets in the stream flow over the same path. Reservation of resources such as network bandwidth and buffer capacity is used to provide constant bit rate and controlled delay characteristics to streams. These are what is needed by multimedia traffic.

ST-II requires every router in the path along a stream to maintain state information, including resources needed, describing all the streams of packets flowing through them. When a stream's virtual circuit is set up, the resources necessary to support it are reserved on each network over which the stream passes. This reservation of resources allows data packets to be forwarded with low delay, low overhead, and a low probability of loss due to congestion, since the reservation is not granted unless the necessary resources are available.

ST-II allows streams with multiple targets, thus supporting conference call applications. The path of the complete stream is a tree as in Figure 1. Streams are created and maintained as a series of "next hops". In the figure, the origin has a single next hop (A), A in turn has a single next hop (B), B has two next hops (C and D), C has a single next hop (Target #1), and D has two next hops (Targets #2 and #3). The stream in the figure carries a four-way conference call.

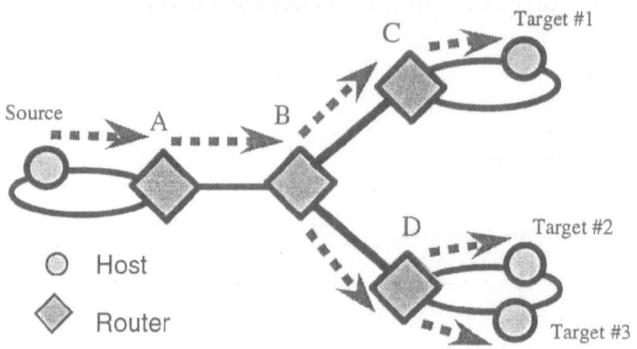

Figure 1. ST-II Stream Implementing Multicast Conference Call

ST-II operates with hop-by-hop routing. Although source routing (selection of the entire route through the network by the originating application) is supported by the protocol, this method was judged not desirable in our network since the burden is then on the application to make the right choices.

Stream setup with hop-by-hop routing operates as follows. Streams are created by using the ST Control Message Protocol (SCMP). The origin of a new stream sends out an SCMP CONNECT packet containing the bandwidth and other characteristics required by the stream, and the IP address of the one or multiple targets of the stream. The initial SCMP CONNECT message is routed using the conventional IP routing mechanism from source to destinations. For each target, the Routing function of the gateway provides the IP address of the next hop to which the packet should be sent, and the device interface out of which the packet must be sent to get to that next hop.

The bandwidth necessary to carry the stream being set up is reserved on the network leading to the next hop. If that network link does not have enough bandwidth available, the call setup fails for that target -- alternate routes are not tried. This process of obtaining a route and reserving bandwidth on the network is repeated for each target of the stream being set up. If there is no bandwidth available to reach any of the targets on the routes provided by the Routing function, the stream setup fails. Thus the choice of route is critical to the success of the stream setup. It is very important for the Routing function to have accurate information about the best route to all targets.

STATIC SPF ROUTING

The correct next hop for each ST-II target is provided to the call setup function by the Routing function. The routing algorithm and protocol implemented in the T/20 router is Shortest Path First (SPF). The "best" route to a target is the route with the smallest cost. In standard static SPF, the cost of each hop is the same, a configured constant which does not change during operation. Thus the lowest cost route is simply the route with the smallest number of hops between source and target. The route is determined by consulting the Forwarding Table, which is built periodically from a tree constructed by the SPF algorithm from the link state database. This database is maintained by flooding routing updates containing the cost, or metric, of each link to every node in the network.

Since the Forwarding Table is built from the SPF tree, each router must know about every topology change in the entire internet. In static routing, there are two ways in which the network topology can change:

1.   A new link may become operational, or an existing link may go down, due to either a hardware failure, or an orderly shutdown of a link.

2.   A new router may become operational, or an existing router may go down. This situation can be treated exactly as case (1) above by viewing the router as the collection of its links which all become operational or non-operational at once.

Each such link state change is sent by the SPF protocol throughout the entire connected internet so that all routers learn about the new topology as quickly as possible. This is a very important operation, since until every router has updated its Forwarding Table to reflect the new topology, packets will be sent to the wrong place or will be sent over a sub-optimal path.

A system which has a link state change sends a routing update packet containing a list of all operational links to all of its neighbors. In turn, each router is responsible for re-transmitting every flooding packet that it receives to all its neighboring routers.

Since every router has the complete network topology in a local database, it can easily construct a Shortest Path tree with itself as root and all other routers as nodes. The "length" of each link is given by its metric. From this tree, it builds the Forwarding Table that indicates which device interface should be used for a given destination IP address. The rapidity with which a router can revise its Forwarding Table in response to a link state change depends on a number of factors. Most important is the processor time that a router can devote to routing tasks such as flooding, building SPF tree, and publishing the Forwarding Table. The more processor time devoted to routing, the faster a router will be able to respond to changes — but the less time it will have (under heavy loads) for other important tasks such as IP forwarding, SCMP processing, and so forth.

THE PROBLEM

In the traditional, static SPF routing detailed above, each path is calculated based simply on the number of hops, since the metric for each link is the same configured-in value. Once the path to a destination is determined, it will not change as long as the network topology remains the same. Therefore, if the bandwidth of any link on the path is saturated through ST-II reservations, future SCMP call setups will be rejected by the ST-II agent even though there might be alternative paths which still have spare bandwidth.

Not being able to use alternative paths is undesirable for a multimedia traffic environment, since the bandwidth requirements are very high, and total available bandwidth on a path is limited. Static routing cannot fully utilize the network resources.

THE SOLUTION

To solve the problem of under-utilization of network resources, we have designed a dynamic routing extension to the SPF algorithm. The method has been implemented in the BBN T/20 router and greatly enhances the network resource utilization. In our scheme, the shortest path from source to destination is calculated based on the available ST-II bandwidth on each participating link.

Following are the somewhat conflicting goals of multimedia Routing in our internet:

• Maximize the chance that a newly placed telephone call will get through.

• Maximize network utilization and load balancing by spreading calls over multiple paths.

• Minimize the total internet resources devoted to telephone calls (don't set up calls over paths containing extra links if they are not really needed).

• Minimize internet resources devoted to routing updates (bandwidth and processing time)

The routing architecture tries very hard to ensure that ST-II connection requests, which proceed hop by hop through the internet, have a good chance of succeeding. That is, at each router along the path of a stream, the network link leading to the next hop selected by the Routing function needs to have enough spare bandwidth to support the call, and this next hop should lead toward a total path with sufficient bandwidth to reach the target. The primary goal, therefore, is to select a path that appears to have sufficient spare ST-II bandwidth so that the connection is likely to succeed.

In order to accomplish this primary goal we have chosen to base the cost, or metric, of each link on the amount of spare bandwidth available on that link for ST-II traffic. In other words, if there is a lot of spare bandwidth available, the metric will be small, and if the link is almost full, the metric will be large.

Routing operates as follows. Each link is periodically sampled for its spare ST-II bandwidth. Weighted averages of the spare ST-II bandwidth and IP bandwidth on each link are maintained. The costs of each link, which are different for IP and ST-II traffic, are calculated based on these weighted averages. Both IP and ST-II metrics in the internet are composed of two parts: a *static* component administered through Network Management, and a *dynamic* component that is load-based and calculated from weighted averages of recent ST-II reserved bandwidth. The static and dynamic components are added to give the total metric for a link. This metric is measured in hops. The smallest metric for a link is 1, indicating that one transmission across a best possible link is required; the weight 1 is given to the fastest possible link in the internet when it is carrying no traffic load. If a given link has a total metric of 2, that link is considered exactly comparable to a path to the same destination consisting of two links, each of which has metric 1. The static weight is administered by Network Management and has different values for different link speeds. It provides a method by which Network Managers can influence the selection of desirable links. That is, a high value for the static weight on a link will discourage the use of that link.

Each router now maintains two separate SPF trees, one for ST-II and one for IP, since their link metrics are in general different. The standard SPF algorithm is used to determine the shortest path routing trees using the updated link ST-II and IP metric values. Each tree is used to build a separate Forwarding Table for that traffic type.

Since link metrics change dynamically according to the amount of ST-II traffic, there is now a third way, in addition to link state or router state going up or down, that the SPF Tree can change: the IP and/or ST-II metric for an existing link may change. Since in SPF every router must be informed about every metric change, and this change may result in re-calculation of the SPF tree and re-building of the forwarding table, the addition of dynamic link state changes increases the amount of processing that Routing must perform. The more frequent the updates, the more accurate will be the routing decisions, but the more time will be spent on routing processing. The frequency of routing updates containing the latest link metric values is well controlled so that the network is not flooded with update packets, but is able to use the latest information for routing as soon as is feasible. Several trade-off were made.

Since the amount of spare bandwidth on a link varies every time a call is setup or torn down, the metric value will change very frequently, causing undesirable fluctuations in routing tree computations. Thus we want the value used in calculating the dynamic portion of the metric to be more complex than simply the spare bandwidth. We estimate the amount of spare bandwidth that is likely to be available very soon on a given link by keeping a moving average of the history of that link's spare bandwidth, and deriving the link's metric from this historical information. The historical average is weighted so recent history has a stronger influence than older history. The second adjustment is to take into account the burstiness of the traffic by including an approximation to the standard deviation to the spare bandwidth. That is, if the traffic is very bursty, the standard deviation will be large, and we want the metric to be larger than if the traffic was very steady so that we will discourage traffic on bursty links. The reason for this is that if a link is very bursty, it is more likely to be full when a call is directed that way.

The metrics are re-calculated every time bandwidth values are sampled. But these new values are not flooded for use in re-calculating the SPF tree and Forwarding Table every time they change. If they were, too many routing updates would be generated. The scheme chosen limits routing updates to the background periodic floods which are generated by every router every 8 minutes, and to "panic" and "recovery" updates. A "panic" update is generated when a link becomes "almost" full, where "almost" is defined by a tunable parameter. This kind of event results in a very large artificial metric being assigned to the link, thus discouraging all new calls from using that link, except those whose only path to the destination is through that link. A "recovery" update is generated when the full link has had enough stream teardowns to allow traffic to go through again. Hysteresis is applied to this panic/recovery process by suppressing the recovery update until a certain amount of time has gone by.

No matter how an SPF flood is triggered, the resulting update packet always contains the currently advertised ST-II and IP metric values for each operational link. There are three ways in which a change in a link's metric can trigger an SPF flood.

1. When an instantaneous snapshot of reserved ST-II bandwidth on a link indicates that the link is near, or over, its administrative maximum. This is the *start* of a "panic" for that link.

2. When the link is in a "panic" and spare bandwidth has been available for at least a certain amount of time t. This is the *end* of that link's "panic."

3. When a static weight W is changed by network management command.

Conditions 1 and 2 ensure that each link on a router can trigger no more than 2 load-based SPF floods within the time limit t. Thus a router with n interfaces will emit no more than $2n$ SPF floods within this interval, because of bandwidth panics. Every update will include the current metrics for *all* links.

The basic quantities used when calculating the dynamic component of IP and ST-II metrics for a given link are: the total capacity of the link, the maximum allowed ST-II capacity of the link, and the currently reserved ST-II bandwidth on the link. Both IP and ST-II metrics are derived from the ST-II reserved bandwidth on a link. There are two major reasons for basing both metrics on a link's ST-II reserved bandwidth:

• Basing metrics upon reservations for ST-II virtual circuits tends to give greater network stability than basing metrics upon observed IP traffic, since streams remain the same even when routing indicates that new traffic should be set up along different links. IP traffic, on the other hand, is routed anew for each packet. So fluctuations in IP routing result in rapid fluctuations of traffic levels on different paths. (See Bertsekas and Gallagher, Section 5.2.5, 1992, and Khanna and Zinky, 1989, for analyses of this phenomenon.)

• Basing metrics upon observed ST-II traffic, rather than reservation, could lead to cases where bandwidth is underutilized most of the time and thus overbooked occasionally, leading to dropped packets. For instance, if the telephony application suppresses packets of silence, many streams might not carry packets for fairly lengthy intervals. However, they will need the bandwidth when the speaker begins to speak; and a routing strategy that encourages overbooking of links may lead to dropped voice packets if too many speakers attempt to speak at once. In short, the routing architecture must honor the reservations guaranteed by an ST-II stream.

IP metric calculations are analogous to ST-II metric calculations for a link, with some differences. The spare bandwidth used in the calculations is based on the total capacity of the link rather than the ST-II capacity. The values of the constants used in the metric calculation formulas are possibly different for IP and ST-II. Finally, IP "link full" situations do not trigger routing updates.

EXPERIMENTAL RESULTS

The addition of the dynamic routing scheme to standard SPF routing does result, in our test laboratory, in increased utilization of network resources. Under heavy ST-II loads, multiple paths through the network are used. However, this improvement in network utilization comes at the expense of CPU time spent on processing a heavy load of routing updates. Experiments on many different network configurations, with many different traffic loads, using many different combinations of Routing parameters, are being tried. Some of the results are given here.

Since the network needs to work in a 500-node network environment, and only a handful of routers were available for testing, a simulator was written to simulate the behavior of the large network. The output from this simulator is a stream of routing updates which appear to have been generated by the virtual routers. The simulated update stream is injected into the network formed by real BBN T/20 routers and real-time measurements are taken on the T/20 to monitor the performance. In this way we are able to study the effects of a heavy routing update load on the routers in a combined environment of both virtual and real network.

An important issue we have to investigate is whether or not the T/20's CPU could stand the load of routing traffic in such a real-world scenario, i.e., what is the percentage of CPU time spent on routing processing in a typical 500-router network. The more routing updates, the more time the router has to spend on processing them, at the expense of other work the router should be doing such as forwarding IP packets or setting up new ST-II streams.

Our analysis has shown that the total number of routing updates is dominated by the "panic" rate in the network and our experiment tells us that in a network of 300 nodes, the CPU shows no signs of strain. Therefore, we will focus on the impact of different panic rates on the CPU time in a large 500 node network. The following experiment is set up:

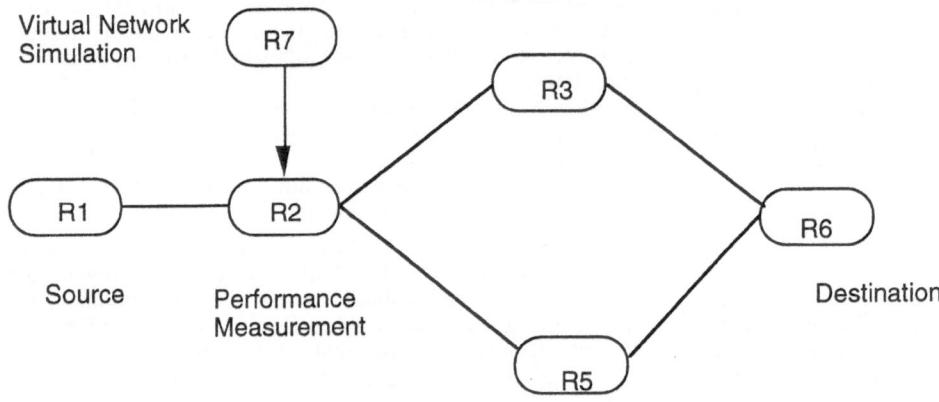

Figure 2. The network setup for routing scalability study

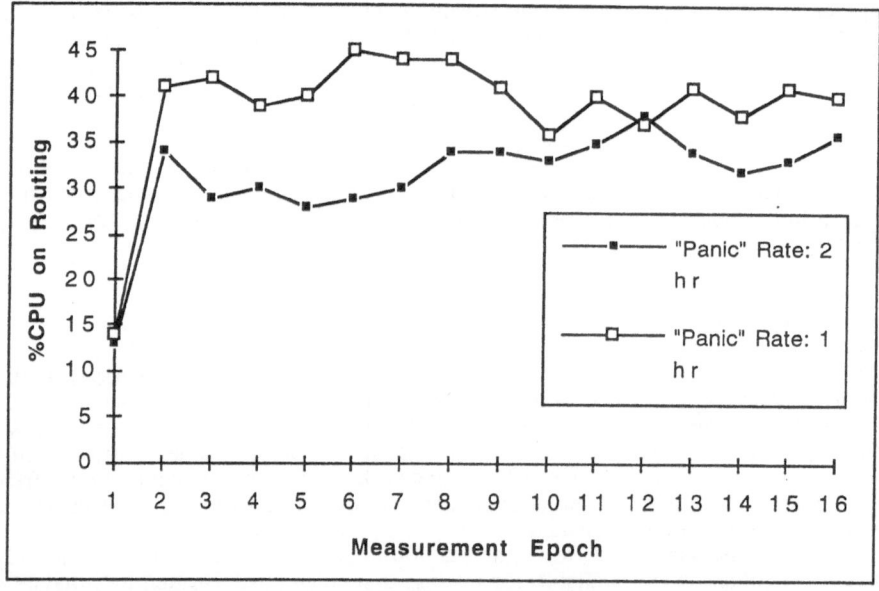

Figure 3: CPU Utilization under Different "Panic" Rates in a 500-node network

ST-II streams are sent from R1 through R2 to R6. The router R7, attached to R4, is used as the "virtual network" simulator of a 500 node network from which a stream of routing update packets is injected into the real network. We control the average number of updates per second in the virtual network by adjusting the panic rate. All the real-time measurements are conducted on the middle router, R2. Figure 3 shows the CPU time occupied by routing processing, with panic rate being on the average once per hour per link and once per two hours per link.

Each data point in the diagram is the average percentage of CPU time spent on routing processing for the previous five minutes. We can see that if the average link panic rate is once every two hours, then the CPU on R2 uses about 30% of its time on routing processing while it runs 40% of the time for routing if the panic rate increases to once every one hour.

The results show us that the SPF routing traffic load in a large-size packet switch network using dynamic routing could pose a problem for the CPU. Using SPF, every "panic" or "recovery" routing update caused by heavy ST-II traffic on any link in the 500 node network must be processed by every router in the network, and all 500 Forwarding Tables must be re-built. Thus, the introduction of dynamic routing updates in a large network adds greatly to the processing load. This problem could be mitigated by breaking up the large network into relatively independent smaller subnetworks and adopting a hierarchical routing protocol. Since the dynamic routing scheme can be implemented without major changes to existing routing protocols, its use in a hierarchical routing protocol is feasible and valid. This scheme then makes possible the benefits of increased network resource utilization, as shown below.

We test our dynamic routing scheme on a simple four node network using T/20 routers. The topology is like the following in Figure 4:

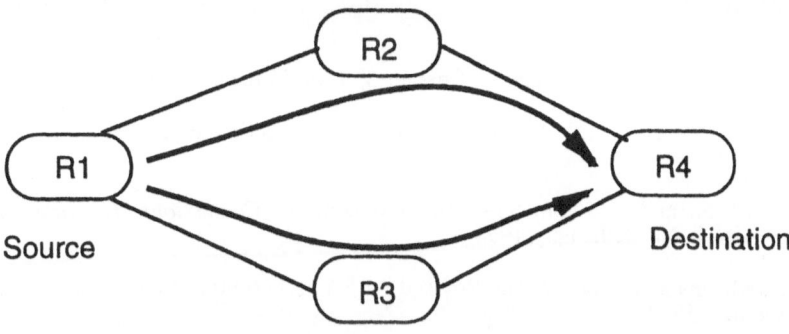

Figure 4: Simple Test Network Topology

The network is configured as follows: each duplex link, e.g. R1-R2, R1-R3, R2-R4, R3-R4, has a bandwidth of 160,000 bits/sec in both directions. ST-II streams are set up from R1 to R4 and each stream has a bandwidth requirement of 16,000 bits/sec. The "panic" threshold for each link is 16,000 and "recovery" threshold is set to be 32,000. The time hysteresis for the next update is one minute. This means that the link would go into "panic" situation once the spare bandwidth is only big enough to hold one more stream and the link would recover from the situation if there is room for two more streams and the time has passed one minute since the panic. In both situations, the relevant routers send out update to inform other routers about the link conditions.

There are no ST-II streams in the system at the beginning. As we start setting up the very first streams on R1, the system chooses R1-R2-R4 as the best route. After nine streams, this route is in "panic" mode with the maximum metric value being set for both link R1-R2 and R2-R4. The router R1 and R2 issue "panic" updates immediately. The router R1 recalculates the best route from R1 to R4 and choose R1-R3-R4. Therefore the next nine streams go through this newly selected route until the system once again reaches "panic" on R1-R3-R4. This time, the system's *total* capacity is saturated and could only take one more stream.

Next we tear down the streams one by one with an interval of about 3 minutes starting from those on route R1-R2-R4. When there are seven streams left, router R1 and R2 both generate a routing updates indicating that the links R1-R2 and R2-R4 have "recovered" with the latest calculated link metric values.

The above simple experiment demonstrated that our scheme could increase the network capacity by taking advantage of the bandwidth available on idle links and direct ST-II calls through them.

CONCLUSIONS

We have shown that use of a dynamic metric based on ST-II bandwidth reservations in standard SPF increases the network utilization and reduces the call setup failure rate. A carefully designed update strategy keeps the update overhead under control in a medium-size network. In a large network requiring hierarchical routing, the scheme is also applicable,

ACKNOWLEDGMENTS

We wish to thank Chip Elliott and Martha Steenstrup of BBN for their contributions to the dynamic routing scheme explained here. Discussions were also held with John Zinky, Jim Cervantes, and Ravi Sastry of BBN.

This work was performed under contract to Computing Devices Canada, Ltd., in support of the Iris battlefield communications system being developed for the Canadian Land Forces. See Elliott and Hill, 1994, for a complete description of this network.

REFERENCES

D. Bertsekas and R, Gallagher, Chapter 5, *Data Networks*, Prentice-Hall, 1992.

C. Elliott and A. Hill, An Internet for Tactical Multimedia Communication, internal report, 1994.

C. Elliott and C. Lynn, ST-II Implementations, *Connexions, The Interoperability Report*, Vol. 8, No. 1, January 1994.

A. Khanna and J. Zinky, The Revised ARPANET Routing Metric, *ACM SIGCOMM Proceedings*, 1989.

C. Topolcic, S. Casner, C. Lynn, P. Park, K. Schroder, *Experimental Internet Stream Protocol*, Version 2 (ST-II), RFC 1190, October 1990.

RESOURCE OPTIMIZATION IN VIDEO-ON-DEMAND NETWORKS

Jeong-dong Ryoo and Shivendra S. Panwar

Department of Electrical Engineering
Polytechnic University
Five Metrotech Center, Brooklyn, NY 11201
ryoo@photon.poly.edu, panwar@kanchi.poly.edu

Abstract

In this paper, we address the problem of designing a cost-minimized network for *Video-On-Demand* (VOD) services satisfying given customer demands and *Grade-Of-Service* (GOS) requirements. The GOS is expressed as an acceptable level of blocking for customer requests.

We present results for an assumed cost function for video servers and links. We then provide a solution that can minimize the network cost by means of the selection of nodes in which video servers will be placed, the service and storage capacities of these video servers, and the allocation of the link capacities that meet traffic requirements.

1 Introduction

It is widely considered that *Video-On-Demand* (VOD) will become an important residential service as a substitute for current home entertainment and information services. VOD services offer instant access to a video library of educational and entertainment programs, as well as communication services. Developing this new information delivery infrastructure at minimum cost requires considerable planning and effort. In order to support fully interactive VOD service, a two-way switched broadband network, which provides an independent connection to each user using a high data rate (e.g., about 4 Mbps for a Constant Bit Rate (CBR) MPEG-2 compressed video stream for a broadcast quality movie), is needed. Considering initial costs in the early deployment stage, a *Hybrid Fiber/Coax* (HFC) transport architecture is regarded as a promising option

---

*This publication was developed under the auspices of the Polytechnic University CAT in Telecommunications and Distributed Information Systems, a New York State Center for Advanced Technology supported by the New York State Science and Technology Foundation, and also the NFS under Grant NCR-9115864.

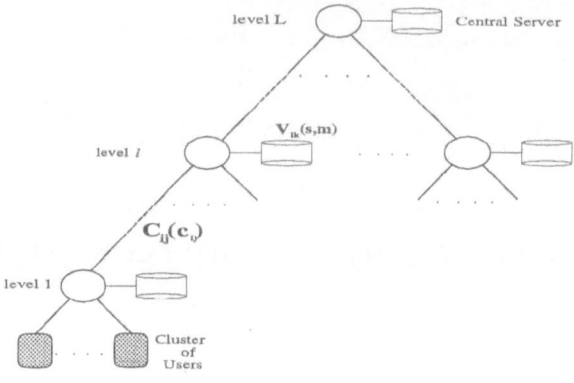

Figure 1: Logical network topology for VOD

for the delivery of interactive broadband services. Given that broadband coaxial cable access is available to over 93 percent of the homes in the United States [5], it gives cable operators the opportunity to play an important role in interactive broadband service. Migration from the current broadcast video service to real-time interactive video service creates the need for a change in the cable network structure and the introduction of switching capability.

There are a few papers dealing with network design issues related to VOD. Nussbaumer, Patel, *et al.* evaluated the effects of program caching and stream sharing on the bandwidth and storage requirement of the VOD network [4]. Bisdikian and Patel addressed the issues on movie allocation in distributed VOD systems [1]. Our approach differs from these in that an acceptable blocking probability is explicitly taken into account in the network design. In addition, as a result of this approach, the network design incorporates the statistical multiplexing gain made possible by using shared resources.

In this paper, we address the problem of designing a cost-minimized network for VOD services satisfying given customer demands and *Grade-Of-Service* (GOS) requirements. The GOS is expressed as an acceptable level of blocking for customer requests. Request blocking arises when there is not enough bandwidth in the network, or the required video file cannot be retrieved from the video server. The latter results from a situation when the video server reaches the limit on the number of requests that can be supported simultaneously, or the requested video file is being accessed by another user and is therefore not available.

We present results for an assumed cost function for video servers and links. We then provide a solution that can minimize the network cost by means of the selection of nodes in which video servers will be placed, the service and storage capacities of these video servers, and the allocation of the link capacities to meet traffic requirements.

2 Network Model

Our logical multimedia network topology is illustrated in Figure 1. Each node of the tree consists of a potential site for a video server. The leaves of the tree represent a cluster of users. The objective of the design is to minimize the overall cost, which is the sum of the costs of video servers and link bandwidths.

$$J^* = min\{\sum_{i,j} C_{ij}(c_{ij})Y_{ij} + \sum_{l,k,s,m} V_{lk}(s,m)X_{lk}\} \qquad (1)$$

The link cost, $C_{ij}(c_{ij})$, is the cost of a link with capacity c_{ij}, connecting locations i and j. $V_{lk}(s,m)$ represents the cost of a video server that is placed at location k, in level l, with service capacity s (i.e. the maximum number of simultaneous requests that can be handled), and storage capacity m. Y_{ij} and X_{lk} are indicator functions, indicating the presence or absence of a link between locations i and j or a server for location k on level l, respectively.

In this paper, preliminary results are obtained for a balanced tree, or in degenerate cases, a forest of balanced trees, all the links at any particular level having the same capacity. Also, symmetric traffic requirements are assumed, i.e., each cluster of users has the same demand characteristics for movies. A constraint is imposed on the overall blocking probability.

Let us assume that the VOD system offers M movies to the customers. Let p_n be the probability that an incoming request is for the nth video file, where $n = 1, 2, 3, \ldots, M$. Without loss of generality, we assume that the video files are ordered by popularity, i.e.,

$$1 > p_1 \geq p_2 \geq \ldots \geq p_M > 0. \qquad (2)$$

Note that we keep multiple copies, $C_{F,n}$, of the nth file to satisfy the demand for that file in a certain video server, as well as in all the video servers at the same level. Under this assumption, more popular movies are more costly to store, since multiple copies have to be provided to satisfy demand. According to the given demand characteristics of video files, the video files are distributed among various levels. We make the assumption that movies are placed, starting with the highest level server, in inverse order to their popularity. Thus by placing the most popular movies closest to the users, the communication cost can be minimized.

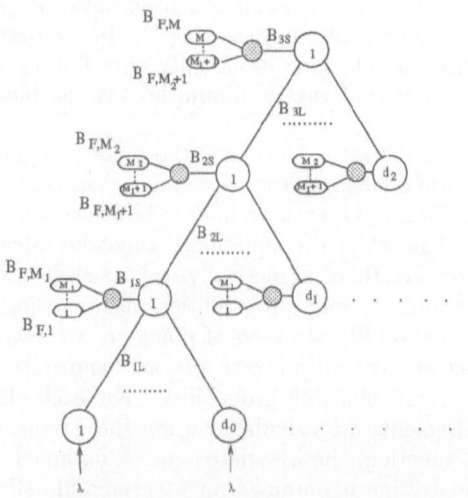

Figure 2: Three level balanced tree

An example of a three level balanced tree is shown in Figure 2. Each video sever in level 1 stores the most popular movie to the M_1th most popular one. Its required storage capacity is $\sum_{n=1}^{M_1} C_{F,n}$. The $(M - M_{L-1})$ least popular movies are stored in the root server. Later, for a given M, we will find optimal values of M_i's that give the minimum cost. Let B_{lL} be the blocking probability of a link between level $l-1$ and l. B_{lS} is the probability of the blocking that occurs due to lack of internal bandwidth of a video server in level l. Let $B_{F,n}$ denote the blocking probability of the nth file, which occurs when all the copies of the requested movie are being accessed by users. d_{l-1} denotes the number of branches under a node in the level l of the tree, where $l = 1, 2, \ldots, L$.

For the example of a three level balanced tree, the constraint could be expressed as follows for a maximum blocking probability, $\overline{B}$,

$$
\begin{aligned}
P_{block} \;=\; & 1 - \sum_{n=1}^{M_1} p_n (1 - B_{1L})(1 - B_{1S})(1 - B_{F,n}) \\
& - \sum_{n=M_1+1}^{M_2} p_n (1 - B_{1L})(1 - B_{2L})(1 - B_{2S})(1 - B_{F,n}) \\
& - \sum_{n=M_2+1}^{M} p_n (1 - B_{1L})(1 - B_{2L})(1 - B_{3L})(1 - B_{3S})(1 - B_{F,n}) \quad (3)
\end{aligned}
$$

$$
\leq\; \overline{B} \quad (4)
$$

assuming that, for a given service request, the blocking events on the required path are all independent. For the example considered in the next section, we shall assume that $B_{F,n}$ is the same for all movies at a given level and that the number of copies of each movie, $C_{F,n}$, is set to achieve this common blocking probability.

3 Computation

In order to determine the overall blocking probability, we could have considered a Markov model of the network, which gives the exact solution. However, it is impractical to perform the calculation even in a moderate sized network. As an approximation method, we adopt the Erlang fixed point solution [3]. In the process of evaluating the various link probabilities, the blocking probability on a link is computed by Erlang's loss formula with an offered load that is diminished by the blocking probabilities of other links along the path.

When all the nodes in the same level have the same number of branches and the incoming traffic at each leaf node has the same demand characteristics, $M+2L$ nonlinear equations, consisting of Erlang B functions, have to be solved for a L-level balanced tree with M kinds of files. The set of the equations cannot be solved analytically, which means that we have to solve them numerically. There exist several methods to find the correct values for those blocking probabilities, such as the repeated substitution method, or Newton's method [2]. Instead of doing so, we searched all feasible sets of blocking probabilities of link, video server internal bandwidth, and video file under the constraint of the overall blocking probability. For each element of the feasible set, the capacity requirements are calculated using the Erlang fixed point equation. With an assumed cost function, the associated cost is obtained. After calculating all the elements, an optimal value is obtained for a specific file allocation. This is then performed for all possible file allocations to determine the lowest cost network.

In figures 3 and 4, we show the computational result for the two level tree case, with $M = 100, d_1 = 3, d_0 = 4, \lambda = 60$ erlangs, and $\overline{B} = 0.1$. The probability p_n that a service request is for the nth video file is assumed to be $P_n = \frac{G}{n}$, where $G = \left(\sum_{n=1}^{M} \frac{1}{n}\right)^{-1}$. This is a standard movie popularity model [4]. The cost function is assumed to be linear

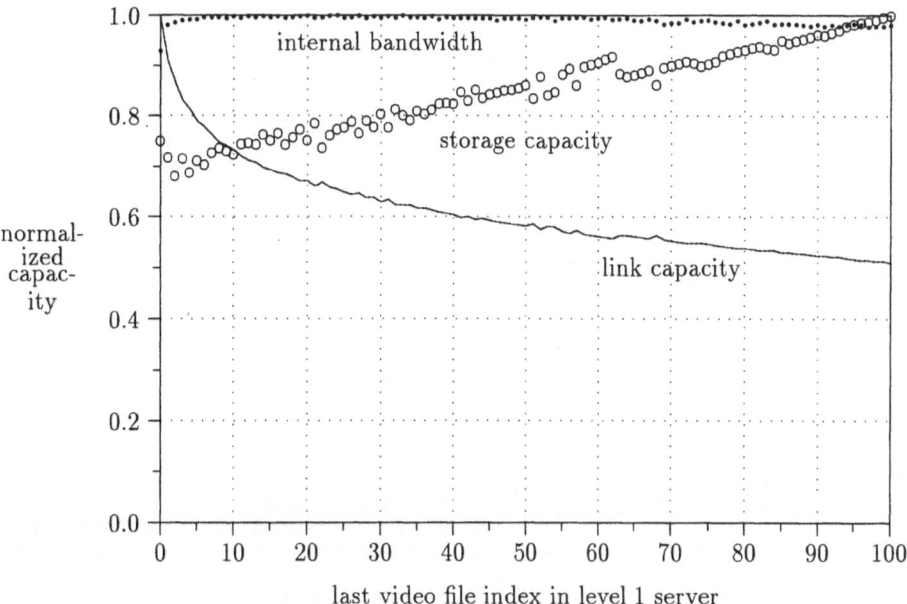

Figure 3: Normalized capacities of link, storage, and internal bandwidth, versus last video file index in level 1 server

with coefficients 1 for a channel in the video server, 7 for a video file, and 37 for a video channel on a link.

Figure 3 shows the normalized values of the total link capacity, the total internal bandwidth of all the video servers in the system, and the sum of the video file storage capacities, as a function of the file allocation. The largest values of the total link capacity, internal bandwidth, and video file storage capacity are 1452, 738, and 1653, respectively. The total link capacity decreases as more video files are placed at the video server closer to the lowest level of the tree. On the other hand, the total storage demand decreases as video files are placed in the root of the tree. This situation can be explained as an example of statistical multiplexing gain, i.e., when the traffic intensity is increased, for a fixed blocking probability, the number of channels required increase less than proportionately. One of the interesting aspects of this example is that the aggregate internal bandwidth (the number of channels) required for all the video servers is almost the same for different video file allocation schemes.

The effect of different video file allocation schemes on the total cost is shown in figure 4. In this two level tree example, the optimal cost is obtained when a video server in Level 1 contains video files ranging from the most popular one to the 98th most popular.

4   Conclusions and future work

Due to the independent interactive capability of VOD as well as the large video bandwidth requirement, one of the main problems in a VOD network is the huge amount of total aggregate link bandwidth required. Also, a massive amount of storage capacity and processing resources are required in order to provide all users with satisfactory services. As a way to overcome this situation, we considered the problem of resource optimization, which can be accomplished by means of assigning link capacity, sizing the video servers and assigning files to each of them. Arbitrary cost functions can be

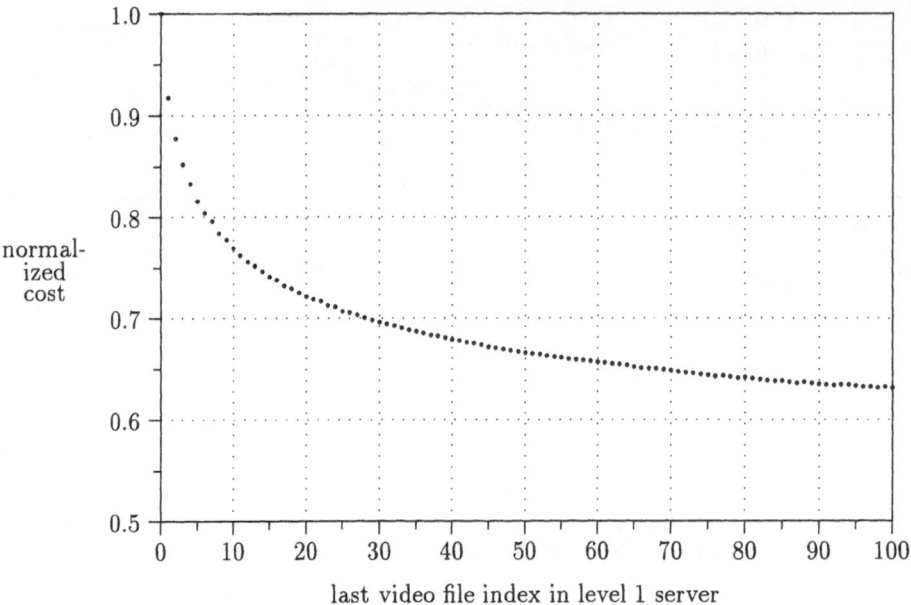

Figure 4: Normalized cost versus last video file index in level 1 server

used for the exhaustive search technique that we have used.

This work is continuing in the direction to find an optimal video file allocation scheme efficiently without exhaustive search. Even for a given file allocation scheme, finding an optimal set of capacities takes a lot of computation time. Thus, we are considering the following optimization problem as a way to speed up the computation.

For a given video file allocation scheme, let $\mathcal{I}$ denote the set of indices for the links present in the network. Let $\mathcal{J}$ denote the set of indices for the nodes where video servers are placed. The optimization problem in equation (1), can be formulated as

$$min\{\sum_{i\in\mathcal{I}} C_i(c_i) + \sum_{j\in\mathcal{J}} V_j(s_j, m_j)\},\tag{5}$$

with constraints

$$
\begin{array}{rcll}
L(\underline{c}, \underline{s}, \underline{m}) & \leq & \overline{B} & (\lambda) \\
c_i & \geq & 0 \quad \forall i \in \mathcal{I} & (\underline{\alpha}) \\
s_j & \geq & 0 \quad \forall j \in \mathcal{J} & (\beta) \\
m_j & \geq & 0 \quad \forall j \in \mathcal{J} & (\gamma)
\end{array}
$$

where $L(\underline{c}, \underline{s}, \underline{m})$ is the overall loss probability with link capacities, c_i, internal bandwidths of video servers, s_j, and storage capacities of video servers, m_j, for all i and j. λ, $\underline{\alpha}$, β, and γ are the vectors of multipliers corresponding to the blocking and positivity constraints.

When we assume $c_i \neq 0, s_j \neq 0, m_j \neq 0$, for all $i \in \mathcal{I}, j \in \mathcal{J}$, then at the optimal solution, $(\underline{c}^*, \underline{s}^*, \underline{m}^*)$, we have $\alpha_i = \beta_j = \gamma_j = 0$. The first-order Kuhn-Tucker conditions are reduced the following condition

$$\lambda = -\frac{\partial C_i(c_i^*)/\partial c_i}{\partial L(\underline{c}^*, \underline{s}^*, \underline{m}^*)/\partial c_i} = -\frac{\partial V_j(s_j^*, m_j^*)/\partial s_j}{\partial L(\underline{c}^*, \underline{s}^*, \underline{m}^*)/\partial s_j} = -\frac{\partial V_j(s_j^*, m_j^*)/\partial m_j}{\partial L(\underline{c}^*, \underline{s}^*, \underline{m}^*)/\partial m_j} \forall i \in \mathcal{I}, j \in \mathcal{J}.\tag{6}$$

Expressed in words, this has the intuitively appealing interpretation that at the optimum allocation, the incremental cost per unit reduction of loss is equal for any resource (any link capacity, any video server capacity, or memory) in the network. We are investigating various numerical methods and approximations to solve this problem by computing the value of the multiplier, λ, at the optimal point.

References

[1] Chatschik C. Bisdikian and Baiju V. Patel, "Issues On Movie Allocation in Distributed Video-on-demand Systems," *ICC '95*, pp. 250-255, 1995.

[2] André Girard and Yves Ouimet, "End-to-End Blocking for Circuit-Switched Networks: Polynomial Algorithms for Some Special Cases," *IEEE Trans. on Comm.*, Vol. COM-31, No.12, December 1983, pp. 1269-1273.

[3] F. P. Kelly, "Blocking Probability in Large Circuit-Switched Networks," *Advances in Applied Probability*, vol.18, pp. 473-505, 1986.

[4] Jean-Paul Nussbaumer, Baiju V. Patel, Frank Schaffa, and James P.G. Sterbenz, "Networking Requirements for Interactive Video on Demand," *IEEE J. Selected Areas in Comm.*, Vol. 13, No. 5, June 1995, pp. 779-787.

[5] Andrew Paff, "Hybrid Fiber/Coax in the Public Telecommunications Infrastructure," *IEEE Comm. Magazine*, April 1995, pp. 40-45.

END USER RECEPTIVITY: AN IMPORTANT FACTOR IN THE DESIGN OF A DISTRIBUTED MULTIMEDIA SYSTEM

Ronnie T. Apteker[1], James A. Fisher[2], Valentin S. Kisimov[1], and Hanoch Neishlos[1]

ronnie@is.co.za, 144jafs@muse.arts.wits.ac.za, valentin@concave.cs.wits.ac.za, hanoch@concave.cs.wits.ac.za

[1]Computer Science Department
[2]Psychology Department
University of the Witwatersrand
Wits 2050, Johannesburg, South Africa

INTRODUCTION

The computer distorts the act of viewing motion pictures from a screen as it has traditionally been perceived. It is significant that this facility is now called a multimedia application, and not a television channel. In essence, a virtual television (a multimedia application window) is constructed, which when connected to a data network, can be programmed or scheduled, similar to the way that television networks program their line-ups. Putting computing into a television was an achievement, but putting television into a computer is quiet a different matter.

Multimedia is in the mainstream. Today's challenge is to use it effectively. With the widespread deployment of computer networks all around the globe, today's underlying requirement is inter-connectivity. Everyone wants to be connected to everyone else. Distributed multimedia applications are the goal, and the problems are those concerning the large volumes of data and network bandwidth constraints that are characteristic of continuous media delivery. A typical digital video with a frame size of 640 x 480 pixels and 3 byte colour depth requires 27 Mbps of network bandwidth. If we have thousands of users on a network each demanding access to a multimedia server (video on-demand) we need transmission speeds well in the gigabyte per second range pushing the limits of optical fibres. Brute force is always an approach. Spend more on high bandwidth networks and support more users. But the growth of these new environments and increasing level of user acceptance drives us to research new ways to building distributed multimedia systems capable of supporting the expanding user community. A hundred year ago Marconi invented a method to enable people to hear audio transmission, and now we are seeking a practical way to extend this into the exciting world of multimedia. Still, 100 years later, we do not have a concrete mechanism to enable everyone around the world to receive and interact with multimedia data all at the same time.

Multimedia Communications and Video Coding
Edited by Y. Wang *et al.*, Plenum Press, New York, 1996

The current view and practice of QOS is one of technical resource management and the system parameters surrounding this management. We argue that this narrow technical focus does not address the very human qualities that constitute a user's concept of quality. This is particularly so in the context of the rich potential for human interaction with multimedia applications and digital technology. Our concern here is to promote a concept that has become critical to our thinking of QOS; the concept of "receptivity". Receptivity, as we conceive it, is a human disposition to embrace and appreciate technology, based upon an acquired knowledge and developed set of expectations surrounding the potential and value of the technological possibilities for the user. In short, receptivity in our scheme of things is a composite of appreciation, understanding, acceptability and expectation on the part of the user.

To our thinking QOS is the logical product of the interaction between the ingredients of receptivity and the technical parameters of a distributed multimedia system. Our aim throughout the paper is to use this idea to incorporate these human feelings not only onto the design of distributed multimedia systems but to ensure that the concept QOS reflect these human qualities as well as the evident technical ones.

QOS therefore encompasses technical system parameters as well as end user receptivity. Through a heuristic process of learning we can gain knowledge as to how compromises in end user receptivity can make a difference on technical parameters in the system and on QOS overall. We develop a model for the purpose of acquiring knowledge and it is against this framework that QOS is revisited as a system consisting of technical parameters and human parameters. A distributed multimedia system designer can utilise this model by acquiring knowledge though heuristics making way for greater flexibility when implementing physical system parameters in the realisation of a working system.

At this stage it is important to stress that our concept of user receptivity is a holistic one; one that is not tied to the exact measurement of, say, milliseconds saved in an interactive task of modelled trade-offs in lag and error rate [1]. These specific components and design guides are. of course, available in the engineering psychology and cognitive ergonomics literature. but our concept is not directed at the piecemeal assembly and addition of such data, but rather to sketch a more dynamic situation posed by multi-tasking with multi-channel tools.

The objective of this paper is to develop a model to aid in the development of a distributed multimedia system. The model presented makes a departure from traditional technical specifications by re-examining the concept of QOS to include end user receptivity. The goal of any system in this field is to maximise end user acceptance whilst minimising system resource utilisation. Our model lends itself to this goal by exploring more than technical issues alone. Our model incorporates the idea of a knowledge acquisition process whereby a system designer can learn when to manipulate technical parameters according to knowledge gained with respect to end user receptivity, and hence allowing for better utilisation of existing resources.

A SIMPLIFIED SYSTEM VIEW: DISTRIBUTED MULTIMEDIA AND QOS

A distributed multimedia application is a system whereby multimedia data is taken as the input, processed, and delivered as a multimedia session as the output [2,3,4]. QOS is traditionally associated with the multimedia application and is measurable in terms of system parameters, for example, frame rate delivered, etc. By introducing heuristic knowledge QOS can be separated out from the physical system so that it is seen to comprise human parameters as well as technical parameters.

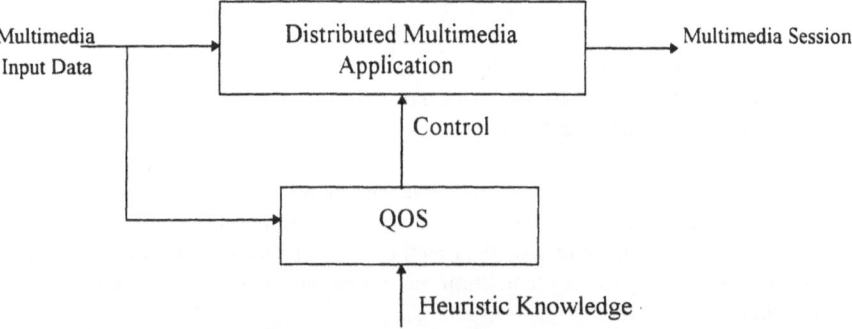

Figure 1. A distributed multimedia system

In Figure 1, "control" represents those algorithms, processes, etc. that constitute the physical characteristics of a distributed multimedia system. QOS is viewed as the collection of these physical characteristics together with heuristic knowledge pertaining to end user receptivity. The functionality of the system is such that an appropriate algorithm (say, colour depth control) will be manipulated when the system knows that end user receptivity will not be severely effected by a change in a specific resource (say, colour) and hence system resources are saved and more users can be supported. The presence of heuristic knowledge serves system designers by allowing them to select appropriate controls.

The input to a distributed multimedia system typically is the multimedia data; the output is the multimedia session. The controls are technical processes interacting with heuristic knowledge. With the rich variety of input data available (multimedia information can be

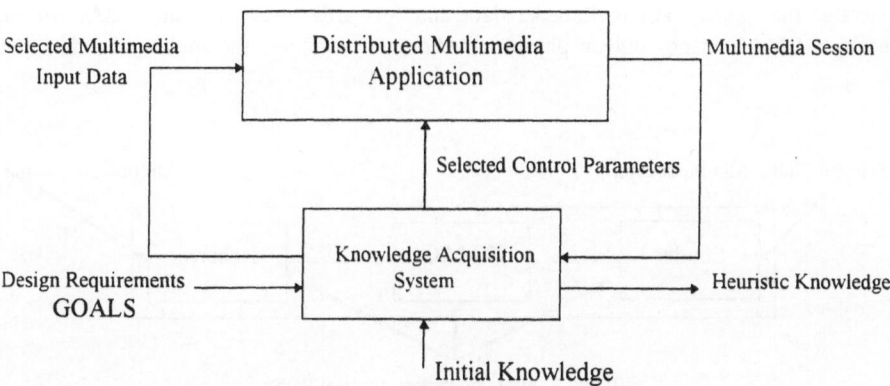

Figure 2. Knowledge Acquisition Model

represented by sports, talk shows, music videos, news, and so on) it should be clear as to the vastness of the control space. In short, the value of heuristic knowledge stems from its use as an effective method for exercising differing controls depending on the multimedia input data.

KNOWLEDGE ACQUISITION MODEL

The model is based on the following principle: collect data through active experimentation, recognising patterns in this data, and then formulate heuristic knowledge through inferences from these patterns.

Although inferred from experimental data, the knowledge derived is generally heuristic rather than empirical because : 1. The procedures of experimentation are designed using heuristic principles, and not only statistical methods. 2. The knowledge rules are inferred from statistical analysis, pattern recognition, neural networks, etc. [5]. 3. The process is one of learning.

Heuristic knowledge is always goal-specific because points 1-3 are achieved with the goal being the primary objective. The primary goal is the production of heuristic knowledge that ideally characterises the optimal set of parameters that satisfies the goal over any given multimedia application, i.e., such values of the system that simultaneously maximise the end user receptivity and minimise system resource utilisation.

Figure 2 illustrates the experiment. The enormous variety of multimedia data (input) and the various controls makes the application and control spaces too complex for an exhaustive acquisition of knowledge. Instead, a representative selection is made which constitutes the application and control spaces. The output of active experimentation is heuristic knowledge.

A DISTRIBUTED MULTIMEDIA SYSTEM: 3 CO-OPERATING BLOCKS

Our model splits the distributed multimedia system into 3 co-operative blocks : Sender, Receiver and the Continuous Media Active Channel (CMAC). These blocks are linked through 2 paths: control and information. The control path is a message stream used to manage the information path that flows between the 3 co-operative blocks. The Sender generates the appropriate multimedia data and forwards it though the CMAC to the Receiver. CMAC is not only a physical medium, it examines the multimedia data and

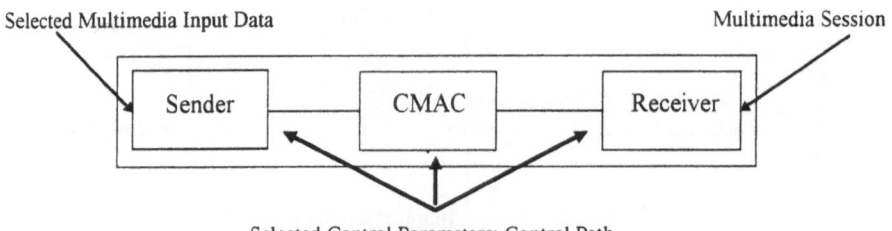

Figure 3. A distributed multimedia system: 3 co-operating blocks.

executes a policy for transmission. An appropriate protocol strategy is incorporated as CMAC's function. The CMAC uses the data that it obtains via the control path and uses this information in its feedback strategy by learning how to better utilise the multimedia data and application data which is carried in the information path.

For each of the co-operative blocks we need to define a set of end user related operational parameters. Applying these parameters to our model we will be able to suggest a set of technical parameters [6] for designing a distributed multimedia system. In this paper we present an initial set of parameters for the co-operative blocks. The purpose of these initial set of parameters is to give a basis of a mainstream set of models and to describe a method for their evolution. The initial set of operational parameters for the Sender comprises: data compression algorithm (lossy versus lossless compression), temporal resolution (frames per second, audio sampling rate), spatial resolution (bit per pixel, bits per colour depth, audio/mono sound), level of Sender/Receiver interaction, and expected network bandwidth. For the Receiver, the initial set of operational parameters include: end user acceptance specifications of picture and audio quality, relation between multimedia data and other application data expressed as a function of the multimedia data (main, supporting, complementary), duration (continuous or pulse) of the multimedia data in relation to other application data, and the role of audio in the multimedia data (complementary, audio oriented, high quality audio). The operational parameters for the CMAC consists of: multimedia and application data transmission priorities, maximum data loss, range of network throughput requirements, minimum spatial and temporal requirements.

To describe the taxonomy of human parameters is beyond the scope of this paper. Rather, we promote the concept of "receptivity" - a composite of appreciation, understanding, acceptability and expectation on the part of the user. A taxonomy would be to break these into quantitative fields, which could include, importance of auditory message, importance of visual message, importance of session context, and so on. A simple scale would be to rate these on "low" and "high" values [7], for example, a talk show would have high auditory message importance and low visual message importance.

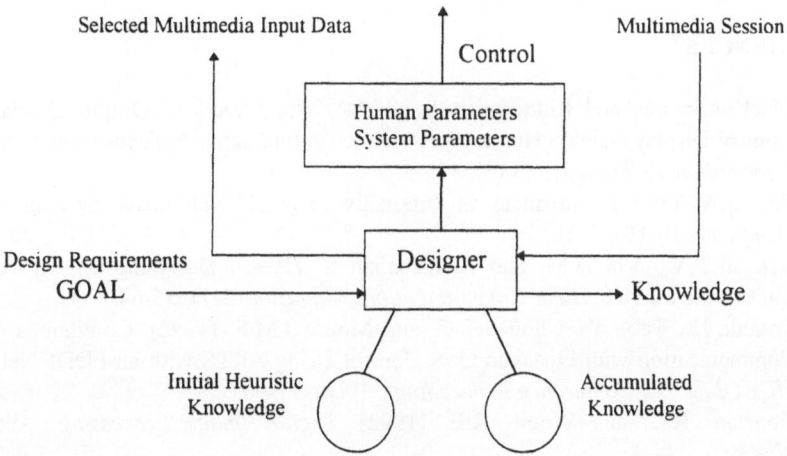

Figure 4. The modelling process: knowledge acquisition

It is essential here to accept the apparent looseness and subjectivity of these human parameters. There can be no truly objective measure of, for example, the importance of the session to the user in the context of wider activity and intention, but most users are able to illustrate this transient state on even the simplest of psychological indicators. Thus an indicated position of a single line, anchored at each end to depict high and low importance, allows the easy conversion to a position on a scale of unitless proportion (0 to 1 or 0 to 100). With n dimensions these can be modelled, using at the lowest level of structuring comparative rank profiles or for more complex structures multi-dimensional scaling techniques, to form a succinct summary of a user's human parameter state or disposition.

The principle objective is to load QOS with knowledge. Figure 4 illustrates how the designer can select different multimedia input data and capture session knowledge with respect to end user receptivity. For all intents and purposes, QOS is an holistic encompassing of technical and human attributes and should guide the system in the control process.

A study was performed that illustrated how a compromise in frame rate [8] lead to a saving in system resources without a severe degradation in end user receptivity over a variety of different multimedia inputs. This simple experiment showed how a change in a technical characteristic could be implemented without the obvious fear of QOS degradation. Different technical parameters can also be experimented with in order to establish their effect on end user receptivity with the consequence of building heuristic knowledge.

CONCLUSION

Designing a distributed multimedia system requires careful consideration of parameters pertaining to physical limitations of existing resources in hardware and software. But, other important parameters are those describing the way multimedia data (video and audio) is presented to the end user. As with physical hardware limitations, the design of a multimedia system should assume some compromises in the end user acceptance parameters.

REFERENCES

[1]   MacKenzie I.S. and Riddersman S. (1994). The Effects of Output Display and Control-Display Gain on Human Performance in Interactive Systems. *Behaviour and Information Technology*, 13 (**5**), 328-337.

[2]   Fox E.A. (1991). Advances in Interactive Digital Multimedia Systems. *IEEE Computer*, 10(**10**), 9-21.

[3]   Rangan P.V., Vin H.M. and Ramanathan S. (1992). Designing an On-Demand Multimedia Service. *IEEE Communications Magazine*. 3(**7**), 56-64.

[4]   Tokuda H., Tobe Y., Chou S.T.C. and Moura J.M.F. (1992). Continuous Media Communication with Dynamic QoS Control Using ARTS with an FDDI Network. *SIGCOMM '92 Conference Proceedings*, 10(**4**), 11-98.

[5]   Gonzalez R.C. and Woods R.E. (1992). *Digital Image Processing*. Addison-Wesley.

[6] Nahrstedt K. and Steinmetz R. (1995). Resource Management in Networked Multimedia Systems. *IEEE Computer*, 28 (**5**), 52-63.

[7] Edwards A.L. (1960). *Experimental Design in Psychological Research*. 2nd Edition. Holt, Rinehart and Winston, New York.

[8] Apteker R.T., Fisher J.A., Kisimov V.S. and Neishlos H. (1995). Video Acceptability and Frame Rate. *IEEE Multimedia*, to be published this year.

SYNCHRONIZATION ISSUES ON SOFTWARE MPEG PLAYBACK SYSTEMS

Gabriel Zuniga and Ephraim Feig

IBM T.J. Watson Research Center
Yorktown Heights, NY 10598

INTRODUCTION

Software based MPEG[1,2] playback systems on today's personal computers suffer from insufficient CPU capabilities. On top of these limitations are further impediments caused by the system memory hierarchy.

Current operating systems lack hard real-time support and it is not possible to allocate a fixed amount of the CPU for processing fixed functions. So even if the CPU is powerful enough, software based MPEG systems could find themselves for some periods of time with insufficient CPU for processing. This indeed happens when the CPU is simply not powerful enough for the task, or even if it is, when other applications or drivers are running concurrently.

Invariably this leads to frame dropping strategies, whereby some of the MPEG video is not decoded and displayed. Generic techniques have a hard time dealing with MPEG playback. Such techniques would naively determine if there is enough time to process a frame at a given instance, and if not would drop that frame. Unfortunately, in MPEG not all frames are created equally. B frames can be dropped at will; but dropping I or P frames can cause many future frames to be dropped in turn. For example, if a sequence contains groups of pictures lasting a whole second which start with an I frame followed by a B frame and then 14 more P-B couplets (such GOPs are quite typical of videos in Compton's MPEG Encyclopedia), then dropping the I frame would mean that the remaining full second of video is also dropped. In fact, in this case, any decoder which cannot process at least 15 frames per second will have to drop at least 1 P frame, and this will cause visually degrading timing artifacts.

The present paper reports on two different approaches to solve the frame dropping problem in two environments. The first more aggressive solution was implemented on IBM's PowerPC running the personal AIX operating system. The CPU was a 66 MHz 601, with 32KB L2 unified (data and instruction) cache and 32MB memory.

The second solution was implemented on Pentium based IBM Aptiva PCs running Microsoft Windows 3.1. The CPUs ranged from 60MHz to 133MHz, some without L2 cache, others with 128KB L2 cache, and each with 8MB memory.

At the time of implementation, there was no hardware support for color space conversion and scaling on the PowerPC, so color conversion and scaling were done in software. Also, on the PowerPC, MPEG audio decoding was done by the CPU, whereas on the Aptiva it was done by an Mwave DSP. As will be seen, the dominant factors which determined the choice of strategies were the amount of available memory and the availability (or lack) of efficient multitasking.

REQUIREMENTS

The frame dropping scheme should fulfill the following requirements:
- Display every frame on time.
  Each frame should be displayed according to its presentation time-stamp (PTS). Otherwise the displayed video will produce unpleasant visual artifacts and lose synchronization with the audio.
- If a frame is decoded then it should be displayed.
  Some schemes decode frames that that are not displayed in order to avoid dropping all the frames depending on it. We avoid this inefficiency.
- CPU utilization should be maximal.
  The decoder should utilize as much CPU as is available. It should not stay idle because of dependencies between frames. For example, we avoid dropping early P frames (in GOPS with several P frames); otherwise, all subsequent frames until the next I frame will be dropped, and the CPU will be idle for a significant time period.
- Avoid dropping I and P frames when possible.

PowerPC IMPLEMENTATION

The target system is based on the PowerPC 601 processor with 32M of memory and 256K L2 cache running the personal AIX operating system (IBM's UNIX implementation). Mpeg audio and video decoding are done by the CPU, as well as color-conversion and dithering to an 8-bit display.

Audio/video synchronization is achieved by using the audio and video "Time Stamps" provided in the Mpeg sequence. The audio acts as the master clock and the video adjusts itself to the audio by dropping frames if necessary.

AIX's multitasking and efficient memory management allow us to use pipeline techniques to solve the frame dropping problem. Each pipeline stage is implemented by a different AIX process. The video decoder module has two pipeline stages. In one stage the video is decoded and in the other stage the video is displayed; the latter stage includes color conversion and dithering. An additional pipeline stage was used to perform file input and audio and video demultiplexing. The decoder architecture is showed in figure 1.

The decoding stage decodes as many frames as it can from a complete GOP. Concurrently, the display stage performs color conversion and dithering, and displays the frames decoded during the previous decoding stage. These are displayed in sync with the audio according to their time stamps. This means that we are always decoding all the frames from a GOP that can be decoded before displaying any of them.

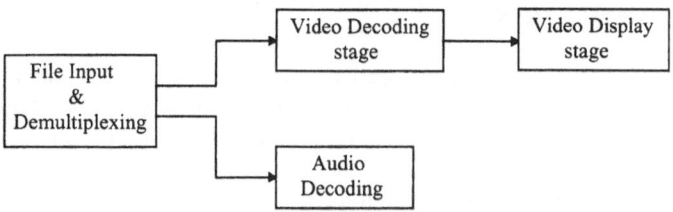

Fig 1. PowerPC Mpeg decoder architecture.

The decoding stage proceeds as follows:

1. Scan the input video stream from the current I frame to the next one.
   This is typically encoded as a GOP (Group Of Pictures), though not necessarily. In this paper GOP will always refer to this structure.
2. For the sake of concreteness, we consider as sequence with an IPB structure
   I B B P B B P B B P B B
   every half a second; this is very typical. Let us also assign indices to the P and B frames so that the above sequence is
   I $B_{1,1}$ $B_{1,2}$ P_1 $B_{2,1}$ $B_{2,2}$ P_2 $B_{3,1}$ $B_{3,2}$ P_3 $B_{4,1}$ $B_{4,2}$
   Prioritize the frames as follows (earlier entries in the sequence have higher priority)
   I P_1 P_2 P_3 $B_{1,1}$ $B_{2,1}$ $B_{3,1}$ $B_{4,1}$ $B_{1,2}$ $B_{2,2}$ $B_{3,2}$ $B_{4,2}$
3. Decode frames in the prioritized order.
   The order was designed to ensure that when a predicted frame is to be decoded the reference frames used for the prediction are already decoded.
4. When the display stage of the pipeline signals that it finished displaying the frames from the previous GOP and it is time to display the current GOP, then the decoding stage stops decoding frames from the current GOP and proceeds to process the next GOP.

The display stage performs the following:

1. Manage a queue of decoded frames in display order. When a frame is decoded it is inserted into a queue waiting to be displayed.
2. When is time to display the next frame in the queue, it is dithered and displayed.

This scheme is very flexible in the sense that it allows prioritized frames to be decoded in an order different from the display order. It will never decode a B frame if a following P frame in the same GOP is not, even in the case that the CPU load changes. The load of the CPU can change by the CPU requirements of other programs running in the system.
Furthermore, different B frames are given different priorities. This allows for great maneuverability in our frame dropping strategy, and also assures a smooth looking display even a very low frame rates, as long as no P frames are dropped. In extreme cases where P frames need to be dropped, our prioritization chooses to drop the frames that lead to the maximum frame rate possible.

APTIVA IMPLEMENTATION

The target environment is a Pentium based PC with graphic acceleration for color conversion and display, and an Mwave DSP for decoding audio. The system is configured with only 8M of memory. Schemes like the previous one for PowerPC, which required large memory, have to be avoided. In order to keep memory requirements as low as possible, frame buffering should be minimal and frame dropping decisions need to be made on the fly. We have kept to our basic principles of dropping B frames first and then only "last" P frames in a GOP. In this different strategy I frames and B frames are always displayed according to their time stamps, but in very low frame rates P frames may be displayed slightly out of sync. This decision is acceptable because in the case that we must drop P frames there is a visually perceptible discontinuity which can be minimized by displaying P frames at slightly different time stamps. The penalty incurred is a jump in the motion, when last displayed P frame is very close in time to the next displayed I frame.

Define the following terms:

Tdisp(frame) - is the time when "frame" has to be displayed.

This time is computed from the video PTS values and the video frame rate.

Tdec(frame) - is the time when "frame" has to be decoded and ready for displaying.

For B frames Tdec(B frame) and Tdisp(B frame) are equal, but for I and P frames these values can be different.

Consider for example the sequence of displayed frames

$$B_{-2} \, B_{-1} \, I_0 \, B_1 \, B_2 \, P_3 \, B_4 \, B_5 \, P_6 \, B_7 \, B_8 \, P_9 \ldots$$

The decoding sequence, assuming all these frames are indeed decoded and displayed, is

$$I_0 \, B_{-2} \, B_{-1} \, P_3 \, B_1 \, B_2 \, P_6 \, B_4 \, B_5 \, P_9 \, B_7 \, B_8 \ldots$$

The I and P frame have to be decoded considerably earlier than their display. For I and P we set Tdec(frame) = Tdisp(previous I or P frame). That is, when is time to display an I or P frame the previous I or P frame must already be decoded.

Time - is the current time.

This time is synchronized to the audio real-time clock.

Pred(I), Pred(P), Pred(B) - are the predicted decoding times for I, P and B frames respectively. These times are predicted from previous decoded frames of similar types.

The compressed video stream is scanned ahead to be able to get the position information of the next frames in the current GOP; the buffering requirement for this is minimal.

The frame dropping decision is done on the fly according to the following rules:

- a B frame will be decoded only if it could be displayed on time and the next P or I frame (whichever comes first) could be decoded on time. Using our notation,

```
if ( Time + Pred(B) < Tdisp(B frame) &&
     Time + Pred(B) + Pred(P) < Tdec(next P) )
then decode frame
else drop frame
```

- a P frame will be decoded only if the next I frame could be decoded on time.

    if (Time + Pred(P) + Pred(I) < Tdec(next I))
    then decode frame
    else drop frame

This rule handles the (unpleasant) case where a P frame must be dropped. If it is time to start decoding the next GOP, all the remaining frames from the current GOP will be dropped.

- an I frame will be decoded only if the next I frame could be decoded on time.

    if (Time + Pred(I) + Pred(I) < Tdec(next I))
    then decode frame
    else drop frame

If for some reason the video decoder got starved from CPU and the video fall behind the audio, this rule will allow to drop whole GOPs in order to allow a fast video update.

It is possible that even though there is sufficient time to decode all I and P frames in a GOP, the on-the-fly decisions of the decoder will decode too many B frames, thereby causing P frames to be dropped. However in our scheme this is very unlikely; this is borne out in our actual implementation.

The performance of this scheme heavily depends on the values of the time predictions Pred(I), Pred(P), Pred(B). These values must be continuously updated in order to accurate reflect the varying decoding time statistics of the video. The advantages of this scheme are its simplicity and low memory and cpu requirements. It is implemented in IBM's Aptiva SoftMpeg product, achieving very good results and smooth looking video playback under different load conditions.

CONCLUSIONS

The bottom line in MPEG playback is perceptual quality. In both environments we achieve much better quality than existing players such as AVI or Indeo. The main reason for this is that MPEG at CDROM rates of 1.5 Mb/sec provides very good quality SIF resolution video at full frame rates (24 - 30 frames per second). Our contributions are enabling software MPEG to run at a reduced frame rate using frame-dropping strategies while providing real time synchronization between the audio and video, and which intelligently decide which frames to drop so as to minimize visual degradation due to jumpiness. Coupled with our previous work on fast algorithms[3,4], we are able to provide a sufficiently high frame rate, well synchronized, so that the overall perceptual quality is pleasing.

REFERENCES

1. MPEG proposal package description, *Document ISO/WG8/MPEG/89-128*, July, 1989.
2. LeGall, MPEG: a video compression standard for multimedia applications, Communications of the *ACM, vol. 34, pp. 46-58*, April 1991.
3. Feig and S. Winograd, Fast algorithms for the Discrete Cosine Transform, *IEEE Trans. Signal Processing, Vol. 40 No. 9, pp. 2174-2193*, Sept. 1992.
4. E Feig, A fast scaled DCT algorithm, *Proc. SPIE Image Processing Algorithms and Techniques, Vol. 1244, R. J. Moorhead and K. Pennington, Ed.*, February 1990.

ADAPTIVE SYNCHRONIZATION
IN REAL-TIME MULTIMEDIA APPLICATIONS

Changdong Liu, Yong Xie, Myung J. Lee, and Tarek N. Saadawi

Electrical Engineering Department
The City College of the City University of New York
New York, NY 10031
mjlee@ee-mail.engr.ccny.cuny.edu

INTRODUCTION

Important work on synchronization has been done by many researchers in many aspects. Although some synchronization schemes perform pretty well under certain conditions, they reveal limitations under other conditions. For example, Alvarez—Cuevas, et al. provide an adaptive technique to achieve minimal voice delay, but require special measurement packets to estimate the mean network delay at the beginning of each silence interval.[1] Escober, et al. intend to give general solutions to the synchronization problem, but they are based on synchronized network clocks and assume that messages are delivered reliably at least for initialization.[2] Katseff and Robinson's work copes with short-term network congestion, but applies only to delivering stored multimedia to a user.[3] The work done by Ramanathan and Rangan minimizes perceptible degradations in the quality of media playback, but needs feedback.[4] Zarros' solution applies to a wide spectrum of applications, but needs pre-collection of network statistics.[5] In contrast, our synchronization mechanism is adaptive to network changes, works with any underlying delay distribution, and does not need a global clock or synchronized clocks, while it provides optimal delay/buffering for given QoS requirements (delay, jitter, and loss ratio).

The synchronization issues we discuss here are at the application level. In the context of TCP/IP, it is above TCP/UDP sockets.

MULTIMEDIA SYNCHRONIZATIONS

We address multimedia synchronization problem in two aspects: intramedium synchronization and intermedia synchronization. For real-time applications like teleconference, absolute delay and delay jitters are major concerns. Even when only a single medium is involved, the network inevitably introduces some variations in the delay of the delivered objects. At the receiver side the application attempts somehow to faithfully recover the object. Faithfully recover in the context of synchronization means that the temporal relationship among data at the source is preserved when they are played back at the sink. How to do it is the very question that should be answered by the synchronization mechanism. For intramedium synchronization the most intuitive solution may be to buffer the incoming data and then replay them at some fixed offset delay from the original generation time at the source so that every object experiences an equalized delay. This fixed offset delay must be larger than the maximal possible delay that any data may experience. In this sense synchronization is nothing but adding the controllable delay.

Our principle to solve the synchronization problem is to compromise between fully deterministic guarantees and best-efforts. Referring to Figure 1, define the playback point as the point in time which is offset from the original generation time by a certain delay. To faithfully recover the original object, we may use a fixed offset delay larger than the maximum end-to-end delay (the line F(t) in Figure 1) to avoid QoS degradation from late objects provided the channel is perfectly reliable. However, this is an expensive solution especially when the tail of the delay distribution is thin and long, which means only a few objects experience extremely large delay. Also, large end-to-end delay is something should be avoided in real time applications. If the channel is not reliable and/or the user can tolerate data loss, a maximal offset may be set such that the data arriving before its associated playback point can be used to reconstruct the object and the data not arriving by the playback point is considered lost even it may arrive at a later time. When the application affords a certain data loss, we may find the minimal playback point that guarantees the loss ratio below a given limit, as indicated by line R(t) in Figure 1. Note that objects are lost. To do this some *a priori* information like characteristics of the traffic and statistics of the network delay and jitter is appreciated. However, in real life, it is hardly feasible to acquire these kinds of information. Approximate statistics may be deduced from actually measured network delay in the recent past, which normally involves time stamping every object at the source and registering arrival time of each object at the sink, resulting in large computational overhead. Factors such as clock offset and frequency drift of clocks, and the number of samples upon which the statistics are derived affect both the accuracy of the statistics and its cost.

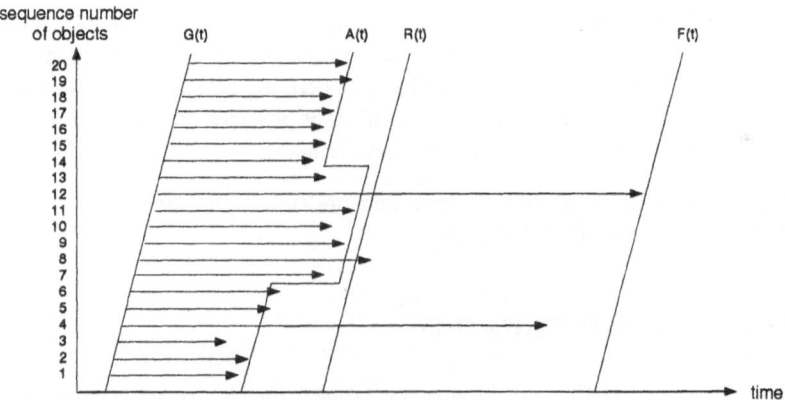

Figure 1. Playback times set up based on different criteria. G(t): objects generation time; A(t): adaptive real-time playback; R(t): real-time playback; F(t): faithful playback.

By carefully examining the synchronization problem in perspective, we make two arguments: First, the strongest temporal relationship among objects exists between adjacent ones since correlation between any two objects decreases as the time interval between them increases. Therefore, it is meaningless to retain temporal relation for two objects which are far apart in time. Second, small error in restoring the temporal relationship is tolerable due to the human perception. Actually, because so many random factors are involved, introducing some errors in synchronization is inevitable.

If the constraint of *being faithfully recovered* can be relaxed, we can exploit a lot from the two arguments made above. From the first: instead of using a unique reference time point, the application can adjust it from time to time. From the second: instead of dropping all the data arriving after the playback point (which creates incomplete signal) the application may still accept those data that arrive *a little bit* after the playback point (e. g. the object with respect to $R(t)$ in Figure 1) for recovering the object even if the object is somewhat distorted. This is useful since obtaining something is better than nothing in majority of applications like packet voice. For example, in teleconference the users generally favor cadence altered voice over the choppy, incomprehensible one. Our solution to the synchronization problem is: *to find the optimal playback point based on which the original object can be* piecewisely *faithfully recovered with acceptable distortion while the loss ratio is kept under a given limit.* "Optimal" is in the sense that while the QoS is ensured, the overall average end-to-end delay, consequently the buffer requirement, is minimal. "Piecewisely" implies that instead of always using one fixed offset delay an *adaptive offset delay* is employed. Line $A(t)$ in Figure 1 features this adaptability. Initially, the playback point could be better set with certain knowledge of the network conditions and dynamically adjusted thereafter. However, since knowing network conditions to any extent is expensive or even impossible in some situations, the adjustment mechanism should be intelligent enough to quickly converge playback point to the optimal state even with an arbitrary initial set. This adjustment should not be based on the worst case assumptions on the network and traffic behavior, rather is computed with proper predictions in response to the actual network delay in the *recent past*. To recover the object as completely as possible, this adjustment should respond to network degradation quickly to avoid serious object loss and rather be conservative to network improvement to avoid excessive surging in playback.

ADAPTIVE SYNCHRONIZATION ALGORITHM

The proposed synchronization algorithm conducts buffering through a *PlayBack Clock* (PBC), three event-counters, and an associated controller. The PBC is a virtual clock that emulates the clock at the sender site. The motivation to have the PBC is that once the source clock can be reproduced at the sink, the synchronization problem is readily solved.

The synchronization scheme requires the user specify the maximum acceptable synchronization error, E_{max}, (the difference between the generation time intervals at sender and playback time intervals at receiver of any adjacent objects), the maximum acceptable jitter, J_{max}, (variance of the synchronization error), and the maximum acceptable object loss ratio caused by synchronization measures,[*] L_{max}. At the sender each object carries a time stamp indicating its generation time in reference to the *Source Clock* (SC). At the

---

[*] It is possible to detect objects loss in the network, and there are also some countermeasures against excessive loss like asking the sender slow down. However, those congestion control issues are out of the scope of this paper and therefore not discussed here.

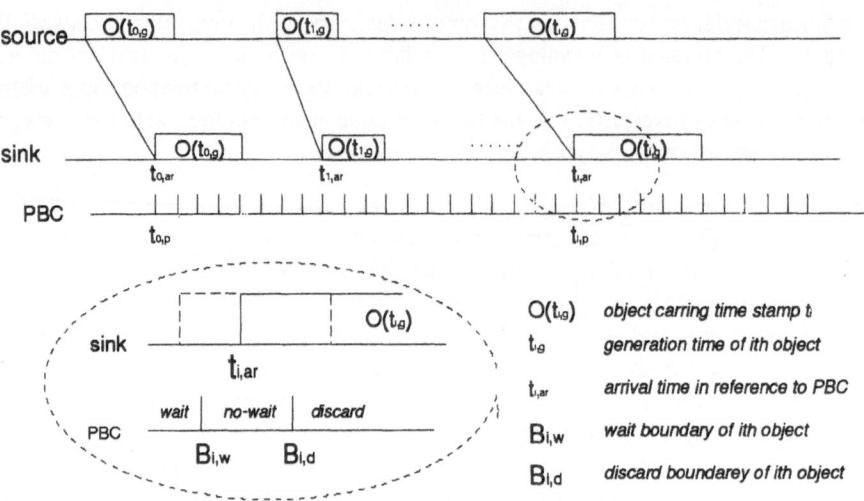

Figure 2. Temporal relations among sender, receiver and PBC.

receiver a PBC and three event-counters, namely *wait counter* C_w, *no-wait counter* C_{nw} and *discard counter* C_d, each with threshold T_w, T_{nw} and T_d respectively, are maintained.

The PBC is triggered by the arrival of the first object and its initial time is set to the time stamp carried by the first object, and then ticks with the receiver clock. At this time the absolute offset between SC and PBC equals the end-to-end delay of the first received object. Due to the stochastic behavior of the object delays, the PBC must be adjusted since then. The vicinity of an object's arrival time in reference to the PBC time is partitioned by the *wait boundary* (B_w) and *discard boundary* (B_d) into three regions: *wait* region, *no-wait* region and *discard* region (refer to Figure 2). These two boundaries are defined for each object carrying time stamp $t_{i,g}$ as

$$B_{i,w} = t_{i,p} = t_{i,g}$$
$$B_{i,d} = B_{i,w} + E_{max} \tag{1}$$

where $t_{i,p}$ denotes the scheduled playback time in reference to the PBC, which equals to $t_{i,g}$ in value. The arrival time $t_{i,ar}$, in reference to the PBC, of the object carrying time stamp $t_{i,g}$ may fall into one of the three regions with respect to its associated two boundaries. The synchronization algorithm conforms to the following rules:

(i) If the object with time stamp $t_{i,g}$ arrives before $B_{i,w}$ ($t_{i,ar}<t_{i,p}$, within the wait region), then it will be played back at $t_{i,p}$ (waiting until the wait boundary).

(ii) If the object with time stamp $t_{i,g}$ arrives after $B_{i,w}$ but before $B_{i,d}$ ($t_{i,p}<t_{i,ar}<t_{i,p}+E_{max}$, within the no-wait region), then it is played back immediately.

(iii) If the object with time stamp $t_{i,g}$ arrives after $B_{i,d}$ ($t_{i,ar}>t_{i,p}+E_{max}$, within the discard region), then it is discarded.

The basic synchronization algorithm is:

Step 1. Initiate the PBC upon receiving the first object and set the initial PBC time equal to the time stamp carried by this object.

Step 2. Upon receiving the ith object, compare its time stamp, $t_{i,g}$, with its arrival time (the current PBC time), $t_{i,ar}$. If $t_{i,g} > t_{i,ar}$, increase the wait counter by one and do not playback the object until $t_{i,p}$. Else if $t_{i,g} \leq t_{i,ar} < t_{i,g}+E_{max}$, increase the no-wait counter by one and playback the object immediately. Otherwise (i. e. $t_{i,ar} \geq t_{i,g}+E_{max}$) increase the discard counter by one and discard the object.

Step 3. Check if the most recently increased counter overflows. If no, go to Step 2. Otherwise continue.

Step 4. When the no-wait counter or the discard counter overflows, check whether the wait counter is full. If no, decrease the PBC: $PBC(t) = PBC(t) - \Delta$. Otherwise go to Step 2.

When the wait counter overflows, check whether the no-wait counter is full. If no, increase the PBC: $PBC(t) = PBC(t) + \Delta$. Otherwise go to Step 2.

This algorithm functions a typical low-pass filter transforming individual object's arrival behavior into the events distribution. Therefore, the thresholds of the three counters are critical to the performance of this synchronization algorithm. Their summation determines the bandwidth of the filter, and consequently the sensitivity of the synchronization control. The ratios among them are to be determined with certain knowledge of the underlying network delay distribution. Nonetheless, we can compute T_{nw} with the worst case scenario without any *a priori* information.

Careful readers should have already realized that our synchronization algorithm is inherently immune to clock frequency drift since its effect can be treated as one of the contributing factors to network delay jitter. To see this, assume network delay has no jitter. If the clock at the sender were synchronized with the one at the receiver, the receiver should always receive an object carrying time stamp $t_{i,g}$ exactly at $t_{i,p}$. When the clock at the sender is faster than the one at the receiver, the receiver should find the objects always fall into the wait region, i. e. $t_{i,g} = t_{i,p} > t_{i,ar}$. This would soon cause the wait counter go beyond its threshold while the no-wait counter and discard counter remain empty. Once this happens the PBC would be increased to track the clock at the sender. Should the clock at the sender be slower than the one at the receiver, it would be handled accordingly. Moreover, the objects received out of sequence will be put in order automatically as they go through the synchronization process.

There are a couple of choices in determining the amount of adjusting the PBC. The simplest is linear interpolation using the ratio of the counter contents. Using more sophisticated rules like soft threshold, e. g. use of a fuzzy controller or neural controller, will improve the performance further at the expense of the computational overhead.

APPLICATIONS

Applying our synchronization algorithm to intramedium situation is straightforward, and extending to intermedia situation is also easy. Assume a group of N media to be inter-synchronized. When the N media try to reach their own intramedium synchronization individually, N PBC's and N sets of counters are needed. The decision of adjusting each PBC is given by its associated set of counters. The adjustment for different PBC's are most likely different. When intermedia synchronization is engaged, all the N PBC's are compared, and the slowest one will dominate the playback of all media in the synchronization group.

This intermedia synchronization scheme allows the application flexibly specify which media are subject to intermedia synchronization and conveniently switch a selected medium

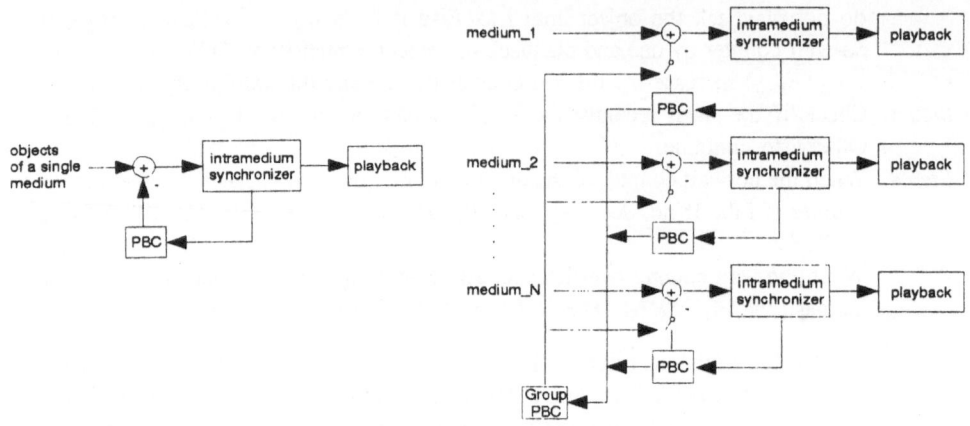

Figure 3. Implementation of intramedium synchronization and intermedia synchronization.

in and out an intermedia synchronization group. This functions are useful in many scenarios like teleconference.

EXPERIMENTS

The adaptive synchronization algorithm has been applied to a multimedia conferencing system in our lab for both intramedium and intermedia synchronizations. The experimental results are encouraging even in changing network conditions.

Our experimental network testbed consists of an 100 Mbps FDDI as backbone, 3 servers (Sun workstation 4/330) connected to FDDI, two 10 Mbps Ethernet segments, each connecting several Sun IPC's and SLC's. Each of the hosts taking part in the multimedia experiment resides on a different Ethernet.

Audio and Video Traffic

We used UDP to transfer video and audio messages and measured delays for audio and video objects for 16.7 minutes (1000 seconds) between two multimedia hosts while the microphone at the sender keeps on picking ambient audio signals and the video camera keeps on capturing a still scene and occasionally passing-by people. Under this condition, delays are mainly caused by processing overhead. We measured the delays experienced by audio and video objects, from the time when an object is created to the time when the object is arrived at the receiver before synchronization (hereafter network delay). For audio, the average delay is $\eta^a = 6.9\ ms$ and the variance is $(\sigma^a)^2 = (3.6\ ms)^2$. For video, the average delay is $\eta^v = 69\ ms$ and the variance is $(\sigma^v)^2 = (10.7\ ms)^2$.

As for the user provided QoS, we assume L_{max}=0.01, E_{max}=2 ms, and J_{max}=(0.8 ms)^2. We chose T_w=800, and calculated T_d=8, and T_{nw}=200. After applying the intramedium synchronization algorithm, the average end-to-end delay of audio is $\eta^a_{intra} = 10\ ms$ and its variance is $(\sigma^a_{intra})^2 = (0.75\ ms)^2$. The average end-to-end delay of video is $\eta^a_{intra} = 85\ ms$ and its variance is $(\sigma^a_{intra})^2 = (1.4\ ms)^2$.

The synchronization algorithm works equally impressive when the network condition changes. We add background traffic by having another set of hosts keep sending dummy files to the communicating multimedia hosts for 5 minutes. Figure 4 shows the network delay experienced by the audio objects under this condition with wait and discard boundaries indicated. This time, network delays caused by queuing at different layers along the transmission path became significant. During the period with the added background traffic both the minimal delay and delay jitter are increased. The intramedium synchronizer nicely tracks these changes as we expected: when the network delay increases the synchronizer follows up the jump very quickly (catch up with one jump), when the network condition improves the synchronizer responses with a conservative manner (reduce in two steps). Figure 5 shows the delays after the synchronization (i. e. the end-to-end delay). As evidenced by comparing Figure 4 and Figure 5, the variance of the delay is significantly reduced.

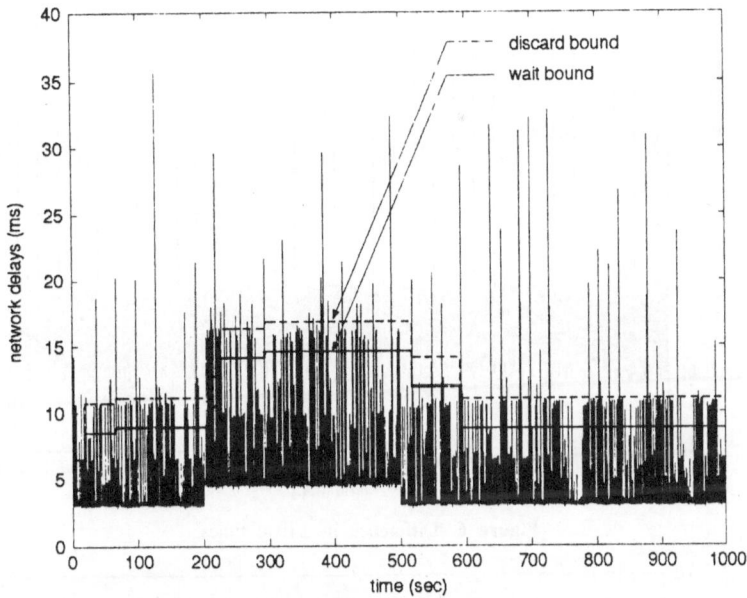

Figure 4. The delay behavior of audio objects when the network condition changes. The solid line is the wait boundary and the doted line is the discard boundary.

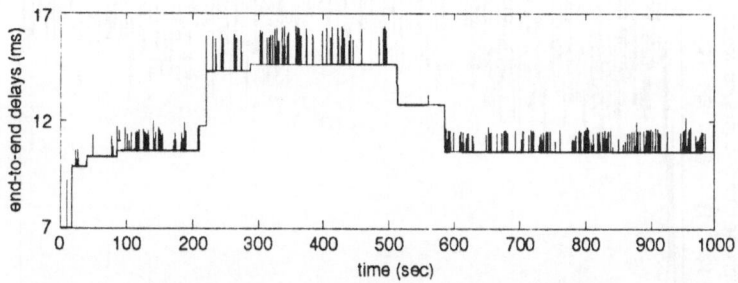

Figure 5. The end-to-end delay of audio objects when the network condition changes.

Intermedia Synchronization between Audio and Video

Audio and video objects are not generated exactly at the same time. Thus, there will be a inherent offset existing between the audio and video objects adjacent in time. The intermedia synchronization error is defined with respect to a video object and the adjacent audio object immediately ahead of it in time. Denote the generation times of this pair of objects, between which a temporal relationship is to be kept, as $t^v_{m,g}$ and $t^v_{n,g}$, and the arrival times of this pair of objects as $t^a_{m,ar}$ and $t^v_{n,ar}$. The temporal distortion after the network but before the synchronization is

$$\Delta t_{m,n} = \left(t^v_{n,ar} - t^a_{m,ar} \right) - \left(t^v_{n,g} - t^a_{m,g} \right). \tag{2}$$

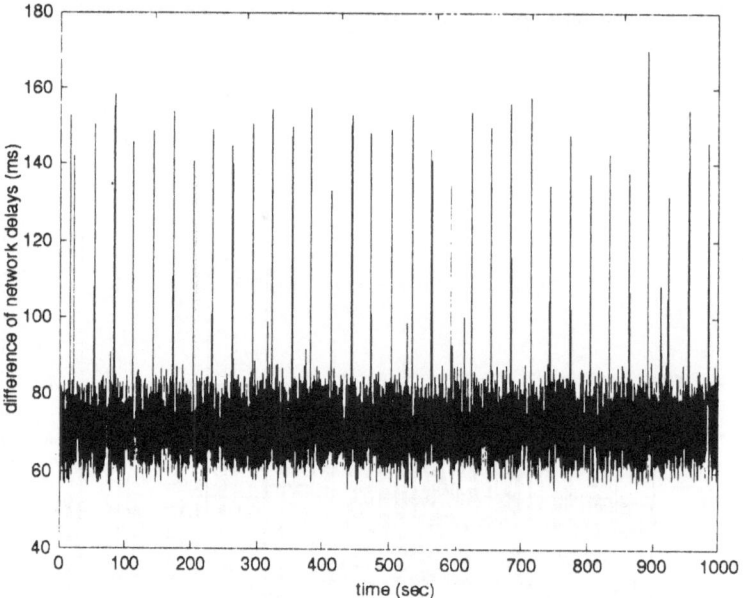

Figure 6. Difference in arrival times.

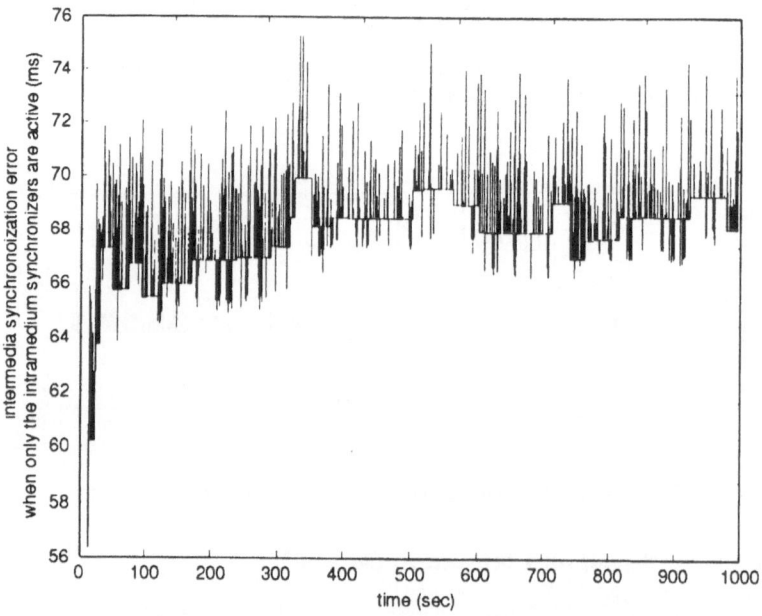

Figure 7. Asynchronousness between audio and video at afte the intramedium synchronizers but without the intermedia synchronizer.

154

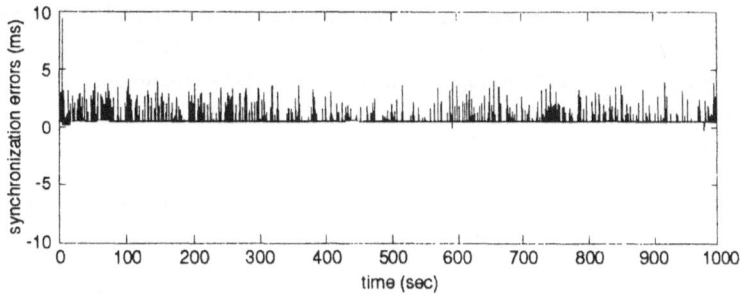

Figure 8. Intermedia synchronization errors when the intermedia synchronizer is active.

The intermedia synchronization error is defined as

$$\epsilon_{inter} = \left(t^v_{n,p} - t^a_{m,p}\right) - \left(t^v_{n,g} - t^a_{m,g}\right) \tag{3}$$

Figure 6 shows the $\Delta t_{m,n}$, as defined in Equation 2, for the arrivals of the audio and video objects. Figure 7 shows the intermedia synchronization error, as defined in Equation 3, between audio and video objects after they having passed through their individual intramedium synchronizer, but without intermedia synchronization. The magnitude and variance of the difference in arrival times have been reduced because the individual intramedium synchronizers discarded objects with abnormally long delay and smoothed each stream. Once the intermedia synchronizer is activated, the performance becomes remarkable, as noticed in Figure 8. All synchronization errors are less than 10 ms and the variance is within $(2 \text{ ms})^2$.

CONCLUSION AND FUTURE WORK

Major issues in the multimedia synchronization are discussed in this paper. Experimental results assured that the proposed synchronization algorithm is efficient and adaptive so that it may be used in many applications. In particular, the proposed adaptive synchronization algorithm features: adaptive to network condition changes, applies to general traffics of both periodic and non-periodic, no needs for a global clock, immune to clock frequency drift and object loss, optimal playback which satisfies QoS requirements with minimal buffer.

Extension to interparticipants synchronization algorithm and more rigorous performance modeling are to be studied. Transporting our multipoints teleconference system to an ATM LAN is underway.

REFERENCES

1. F. Alvarez-Cuevas, et al., "Voice Synchronization in Packet Switching Networks", *IEEE Network*, September, 1993.
2. J. Escober, et al., "Flow Synchronization Protocol", *IEEE/ACM Trans. on Networking*, Vol. 2, No. 2, April, 1994.

3. H. Katseff and B. Robinson, "Predictive Prefetch in the Nemesis Multimedia Information Service", *Proc. ACM Multimedia '94*, October, 1994.

4. S. Ramanathan and P. Rangan, "Adaptive Feedback Techniques for Synchronized Multimedia Retrieval over Integrated Networks", *IEEE/ACM Trans. on Networking*, Vol. 1, No. 2, April, 1993.

5. P. Zarros, et al., "Statistical Synchronization Among Participants in Real-time Multimedia Conference", *Proc. of IEEE Infocom '94*, pp912–919, Toronto, CA, 1994

A SYNCHRONIZATION MECHANISM FOR MULTIMEDIA PRESENTATION

Ying Cai[1] and Grammati E. Pantziou[2]

[1]Computer Science Department, University of Central Florida
Orlando, FL 32816-2362
cai@cs.ucf.edu
[2]Computer Technology Institute, P.O. Box 1122
26110 Patras, Greece
pantziou@cti.gr

Abstract

The Petri Net-based synchronization models impose strict synchronization requirements among multimedia objects. This may introduce unecessary delays, and may increase the presentation time. In this paper, we propose a hybrid model that uses a Petri Net-based model to describe strict synchronization requirements among multimedia components, and a lower-level mechanism to describe *weak* synchronization requirements, in the case that strict synchronization is not necessary. Skipping of objects which arrive late, is modeled. The model supports both inter- and intra-stream synchronization, it is simple to use, and may reduce the delays, and also, the size of the buffers needed to handle slow streams.

1 Introduction

One of the most important requirements of a Distributed Multimedia Information System is the composition of multimedia data from distributed sites and the presentation of the composite multimedia objects to the user. A composite multimedia object usually has specific temporal relationships among the different types of multimedia components. The temporal relationships among different multimedia objects should be explicitly formulated for ensuring proper scheduling of the synchronized presentation. Evaluating the temporal relationships of the multimedia objects, and scheduling their presentation is known as *synchronization.*

Multimedia data is scheduled for transmission according to its media-specific Quality-of-Service (QoS) parameters. However, due to nondeterministic behavior of the network, random network delays should be taken into account while generating a mechanism for synchronization in order to preserve the timing presentation constraints

Multimedia Communications and Video Coding
Edited by Y. Wang *et al.*, Plenum Press, New York, 1996

of the multimedia data. Several methods have been suggested in the literature to address the problem of modeling the temporal relationships of the multimedia components of a presentation as well as specifying the mechanism for synchronization over a network in a distributed environment (e.g., [3, 7, 2, 4, 5, 9, 6, 8, 10, 11]).

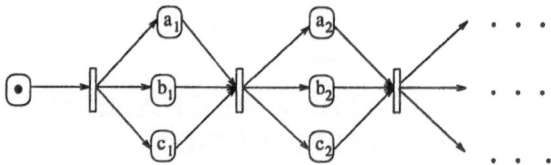

Figure 1: An example of a Petri Net-based specification of the synchronization requirements among the objects A, B, and C.

Synchronization is in general of two types: Inter-stream and intra-stream synchronization. Inter-stream synchronization is required where there are temporal relationships between two (or more) multimedia components of a presentation (e.g., audio and images in the case of a video presentation). Intra-stream synchronization is a low level synchronization and is concerned with delivering each object of a multimedia component in time to meet its playout deadline by smoothing out network delays and jitter in this multimedia stream. Note that buffers of appropriate size need to be used to handle streams in which the packets do not arrive at the destination in time to meet their respective playout deadlines, and therefore, correction actions need to be taken.

1.1   The problem and our proposal

A number of modifications of earlier Petri Net models (e.g. Timed Petri-Net (TPN) [1]) have been suggested for both types of synchronization. We mention here the Object Composition Petri Net (OPCN) model [3], and the eXtended OCPN (XOPCN) model [7]. In these models, strict synchronization points are defined among different multimedia objects and therefore, the temporal synchronization is guaranteed. For example, consider a multimedia presentation that involves three multimedia components $A = \{a_1, a_2, \cdots\}$, $B = \{b_1, b_2, \cdots\}$ and $C = \{c_1, c2, \cdots\}$ where a_i, b_i and c_i, $i \in \{1, 2, \cdots\}$, are objects of the multimedia components A, B, and C, respectively (e.g., if A is the audio component in a video presentation then a_i is an audio sample). Suppose that using a Petri Net-based model we specify strict synchronization points among the objects a_i, b_i and c_i, for $i \in \{1, 2, \cdots\}$. Then a_{i+1}, b_{i+1}, and c_{i+1} can be presented only after the presentation of a_i, b_i, and c_i has been completed, for each $i \in \{1, 2, \cdots\}$ (see Figure 1).

However, since a_i, b_i, and c_i may come from different sources in the network, it is hard to guarantee that these objects will arrive at the presentation cite simultaneously. Even for the same source, the different packets may arrive at the destination out of the sending order, because they may be transmitted via different network routes. For example, it is possible for a_2 to arrive before a_1, although a_1 was sent before a_2. Therefore, a Petri Net-based model may introduce delays in the presentation due to the fact that the presentation of the next triple of objects a_{i+1}, b_{i+1}, and c_{i+1} is delayed until all of a_i, b_i, and c_i are presented. On the other hand, strict synchronization as it is guaranteed by the a Petri Net-based model, is not always necessary. For example, in the case of video presentation where we have image, audio, and text components, we may need strict synchronization between images and text but we may not need it between images and audio. Therefore, we may reduce the delays, and also, the actual presentation time by allowing a type of "weak" synchronization among components that do not require strict synchronization.

In this paper, we propose a hybrid model that uses a Petri Net-based model, and a lower level register-based mechanism to provide both strict synchronization and weak synchronization (in the cases that strict synchronization is not necessary). More specifically, a two-level model is proposed where at the first level, a Petri Net-based model is used to describe strict synchronization requirements among multimedia components and/or objects, while at the second level a lower-level mechanism is used to describe weak synchronization requirements. Skipping of objects which arrive late, is modeled. The model supports both inter- and intra-stream synchronization, it is simple to use, and may reduce the size of the buffers needed to handle slow streams.

2 The proposed synchronization mechanism

In this section we describe our mechanism for synchronization. We first give some terminology. We say that the *flexibility* of the multimedia component A with respect to (w.r.t.) the multimedia component B is n, if the currently presented object of A may precede the currently presented object of B by at most n time units. If the flexibility of A w.r.t. B is zero, and the flexibility of A w.r.t. B is zero, then the multimedia components A and B are *strictly synchronized*. As an example consider three multimedia components $A = \{a_1, a_2, \cdots\}$, $B = \{b_1, b_2, \cdots\}$ and $C = \{c_1, c_2, \cdots\}$, where a_i, b_i, and c_i, $i \in \{1, 2, \cdots\}$, are multimedia objects of the components A, B, and C, respectively. Suppose that the flexibility of A w.r.t. B and C is one, the flexibility of B w.r.t. A and C is zero, and the flexibility of C w.r.t. A and B is zero. Then, if the presentation of b_i, c_i, a_{i+1} is completed, a_{i+2} can go ahead. But a_{i+3} must wait until b_{i+1} and c_{i+1} are done. Since B and C are strictly synchronized, they cannnot precede each other, and also they cannot precede over A.

We associate each object of a multimedia component with a unique integer identification, called *Object Indetification (OI)*. The playout durations of objects of different multimedia components may be different, and these durations are supposed to be preknown. If an object a follows an object a' in the multimedia component, then the OI of a is greater than the OI of a'. I.e., the OIs of the objects follow the timing order of the objects in the multimedia component. In our example, we assume that the OI value of a_i is i. Each object α in a multimedia component A, is also associated with a variable, storing the *Beginning Time Unit (BTU)* of α, i.e., in how many time units after the start of the presentation of the first object of A, the presentation of α should start. If the BTU of α is t, then the presentation of α should start t time units after the start of the presentation of the first object of A.

We associate each multimedia component with two registers. One is called *Proceeding State Register (PSR)*, showing how many time units the stream has been played. The other is called *Object Identification Register (OIR)* and keeps the most recently presented object of the component. The information kept by these registers is used to schedule the playout times of the objects of the multimedia component.

A Petri Net-based model is used to define at a first level, the temporal relationships among related multimedia components. In this way, strict synchronization requirements for certain multimedia objects are imposed by the firing rules of the Petri Net-based model. Between points of strict synchronization, as they are specified by the Petri Net-based model, lower-level mechanisms are activated to impose weak synchronization requirements among different objects of the multimedia components. For each one of the multimedia components involved in the presentation, a procedure is invoked parameterized with the flexibility of the multimedia component with respect to the other components involved in the strict synchronization requirements imposed by the Petri Net-based model, at the first level of synchronization. To be more specific, consider the following case. Let PSR_A (OIR_A) be the PSR (resp., OIR) of the multimedia component A. Suppose that using a Petri Net-based model, we specify that the components A, B and C will be presented simultaneously (i.e., we have strict synchronization requirements among A, B, and C). For the sake of simplicity, let the flexibility f of A w.r.t. B be equal to the flexibility of A w.r.t. C. Suppose also that objects of A can be skipped when they arrive late. Then, when an object α of A arrives, the following procedure is activated.

if $(OIR_A > OI$ of $\alpha)$ **then** discard α; /* skip α since some following object has been presented */
else
    minPRS=$min\{PRS_B, PRS_C\}$; /* the least value of PRS in all other
    multimedia components that are strictly synchronized with A */
    $PRS_A = BTU$ of α;
    $OIR_A = OI$ of α;
    while (the presentation of α is not completed)
        while $(PRS_A <=(minPRS+f))$ /* the object is being presented */
            playout one time unit of the object;
            minPRS=$min\{PRS_B, PRS_C\}$; /*minPRS may change each time unit*/
            $PRS_A + +$;
        if $(PRS_A >(minPRS+f))$ **then** wait one time unit;
        minPRS=$min\{PRS_B, PRS_C\}$;

Recall that in our example the flexibility of the component A w.r.t. B and C is one. Therefore, when a_3 arrives, first we check if any following object (e.g., a_4) has been presented. If yes, then a_3 should be skipped. Otherwise, we check whether a_3 proceeds too many time units or not, by checking the PRS of B and C. If yes, a_3 needs to wait. Otherwise, a_3 is presented while OIR_A and PRS_A are updated.

If the flexibility of a multimedia component A w.r.t. a component B is greater than zero, and the flexibility of B w.r.t. A is also greater than zero, then in the case that there are delays in both streams, a weak type of synchronization that exploits the flexibility of the components, may reduce the time of the whole presentation. To see that consider the following example. Let A consist of four multimedia objects a_i, $i \in \{1, 2, 3, 4\}$, with duration of one time unit each, and let B consist of five multimedia objects b_i, $i \in \{1, \cdots, 5\}$. Suppose that the flexibility of A w.r.t. B is 2 and the flexibility of B w.r.t. A is also 2. Consider that the objects a_i, $i \in \{1, 2, 3, 4\}$, arrive at the time units 1, 2, 3 and 7 respectively, while the objects b_i, $i \in \{1, 2, 3, 4, 5\}$, arrive at the time units 1, 3, 4,5 and 6, respectively. It is easy to see, that using our weak synchronization mechanism, the playout of the objects will have been completed after 7 time units, while in the case that we specify strict synchronization between the objects a_i and b_i, then 8 time units are required.

A procedure simpler than the one used above for inter-stream synchronization, may be used for intra-stream synchronization. We say that the *flexibility* of a multimedia component A *w.r.t. itself* is n if the OI of the currently presented object may proceed the OI of the last presented object of A by at most n. When an object a of A arrives, then the value of the OIR of A may be used to decide whether a will be presented or not. For example, let the flexibility of A w.r.t. itself be 4. Suppose that a_{89} arrives, and the last presented object of A is a_{84}. Then, the presentation of a_{89} is blocked until one of a_{85}, a_{86}, a_{87} or a_{88} arrives. Different policies may be implemented to handle slow streams. We mention here the blocking policy, and the restricted blocking policy (see [7]). These policies should be appropriately modified to take into consideration the flexibility of the multimedia component w.r.t. itself.

Note that the proposed synchronization mechanism may reduce the size of the buffers required to handle slow streams or streams whose packets arrive out of the sending order. This is achieved since the mechanism reduces the delays in the presentation of multimedia objects, and therefore, the time period during which the multimedia objects remain in the buffers.

References

[1] J.E. Coolahan Jr. and N. Roussopoulos, "Timing Requirements for Time-Driven Systems Using Augmented Petri Nets", *IEEE Trans. Software Eng.*, vol. 9, Sept. 1983, pp 603-616.

[2] W.H. Leung, T.J. Baumgartner, Y.H. Hwang, M.J. Morgan, and S.C. Tu, "A Software Architecture for Workstations Supporting Multimedia Conferencing in Packet Switching Networks", *IEEE Journal on Selected Areas in Communications*, April, 1990, pp. 380-390.

[3] T.D.C. Little and A. Ghafoor, "Synchronization and Storage Models for Multimedia Objects", *IEEE Journal on Selected Areas in Communications*, April, 1990, pp. 413-427.

[4] T.D.C. Little and A. Ghafoor, "Multimedia Synchronization Protocols for Broadband Integrated Services", *IEEE Journal on Selected Areas in Communications*, Vol. 9, No. 9, Dec. 1991, pp. 1368-1382.

[5] C. Nicolaou, "An Architecture for Real-Time Multimedia Communication Systems", *IEEE Journal on Selected Areas in Communications*, Vol. 8, No. 3, April 1990, pp. 391-400.

[6] B. Prabhakaran and S.V. Rangan, "Synchronization Models for Multimedia Presentation With User Participation", in Proc. *ACM Multimedia*, 1993, pp. 157-166.

[7] N.U. Qazi, M. Woo, and A. Ghafoor, "A Synchronization and Communication Model for Distributed Multimedia Objects", in Proc. *ACM Multimedia*, 1993, pp. 147-155.

[8] S. Ramanathan and P.V. Rangan, "Adaptive Feedback Techniques for Synchronized Multimedia Retrieval over Integrated Networks", *IEEE Transactions on Networking*, April 1993.

[9] D. Shepherd, D. Hutchinson, F. Garcia, and G. Coulson, "Protocol Support for Distributed Multimedia Applications", *Computer Communications*, Vol. 15, No. 6, July 1992, pp. 359-366.

[10] D. Shepherd and M. Salmony, "Extending OSI to Support Synchronization Required by Multimedia Applications", *Computer Communications*, Vol. 13, No. 8, September 1990, pp. 399-406.

[11] R. Steinmetz, "Synchronization Properties in Multimedia Systems", *IEEE Journal on Selected Areas in Communications*, Vol. 8, No. 3, April 1990, pp. 401-412.

ARCHITECTING VIDEO-ON-DEMAND SYSTEMS: DAVIC 1.0 AND BEYOND

Alexandros Eleftheriadis

Department of Electrical Engineering
Columbia University
New York, NY 10027
eleft@ctr.columbia.edu

INTRODUCTION

The culmination and convergence of several different technologies, including video compression and networking, are today making possible the design (and soon the deployment) of commercial interactive video services. Several commercial trials have been announced, are in progress, or are already completed. The multi-disciplinary nature of video-on-demand services, which includes content providers, service providers, network providers, cable TV distribution companies, as well as the computing and consumer electronics industries, has made it extremely difficult to concentrate within a single organization the expertise needed to design and implement such services. Attesting to that is the significant activity that has been created within the industry in terms of alliances and partnerships.

As with any prospective service of such anticipated scale, it is imperative to ensure that open and interoperable systems are used. This not only alleviates the problem of having to design vertical solutions, but also helps to stir healthy competition (technical and economic) within well-defined sub-system domains (e.g., among content providers or network providers). Most importantly, it allows for—and actually encourages—a fast-paced evolutionary path by easing the process with which innovations and new research results can be incorporated: having well-defined interfaces enables the substitution of existing subsystems with more sophisticated or better performing ones, without having to redesign the whole system.

We discuss the design of Video-on-Demand (VoD) systems, focusing on the efforts of the Digital Audio-Visual Council (DAVIC[1]). DAVIC, established in June 1994, is an association of more than 200 industrial and academic organizations from around the world, pursuing the definition of interfaces and protocols for interoperable VoD systems. DAVIC's scope is extremely broad, addressing the complete spectrum of vertical (bottom-up) and horizontal (end-to-end) specifications. It primarily performs a "systems integration" function, interconnecting several different technologies to create a complete and coherent system. As such, it attempts to distill all current experience about the design of VoD systems (actively monitoring and/or collaborating with related activities, such as ISO, ITU, the ATM Forum, IMA, and the IETF), as well as create solutions for problems that still remain unresolved. As a result, several recent developments in standardization activities (both official and industry-based) have been incorporated. This includes the ISO/IEC MPEG-2 and MHEG-5 specifications, ATM Forum and ITU-T specifications on ATM, the recently designed IP Version 6 (most often referred to as IPng, for "next generation"), CORBA 2.0, etc.

DAVIC expects to release its first set of specifications (DAVIC 1.0) in December 1995. A common and very important theme in DAVIC is to define a *single* solution for each functionality desired, thus eliminating the need to support multiple (and typically conflicting) specifications.

VoD is just one from a number of possible applications that can be supported from a video-capable communication system. In fact, by proper provisioning, a large number of different applications can be accommodated without overburdening the implementation. Furthermore, an open approach allows the introduction of new services that may not have been anticipated by the system's designers.[2,3]

The structure of the paper is as follows. We first give an overview of a System Reference Model, that defines the basic components of a VoD system. We then describe in detail each of these components, namely the server, the delivery system (core and access networks), and the subscriber terminal, as well the necessary protocols operating between them. We conclude by discussing media encodings and future directions of the DAVIC effort. The description is necessarily brief, and it should be noted that it is based on the current draft status of the DAVIC specifications; the reader is encouraged to consult the actual text of the specifications for more detailed and up-to-date information.

SYSTEM REFERENCE MODEL

The notion of a VoD system is quite broad, and has been used to denote quite different application domains. For our purposes, VoD indicates an interactive digital audio-visual service with a prevalent video component. Two fundamental design problems are the delivery of high-quality digital video to the end-user, and the handling of interactivity. What makes them particularly challenging is the desire for scalability, i.e., to be able to efficiently support a large number of users, as well as the interoperability constraint. The latter mandates the definition of several different and independent subsystems, the specification of their interfaces, and the verification of their interoperation. The interoperability constraint can be applied at all levels, from inter-regional systems (across regulatory domains), to individual protocol components.

The DAVIC system architecture is based on four basic entities: the end-user, the delivery system provider (or network provider), the service provider, and the content provider. The end-user interacts with the system using a subscriber terminal, or Set-Top Unit (STU). STU connectivity is provided by the delivery system provider using a possibly tiered approach: a core network that forms the backbone of the delivery system, and an access network that connects the STU to the core network (typically covering the last few miles). This tiered architecture is necessitated by the design of the physical networks that are already available to end-users (cable TV coax and telephone twisted pair), although an end-to-end ATM solution has distinct advantages.[4] Systems that can be directly connected to the core network (e.g., computers using high-speed connections) can have a much simpler architecture, incorporating the access network functionality (service selection etc.) at the end-system. Finally, the service providers offer access to stored or live content, while the content providers generate the programming material provided by the service.

Within each system, and across all of them, we can distinguish a number of different information flows and distinct interfaces. Figure 1 indicates the overall architecture, in the form of DAVIC's System Reference Model (SRM). The figure depicts a number of different information flows (S1–S4) and associated interfaces or reference points (denoted by A). The latter designate the entire protocol stack at the particular interconnection point they refer to. In the following sections each of the components is described in more detail. Our interest is the interfaces and protocols between components and not the component implementation details.

In order to properly define the operational aspects of the system, it is important to identify a common set of functionalities that are needed to implement a desired set of applications. The initial set of applications targeted for support in DAVIC 1.0 includes movies-on-demand, teleshopping, broadcast, near video-on-demand, delayed broadcast, games, telework, and karaoke-on-demand; the common functionalities needed to support these functions have been taken into account in the overall design of the system.

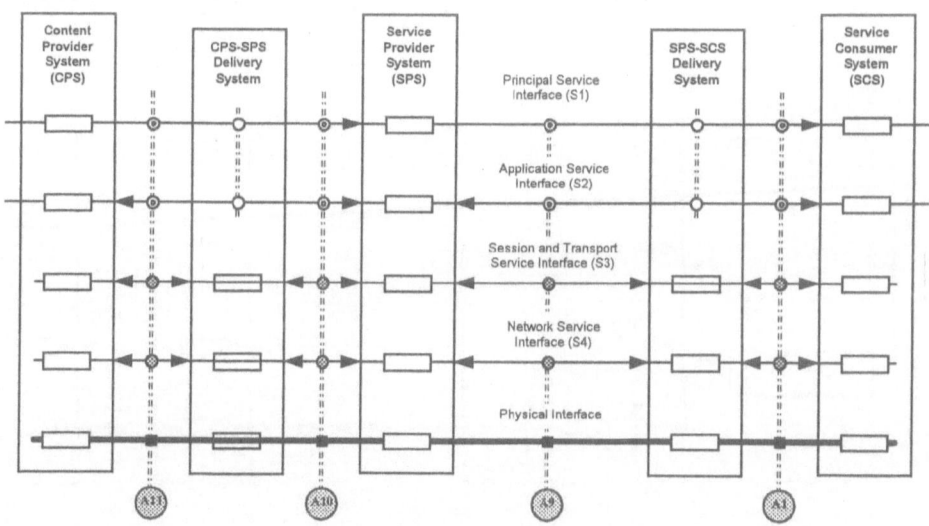

Figure 1. DAVIC System Reference Model. Also shown are interfaces (A1, A9, etc.), and information flows (S1–S4) across system components.

SERVICE PROVIDER SYSTEM

The server has four basic responsibilities: service gateway functionality, stream services, application services, and content services (Figure 2). Gateway functionality is a "brokerage" service that provides the means for registering and decommissioning services, allows the client (STU) to discover the existence of a service (e.g., an application for accessing movies-on-demand), and establishes and manages sessions. The stream services provides the repository and source for streams, i.e., content bitstreams. This service is initiated from the service gateway, when a content selection has been made by the user (e.g., to playback a movie). The application service is the key vehicle with which application functionality is provided, and an important source of added-value by individual service providers. Finally, the content services are responsible for content management, including loading and unloading between the content provider and service provider, as well as between service providers. These core services can be used to define additional ones, such as client profiling, file , and download services.

The application service interface (S2 flow) is based on an object-oriented architecture, centered around OMG UNO[5] (Universal Networked Objects) 2.0 and the default Internet Inter-ORB Protocol (IIOP). The session service interface (S3) is based on DSM-CC U-N (ISO/IEC 13818-6, Digital Storage Media Command and Control, User to Network part). Hence OSI Layers 4 and 3 for these interfaces utilize TCP and UDP over IP. The principal service interface is based on the MPEG-2 TS (Transport Stream) delivery. More details about protocol specifics are given later on.

All of the server internals are purposely left unspecified, since they do not affect interoperability (the only needed specification is for the interfaces exported across the A9 interface as shown in Figure 2). In addition, due to the challenges involved in server design, they are highly platform-specific.

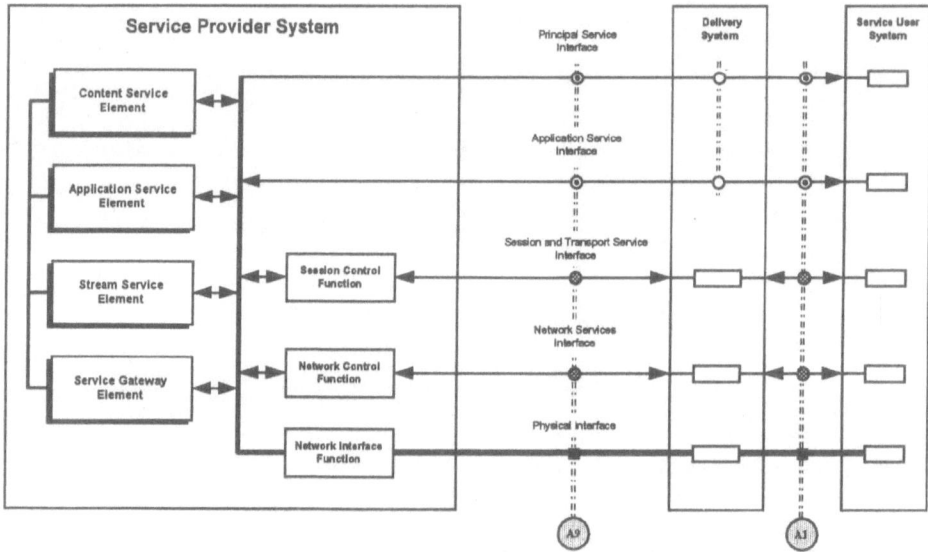

Figure 2. Service provider system architecture. Internal interfaces (those that are not visible through the A9 point) are left unspecified, as they are implementation-dependent and do not affect interoperability.

DELIVERY SYSTEM

Delivery systems can include both physical media-based (e.g., CD-ROM) as well as networked systems. In DAVIC 1.0 only networked systems are considered, for both wired (e.g., CATV) and wireless (terrestrial and satellite broadcast) networks. The delivery system architecture follows a potentially tiered approach that includes a core network that forms the backbone of the system and an access network that bridges client STUs with the core network. The overall architecture is shown in Figure 3.

Core network switching and multiplexing is based on ATM. The Access Node (Figure 3) has the responsibility of adapting between the ATM-based core network and the specifics of the access network used. DAVIC provides support for a variety of access network types, including Asymmetric and Very high speed Digital Subscriber Lines (ADSL,

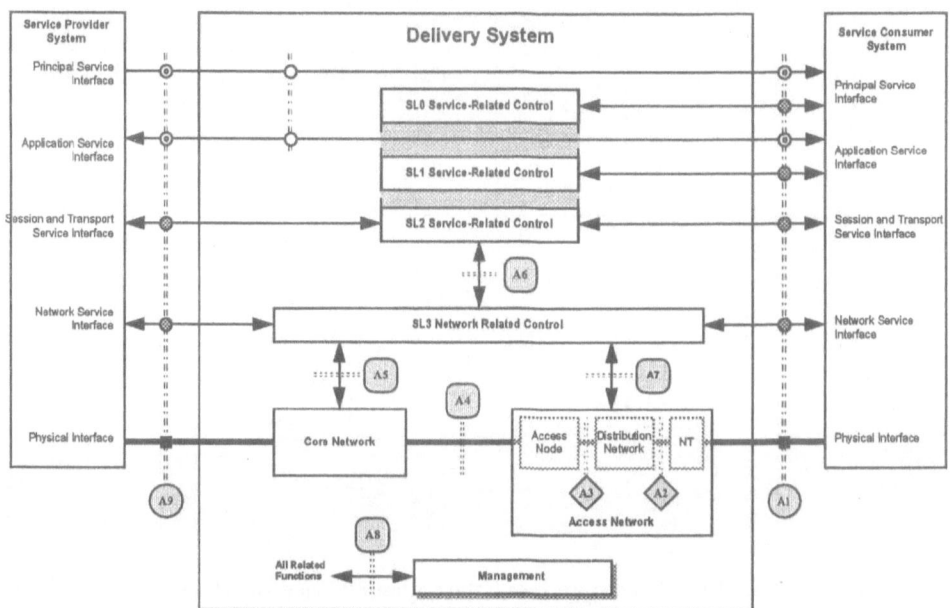

Figure 3. Delivery system architecture. Both core and access networks are included. The access network involves an access node for interconnection with the core network, the distribution system, and network termination (NT). NT may be either active, or passive if the actual termination is performed at the STU. Note that the A2 and A3 interfaces will not specified by DAVIC.

VDSL), Fiber-to-the-Curb (FTTC) and Fiber-to-the-Home (FTTH), as well as ATM-based and non ATM-based Hybrid Fiber Coax (HFC).

Satellite systems can function both as core networks and access networks; in the former case they have to comply with the A4 interface; while in the latter with the A1 interface. Obviously, in order to provide user control, satellite delivery systems have to be augmented with appropriate reverse channels; the mechanism for this is not yet determined by DAVIC.

Regarding to the in-house network, i.e., the part between the NT and the STU(s), two options exist. In an active NT configuration, a bus topology can connect several STUs to the same connection; in a passive configuration (where the STU provides termination functionality) the in-house network becomes more complex and is not specified.

Network-related control (S4 information flow, at the Network Service Interface) is on ITU-T Q.2931, as these are public networks. The access network beyond the A4 interface can be based either on ATM or on MPEG-2 TS delivery. A mapping function is then necessary in order to assign virtual paths/channels to MPEG-2 TSs. With respect to network management functions, two candidates under consideration are CMIP defined by ITU-T M.3010 and SNMP defined by RFC1157. The former is likely to be used for the core and access networks, while the latter is preferred by end service providers.

SERVICE USER SYSTEM

The STU is arguably one of the most interesting components of a VoD system, as it represents a significant percentage of the opportunities for added value for service and content providers. It is also the only component (hardware-wise) that belongs to the domain

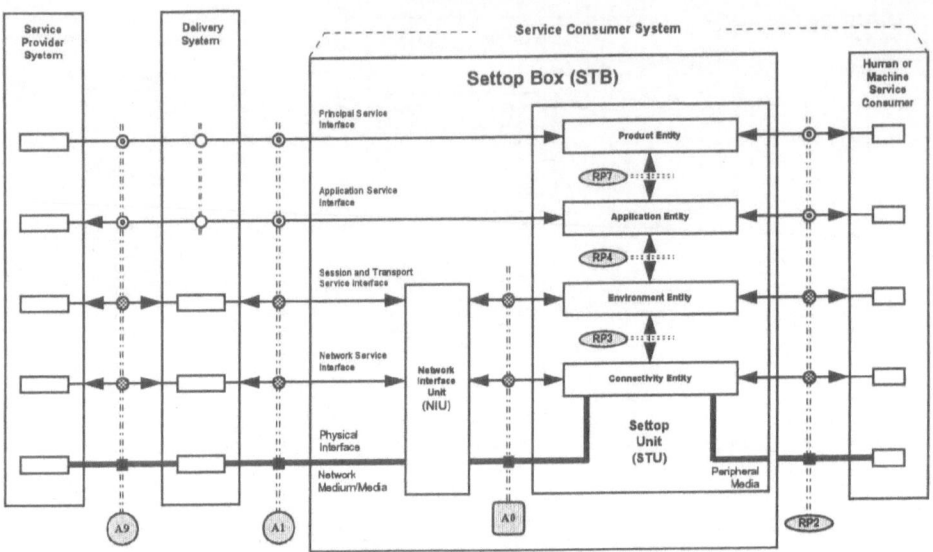

Figure 4. Set-top Unit (STU) Architecture. Note that the actual Set-Top device also includes the Network Interface Unit (NIU), to form the Set-Top Box (STB). To facilitate application interoperability, the architecture defines internal STU interfaces, at Reference Points (RPs) 3 and 4.

of consumer electronics: users will be able to select among several different models and obtain the one of their choice.

The STU accepts a signal compliant to the A1 interface (high-speed downstream channel, Figure 4) through the physical interface. The Network Interface Unit (NIU) then processes the signal based on the type of A1 interface used, for example QAM demodulation for HFC or satellite networks, and then passed through the A0 interface (if one exists). The high-speed data is processed by an MPEG-2 demux, with any relevant control messages (possibly out-of-band) passed on to the appropriate entities, as shown in Figure 4. Conditional access functionality is also provided at the STU/NIU, but its specification is not included in DAVIC 1.0.

In order to facilitate interoperability at the application level, a standard application interface has been adopted. Between the RP3 and RP4 reference points we have the STU Operating System (OS), on which a Run-Time Engine (RTE) resides (these form the Environment Entity). The RTE exports a well-defined interface across RP4, to be used by application developers. The application itself is based on MHEG-5[6] (residing on top of RP4), which was designed to support the distribution of interactive applications across platforms of different types and brands. MHEG-5 is actually a subset of MHEG-1, in which platforms with minimal resources are taken as primary implementation targets. This portable application format is stored at the server, for downloading to the STU. In order to allow for application-specific functionality, extensions to the OS (and thus the RTE) as well as the application code are allowed. This provides platform-independent content representation, without sacrificing flexibility or the capability for creative differentiation. Note that MHEG-5 does not provide a scripting language, but it allows using one to perform more complicated tasks from the generic ones the standard supports. MHEG-5 object communication between the server and the STU is performed using the DSM-CC U-U (User-to-User) specification. Although the mapping of object and content references is defined in DAVIC 1.0, that of API actions is not (although work is under way to ensure that sufficient functionality is provided to support high-level API run-time facilities).

PROTOCOL CONFIGURATIONS

The preceding discussion described the individual architecture of each of the components of a VoD system, as conceived in the DAVIC model. The glue that interconnects these components are the individual protocols that operate between them. Due to the large number of interfaces and variations thereof (due to the multiplicity of access networks), we cannot exhaustively list them all here. We thus provide only indicative examples only for the server.

The basic stack structure is centered around ATM/AAL5, MPEG-2 TS, and TCP, UDP, and IP. The ATM Forum's 5/8 mapping is currently used for mapping MPEG-2 TS packets to AAL5 PDUs, but the subject is far from being a closed issue.[7] ATM is used within the core network and, depending on the access network, may extend end-to-end. MPEG-2 TS is used for the downstream channel (from the server to the STU), carrying audio-visual information as well as other data in the private data part of the bitstream. IP is used for application command and data information transfer (e.g., DSM-CC U-U), as well as network management (SNMP). Figure 5 indicates, for example, the protocol stack to be used for the S1 information flow (downstream traffic) across the A9 interface (server to core network); note that various physical layers are acceptable. Figure 6 shows the corresponding stack for the S2 information flow, consisting of application command and data information. Similarly, Figures 7 and 8 depict the protocol stack for the S3 (session control) and S4 (network control) information flows.

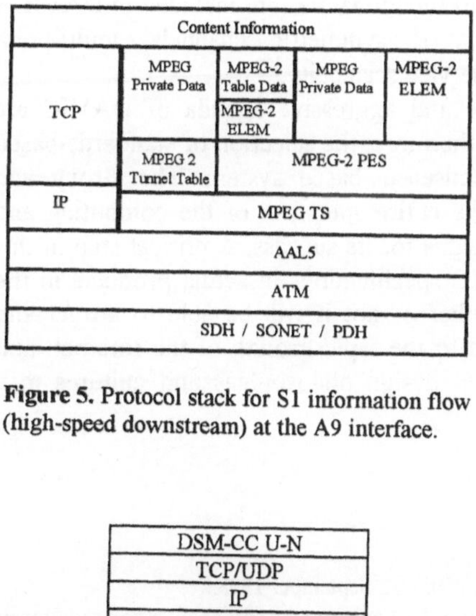

Figure 5. Protocol stack for S1 information flow (high-speed downstream) at the A9 interface.

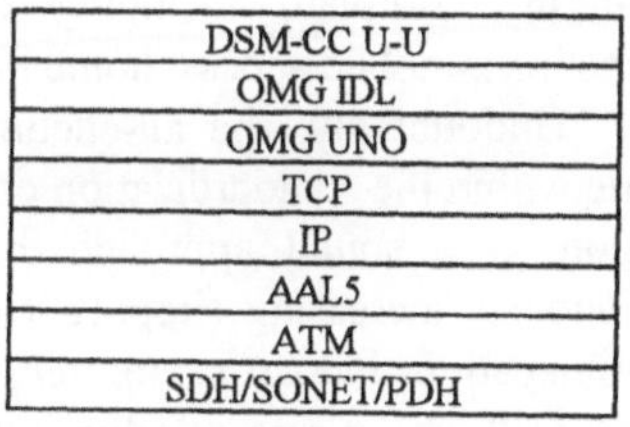

Figure 6. Protocol stack for S2 information flow (application command and data transfer) at the A9 interface.

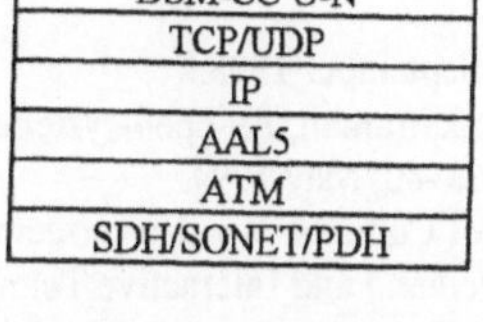

Figure 7. Protocol stack for S3 information flow (session control) at the A9 interface.

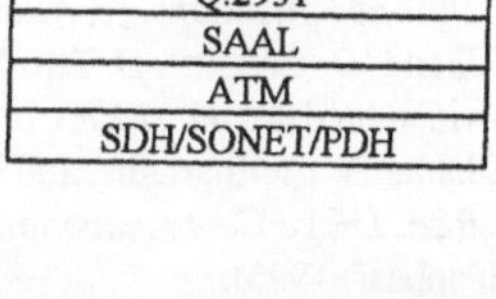

Figure 8. Protocol stack for S4 information flow (network control) at the A9 interface.

Protocol stacks for other corresponding interfaces are similar, with modifications primarily to the lower layers. For example, for an ADSL-based access network the only difference in Figure 5 would be the substitution of the physical layer.

CONTENT REPRESENTATION

Content representation for multimedia information is based, as mentioned before, on the MHEG-5 specification. For monomedia encoding, DAVIC uses the following formats: MPEG-2 for video (MP@ML) and still picture encoding, MPEG-1 Layers I and II for single and dual channel audio (32, 44.1, and 48 kHz), DAVIC-specific formats for linear (PCM) audio and graphics bitmaps, and Unicode (ISO/IEC 10646) for character encoding (with the Latin-1 subset being mandatory). With respect to graphics, the specified format is error-resilient and extensible, and supports: CLUT1, 2, 4, 8 and RGB16 formats; square, 4:3, and 16:9 pixel aspect ratios; transparency; 50% translucency; and full or half-screen resolutions.

BEYOND DAVIC 1.0

The preceding discussion only described some of the key features of the architecture of a DAVIC system. Several components were omitted, including security, local session and initialization protocols between the STU and the access network, usage information protocols (for billing etc.), as well as profile and level delineation. DAVIC is already looking forward, and several issues have been identified for coverage in specifications after version 1.0. These include definition of the A10 interface (between content provider and server), server MIB and management, satellite return channel specification, A0 and A1 connectors and STU data ports, scrambling, key management, authentication, and copyright enforcement, 3-D graphics, etc. More sophisticated functionalities are included in DAVIC's long-term workplan, including distributed servers, symmetric channels, multi-point connections, mobile access, home networks, and service scalability.

Undoubtedly, the all-encompassing scope and aggressive agenda of DAVIC are unique within the standardization community. Furthermore, the selection of standards-based solutions is a sound approach in building a consensus-based system. The strong and continuously increasing support it enjoys from the entire spectrum of the computing and communications industries are very encouraging signs for its success. A critical step in this respect will be the rate of adoption of the DAVIC specifications in actual products in the months to come. Another key issue will be whether or not it will be able to attract and absorb recent developments that have contributed to the rapid growth of the Internet, and provide a unified delivery platform. The different design philosophies and cultures may prove critical in this respect.

REFERENCES

1. DAVIC 1.0 Specifications, Revision 4.0 (draft), Hollywood, CA, September 1995.
2. Y. H. Chang, D. Coggins, D. Pitt, D. Skellern, M. Thapar, C. Venkatraman, An open-systems approach to video on demand, *IEEE Comm. Mag.*, Vol. 32, Nr. 5, pp. 68–80, May 1994.
3. S.-F. Chang, A. Eleftheriadis, and D. Anastassiou, Development of Columbia's video-on-demand testbed, *Image Communication*, Special Issue on Video-on-Demand and Interactive Television, (to appear, 1995).
4. T. Kwok, A vision for residential broadband services: ATM-to-the-Home, *IEEE Network. Mag.*, Vol. 9, Nr. 5, pp. 14–28, October 1995.
5. Object Management Group, Universal Network Objects Version 1.0, Common Object Request Broker Architecture (CORBA) Version 2.0.
6. ISO/IEC CD 13522-5 (Committee Draft), Multimedia and Hypermedia Experts Group (MHEG), July 1995.
7. S. Dixit and P. Skelly, MPEG-2 over ATM for video dialtone networks: issues and strategies, *IEEE Network. Mag.*, Vol. 9, Nr. 5, pp. 30–40, October 1995.

BUSINESS VIDEO SERVER: THE LAN/WAN CONNECTIONS

Mon-Song Chen, Shiow L. Huang, Chris D. Song, and Sharon H. Chen

InfoValue Computing, Inc.
P.O. Box R
Goldens Bridge, NY 10526
(914) 232-6642; (914) 232-1970 (fax)
monsong@ix.netcom.com

1. INTRODUCTION

Today most PCs can do standalone multimedia computing, i.e., reproducing good quality motion video and audio from digital files stored in local hard disk or CD-ROM. The challenge for the information technology industry is to enable PCs to do networked multimedia computing in the near future.

There are two types of networked multimedia computing: synchronous and asynchronous. With synchronous computing, users can interact with each other via real-time exchanges of live motion video and audio, such as in video conferencing. The primary technical challenges in synchronous computing are the real-time compression and collaboration technologies[1,2,3]. With asynchronous computing, users can instantaneously retrieve and playback video or audio from multimedia databases stored in specialized servers. The primary technical challenge in asynchronous computing is the technology to store, manage, and deliver multimedia content over computer networks. Such a technology is also commonly referred to as the video server technology.

A video server is essentially a specialized file server. Like a file server, a video server facilitates sharing of video and audio content for workgroups. Unlike a file server, a video server must ensure that video and audio contents are delivered with strict isochronous timing to the requesting clients. A file server, however, is sensitive only to the latency, which is the period between start and completion instants, but not the timing in between.

A lot of interests on video servers come from the video on demand (VOD) field trials taking place in the last two years. Many projects have proposed various techniques to increase the overall system capacity by, for example, optimizing disk arm scheduling[4,5,6], system architecture and service scenarios[7,8].

Video is also an important medium for business computing. Business video (BV) and entertainment video (EV) share the same video technologies, such as compression methods and formats. But the ways they are used and consequently their system designs are quite different, however. The objective of this paper is to present a complete discussion about the BV server. The structure of the paper is as follows. In Section 2 we will introduce the BV servers through a couple of application examples. In Section 3 we will present a system model for BV servers, followed by the comparisons between BV and EV servers. In Section 5 we will discuss the issue of real-time transport and describe a new client/server transport protocol. Concluding remarks will be in Section 6.

2. BUSINESS VIDEO APPLICATIONS

Video is an important medium to business computing in many ways. For media industry companies, for example, video is their main business. For education and training, video-enriched multimedia titles are becoming important tools. For these and many other applications, video servers can be used to significantly improve the efficiency by extending the standalone operations to workgroup oriented client/server computing. Two examples are briefly described below to illustrate the benefits and system requirements of such video server based network computing:

Corporate Training And Education

Video-enriched multimedia titles are increasingly being used to assist in training and education. These titles are mostly stored in CD-ROMs and distributed to the necessary employees. As the number of CD-ROM titles increases, management and administration, e.g., ensuring versions are consistent, quickly become complicated tasks. Also CD-ROM, as a storage medium, is less flexible and more expensive when the contents need to be updated frequently.

The alternative is to organize the video-enriched titles on video servers and let workgroup users perform need-base retrievals. In such a client/server configuration, the tasks of management and administration are easier and more direct. Changes are instantly effective and incur no cost.

Additionally individualized on-line training can be developed to reduce the overhead and constraints of the conventional classroom oriented training programs. In conventional training programs, classes are organized based on employees' needs, as perceived by the program coordinator, and time, space, and instructor's availability. The effectiveness of the classes vary depending upon instructors' quality and enthusiasm, and the intended employees' availability and concentration. Such programs are inflexible, and have significant recurring cost.

Video servers facilitate individualized on-line training, which essentially transforms training into an integral part of desktop computing. Since video-enriched training titles are stored in video servers, employees can access the lessons they need when they need them at the their own desks. This reduces the expense and the problems of conventional training classes, such as determining the specific training needs of each individual, arranging for classroom space and scheduling. In addition, productivity will increase since learning is immediately applied.

Video Kiosks

Kiosks are effective tools to deliver information to public audiences, such as in shopping malls or convention centers. A Kiosk is essentially a PC with touch sensitive screen and information organized for easy navigation. Most kiosks today are standalone PCs and have little or no high quality motion video, because of the difficulties in maintaining and updating the contents.

The situation can be greatly improved by networking kiosks with one or more video servers. In such a client/server configuration, the kiosks need not store any video content. When requested by the users, the kiosks retrieve the requested video clips from the video servers and perform real-time playback. Hence the system administrators need only maintain and update the collection of contents in the video servers.

3. SYSTEM MODEL OF BUSINESS VIDEO SERVER

In this section we will present a system model, depicted in Figure 1, for the business video servers, and discuss the critical requirements.

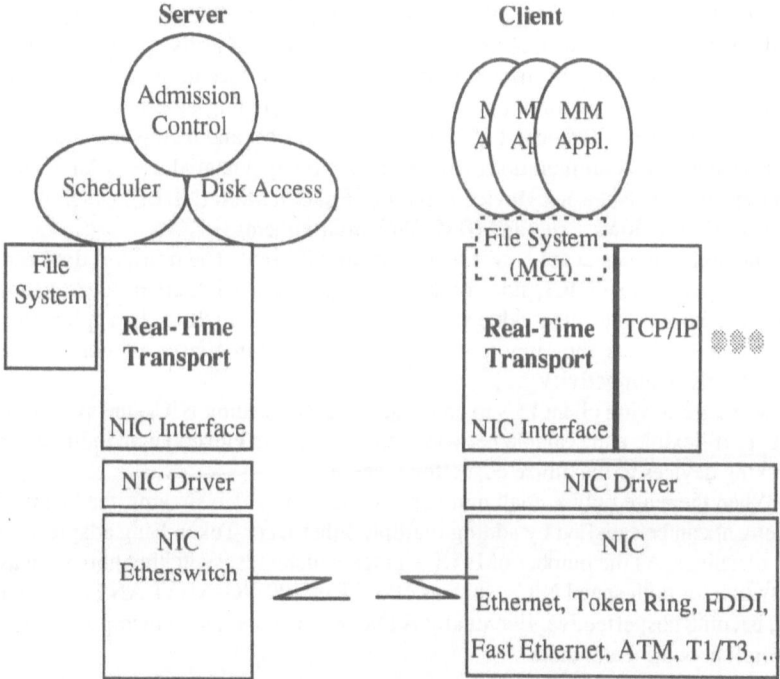

Figure 1. System Model For Business Video Server

The model is based on the principle that business video must fit into existing business computing. It must support existing multimedia applications without code modifications, and the standard application programming interface (API) for ease of developing new applications. Its administration and management procedures must be integrated or compatible with the existing operations. Lastly but not least importantly, it must have low entry cost and strong scalability.

Transport

To deliver isochronous video streams, new real-time transports are needed as conventional transports such as TCP/IP and SPX/IPX address only the latency issue, but not the strict isochronous timing requirements. There have been several important proposals for real-time transport protocols, e.g., the Tenet Suit for U. C. Berkeley and RSVP from IETF. We will defer protocol discussions till Section 5, and concentrate for now on how a real-time transport should fit in the system model.

The real-time transports should have the right interfaces so that it can fit into most existing environments. As depicted in Figure 1, a good model is that it provides file system interface in the upper layer, and interfaces with network interface cards (NICs) in the lower layer.

With file system interface, the real-time transport can be embedded in a redirector so users can access files in servers in the exact same way as files in local hard disks. The choice of file system interface is based on the fact that most applications use file system commands to get video data for their processing such as video decompression.

There are alternatives to the file system interface if there are higher layer multimedia interfaces. An example is the **Media Control Interface (MCI)** in Microsoft's Windows (including Windows For Workgroups, Windows 95, and Windows NT) and IBM's OS/2 environments. Since most, if not all, multimedia applications interface with MCI, instead of with file system directly, the real-time transport can use MCI as its upper layer interface.

The real-time transport must have a lower layer interface to inter-operate with any existing or emerging network interface cards (NICs), including Ethernet, Token-Ring, FDDI, Fast Ethernet, ATM, and T1/T3. Otherwise the cost and inconvenience of changing adapters or hardware configurations will intimidate many potential users. An example of the NIC interface is the **Network Device Interface Specification (NDIS),** which is supported in Microsoft's Windows series and IBM OS/2 environments.

The requirements at the server machines are different. The transport does not have to inter-operate with most NICs, since quite often high speed NICs are needed to support multiple concurrent out-going video streams. Also, since no multimedia applications will run on the server machines, the support of file system or MCI interfaces are not necessary.

Network Connectivity

Besides allowing client PCs to continue use their existing NICs, the system should also support flexible and scalable network connectivity, and leverage on industry standard networking devices to maximize cost effectiveness.

When there are only a small number of concurrent video streams, the bandwidth requirement can be satisfied by adding multiple Ethernet or Token Ring adapters in the server machines. As the number of LAN segments increases, switching hub, such as Etherswitch, or high speed NICs, such as 100 BT or 100 VG-ANYLAN Fast Ethernet or FDDI, become cost effective. Eventually ATM switches should be incorporated to expand the overall network bandwidth.

Client

At a client machine, the real-time transport should co-exist with other transports, such as the popular TCP/IP, which may be needed by other applications. The multiple transport configuration is supported in most PC and workstation platforms.

Server

At a server machine, the three main tasks directly involved in serving isochronous video streams are:

- admission control (AC): responsible for granting or rejecting client's requests, by considering the utilization of storage, network, CPU, and other resources.
- disk access (DA): responsible for retrieving relevant video data from disk storage for in-time delivery to the clients.
- scheduler: responsible for delivering video data to the requesting clients with isochronous properties, and react to client's commands, such as play, pause, fast forward, fast reverse, or stop.

Finally the video servers should co-exist with other file or application servers, as shown in the enterprise video/data network depicted in Figure 2. A file server serves conventional data to the clients, and a video server serves both video and data to the clients. Because of bandwidth considerations, most clients only go through LAN connectivity, and only a few remote clients go through TCP/IP WAN connectivity. Etherswitch or similar bandwidth expansion devices are used to expand the LAN bandwidth. Routers are used to interface to wide area networks.

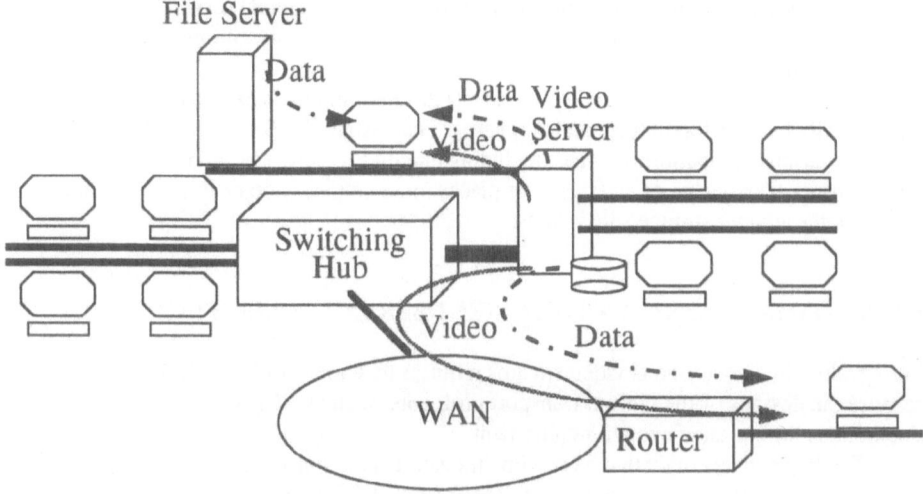

Figure 2. Enterprise Video/Data Network

4. COMPARISONS OF BUSINESS VIDEO AND ENTERTAINMENT VIDEO

Business video (BV) and entertainment video (EV) systems are similar in general concepts, but have many different system requirements. Since there are a lot of interests on video on demand (VOD) systems, we will compare these two types of systems in this section.

The differences between BV and EV systems mainly come from the differences in user's expectation. With BV systems, users expect the same types of performances and procedures as their business computing, while with EV systems the expectations are gauged by user's experiences of using VCR.

The specific differences are:

- **Client:** EV clients are new and dedicated set-top boxes, while BV clients are existing computers. BV applications will be used together with many existing data applications. Hence BV clients have the additional compatibility requirements, including computer hardware platform, operating system, and network interface.
- **Network:** EV systems like VOD typically have high and/or dedicated bandwidth connectivity between clients and servers, while BV systems generally must leverage on existing corporate networking infrastructure. The latter means conventional LANs, such as Ethernets and Token Rings, which usually are limited in bandwidth and do not have any bandwidth guarantee.
- **Interactivity:** The video clips for BV systems generally range from a few tens seconds to a few minutes, which are much shorter than 90 minutes to 2 hours movies. In addition, users of BV systems generally are more inclined to manipulate the playing of video, while users of EV systems tend to passively watch the video.
- **Server:** the throughput requirements in EV systems, such as VOD systems, are much higher than those of BV systems. A VOD server typically needs to serve

hundreds or thousands of households, while a BV server of a workgroup in many situations only need to serve up to a few tens concurrent streams. Consequently BV servers should be designed to leverage on industry standard storage subsystems to maximize the cost effectiveness, while VOD servers may have to have specially designed storage sub-systems to satisfy the throughput requirements.

In summary, BV and EV systems are very different in user's expectation, and consequently have different technical emphases. The design of a BV server requires more efforts in various compatibility issues, including client PCs, networks, and operational environments. The design of an EV server places more emphasis on enabling the required very high throughput storage subsystems.

5. REAL-TIME TRANSPORT FOR CLIENT/SERVER COMPUTING

Delivering isochronous video streams requires new real-time transport protocols, because the designs of the current transport protocols, such as TCP/IP, did not include isochronous timing as an explicit requirement.

The key enhancement that a real-time transport should provide is the quality of service (QoS) guarantee. In wide area networks, RSVP from IETF appears to be most prominent among many proposals. RSVP is a mechanism for users to set up a communicating path with guarantee bandwidth before starting data delivery. Once setup, the relevant routers and switches will ensure that the data stream will be delivered with guaranteed QoS.

Similarly in local area networks there are proposals to incorporate priority into the existing multi-access schemes, such as the priority Token Ring and VG-ANYLAN. But none has emerged as the prominent leader.

These real-time protocols are designed to provide generic transport services, and are peer-to-peer in nature. They are useful for any networked multimedia computing, but not necessarily optimized for video server systems, which are asymmetric client/server configuration. In the rest of this section we will describe an optimized transport to be used by video servers.

In any client/server system, the server is the hot spot and usually the bottleneck of the system. In video server systems, for example, the most important performance measurement is the number of concurrent video stream a given server hardware platform can support. To increase the capacity of a video server, one way of doing it is to reduce the server's processing load in serving each client.

An effective approach is to let the video server anticipate the needs of each client, and make the just-in-time delivery of the most relevant data accordingly. This approach is sometimes referred to as the Push model, in contrast to the usual Pull model in which the server passively responds to client's requests.

Push model delivery in business video servers is more complicated than in other systems, such as VOD servers. This is because business video servers must deal with very heterogeneous environments, including variations in:

- client configurations, e.g., microprocessors, network interface, memory sizes, and operating systems,
- network connectivity, e.g., Ethernet, Token Ring, or FDDI,
- video formats, e.g., Indeo, Cinepak, QuickTime, MPEG-1, MPEG-2, or M-JPEG,
- video decompression, e.g., software-only or specialized hardware, and whether or not data integrity is required
- usage patterns, e.g., degree of interactivity

To overcome the complexity, we have designed a hybrid Push & Pull (PP) model. The server smoothly pushes data out as much as possible, based on the appropriate information about the video format and client's needs. The client, in the meantime, dynamically monitors the user's behaviors and relays significant changes to the server. For example, if data integrity is not a critical requirement in a particular client, the server, once informed, can avoid doing error recovery and consequently reduce the processing load.

A complete design and implementation of the PP model has been developed by InfoValue Computing in its patent pending QVS Transport Protocol (QVSTP). The protocol is used in the **Quick Video Server (QVS)** product. QVS is a client/server software product, consisting of QServer and QClient. QServer runs under Windows NT and turns standard PCs into powerful video servers. QServer can be installed on any hardware platform which supports Windows NT, including 486 and Pentium processors, PowerPC, SMP, Alpha and MIPS. With off-the-shelf storage and network components, a single PC can serve many, e.g., 40 or more, concurrent video streams. QClient enables PCs to playback video from QServer in real time. QClient works with existing operating system platforms, and requires no hardware changes. Existing multimedia applications continue to run without any modification.

6. CONCLUSIONS

The objective of this paper is to present a complete discussion about business video servers. This is accomplished by describing business video server's characteristics through a couple of application examples, and then presenting a system model for business video servers. Since there are a lot of interests on video on demand systems, we have also compared the differences between these two types of video servers. Finally we have discussed the issue of real-time transport protocols and presented a client/server transport specially optimized for video server. The transport is implemented in the video server product, **Quick Video Server (QVS)**, developed by InfoValue Computing, Inc.

REFERENCES
1. H. Vin, M.-S. Chen, and T. Barzilai, Collaboration management in DiCE, The Computer Journal, Vol. 36, No. 1, 1993
2. M.-S. Chen, Z.-Y. Shae, D. Kandlur, T. Barzilai, and H. Vin, A multimedia desktop collaboration system, Globecom'92
3. Z.-Y. Shae and M.-S. Chen, "Video Mixing in JPEG Compressed Data Domain," (Invited Paper) Image System Journal
4. M.-S. Chen, D.D. Kandlur, and P.S. Yu, "Optimization of he grouped sweeping scheduling (gss) with heterogeneous multimedia streams," Proc. 1st ACM Int. Conference on Multimedia, ACM
5. P. Yu, M.-S. Chen, and D. D. Kandlur, Design and analysis of a grouped sweeping scheme for disk scheduling to support multimedia applications, Multimedia System Journal
6. P. Rangan, H. Vin, Designing file systems for digital video and audio, Proc. ACM Symposium on Operating Systems Principles
7. D.D. Kandlur, M.-S. Chen, and Z.-Y. Shae, Designing a multimedia server, IEEE and SPIE High Speed Networking and Multimedia Computing Conference, Feb., 1994, San Jose, CA
8. W. Sincoskie, System Architecture for Large scale video on demand services, Computer Networks ISDN Syst. 22:155-162.

THE BLOCK ALLOCATION PROBLEM IN VIDEO SERVERS

William Tetzlaff

Robert Flynn

IBM T. J. Watson Research Center
Post Office Box 704
Yorktown Heights, NY 10598

Polytechnic University
36 Saw Mill River Road
Hawthorne, New York 10532

Introduction

Recent advances in technology have made possible the development of inexpensive multimedia display stations [1]. Rangan [2] noted that the cost of storing the material will require a high degree of reuse and sharing of video source material. This leads to the creation of large servers that serve large numbers of users [3]. These server systems use computers to store and retrieve compressed digital video data which they then transmit over digital networks to client stations.

Given that the video servers ultimately serve a home consumer market, end user cost is a key factor. This is to be contrasted with the size of the data sets that the server deals with. The typical video considered here is, when looked at as a file, very large. Given the low cost and large size, the products and the users expect a significant reuse of the videos (files). Video servers will be used in telephone offices and cable company head-ends to feed the local distribution system. They will also be used through wide area networks to provide video through multiple local distribution networks.

The primary storage medium for video servers are rotating magnetic disk drives. In normal computing systems occasional long queue lengths and the attendant long service time is accepted. In video systems long delays are unacceptable because the video information must be displayed in real time. Computing systems are planned based on random arrivals of I/O requests. Once a video has been started the times of all of the subsequent requests are predetermined. An effective allocation strategy could take advantage of this characteristic to distribute the load across the devices. This paper will investigate potential algorithms and how they interact with the workload and system size. Relatively constant interarrival times of read requests can result in high utilization with relatively constant service time.

Video Server Environment and Terminology

Any interruption in the timely supply of data to the client results in *jitter* that is both visible and audible to the viewers. An uninterrupted supply of data to the client systems is required for continuous display of video data. A continuous play of digital video to a viewer is frequently called a *stream*.

Movies on demand and *Interactive video* are different video server applications. In Movies on Demand, pausing, video reversing and repositioning present systems software with challenges. Existing data from movie theaters and video store rentals suggests a strong tendency toward viewing of a small number of movies. The conventional view is that movies on demand to homes will have a similarly skewed distribution toward a very few movies. Assumptions usually call for the single most viewed movie to account for from ten to twenty percent of the viewers. In Interactive Video (education, shopping, games etc.) the video sequences are shorter and there is a need for short messages that directs, in a scripted manner, the sequence of video accesses played to the viewer (buying an object from the interactive video provider.) The individual video clips may also be separated by long periods during which there is no video or the last frame being shown. Observations of students using video disk course material show that full motion video is only viewed about twenty percent of the elapsed time.

Multimedia Communications and Video Coding
Edited by Y. Wang *et al.*, Plenum Press, New York, 1996

Disk storage is the primary medium for storing content. Current disks (one to two Gigabytes) have the space to hold one or two movies and have the throughput to allow 5 to 20 (on the order of 10) viewers concurrent access to the material. The tendency for a small amount of the material to be frequently viewed makes dedicating disks to particular movies undesirable. Some disks will exceed playout capacity so the movie will need to be replicated on many additional disks. Other movies will be so infrequently viewed that only a tiny fraction of the available throughput will be used. *Striping* is a technique that is normally used to provide very high data throughput interleaving sequential material across multiple devices. Striping is useful in video servers in order to provide the needed throughput to material concurrently viewed by many people. In addition it can be used to mix high use and low use content and spread it across multiple devices in order to make full use of both the storage space and the bandwidth.

Compression technology is essential to the success of storing video information digitally. The compression reduces the bits needed to store the same material by about two orders of magnitude. Part of the reduction is due to *lossy* compression in which some information is lost. The *compression ratio* is defined as the uncompressed size divided by the compressed size. If *Constant Bit Rate* (CBR) is used then the compression ratio is constant across multiple frames, thus buffering needed within the decoder is reduced and resource management is simplified. The same visual quality can be obtained with *Variable Bite Rate* (VBR) compression while using twenty to thirty percent fewer bits. This study addresses the resource requirements of fixed and variable rate compression.

A block size of 256K bytes is in the reasonable range because it gives a disk data transfer efficiency of about 75 percent for contemporary disks. The *data transfer efficiency* is defined as the time spent transferring data divided by the elapsed time that the disk is in use. Large block sizes are preferred because they amortize the cost of the seek and rotation across a larger number of bytes and thus increase the data transfer efficiency. A 2 hour movie compressed at 1.5 Mbits per second will contain over 5000 blocks of 256 Kbytes which will require 1280 Gigabytes of disk storage.

Block Allocation Problem

Once the decision has been made to stripe across multiple disk drives the next decision has to do with the strategy for selecting the order in which disks will be used. The block allocation problem becomes a scatter function or hashing function problem when one realizes that there is no need to linearly assign the blocks to the disks (e.g. blocks 1,2,3, ... m to drives 1,2,3, ... $n - 1(\bmod n)$) corrected only by error detection and correction considerations. The scatter function used may be different for different movies, in addition. This study explores several striping strategies for sequencing the remaining blocks across the available devices.

The *scan* algorithm causes blocks to be allocated in a fixed, linear order across all of the disks. Since movies consist of over 5000 blocks they will sequence through the disks many times. All movies are given the exactly the same sequence of disks to repeat until all of the blocks of a movie are stored. Thus the activity will be guaranteed to be uniformly distributed across all devices. The dynamic behavior should also be uniform since viewers will all be sequencing disk reads in the same order through the devices.

The *Individual Permutation* (IP) algorithm chooses a particular sequence through the disks for EACH movie that is stored. Once all of the disks have been used the same sequence is repeated. The IP algo-

Disk 1	Disk 2	Disk 3	Disk 4	Disk 5
A1	A3	A2	A4	A5
A6	A8	A7	A9	A10
A11	...	...	...	...
B5	B1	B3	B2	B4
B10	B6	B8	B7	B9
...	B11	...	...	...
C4	C5	C1	C3	C2

$A=(1,3,2,4,5)$
$B=(2,4,3,5,1)$
$C=(3,5,4,1,2)$

Figure 1. Individual Permutation (IP) allocation.

rithm is similar to scan in that both methods used fixed permutations. In IP, the permutation is fixed, movie to movie, but repeats, cyclically depending on the number of drives. Scan uses one permutation for all movies and that permutation is the linear order permutation. (IP allocation will also assure that, in the long run, all of the device activity is uniformly distributed across the devices. In the short term the different paths taken by each movie will not have the same device uniformity that scan has.)

The **Random Block Permutation** (RBP) or movie specific block permutation method is a block scatter method in which a random path through the disks is taken until all of the blocks have been stored. The device numbers are chosen is such a way that there will be equal use of all devices for each movie. The order within the list is then randomized. This algorithm will have the long term uniformity in device utilization. However, the short term will not be uniform, and should have at least the variability of IP.

Zipf Distributions

Zipf distributions are frequently used to express the probability of selection of a particular object from a fixed number of objects where there is a skew toward some of the objects. A pure Zipf (Figure 3) distribution has a single parameter, the number of discrete objects to be selected from. In a Zipf like distribution a second parameter, theta, is added to specify the skew. If theta is zero the distribution is Zipf, if it is one it is uniform. Relative probabilities are converted to probabilities by normalization. We, like Dan [4], made use of a Zipf like distribution with a theta of .271 because it closely matched a published video store rental distribution [5] of 92 movies.

Disk Modeling and Scheduling

For this study we assume a high performance disk that is capable of providing service to multiple clients simultaneously. The disk modeling is based on work by Ruemmler and Wilkes [6]. The disk character-

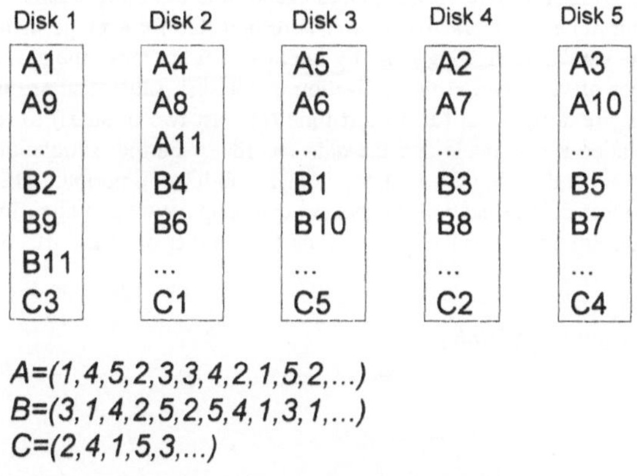

Disk 1	Disk 2	Disk 3	Disk 4	Disk 5
A1	A4	A5	A2	A3
A9	A8	A6	A7	A10
...	A11	...	...	...
B2	B4	B1	B3	B5
B9	B6	B10	B8	B7
B11	...	...	...	...
C3	C1	C5	C2	C4

$A=(1,4,5,2,3,3,4,2,1,5,2,...)$
$B=(3,1,4,2,5,2,5,4,1,3,1,...)$
$C=(2,4,1,5,3,...)$

Figure 2. Random Block Permutation (RBP) allocation.

Zipf Distribution with parameter n

Frequency   For i = 1 to n $F_i = \dfrac{1}{i}$

$F_1 = \dfrac{1}{1}$ $F_2 = \dfrac{1}{2}$ $F_3 = \dfrac{1}{3}$ ... $F_n = \dfrac{1}{n}$

Zipf Like Distribution with parameters n and theta

Frequency   For i = 1 to n $F_i = \dfrac{1}{i^{1-Theta}}$

Probability

Prob{selection=i} = P_i For i = 1 to n $P_i = \dfrac{F_i}{\sum\limits_{j=1}^{n} F_j}$

Figure 3.

istics similar to the ones used in [6] are used in this study because they represent a contemporary high performance disk. Disk service time will be the sum of seek time, rotational latency and data transfer time. Maximum seek time is 25 milliseconds, average latency is 7.5 milliseconds, and the time to read 256K bytes of video is 69 milliseconds. This is composed of 64 milliseconds for transfer time at 4Mbytes per second, plus 2 milliseconds head settling time and 3 milliseconds for minimum fixed seek time.

Disk scheduling algorithms are somewhat different in video systems. First the steady rate of reading of blocks for one client is well below the disk throughput capacity. If a new read operation is queued for the next block as soon as the previous one has been read then there will appear to be very long queue lengths against the devices. Second the equal service times for block playout for movies leads to constant interarrival times for disk requests on behalf of each viewer. This may lead to more uniform interarrival times of disk requests which may allow higher utilization without jitter.

The key to scheduling the I/O operations is to be sure that they will complete by the time the data is needed. This suggests using Earliest Deadline First (EDF) device scheduling. Whenever a block is emptied by a write to the network the block is free and a new request is created and queued against the appropriate device in order by deadline. This study assumes the use of two buffers for each viewer. When a new request comes into the system it does not have a deadline, but the device is known. It is more important to assure that existing viewers not be subjected to jitter than it is to have a fast start of new material. The system should be cautious about when to initiate a new I/O request so as not to make other requests late by examining the queue looking for a gap that the request could fit into. We define the *jitter rate* as the number of jitters (or late blocks) per viewer per hour. This definition allows for easy comparison of jitter rate for different capacity servers.

Simulation Operation

The magnitude of the workload of the simulations is determined by the number of viewers that are watching content. Each movie viewer places exactly the same incremental workload on the disk subsystem. In the long run each interactive viewer is responsible for average incremental workload on the disk subsystem. The pauses in viewing done by interactive users make the absolute workload per user lower. Thus the number of viewers and the utilization of the disk subsystem are linear functions of each other. The simulations are always run for a short time (at least two minutes) to let the simulation stabilize. A two hour simulation run is made and the data recorded. Further simulations are done by increasing the number of viewers and giving the system another stabilization period. The objective is to find out how many users or what disk utilization the system can support without significant jitter or long initial delay times. However, different server sizes, meaning the number of disks, affects the number of users that can be supported.

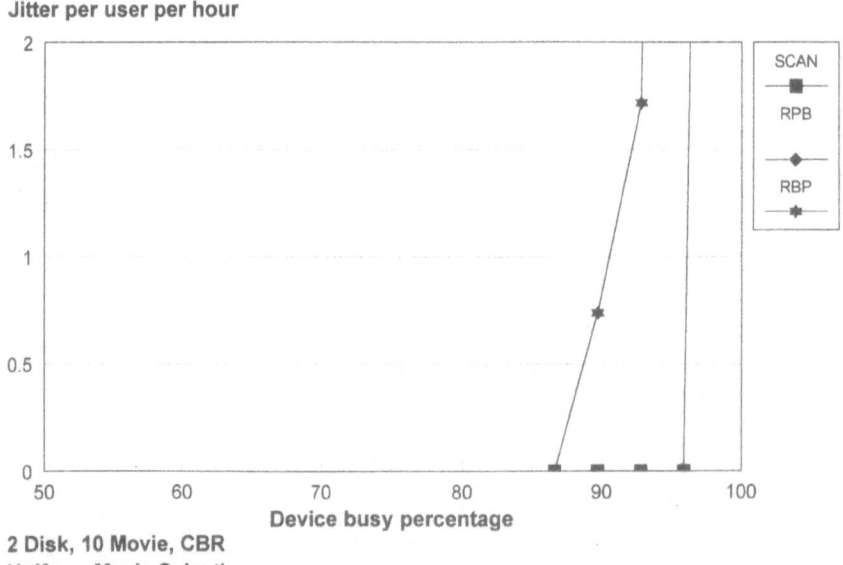

2 Disk, 10 Movie, CBR
Uniform Movie Selection

Figure 4. Jitter rate for 2 disk system.

The disk utilization for each user is a constant. If a particular block allocation algorithm allows higher utilization without jitter it means that more users can be supported without jitter. Since the objective is to make the highest utilization of the disks that is possible the graphs show either the jitter rate or the maximum initial delay as a function of average device utilization. The device utilization proportional to the number of viewers. Device utilization is defined as the percentage of elapsed time that devices are busy, which includes fixed overhead, seek, rotation, and data transfer time.

In running the simulations we found that small jitter rates began well before the systems collapsed into high jitter rates. It is assumed that some amount of jitter will probably be tolerated because the total elimination is expensive. In this paper a system will be considered *overloaded* if the jitter rate goes above .5 jitters per user per hour. This is about one jitter in a full length movie. We chose this for the upper limit for two reasons. One is that our observation was that it was usually near the knee of the jitter curve (Figure 4) and that a small load increase will result in a large jitter rate increase. The second reason is that at most one jitter per movie seems to be a reasonable level of quality for viewing.

Initial Investigations of Server Sizes

The first set of simulations were done to understand of the block allocation algorithms on systems of different sizes. For this part of the study Constant Bit Rate (CBR) compression and a movie workload was assumed. Systems of two, five, ten and fifty disks are simulated.

Two disk systems

Simulations of two disk systems (Figure 4) shows the value of either scan or IP over RBP placement. The fact that IP and scan appear to give the same performance results from having two devices so the only orderings possible are (1,2) or (2,1) which are ultimately the same.

Ten disk systems

Simulations of ten disk systems (Figure 5) again shows a further decline in jitter free device utilization when RBP is used. This is because the risk with RBP is that all, or many, viewers will need the same device at the same time. It also shows divergence between scan and IP.

Fifty disk systems

Simulations of fifty disk systems (Figure 6) are of a system that is more appropriate to a commercial video server. The difference between scan at about 90 percent device utilization and IP at about 77 percent

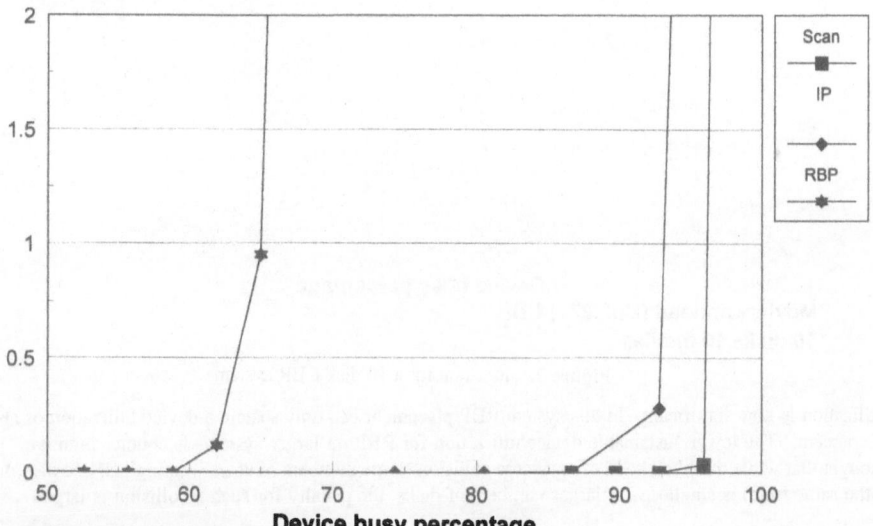

10 disk, 10 movie CBR
Uniform movie distribution

Figure 5. Jitter rate for 10 disk system.

Jitter per hour per user

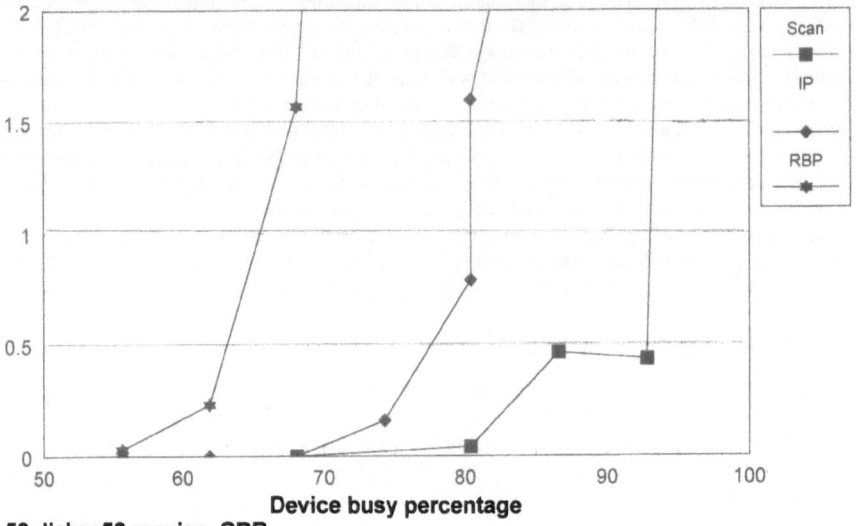

50 disks, 50 movies, CBR
Uniform movie selection

Figure 6. Jitter rate for 50 disk system.

Jitter per hour per user

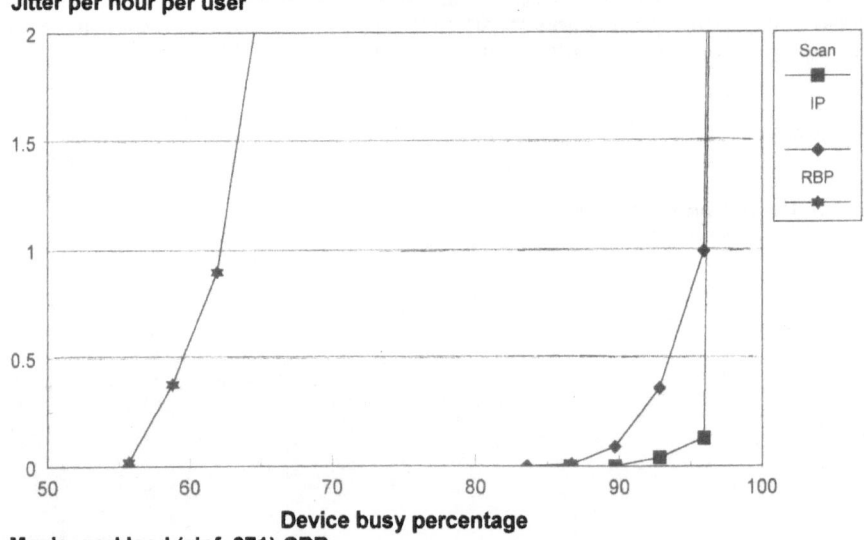

Movie workload (zipf .271) CBR
10 disks 10 movies

Figure 7. Jitter rate for a 10 disk CBR system.

utilization is now significant. In this system RBP placement can only sustain a device utilization of about 63 percent. The lower sustainable device utilization for RBP on larger systems is counter intuitive. It is thought that while the likelyhood of an access collision (large numbers of users wanting just one disk drive a the same time) is smaller with larger numbers of disks, the penalty for such a collision is larger.

Effect of Workload and Compression

The two compression types (VBR and CBR) and the two workloads (Movies and interactive) provide four environments in which to explore the effectiveness of the three allocation algorithms. A moderately skewed (theta of .271) is used in all of the following simulations. In the this section we will show simulation results and comment on the results for all four combinations of workload and compression. Maximum initial delay is generally acceptable whenever jitter is also acceptable (below .5 per user per hour). In any case where this is not true it will be specifically mentioned.

Movies with CBR Compression

A movie workload with CBR compression is the most favorable workload for a video server because very long sequences of video are shown and because all of the viewers consume material at a constant rate. This is the workload that is able to run the devices at the highest utilization without introducing unacceptable jitter.

Movies with VBR Compression

Variable Bit Rate compression is simulated by keeping a list of block play times for each block in a piece of content. Every viewer uses the same list of block play times, but will be in a different place within the list. For these simulations the same mean block play time of 1.365 seconds, for a 256 Kilobyte block, is used. The individual block play times are randomly generated with a uniform distribution about the mean with a range of .737 seconds. Baugher [7] showed peak to mean data rate ratios for 256 Kilobyte blocks to average 1.27. The use of a uniform distribution from 27 percent below the mean to 27 percent above the mean, or a range of .737 seconds, conforms to the Baugher observation.

The introduction of variable play times to the blocks introduces yet more variability into the workload. For the RBP placement it means yet lower, in fact the lowest of anything tried and hence fewest viewers served, utilization with acceptable jitter. With a fifty disk system (Figure 9) a device utilization of only fifty percent is possible.

The use of VBR encoding with the scan algorithm (Figure 9). is not desirable. Scan is very good for CBR long running material because the load is very constant and the viewer content progresses at constant rate from device to device. With VBR material the viewer content may bunch up and the congestion will move from device to device until it clears. The overall performance of the system becomes erratic under higher loads. The system will go in and out of overload through time (Figure 10).

Interactive with CBR Compression

Interactive workloads are simulated using a random clip size that averages ten second and is uniformly distributed across a ten second range. Each clip is followed by a user think time (pause) that is also uniformly distributed across a ten second range with a mean of ten seconds.

Jitter per hour per user

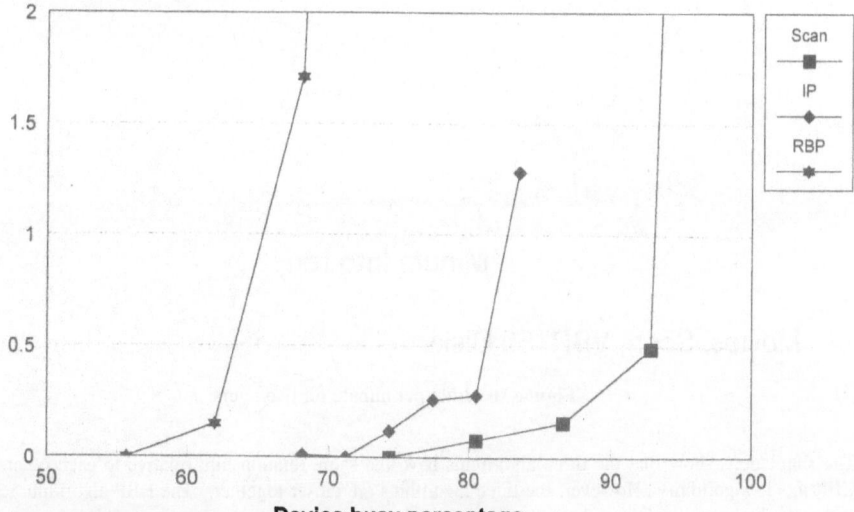

50 disks, 50 movies
Movie workload (zipf .271) CBR

Figure 8. Jitter for a 50 disk CBR system.

185

Jitter per hour per user

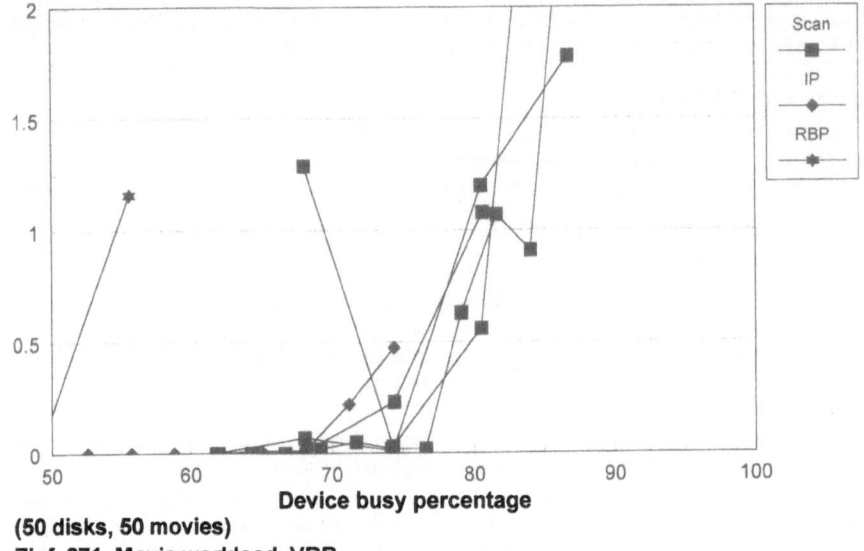

(50 disks, 50 movies)
Zipf .271, Movie workload, VBR

Figure 9. Jitter for a 50 disk VBR system.

Jitters per minute

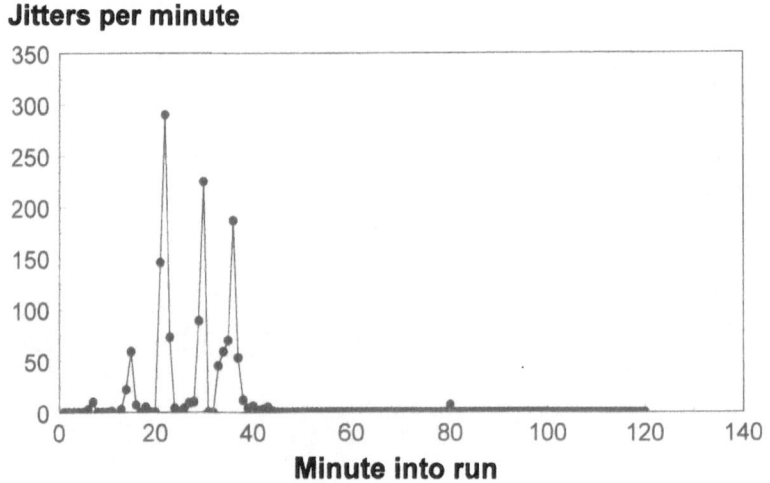

Movies, Scan, VBR, 50 Disks

Figure 10. Jitter per minute for two hours.

The simulations show that the three algorithms have the same relationship, relative to each other as the CBR movie algorithms. However, the three algorithms are closer together. The RBP algorithm achieves somewhat higher utilization and the scan algorithm has lower utilization (fewer sustainable users).

The introduction of more starts and stops introduces more variability into the system. The pauses also mean that the number of active users could potentially range from no users to all users. It would be intuitive to think that lower device utilizations would be necessary to prevent jitter. The actual situation is

Jitter per hour per user

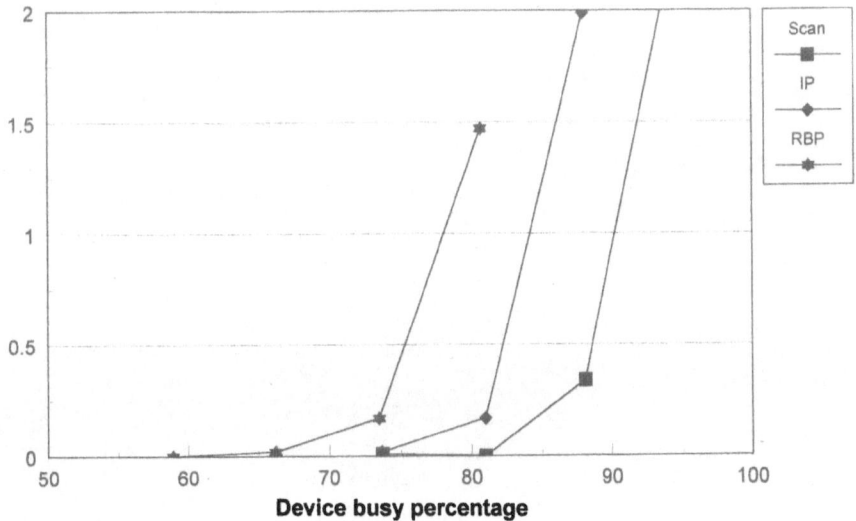

10 disk 10 movie
Zipf .271, Interactive workload, CBR

Figure 11. Jitter for a 10 disk interactive CBR system.

Jitter per hour per user

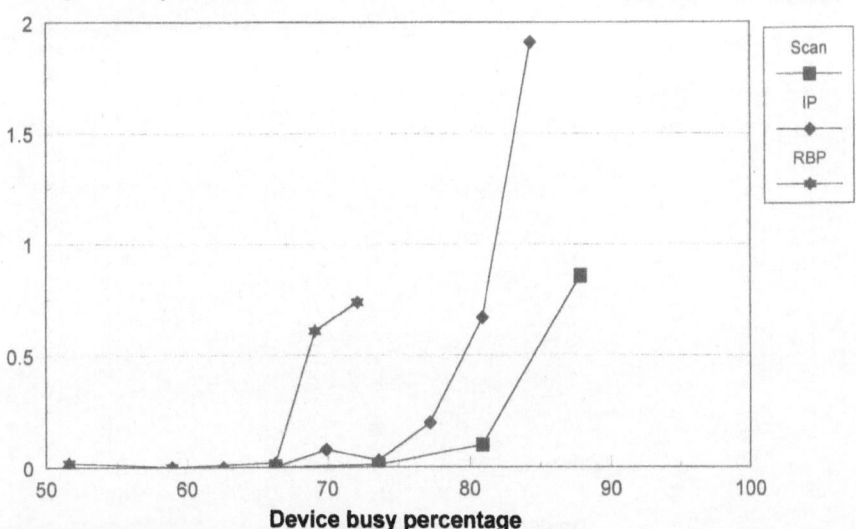

50 disks, 50 movies
Zipf .271, Interactive workload,CBR

Figure 12. Jitter for a 50 disk interactive CBR system.

Jitter per hour per user

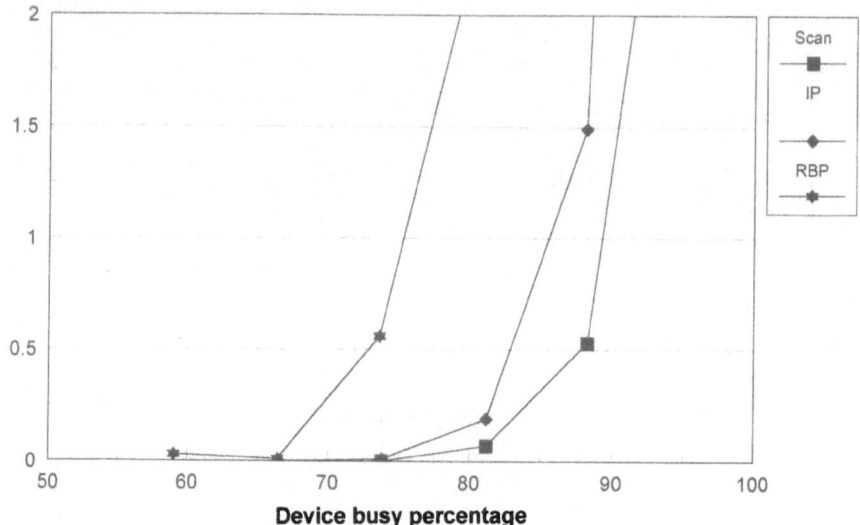

10 disk 10 movie
Zipf .271, Interactive workload, VBR

Figure 13. Jitter for a 10 disk interactive VBR system.

Jitter per hour per user

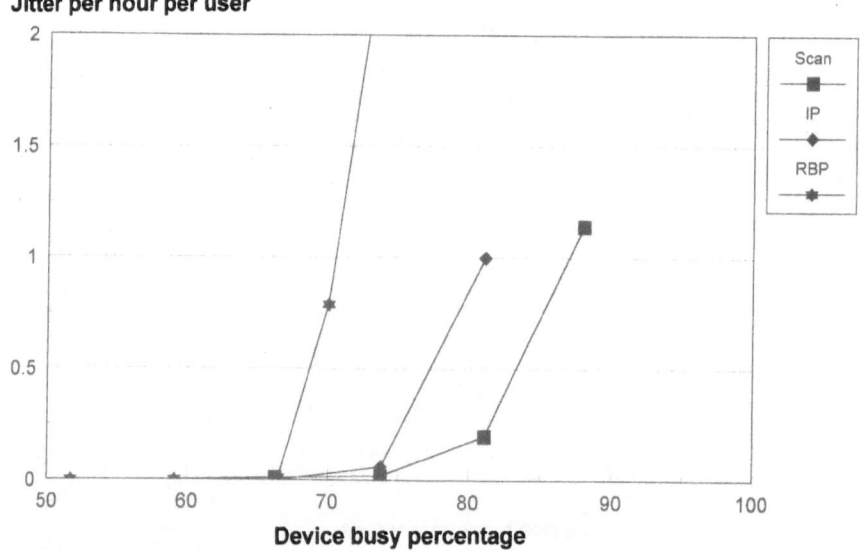

(50 disks, 50 movies)
Zipf .271, Interactive workload, VBR

Figure 14. Jitter for a 50 disk interactive VBR system.

188

counter intuitive because there is a stronger effect in the system. The device scheduling gives first priority to reads for material that is already playing. The initial read is only done if it can be done without making anything that is queued late. Thus the initial plays are delayed whenever a device is temporarily overloaded. This micro level load deferral allows the interactive system to sustain quite high device utilizations. It also validates the need for this sophisticated device scheduling.

Interactive with VBR Compression

The runs representing interactive workload with VBR compressions used the same parameters as the earlier CBR interactive runs. This is the most variable workload in the simulations. Despite the additional workload variation due to the VBR workload the device scheduling defers play starts in order to smooth the workload and produces device utilizations that match the CBR runs.

Conclusions

The RBP block placement algorithm always resulted in significantly lower maximum device utilization. Systems using this algorithm will require many more disks to supply the same throughput. The IP placement can be considered as the best choice if the workload and compression are not known. It is always much better than RBP and it is usually close to scan. With the exception of VBR movies, scan allows the highest maximum device utilizations. The VBR encoding would result in a reduction in throughput requirement and storage requirement of twenty-seven percent. It did introduce variability that required somewhat lower device utilization. However, VBR remains a clear benefit because the bit rate throughput reduction is far larger than the small reduction in maximum device utilization. The combi-

Movies	CBR	VBR
RBP	63	50
IP	81	75
Scan	90	(unstable)75

Interactive	CBR	VBR
RBP	66	66
IP	79	77
Scan	85	84

Device utilization percentage without more than .5 jitters per hour per viewer

Figure 15. Maximum disk utilization with acceptable jitter (10 disks).

Movies	CBR	VBR
RBP	60	60
IP	94	84
Scan	97	84

Interactive	CBR	VBR
RBP	75	73
IP	82	82
Scan	88	87

Device utilization percentage without more than .5 jitters per hour per viewer

Figure 16. Maximum disk utilization with acceptable jitter (50 disks).

nation of deadline scheduling of ongoing streams and the lower priority of initial reads works to good effect. The deadline scheduling is essential to the VBR runs. For CBR systems deadline and FIFO scheduling would be equivalent. The lower priority for initial reads is effective in all systems.

REFERENCES

[1] Rangan, P. V., Kaeppner, T. and Vin, H. M.
    Techniques for Efficient Storage of Digital
    Video and Audio.
    UCSD Technical Report, CS91-209, 1992.

[2] Rangan, P. V., Vin, H. M. and Ramanathan, S.
    Designing a Multi-User Multimedia-On-
    Demand Service.
    UCSD Technical Report, CS92-231, 1992.

[3] Tetzlaff, W., Kienzle, M. and Sitaram, D.
    A Methodology for Evaluating Storage Sys-
    tems in Distributed and Hierarchical Video
    Servers.
    *Proceedings of IEEE 1994 Compcon confer-
    ence*, pages 430-441, 1994.

[4] Dan, A., Sitaram, D. and Shahabuddin, P.
    Scheduling Policies for an On-Demand Video
    Server with Batching.
    *Proceedings of ACM Multimedia 94 Confer-
    ence*, pages 15-21, October, 1994.

[5] Magazine, V. S.
    Video Store Magazine.
    Video Store Magazine, December 12, 1992.

[6] Ruemmler, C. and Wilkes, J.
    An Introduction to Disk Drive Modeling.
    IEEE Computer, 27,3:17-28, March 1994.

[7] Baugher, M.
    The OS/2 Resource Reservation System.
    *Proceedings of Multimedia Computing and
    Networking 1995*, February 6-8, 1995.

A MOVIE-SCHEDULING POLICY FOR VIDEO-ON-DEMAND SYSTEMS

Sameer Dubey,[1] Grammati E. Pantziou,[2] and Nilay Sheth[1]

[1]Computer Science Department, University of Central Florida
Orlando, FL 32816-2362
[2]Computer Technology Institute, P.O. Box 1122
26110 Patras, Greece

Abstract

In this paper we present a policy for scheduling movies in a video-on-demand system which is equipped with the multicast facility. A 2-layer architecture is used where both a master server and a number of local servers are used to serve the users. The policy enables users to select both local and non-local movies, reduces the average waiting time before a user is served, and prevents the replication of movies at each site. Real-time threads have been used for the implementation of the mechanism, and a simulation study has been conducted.

1 Introduction

Recent advances in computing and communication technology have made feasible video-on-demand systems. Basic services that a video-on-demand system provides, include selection from a menu, delivery, and viewing of a movie through interaction with a presentation device. More advanced video-on-demand systems provide the user with individualized control over the presentation. For example, the user is allowed to pause and then restart the presentation at arbitrary times, reposition, forward, and reverse play.

Different issues involved in the design of a video-on-demand system have been studied by different researchers the last few years. Architectural issues have been addressed in [8, 10, 13, 14]. Physical storage organizations necessary for supporting video-on-demand systems have been proposed in [1, 2, 11]. Probabilistic models for the assignment of video data onto a video-on-demand storage hierarchy have been studied in [12, 9].

A typical architecture of a video-on-demand system includes a set of video disks interconnected to a set of servers which route the data to individual users through a high bandwidth network. The user requests for movies arrive at random time intervals and are independent of each other. Due to tight response time requirements, continuous delivery of video data to the user has to be guaranteed by reserving the computing resources as well as the disk I/O bandwidth required for delivery. The computing resources and the disk I/O bandwidth impose an upper bound on the number of sessions that can be simultaneously supported by a server. Therefore, a video-on-demand system that can support a large number of active users, requires very large server capacity.

In the case that a large number of users of a video system do not require dedicated sessions (e.g., pay-per-view, or quasi-video-on-demand), multiple requests for the same movie can be batched together and serviced using a single session. Scheduling policies for on-demand video servers with batching have been proposed by [4]. More specifically, a First Come First Served policy is proposed that schedules the movie with the longest outstanding request, and a Maximum Queue Length policy is proposed that schedules the movie with the maximum number of outstanding requests. In [7] alternative policies have been used that take into account the sum of the waiting time of all users for a movie.

In this paper we present a new policy for scheduling movies in a video-on-demand system which is equipped with the multicast facility. The system supports pay-per-view and quasi video-on-demand, and a number of sessions are reserved to support true-video-on-demand. Our policy differs from the ones that have been previously proposed in the following. First, it is based on a 2-layer architecture where both a master server and a number of local servers are used to serve the users. Therefore, the policy enables users to select both local and non-local movies and additionally, increases the probability that a user will be served and prevents the replication of movies at each site. Second, an appropriate weighting of the user requests is employed which contributes towards decreasing the average waiting time before a user is served, and increasing the fairness of the service.

2   The Movie-Scheduling Policy

The system architecture and a high-level description of the scheduling policy.
The video-on-demand architectural model that we use consists of a remote (or master) server which acts as a master database and a set of n local servers which are connected to the remote server via dedicated lines of fixed capacity. Each local server as well as the remote server can support only a fixed number of movie-streams (sessions). The users also connect to the local servers via dedicated links and make requests to them. Therefore, a 2-layer memory architecture is used (see Figure 1) and each user may select either a movie located to his local storage devices. or a non-local movie located to the master database and served by the master server. The communication network used is equipped with a multicast facility. Thus, the same movie-stream can be sent to more than one users without causing any extra overhead to the server, and therefore, multiple users can participate in a single session.

Movie popularities are calculated during off-peak hours and the distribution of the movies to the local storage devices is done based on those popularities. Therefore, initially each one of the n local disk systems will contain the k_i most popular movies (k_i is the movie capacity of the ith local disk system). A 2-level movie-scheduling scheme is proposed. At the first level, the local servers may multicast movies local to their storage devices, while at the second level, the master server decides on a movie-allocation scheme that allocates both local and non-local movies. The master server is activated when a large enough number of non-local requests has been collated.

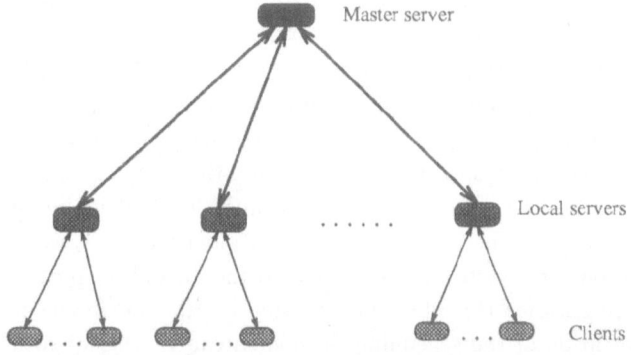

Figure 1: Video-on-demand system architecture

A high-level description of our movie-scheduling scheme follows:

1. All user requests are collated at the local servers.

2. Each local server based upon the movie caching in its storage devices, divides the requests into local and non-local, and gives appropriate weights to them.

3. For each $j \in \{1, \cdots, n\}$, every p_{lj} time units, local server j multicasts local movies to the users. The movies that are sent, are selected according to their weight. The value of p_{lj} varies according to the number of sessions remaining free between two consecutive multicasts.

4. Every p_m time units, $p_m \geq p_{lj}$, for each $j \in \{1, \cdots, n\}$, the following are taking place:

   - All requests (local as well as non-local) appropriately weighted, along with information regarding server capacities for delivering additional video streams and other resource availability, are sent to the master server.

   - The master server evaluates the above information and decides on an allocation strategy which minimizes the average waiting time before a client is served, and which is fair to all movie requests (irrespective of the popularities of the movies).

     According to the scheduling of the movies decided by the master server, local servers multicast local movies to the users, and non-local movies are downloaded from the master server to the appropriate local servers and sent to the users.

The value of p_m varies according to the number of non-local requests made by the users.

The implementation of our scheduling scheme is based on the real-time periodic thread model. The model is based on an explicit declaration of *timing attributes*. A periodic thread i's timing attributes include the start time s_i and the period p_i. A new instantiation of the thread i will be scheduled at s_i and then the activity will be repeated in every p_i. The real-time thread model has been used in [15] to implement dynamic Quality-of-Service control schemes.

The real-time periodic thread model is used as follows in our movie-scheduling scheme: Two types of real-time periodic threads are used: *Local scheduling session threads* t_{lj} with period p_{lj}, for each $j \in \{1, \cdots, n\}$, and a *global scheduling session threads* t_m with period p_m. A local scheduling session thread t_{lj} includes step 3 of the movie-scheduling scheme. I.e., during an execution of the local scheduling session thread local server j multicasts to the users as many local movies as the available resources allow. Each execution of t_{lj} starts p_{lj} time units after the start of the previous execution. The global scheduling session thread t_m includes step 4 of the movie-scheduling scheme. I.e., during an execution of the scheduling session thread t_m the following are taking place: All local servers in parallel send the requests (made to them by the users) appropriately weighted, along with server capacities to the master server. The master server evaluates this information and decides on an allocation strategy (the allocation strategy is described in the sequel). Each execution of t_m starts p_m time units after the start of the previous execution. We assume that $p_m \geq p_{lj}$, for each $j \in \{1, \cdots, n\}$. If the time of an execution of the global scheduling session thread t_m coincides with the time of the execution of one (or more) local scheduling session thread t_{lj}, then t_m is executed, the current execution of t_{lj} is cancelled, and the next execution of t_{lj} is taking place after p_{lj} time units.

Data structures and the movie-scheduling policy. We give now the data structures used by the scheduling session threads as well as the details of the movie-allocation strategies used by the master and local servers. With each user request for a movie we associate a weight reflecting the waiting time of the user for the specific request. If a user's request for a movie is not satisfied by the scheduling decided during the current execution of a scheduling session thread, the request is resubmitted with increased weight, to be satisfied during the next execution. Each local server j keeps "movie-request" information into two data structures $Local\_LS_j$ and $NonLocal\_LS_j$. $Local\_LS_j$ keeps information about requests for movies local to the server j, while $NonLocal\_LS_j$ keeps information about requests for movies located not to the storage devices of server j but to those of the master server. For each requested movie which is local to the server j, $Local\_LS_j$ has one entry with the following information:

- Id_Movie_i: the identifier of movie i;

- $Weight\_Movie_i$: the sum of the current weights of all user requests for movie i;

$NonLocal\_LS_j$ keeps the same information for each requested movie which is not local to the server j.

The master server keeps a global data structure MS with information about each movie requested from some local server while it was not available at that specific server. For each such movie, MS has one entry with the following information:

- Id_Movie_i: the identifier of movie i;

- $Total\_Weight_i$: the sum of the current weights of all user requests for movie i over all the local servers, i.e., $Total\_Weight_i$ is equal to the sum of $Weight\_Movie_i$ overall $NonLocal\_LS_j$ where movie i appears.

- LS_List_i: a list of all local servers which have requests for movie i.

For each local server j, every p_{lj} time units the local scheduling session thread t_{lj} is executed. During an execution the following are taking place:

1. If j has available resources to support one or more sessions, the entries in $Local\_LS_j$ are sorted in decreasing order according to the $Weight\_Movie_i$.

2. Let a_j be the number of the currently available (free) sessions of server j, and let e_j be the number of the entries (local movies that have been requested and not served) in $Local\_LS_j$. Then, the local server j multicasts the topmost $min\{a_j, e_j\}$ movies.

3. The topmost $min\{a_j, e_j\}$ movies are deleted from $Local\_LS_j$.

Note that the period p_{lj} of the local scheduling session thread t_{lj} for the local server j should be such that the number of sessions remaining idle between two successive executions of t_{lj} is minimized, and at the same time, there are no odd executions of t_{lj} (in the case that there are no free sessions). To meet that objective, we use a self-stabilization scheme to determine dynamically the period p_{lj} of t_{lj}. We start with an arbitrary value of p_{lj} that depends on the time of the day (peak hour or not), the number of the active users, and the number of the local movies. If the number of sessions remaining idle between two successive executions is larger than a threshold, then p_{lj} is decreased; if the number of sessions remaining idle between two successive executions is smaller than another threshold, then p_{lj} is increased; otherwise, p_{lj} remains the same.

Every p_m time units the global scheduling session thread is executed. During an execution the following are taking place:

1. For each local server j, if j has available resources to support one or more sessions, the entries in $Local\_LS_j$ are sorted in decreasing order according to the $Weight\_Movie_i$.

2. For each local server j, the entries of $NonLocal\_LS_j$ are used to update the entries of MS.

3. The entries in MS are sorted in decreasing order according to the $Total\_Weight_i$.

4. The following is repeated until all the available sessions have been occupied, or MS and/or $Local\_LS_j$ for each local server j, are empty:

   for the topmost entry in MS

       if the sum of $Weight\_Movie$ of the topmost entry over all the data structures $Local\_LS_j$, where j is a local server in the LS_List corresponding to the first entry of MS which has an available session,

is larger than or equal to $Total\_Weight$ of the topmost entry in MS
then schedule to multicast the movie corresponding to the topmost entry of all
such $Local\_LS_j$; delete the topmost entry of all such $Local\_LS_j$
else schedule to multicast the movie corresponding to the topmost entry of
MS; delete the topmost entry of MS

5. Each local server j with a_j remaining available sessions, and e_j entries (movies) in $Local\_LS_j$, multicasts the topmost $min\{a_j, e_j\}$ movies. Those movies are deleted from $Local\_LS_j$.

Note that the period p_m of the global scheduling session thread t_m should be such that the service is fair enough by frequently taking into consideration in its allocation strategy the non-local requests, and at the same time, the number of odd executions (in the case that there are no enough non-local requests) of t_m is minimized. Therefore the following scheme is used to determine the value of p_m. We start from a value of the period p_m that depends on the time of the day (peak hour or not), the number of the local movies on the different local servers, and the number of active users. If the number of non-local requests (size of MS) at the beginning of the current execution of t_m is larger than a threshold, then p_m is decreased; if it is smaller than another threshold, then p_m is increased; otherwise, p_m remains the same.

Note also that the movie allocation strategy uses the weights given to the movie requests and these weights are heavily dependent on the users' waiting time. This contributes towards reducing the average waiting time before a user is served. Although the strategy's priority is to reduce the average waiting time, at the same time, the strategy is not unfair because the current weight of a movie does not depend only on the popularity of the movie but also on the waiting time of the users requesting the movie. Recall here that minimizing the average waiting time before a user is served, may decrease the fairness of the service.

The user interface. There may be two policies regarding the user interface of a video-on denand system that uses the proposed 2-layer architectural model and movie-scheduling scheme. In the first one the users have a single menu that includes both local and non-local movies. In the second policy, the users have two movie menus to select from. One contains the local movies, and the other contains the non-local movies. Users are aware of the fact that non-local movies may delay more than local ones. Different charging policies may also apply for local and non-local movies.

The second policy regarding the user interface of a video-on denand system, may reduce the number of non-local requests and increase the efficiency of the system. On the other hand, it makes the user interface more complicated with respect to the one of the first policy.

Quasi and true video-on-demand sessions. The proposed movie-scheduling policy applies in the case of pay-per-view and quasi video-on-demand sessions. In the case that we want to provide the users with the capability to pause and then restart at arbitrary times the movies, then a number of efficient methods discussed in [5, 3, 6, 4] may be employed. More specifically, if the user of a multicast stream pauses, then a small amount of buffer can be used for caching the blocks of the multicast stream [6].

If the pause is short, the user can restart without any delay. If the pause is longer, then the request may be batched with an existing stream if a sufficiently close stream exists. In the case that such a sufficiently close stream does not exist, then the user is restarted using a free channel from a set of channels that have been set aside to be used in cases like this one. Such channels can be used to support also a small number of video-on-demand sessions.

3 Simulation study

We developed a simulation program to compare the performance of the proposed 2-layer architecture to a 1-layer architecture. The simulation program models the users as clients making requests to the server (local or remote as the case maybe) and obtaining service based on the available resources. The number of disks on the remote server (1-layer) is equal to the sum of the disks on the local and the remote servers (2-layer). The total number of movie streams supported by the remote server (1-layer) is also equal to the total number of streams supported by the local servers. We assume a uniform batching interval for collating requests. We also define a reneging interval after which a user drops his request. Simulations were carried out for varying remote server capacities, user arrival rates and number of local servers. The user service rate (number of users served as a percentage of requests made) and the reneging rate was measured with respect to time. The 2-layer system performed better compared to the 1-layer system especially for the case of a large number of movies. The improved performance was due to local caching and improved availability of the movie at the local server site.

References

[1] H.-J. Chen and T.D.C. Little, "Physical Storage Organizations for Time-Dependent Multimedia Data", in Proc. *ACM FODO*, (1993).

[2] S. Christodoulakis and C. Faloutsos, "Design and Performance Considerations for an Optical Disk-Based, Multimedia Object Server". *Computer 19*, (1986), pp.45-56.

[3] A. Dan, P. Shahabuddin, D. Sitaram and D. Towsley, "Channel Allocation under Batching and VCR Control in Movie-On-Demand Servers", IBM Research Report, RC 19588, Yorktown Heights, NY, 1994.

[4] A. Dan, D. Sitaram and P. Shahabuddin, "Scheduling Policies for an On-Demand Video Server with Batching", in Proc. *ACM Multimedia*, (1994), pp. 15-23.

[5] A. Dan, D. Sitaram and P. Shahabuddin, "Scheduling Policies with Grouping for providing VCR Control Functions in a Multi-media Server", U.S. Docket No YO993-030, 1994.

[6] A. Dan and D. Sitaram, "Buffer Management Policy for an On-Demand Video Server", IBM Research Report, RC 19347, Yorktown Heights. NY, 1993.

[7] H.D. Dykeman, M.H. Ammar and J.W. Wong, "Scheduling Algorithms for Videotex Systems under Broadcast Delivery", in Proc. *ICC* (1986). pp. 1847-1851.

[8] A.D. Gelman, H. Kobrinski, L.S. Smoot and S.B. Weinstein, "A Store-and-Forward Architecture for Video-On-Demand Service", in Proc. *ICC* (1991).

[9] T.D.C. Little and D. Venkatesh, "Probabilistic Assignment of Movies to Starage Devices in a Video-On-Demand System", Proc. *Network and Operating System Support for Digital Audio and Video (NOSSDAV)*, (1993), pp.204-215.

[10] S. Loeb, "Delivering Interactive Multimedia Documents over Networks", *IEEE Communications Magazine 30*, (1992), pp.52-59.

[11] P.V. Rangan, H.M. Vin, S. Ramanathan, "Designing an On-Demand Multimedia Service", *IEEE Communications Magazine 30*, (1992), pp.56-64.

[12] R. Ramarao, V. Ramamoorthy, "Architectural Design of On-Demand Video Delivery Systems: The spatio-Temporal Storage Allocation Problem", in Proc. *ICC* (1991).

[13] W.D. Sincoskie, "System Architecture for a Large Scale Video-On-Demand Service", *Computer Networks and ISDN Systems 22*, (1991), pp. 155-162.

[14] J. Sutherland, L. Litteral, "Residential Video Services", *IEEE Communications Magazine 30*, (1992), pp.36-41.

[15] H. Tokuda, T. Kitayama, "Dynamic QOS Control based on Real-Time Threads", Proc. *Network and Operating System Support for Digital Audio and Video (NOSSDAV)*, (1993), pp.114-123.

PERFORMANCE AND GUARANTEED QUALITY OF SERVICE FOR AT&T MULTIMEDIA COMMUNICATION SYSTEMS

Dr. Kabekode V. Bhat

Software and Communication Solutions
AT&T Global Information Solutions
Naperville, IL 60566

ABSTRACT

In this paper, we model and analyze the performance of ATM based AT&T multimedia (MM) communication server for server-centric and peer-to-peer scenarios. We present data structures and algorithms and a way to ship the needed MM data from the database to clients in a timely manner avoiding both starvation and overflow. The system level capacity and performance metrics we use are the maximum number of simultaneous video streams of given bit rate supported and the response time for a typical user command respectively. By modeling the processing scenarios using queuing network and accounting for the parallelism within the architectures, we show that the system supports 14, 8 Megabit per second streams in the server-centric case and over 45, streams in the peer-to-peer case. The delays experienced by the data packets at the disk subsystem, bus complexes, and server CPU limit the capacity in the former case and the delays that occur when data packet moves from the disk to ATM adapter via the bus complexes limit the number of streams in the later case. The methodology is general, and is useful in designing, prototyping and configuring servers and provides a basis for realizing guaranteed quality of service in MM communication context.

1. INTRODUCTION

Emerging technologies and quest for quality of life fuel research on MM communication (MMC)[1-6]. Performance and guaranteed quality of service (GQOS), the desirable attributes of these systems are determined by the architecture and technologies used and the number of streams of a given bit rate and the average response time for typical user command are used as metrics for system capacity and performance. Here, extending the processing scenario based modeling methodology[2], we analyze ATM network based MMC server performance to find the number of clients supported on server-centric and peer-to-peer AT&T MMC servers. The MMC server-1 shown in figure 1 is a symmetric multiprocessing computer. It has 1-4 Intel Pentium processors, a 400 MB/second processor-memory bus bridged to two PCI buses. One of the bus supports 4 PCI adapters and the other 2 PCI adapters and 2 EISA adapters for connecting various I/O devices. More I/O slots are obtained by adding PCI buses using bridges.

The dominant bus traffic for MMC is due to the disk and network I/O and the use of separate I/O buses for the disks and network has performance advantages over using a single bus for both traffic. Configurations where one PCI bus is dedicated to continuous media applications while the other to business critical applications are also possible. Growth in the number of clients is provided by the OC-3 connections to an ATM switch that connects to a number of clients. If more connections are desired, either a larger switch or additional switches are used via OC-3 ATM adapters on the PCI bus.

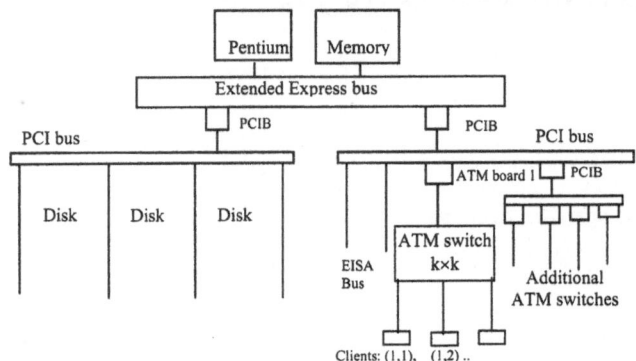

Figure 1. ATM based AT&T MM Server

In the next section, the notations, data structures, the server-centric processing scenario and models for subsystems, response time, and system capacity are presented and illustrated. Section 3 is devoted to the peer-to-peer architecture. Our work is used for performance and bottleneck analysis, configuration design and management for any MMC system in the context of ATM networking and constitutes a basis for realizing GQOS for the MM users.

2. PERFORMANCE MODEL FOR THE SERVER-CENTRIC SCENARIO

Large chunks of MM data from disks are moved to the clients via the server memory and network. Assume that there are n ATM adapters ordered as $\{1 \leq i \leq n\}$. Table 1 gives the model parameters and notations.

The j-th stream in the i-th ATM adapter terminating at *client(i,j)* where the video is displayed at *rate(i,j) Mbit/second* is denoted by stream $S_{i,j}$ and $\{S_{ij} : 1 \leq i \leq n; 1 \leq j \leq nstr(i)\}$ is the set of all concurrent streams supported by the MMC system. The dominant application processing scenario within the system is as follows.

1. A user at *client(i, j)* requests the server (either via ATM network or over the Ethernet, EISA and PCI buses) for a specific video available in the disk storage at *rate(i,j) Mbit/second*.

2. The server recognizes and routes the request through a resource allocation algorithm that tests if the server can meet the QOS requirements of existing clients and the new one. If it can it reads the data from correct disk(s) into the appropriate server buffer group; otherwise the request is given a busy tone.

3. The data delivery manager builds the next chain of buffer pointers for assembling the next AAL5 packet, multiplexing appropriate data buffers associated with the active streams and the new request(s).

4. The data delivery manager notifies the i-th ATM adapter to transfer the next AAL-5 packet sequence. The adapter logic segments and reassembles the packet into standard ATM cells and transmits them over the 155 Mbit/second OC-3 line into the switch. The server buffers associated with the buffer chain pointers are marked ready to be filled with the next chunk of data if any by the server.

Table 1. Notations and model parameters used in the model

Notation	Description	Notation	Description
rate(i,j)	Data rate for $S_{i,j}$ at client(i,j)	T_{ufill}	Delay to fill a RAM buffer from disk
nstr(i)	Total number of streams at the i-th ATM adapter	$T^{\wedge}_{ufill}$	Effective delay to fill a RAM buffer from disk
disk_size	Giga Bytes of storage per disk	T_{seek}	Average seek delay for disk
tseru	CPU service time - upstream	T_{user}	CPU delay to fill a RAM buffer from disk
tserd	CPU service time - downstream	T_{scsi}	SCSI delay to fill RAM buffer from disk
tserap	CPU service time - application	T_{mpci}	PCI bus delay for a request or message
U_{ser}	CPU utilization server-centric case	T_{dfill}	Time to copy RAM buffer to ATM adapter
(i,j)	A pair (i,j), i=1, 2... , n; j=1, 2, .. nstr(i)	T_{dser}	CPU delay for RAM buffer - ATM adapter copy operation
ndisk	Number of disks in the system	Uscsi	Utilization of the SCSI bus
usp	Speed-up due to multiple SCSIs	T_{meisa}	Delay for a message on the EISA bus
$T_{ethernet}$	Message delay on Ethernet and EISA bus	Tserap	CPU delay per buffer_size for application
video_size	Giga Bytes of data for a video	T_{atmc}	Delay at the ATM adapter card
buffer_size(i,j)	Server & client buffer size for $S_{i,j}$ in MB	du	Number of SCSI adapters
C(i,j,1), C(i,j,2)	The two buffers at client(i,j).	$T^{\wedge}_{dfill}$	Effective delay to copy a RAM buffer to ATM adapter
B(i,j,1), B(i,j,2)	(i, j)-th RAM memory buffer pair at the server	R	Average response time for customer command
$T_{disktoatm}$	Disk to ATM buffer copy time	x	Command response time cut off

The switch sends the ATM packet sequence to appropriate set of destination clients, {client(i,q), q = 1,2,.. nstr(i) } via 25 Mbit/sec ATM connections and ATM interface to the clients. For each stream $S_{i,j}$, the server buffer pair (B[i,j,1], B[i, j,2]) facilitates a double buffering scheme. While one of the buffers is being emptied by a DMA operation from the ATM adapter i, the other is filled by the disk subsystem with the next chunk of data from the stream. The server fills the empty buffers with the next relevant data from disks in a round robin fashion so that when a buffer is emptied and more data is needed, the other has it ready. If this were not to happen, client(i,j) will see a service interruption. As in [3], in *upstream*, data for a stream is read from disks via the SCSI, PCI and processor-memory bus into the corresponding empty server buffer. In *downstream,* data from server buffers are multiplexed and assembled into an AAL5 packet and transmitted via the ATM network to the clients. Double buffering is also used at the clients. Figures 2 and 3 show the data flow (where only the i-th ATM adapter dual buffer chains are shown) and the overall buffering schemes respectively.

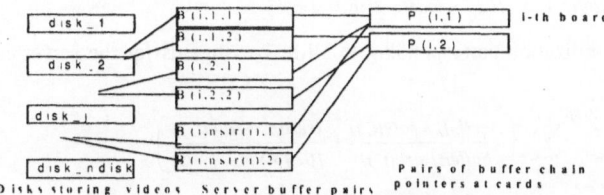

Figure 2. Data flow from disk to ATM adapter

A MMC application need is that the system must feed all active clients with the right data at the rate the *clients need* so that no client starves and no data overflows in any buffer within the

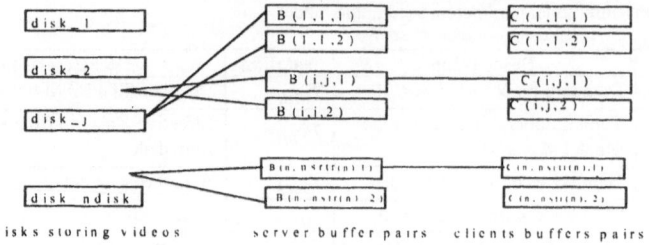

Figure 3. Double buffering - server and client

system. Thus for each $j = 1, 2, ..., nstr(i)$, before the currently being worked client buffer $C(i,j,1)$ is completely empty, the buffer $C(i,j,2)$ should be filled with the next data chunk from $B((i,j,2)$ via the network. Further, $B((i,j,2)$ should be filled from the disk subsystem when the buffer C(i,j,2) needs the next data chunk. Management of upstream buffers and buffer pointer chains is a key server software function necessary for this scheme work correctly. Further, the server system (hardware, software and application) must be able to read the right data from the disk subsystem over the bus complex into server buffer at least at the same rate as that needed by the respective clients. A necessary condition for the scheme to work correctly for $S_{i,j}$ follows.

Assertion 1. *The upstream rate for the stream $S_{i,j}$ must be at least equal to rate(i,j).*

2.1 THE UPSTREAM MODEL

In MMC-1, *ndisk* disks are distributed over *du* SCSI host adapters that are connected to PCI bus and a disk stores all the data for a MM stream. The delay for filling a server buffer from the disk is:

$$T_{ufill} = T_{user} + T_{upci} + T_{scsi} + T_{seek} \tag{2}$$

The values for T_{user}, T_{upci} and T_{scsi} are obtained from a M/M/1 queuing model[4]:

$$latency = \frac{service\_time}{(1 - utilization)} \tag{3}$$

where, *latency* is the response time for a station, *service_time* is the time for servicing the request at the station and *utilization* = *arrival rate × service time* for the station. Next we compute T_{user}, T_{upci} and T_{scsi}.

The server subsystem. *tser*, the CPU service time for shipping a *buffer_size(i,j)* of data of $S_{i,j}$, is the sum of the CPU service times for upstream and downstream processing. For the life of the stream of *(1024×video_size/ rate(i,j))* seconds, there is an application processing overhead of *tserap*. The average server utilization due to processing of $S_{i,j}$ alone is:

$$\frac{tser(i, j) \times rate(i, j)}{buffer\_size(i, j)} + \frac{tserap \times rate(i, j)}{1024 \times video\_size} \tag{4}$$

U_{ser} , the server utilization corresponding to all active streams for the server-centric scenario is given by:

$$U_{ser} = \sum_{i=1,}^{i=n} \sum_{j=1}^{j=nstr(i)} \left(\frac{tser(i, j) \times rate(i, j)}{buffer\_size(i, j)} + \frac{tserap \times rate(i, j)}{1024 \times video\_size} \right) \tag{5}$$

Since the service time for upstream processing of a single buffer at the CPU is *tseru(i,j)*, from equation (3),

$$T_{user} = tseru(i,j)/(1 - U_{ser}) \tag{6}$$

The PCI bus. Since *buffer_size* of data visits PCI bus-1 once in upstream, the resulting average bus utilization is *rate(i,j)/bw_pci*. For all active streams, we have:

$$U_{pci} = \sum_{i=1,}^{i=n} \sum_{j=1}^{j=nstr(i)} \frac{rate(i, j)}{bw\_pci} \tag{7}$$

Using U_{pci} and the bus service time of $(buffer\_size/bw\_pci)$ for shipping a $buffer\_size$ of data in Eq (3),

$$T_{pci} = (buffer\_size/bw\_pci)/(1 - U_{pci}) \tag{8}$$

The disk subsystem. The data throughput rate required is equally shared by the du SCSI buses. The SCSI bus service time to ship a $buffer\_size$ of data is given by $buffer\_size/bw\_scsi$ seconds. The utilization of a SCSI bus denoted by U_{scsi} is given by:

$$U_{scsi} = \sum_{i=1,}^{i=n} \sum_{j=1}^{j=nstr(i)} \frac{rate(i, j)}{bw\_scsi \times du} \tag{9}$$

Thus in view of equation (3) we have :

$$T_{scsi} = (buffer\_size(i,j)/bw\_scsi)/(1 - U_{scsi}) \tag{10}$$

Effect of parallelism. As done in[3], we account for the parallelism (pipe lining and look ahead processing features) in implementing the upstream scenario. A large disk cache eliminates T_{seek} for all sequential disk reads except the first one and DMA facilitates parallelism between CPU processing and data transfers from disks to CPU RAM. Further, data transfers from multiple disks and those from multiple SCSI host adapters across the PCI can be concurrent. These render $\hat{T}_{ufill} <$ T_{ufill}. Next determine $\hat{T}_{ufill}$. Now the maximum number of streams supported is $N = \sum_{i=1}^{i=n} nstr(i)$. For simplicity, assume that each stream has av_rate MB/second and uses server buffers of $buffer\_size$. Since the data visits the PCI bus only once during upstream, no speed up is possible for T_{pci}. Since the memory bandwidth $> bw\_pci$, for $du > 1$, usp, for T_{scsi} due to concurrency at the disk subsystem is:

$$usp = Minimum\{bw\_pci/bw\_scsi, N, du\} \tag{11}$$

For N>1, CPU processing and data transfer from the disks to server is done in parallel. In order to quantify the time saved, denote P_u to be the probability that the CPU does upstream processing while data is moved from the disk to the server (under multiprogramming situation). $P_u = $ (CPU utilization for upstream processing of *all the concurrent streams*)× (utilization of the highest utilized PCI or SCSI-2 bus on the disk to server path. Thus,

$$P_u = \left(\frac{N \times tseru \times av\_rate}{buffer\_size} \right) \times Maximum\left(\frac{N \times av\_rate}{bw\_pci}, \frac{N \times av\_rate}{bw\_scsi \times du} \right) \tag{12}$$

The CPU real time delay while a $buffer\_size$ of data is moved to server RAM from disks at the upstream is $P_u \times T_{user}$. The CPU delay that occurs beyond T_{upci} and T_{scsi}/usp in computing effective delay $\hat{T}_{ufill}$ is $(1-P_u) \times T_{user}$. Thus we have:

$$\hat{T}_{ufill} = (1 - P_u) \times T_{user} + T_{pci} + \frac{T_{uscsi}}{usp} \tag{13}$$

With each disk supporting one stream, in order that *client (i,j)* has no starvation the following must hold.

$$\frac{buffer\_size}{\hat{T}_{ufill}} \geq av\_rate \tag{14}$$

Note that if each disk were to concurrently support m streams at the rate of av_rate MB/second, then the following inequality must be satisfied in order that there is no starvation at any of the corresponding m clients.

$$\frac{buffer\_size}{\hat{T}_{ufill}} \geq m \times av\_rate \tag{15}$$

3.2 ANALYSIS OF DOWNSTREAM PROCESSING

The ATM packet delay within the network < 0.2 milliseconds. The utilization of processors at the switch and ATM boards are designed to be low and the queuing delays here are negligible. Consider the situation when all *clients* stream. Each ATM board serves its active clients simultaneously via the $(k \times k)$ switch. If the server and disk subsystem were able to stream all ATM boards, no clients will starve for data. Assume that the server sends data to ATM board *1, 2, 3, ..n* in a round robin fashion and the buffer $C(1,j,1)$ has been marked ready to be emptied by the *client(1,j)*, for all $0 < j \leq nstr(1)$. The *client(1,j)* for all $0 < j \leq nstr(1)$ empties the filled buffers in exactly *buffer_size(1,j)/rate(1,j)* seconds and during that period, if the system were to function properly, the server must fill the set of buffers: $\{C(p,q,1): p= 2, 3, 4, n; q=1, 2, 3.. nstr(p)\}$ and $\{C(1,j,2): j=1, 2, 3, 4, nstr(1)\}$. Note that these buffers are filled concurrently in the ATM network. The time T_{dfill} that elapses between a request for filling a client buffer and its completion is given by :

$$T_{dfill} = T_{dser} + T_{dpci} + T_{atmc} + T_{switch} + T_{atmc} + T_{dpci} + T_{client}. \tag{16}$$

where $T_{dser}, T_{client}, T_{switch}, T_{dpci}$, and T_{atmc} are the latencies at the server CPU, client CPU, switch, PCI bus and the ATM board respectively. The latencies on the passive devices are deterministic and those at the server and client are determined by the queuing model.

The bandwidth of the link from server ATM adapter to the switch is 155 Mbit/second and that from the switch to a client is 25 Mbit/second. Thus delay for a cell of size 53 Bytes from ATM adapter to the switch is 2.74 microseconds. The cell delay from the switch to the client ATM adapter is 16.96 microseconds. Assume 6 clients per switch so that no clients starve (a pessimistic case). Here ATM cell delay alone is 19.7 microseconds. The ATM switch is designed for *s millions of switching operations per second*. For s = 2, the switching delay per cell is 0.5 microsecond. The delay at the ATM board (corresponding to several cycles at 30 MHz clock rate) is of the order of microseconds and ATM cell delay would under 0.05 milliseconds.

The ATM adapter throughput depends on the AAL protocol and internet protocols used above the AAL. The OC-3 guarantees only $(48/53) \times 155$ (= 140) Mbit per second for ATM cell payload that contains protocols information and the effective MM data rate will be less. With 25% data rate for AAL-5 protocol overheads, the peak MM throughput would be 105 Mbit per second per ATM adapter at the server.

As shown in Figure 2, at each ATM board, there are two buffer pointer chain structures each of which has pointers to the data buffers in the server RAM so that the ATM adapter can set up a DMA operation to pull the multiplexed AAL-5 packet from the server. One of these has pointers to currently emptying server buffers of data associated with the streams serviced by the ATM adapter. The other points to respective server buffers currently being filled from disks. These structures are passed to respective ATM boards so that they can DMA the data buffers and other headers in the right order and perform SAR function on those and ship 53 bytes ATM packets over the OC-3 channel to the switch.. The AAL-5 packets for each adapters are assumed to be formed in a round robin fashion so that at the end of each period, there will be a set of new AAL-5 packets for each board. The server delay for processing a *buffer_size* of data at the downstream is:

$$T_{dser} = tserd/(1 - U_{ser}) \tag{17}$$

The following can be stated.

Assertion 2 A necessary condition for all the clients not to starve is:

$$N \times T_{dfill} \leq buffer\_size \times n/av\_rate \tag{18}$$

Although T_{dfill} is the real time to fill a client buffer from the server RAM, it takes $(N \times T_{dfill}/n)$ seconds to fill all N buffers (one each of the N clients). This is because the buffers of clients associated with a specific ATM adapter board are filled in parallel. Also, the server CPU processing and DMA transfer from server buffers to ATM adapter can be done in parallel. We account these by taking the effective total delay $\hat{T}_{dfill}$ for shipping a *buffer_size* of data to the clients. Let P_d be the probability that the server CPU does processing concurrently with the data transfer from the server to clients. Then,

$$\hat{T}_{dfill} = (1 - P_d) \times T_{dser} + T_{dpci} + T_n \tag{19}$$

where, $P_d = \left(\dfrac{N \times tser \times av\_rate}{buffer\_size} \right) \times \left(\dfrac{N \times av\_rate}{bw\_pci} \right)$ and T_n is the end to end delay at the ATM network. In order that none of the N clients starve at the downstream, the following condition should be met:

$$N \times \hat{T}_{ufill} \leq \frac{buffer\_size}{av\_rate} \qquad \text{and} \qquad N \times \hat{T}_{dfill} \leq \frac{n \times buffer\_size}{av\_rate} \tag{20}$$

2.3 COMMAND RESPONSE TIME AND NUMBER OF STREAMS

R, the mean response time for a play command is the sum of the delays for the request communication to the server, transfer a *buffer_size* of data from disk to the server, application processing at server, the downstream processing and that at the ATM network. R is given by:

$$R = T_{client} + T_{Ethernet} + T_{mpci} + T_{serap} + T_{ufill} + T_{dfill} \tag{21}$$

Response times for other commands such as pause, fast forward etc. can be similarly determined.

If the server has adequate memory and number of CPU (so that these are not system bottlenecks), Eqs. (1-21) are used to find N. Once N is found, Eqs. (5,7,9,21) are used to find component utilization and the response time. N is obtained by solving the following constraints.

1. N < (server memory (MB) - memory needs of software)/(2×*buffer_size*). Assures adequate server RAM.
2. *ndisk* > (Total video data in GB)/*disk_size* in GB. Assures number of disks to store needed videos.
3. N < *ndisk* × ⌊*disk_size/video_size*⌋. Adequacy of disk storage.
4. Equation (20). Avoids buffer starvation at the upstream and lets the clients streaming.
5. N < Minimum{measured PCI bus bandwidth, sum of bandwidths of SCSI buses, n×OC-3 bandwidth×data content, sum of the bandwidths of ATM links to clients)/*av_rate*. Data path hardware bandwidth limitation.
6. $R < x$. Mean response time for a key command must be within x seconds.
7. $N < n \times (k-1)$. Connectivity limitation at the ATM network.

2.4 EVALUATION OF ARCHITECTURAL ALTERNATIVES

The system has 28 disks distributed on 4 SCSI buses on a PCI bus. Current parameters values are: bw_pci = 80 MB/second, bw_scsi = 13.2 MB/second, av_rate = 1 MB/second, $tserd$ = 0.0331 second, $tseru$ = 0.0331 seconds. First we get the value of usp to find P_u. We use Eqs. 2 and 3 to find a conservative value for N.

$$T_{ufill} = \frac{33.1 \times 0.001}{(1 - \dfrac{2 \times 33.1 \times N}{1000})} + \frac{\dfrac{1}{80}}{(1 - \dfrac{N}{80})} + \frac{\dfrac{1}{13.2}}{(1 - \dfrac{N}{4 \times 13.2})}$$

Substituting the above in inequality (20) and solving for N we see that $N >$ 4 and usp = $Minimum\{80/13.4, 4, N\} = 4$. Now using Eq. (13) with $P_u = 0$, in inequality (20), we get:

$$\left(\frac{33.1 \times 0.001}{(1 - \dfrac{2 \times 33.1 \times N}{1000})} + \frac{\dfrac{1}{80}}{(1 - \dfrac{N}{80})} + \frac{\dfrac{1}{13.2 \times 4}}{(1 - \dfrac{N}{4 \times 13.2})} \right) \times N \leq 1$$

The maximum integer value of N satisfying the above inequality is 8. Substituting $N = 8$ in Eq. (12), we get $P_u = (8 \times 0.0331 \times 8/(4 \times 13.2) = 0.04$. Using this value in Eq. (13), we get:

$$\hat{T}_{ufill} = \frac{(1 - 0.04) \times 33.1 \times 0.001}{(1 - 2 \times 0.0331 \times N)} + \frac{\dfrac{1}{80}}{(1 - \dfrac{N}{80})} + \frac{\dfrac{1}{13.2 \times 4}}{(1 - \dfrac{N}{4 \times 13.2})}$$

Using inequality (20) we get:

$$\left(\frac{(1-0.04) \times 33.1 \times 0.001}{(1-0.0331 \times 8 \times N)} + \frac{\frac{1}{80}}{(1-\frac{N}{80})} + \frac{\frac{1}{13.2 \times 4}}{(1-\frac{N}{4 \times 13.2})} \right) \times N \le 1.$$

Solving the above, maximum of $N = 8$ is feasible. Using $N = 8$, and Eqns. (5, 7, and 10), we get:

$U_{ser} = 0.0662 \times 8 = 0.53.$ $U_{pci} = 8/80 = 0.10.$ $U_{scsi} = 8/4 \times 13.2 = 0.15.$

The server CPU delay is a dominant component in this. Let us see the impact of adding a sever CPU on N. This is equivalent to cutting the *tseru* and *tserd* to half its value with 1 CPU. Thus *tseru* = *tserd* = 0.0165 .

$$\left(\frac{16.5 \times 0.001}{(1-\frac{2 \times 16.5 \times N}{1000})} + \frac{\frac{1}{80}}{(1-\frac{N}{80})} + \frac{\frac{1}{13.2 \times 4}}{(1-\frac{N}{4 \times 13.2})} \right) \times N \le 1$$

$N = 14$ satisfies the above. Using $N = 14$ in Eq. (12), we get $P_u = 0.06$. Using this value in Eq. (13) we get:

$$\hat{T}_{ufill} = \frac{(1-0.06) \times 16.5 \times 0.001}{(1-0.0331 \times N)} + \frac{\frac{1}{80}}{(1-\frac{N}{80})} + \frac{\frac{1}{13.2 \times 4}}{(1-\frac{N}{4 \times 13.2})}$$

Using the above equation in inequality (20), we get a feasible $N = 14$. Since at the downstream, an ATM adapter can handle 6 clients, use of 3 ATM adapters and switches can support 14 streams with no starvation.

3. PEER-TO-PEER ARCHITECTURE AND SYSTEM CAPACITY

With peer-to-peer communication, a client sends a request to the server via the network and PCI bus to play a video. The server issues reads from the correct disk(s) into the SCSI adapter buffer. The SCSI adapter transfers a *buffer_size* of data via PCI bus and network into client buffer. Buffer chain pointers computed by server CPU are sent to the ATM adapter which DMA transfers appropriate data buffers from SCSI adapter and the ATM cells are transmitted from the OC-3 connection to the switch as in the server centric case. Each SCSI host adapters have 2 buffers and when one is being emptied the other may be filled by the disk subsystem. MM data does not visit the server memory, server manages data flow from disk to the clients. The PCI bus load here remains the same as that in the server-centric case. The load on the CPU memory bus is halved. CPU will need to process issues of disk reads based on the interrupts from the ATM boards.

The number of clients needing data rate of *rate* MB/second is determined by $T_{dsktoatm}$, the delay to ship a *buffer_size* of data from disk to a client buffer and without considering parallelism we have:

$$T_{dsktoatm} = T_{dcser} + T_{scsi} + T_{pci} + T_{atmc} + T_n \tag{23}$$

Let P_{dc} be the probability that CPU processes while data moves from the disk to ATM adapters. P_{dc} is given by (*CPU utilization*) × (*utilization of the maximum utilized bus amongst EEX, PCI and SCSIs*). Thus:

$$P_{dc} = \frac{N \times tdcser \times rate}{buffer\_size} \times Maximum\left(\frac{N \times rate}{du \times bw\_scsi}, \frac{N \times rate}{bw\_eex}, \frac{N \times rate}{bw\_pci} \right) \tag{24}$$

With parallelism, the effective delay for the data shipping when there are N streams is:

$$\hat{T}_{dsktocard} = (1 - P_{dc}) \times T_{dcser} + 2 \times T_{dcpci} + \frac{T_{uscsifw}}{usp} \tag{25}$$

where, *tdcser* and T_{dcser} are the CPU service time and server delay associated with the shipping of data buffer. Since the data visits both PCI buses, the delay on PCI buses is twice that on a single PCI bus. The delay at the EEX bus is negligible since its bandwidth is more than double that of the PCI bus. With *buffer_size* = 1 MB, *tdcser* = 0.0035 seconds, N *streams* distributed over 4 ATM adapters, we get P_{dc} = 0.0000663N^2, (*usp*= 4). On substitution of these values in (25) we get:

$$\hat{T}_{disktoatm} = \frac{(1 - 0.0000663N^2) \times 3.5 \times 0.001}{1 - N \times 0.0035} + \frac{\frac{1}{13.2}}{(1 - \frac{N}{4 \times 13.2}) \times 4} + \frac{\frac{2}{80}}{(1 - \frac{N}{80})}$$

In order that no ATM adapters starve, $4 \times \hat{T}_{disktoatm} \leq 1$. That is: *0.001[3.5(1-0.0000663N²)/(1-0.0035N) + 18.94/(1-0.0189N) + 25/(1-0.0125N))×4≤ 1*. By trial and error, largest N satisfying all the constraints is 47.

4. SUMMARY AND POSSIBLE DIRECTIONS

Using the processing scenario based queuing network model, the performance and capacity of AT&T MMC system for server-centric and peer-to-peer architectures has been modeled and illustrated. The models take into account of possible parallel processing. Our models and the results are partly validated by laboratory measurements, show that the peer-to-peer scenario increases the system capacity significantly. Our work is applicable for any client-server architecture and applicable in sensitivity analysis or simulation studies to account for hardware and software enhancements to existing systems.

In production systems, some protection against malicious users do exist and bit rates needed by users would be known a priori or dynamically from the applications. Here, a resource manager can maintain up-to-date view of resource utilization or availability. An admission control algorithm uses these information with our model to decide if a request can be granted providing GQOS to all requests. The detailed algorithm is presented in[5]. Future works include extending the approach to massively parallel and distributed architectures, interactive and collaborative MM applications. The results of our analysis are expected to yield simpler admission control schemes on specific systems. The model can be extended to incorporate results on optimal striping and placement of video on the disk subsystem[6]. These will be pursued elsewhere.

REFERENCES

1. T. Suzuki, ATM adaptation layer protocol, IEEE *Communications Magazine 32:4, PP80-83, April 1994.*

2. K. V. Bhat, Performance modeling and analysis for AT&T 3416 based multimedia server architecture, *Proc. Int. Conf. on Telecommunications 95, April 1995.*

3. Y. Chang, D. Coggins, D. Pitt, D. Skeelern, M. Thapar, and C. Venkatraman, An open systems approach to video on demand, *IEEE Communications Magazine, PP 68-80, May 1994.*

4. M. K. Molloy, Fundamentals of Performance Modeling, Macmillan Publishing, New York, 1989.

5. K. V. Bhat, Performance model directed admission control for multimedia enabled servers, *to appear in Proc. Int. Symp. on Communications 1995.*

6. B. Ozden, R. Rastogi and A. Silberschatz, Disk striping in video server environments, *AT&T Bell Lab.Tech. Rep. 113830-950220-01, Murray Hill 1994.*

Acknowledgments. Thanks are due to Dr. B. R. Rowland, Mike Ryder, Catherine Boss, Steve Miller, Larry Pelletier, Dr. Chas Gimarc, Dave Wood of AT&T GIS and Dr. Vijay Kumar and Sunder Ratnavelu from AT&T Bell laboratories for valuable comments.

ARCHITECTURAL TRADE-OFFS FOR IMPLEMENTING VIDEO ENCODERS

Cesar A. Gonzales and Elliot Linzer
IBM T. J. Watson Research Center
Yorktown Heights, NY 10598

INTRODUCTION

Designing VLSI chips to implement video encoders capable of realtime MPEG-2 compression at MP@ML or MP@HL levels of resolution is an extremely challenging task. The computational requirements are enormous and, to date, there are no single chip solutions that support the full richness of functionality available in MPEG-2's Main Profile. As silicon technology continues to evolve, there is no doubt that function-rich, single-chip, MPEG-2 encoders will exist in the future. However, with today's technology we must be content with multiple-chip solutions or, alternatively, with less than full Main Profile functionality in a single chip. If a full-function MP@ML implementation is the goal, a VLSI designer needs to make a number of architectural decisions that impact the flexibility and complexity of the chip or chips, as well as the cost of a system that incorporates them. We distinguish between two approaches to this problem.

In the first approach the emphasis is put on the flexibility to implement multiple levels of encoding performance. In this context performance means video resolution. To accomplish this a video picture is segmented into horizontal stripes of macroblocks and one encoder chip is assigned for each one of the stripes. We refer to this approach as "performance scalable architecture," or PSA for short.

In the second approach, the emphasis is on minimizing the cost of the overall encoding solution at a target level of performance. However, as we will discuss later in this paper, other forms of scalability are also possible here. For that reason, we refer to this approach as "functional scalable architecture," or FSA for short. As opposed to PSA chips, multiple FSA chips will all operate on full video pictures.

Depending on the application at hand, either approach may be more appropriate. In this paper we study the characteristics of both architectural approaches. We point out the advantages and disadvantages in each case. It is not the objective of this paper to conclude that one approach is better than the other, but to guide system designers in evaluating which of these two architectures may be more appropriate for their specific application. Needless to say, there are several other features that also need to be considered in making the right choice, not the least of which is video quality. However, we don't discuss those issues here.

Multimedia Communications and Video Coding
Edited by Y. Wang *et al.*, Plenum Press, New York, 1996

PROFILE AND LEVEL ASSUMPTIONS

The characteristics of the two encoder architectures that we propose to study can be described in terms of the following parameters:

- Flexibility: By this we mean the ability to configure the encoder chips to address different applications.
- Total memory requirement: The MPEG encoding process requires a minimum number of frames stored in memory. The choice of architecture can increase this minimum requirement, thus increasing the cost of the total system.
- Memory bandwidth: This can also impact the system cost as a larger bandwidth translates into more expensive memory, e.g. SRAM, VRAM, or SDRAM. Alternatively, larger bandwidth can also be handled with wider DRAM buses, which can also result in less than optimal utilization of available memory.

To estimate these parameters we need to define more precisely the target encoder. For this study we assume an MPEG-2 encoder with the following features:

- CCIR 601 format and pixel rates
- Full I, P, and B picture encoding support
- M=3
- Support for a motion estimation range of ±64 between reference pictures; horizontally and vertically. This is half of the maximum allowed in MP@ML, but large enough to handle most complex video content, including sports. We will assume a hierarchical motion estimation algorithm so that reasonable computations and memory bandwidth result.

Other elements of MPEG-2's main profile are of lesser importance in determining the memory size and bandwidth requirements of a particular architectural approach.

FSA ENCODERS

A high level block diagram of an FSA encoder is shown in Figure 1. Examples of this type of architecture have been reported in the literature (Matsumura et al, 1995; Armer et al, 1995; Ngai et al, 1995).

Flexibility

There exist many different ways in which the MPEG-2 encoder functions can be partitioned into multiple chips. Most approaches require that all chips work together to implement the encoding function, thus offering little flexibility. Ngai et al (1995) reported one approach in which the MPEG-2 function has been partitioned into three chips, such that the first chip alone is capable of implementing an I-only MPEG-2 compression. Adding the extra chips will increase the functionality of the encoder such that the full complement of three chips implements the full syntax of MPEG-2's main profile. This property is what led us to the name of "functional scalable architecture."

Memory Size

For M=3, a minimum of three input frame buffers are required to store two B pictures, plus the most recent I or P picture. Motion compensation requires the storage of two reconstructed reference pictures. Miscellaneous storage may be required for the output compressed stream and down sized reference pictures used in hierarchical motion estimation. In addition, bandwidth requirements plus the granularity of RAM components will also determine the total requirement for system memory. Separating input memory from reference memory will also introduce inefficiencies.

Typically, a total of 4-6 Mbytes of system memory will be required.

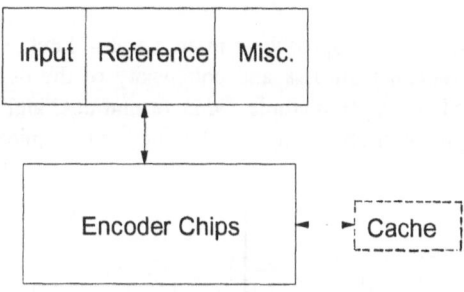

Fig 1. FSA Encoding System

Memory Bandwidth

During one frame period the following memory accesses take place:

- 1 Input frame is stored
- 1 frame is read from storage for processing
- For P and I picture processing, 1 reconstructed frame is stored to serve as reference
- For P pictures, 1 frame is read for motion compensation. For B pictures, 2 frames are read for the same purpose

Thus ignoring motion estimation--regardless of P or B picture processing--a total of 4 frames need to be moved between memory and the encoder chips. Hierarchical motion estimation can be implemented with or without full caching of the search windows. Without full caching, we estimate the number of memory accesses to be about 20 luminance frames per frame period. With full caching, the primary memory access is 2 luminance frames. Since one frame period represents about 15 Mbytes/s, and one luminance frame represents about 10 Mbytes/s of memory bandwidth, the total primary memory bandwidth[1] is approximately:

Bandwidth without full cache: 260 Mbytes/s
Bandwidth with full cache: 80 Mbytes/s

---

[1]   Because of MPEG-2's dual prime, P frames require as much bandwidth as B frames.

In this analysis we have ignored other small contributors to the memory bandwidth such as the compressed bitstream write/read. While 80 Mbytes/s is manageable with a 64 bit data bus and standard DRAM, this implementation will require substantial cache memory which needs to be considered in the total system cost.

PSA ENCODERS

A high level block diagram of a PSA encoder is shown in Figure 2. One example of this type of implementation is described in Wayner (1994).

Flexibility

The main advantage of this approach is in this area. First, a single IC needs to be designed, presumably reducing the risk and complexity of the design task. Second, this component can be used to target multiple video resolutions, and even multiple levels of functionality by scaling the number of chips used in the target application.

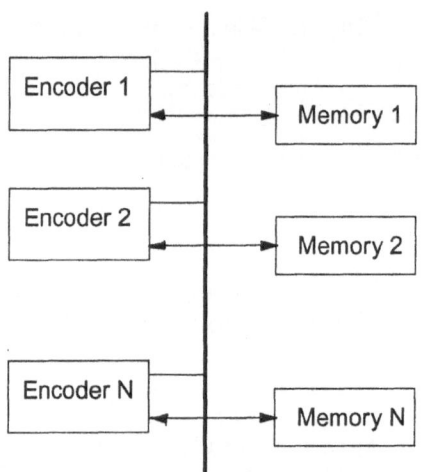

Fig 2. PSA Encoder

Memory size

The primary memory requirements for a PSA encoder are larger than those of an FSA encoder. First the input and reconstruct memory is distributed among a number of memory banks. Depending on the total number of encoder chips, i.e. the number of macroblock segments assigned to each processor, the inefficiencies of the granularity of commercial RAM components can play a very significant role. Another significant source of inefficiency is due to the fact that contiguous memory banks need to store overlapping portions of reference data. This is to allow motion estimation and motion compensation of the top and bottom row of macroblocks in each segment. Assuming:

- N encoder chips
- W*H pixels
- V, motion vector range

Each processor needs $W * H/N + W * (2V + 16)$ pixels for each reference frame. The total reference memory size for all encoder processors can easily be calculated as follows:

N	Mbytes/encoder	Total Mbytes
4	0.6	2.4
8	0.45	3.6
16	0.37	5.9

These totals should be compared against a requirement for 1.2 Mbytes of reference memory in the FSA architecture. As mentioned before, bandwidth requirements and RAM granularity can make matters worse.

PSA systems can incur in 50% or greater size inefficiencies as compared to FSA systems.

Memory Bandwidth

The overall memory bandwidth in a PSA system is increased because of the need for broadcasting the overlapping portions of reference data among contiguous memory banks. On the other hand, this overall bandwidth is then distributed among the number of encoder chips being used. It is conceivable then that a system may be implemented without the need of a full cache and still use reasonable size DRAM memory banks. Again, bus width and DRAM granularity play an important role in determining the final memory requirements.

CONCLUSIONS

We have studied two architectural approaches to the implementation of MP@ML MPEG-2 encoders. Both have interesting advantages and disadvantages in terms of flexibility and overall system cost. Both of these factors need to be considered to determine which chip set is more appropriate in each individual application.

References

T. Matsumara, H. Segawa, S. Kumaki, Y. Matsuura, A. Hanami, H. Yamaoka, R. Streitengerger, S. Nakagawa, K. Ishibara, T. Kasezawa, Y. Ajioka, A. Maeda, M. Yoshimoto, A chip set architecture for programmable real-time MPEG2 encoder, IEEE Custom Integrated Circuits Conference, 1995.

J. Armer, J. Bard, B. Canfield, D. Charlot, S. Freeman, A. Graf, R. Kessler, G. Lamouroux, W. Mayweather, M. Patti, P. Paul, A. Pirson, F. Rominger, D. Teichner, A chip set for MPEG-2 video encoding, IEEE Custom Integrated Circuits Conference, 1995.

A. Ngai, J. Kaczmarczyk, J. Murdock, S. Pokrinchak, VLSI architecture of an I-frame encoder for MPEG-2 video compression, Hot Chips VII, Stanford, CA, 1995.

P. Wayner, Digital video goes real-time, Byte magazine, January, 1994

MODEL-BASED VIDEO CODING
— SOME CHALLENGING ISSUES

Thomas S. Huang   Li-an Tang

Beckman Institute and Coordinated Science Laboratory
University of Illinois at Urbana-Champaign
405 N. Mathews Avenue
Urbana, IL 61801

INTRODUCTION

Most of existing image coding techniques fall into the category of so-called waveform based methods with limited compression factors. In recent years, a new video compression approach using model-based coding techniques arises as more and more researchers seek for very low bit-rate video compression[1]–[4]. The main idea of this approach is to use 3D models of moving objects and derive the motion parameters associated with the model structure, then the original video sequence could be approximately reconstructed using the 3D models and the motion parameters. At present, in most research activities on model-based coding systems, the dominant parts in the video scenes are human faces talking and showing expressions, for example, in videophone and teleconferencing. Therefore, three research topics – the face model, facial motion analysis and facial expression synthesis, are of paramount importance and many researchers in both computer vision and computer graphics have contributed to these themes significantly.

3D Face Models.   This is a widely addressed problem in the computer animation area. Different kinds of 3D face models have been developed. Parke[5] first introduced a parameterized face model to facial animation. Some other types of face models, such as the physics-based models by Terzopoulos et al.[3], the anatomy-based models by Platt and Badler[6] and by Waters[7], were built thereafter.

According to the Facial Action Coding System(FACS)[8], every facial motion could be defined as a linear combination of several independent facial muscle movements called "Action Units"(AUs). The Candide-model[9] was the simplest geometric model which could perform almost all AUs quantitatively. It has been widely used in model-based image coding and some extended versions of this model have been developed[2],[4].

There were also a number of papers dealing with fitting a generic 3D face model

Multimedia Communications and Video Coding
Edited by Y. Wang *et al.*, Plenum Press, New York, 1996

to some specific person's face using several available 2D images, so that a personalized face model could be generated for further application. Some used interactive software and others tried automated ways to do it provided certain constraints were applied to the face images. Akimoto *et al.*[10] proposed a fully automated approach to creating a specific face model for a person based on two orthogonal 2D face images, namely, the front and side profile views of that person, although the constraints were rather stringent.

Facial Motion Analysis. In computer vision, many algorithms of motion estimation and feature tracking[11] could be adopted for analyzing human facial motions, such as optical flow-based algorithms[2],[12],[13], feature-based algorithms[3],[14] and stereo-based algorithms[15].

Facial Expression Synthesis. This issue was extensively studied by many researchers in computer graphics. Several techniques have been developed for facial expression animation[16], such as key framing, speech driven animation and using FACS parameters.

We have been developing a complete human face modeling system and trying to solve general problems related to it. In this paper, we shall describe an approach to synthesizing different human facial motions based on 3D face model and motion vectors using interpolation techniques. As the basis of our model-based video coding system, this method is applied to various video sequences with different facial expressions and some preliminary compression results are shown in this paper. This paper is concluded by discussing some remaining challenging problems related to model-based video coding system.

MODEL-BASED CODING SYSTEM

The complete model-based coding system we are currently developing is shown in Fig. 1.

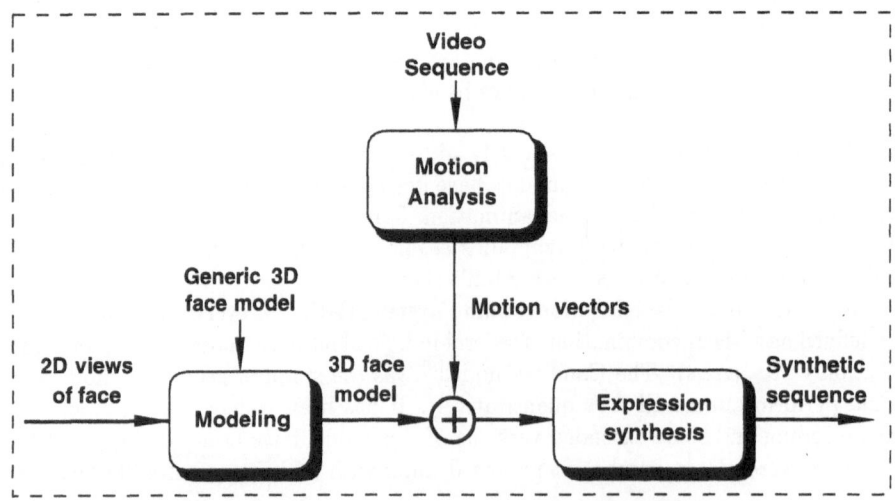

Figure 1. The model-based coding system.

The system consists of three parts. (1) 3D face modeling; (2) motion analysis to obtain facial motion vectors; (3) facial expression synthesis using the 3D face model and the motion vectors.

3D Face Modeling

The 3D face modeling is done by fitting a generic 3D wire-frame face model to several 2D views of a person's face images. The generic 3D face model is a parameterized wire-frame model. The face surface is approximated by triangular meshes. We are using an extended version of the Aizawa-model[17]. Currently it consists of 353 vertices and 578 triangles. An interactive human face modeling system has been developed to create 3D face model for a particular person. Several key vertices on the generic face model are first interactively mapped onto the face images. The rest of the vertices are located based on these key vertices using a set of anthropometrically designed interpolation functions. By this process, we can obtain a 3D wire-frame face model which exhibits the face shape of the person we are modeling. Moreover, the intensities are taken from the 2D face images and mapped to the 3D face model by texture mapping. In this way, a complete 3D face model which possesses both the shape and the texture of the person's face is built. Fig. 2 shows the face models and the fitting procedure.

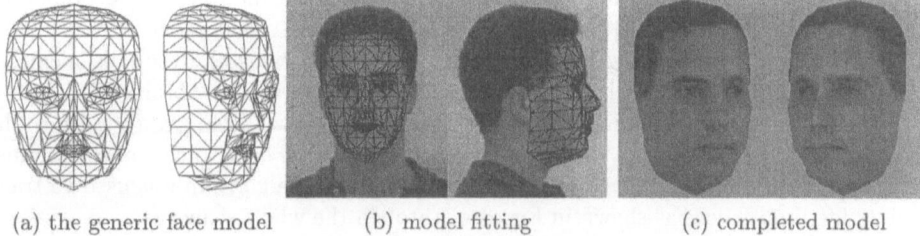

(a) the generic face model   (b) model fitting   (c) completed model

Figure 2. The modeling procedure.

Facial Motion Analysis

The facial motions in video sequences are represented by motion trajectories of a number of key feature vertices. Unlike FACS-based systems, our system will directly use these motion parameters to synthesize facial expressions instead of converting them into combinations of several facial action units. The latter seems impractical at the moment since it is not quite clear how a specific facial expression can be represented by a combination of action units. The motion trajectory of each feature vertex is estimated using a feature tracking algorithm[18] which uses a motion-based template matching strategy to find the motion of a chosen point. The motions of non feature vertices are derived using the interpolation functions as in the face modeling process. The motion trajectories are further linearized into several line segments. Fig. 3 shows a smiling sequence and the estimated motion trajectories for all model vertices.

Facial Expression Synthesis

We now have both the 3D face model and the facial motion parameters. We are ready to reconstruct the original facial expressions on the face model. The main idea of facial expression synthesis is to dynamically deform the 3D face model using the

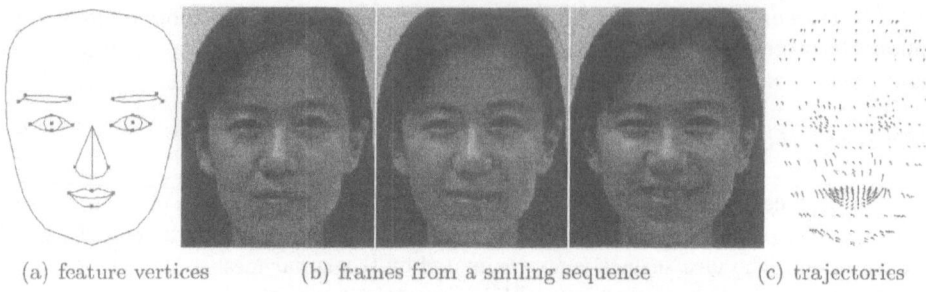

| (a) feature vertices | (b) frames from a smiling sequence | (c) trajectories |

Figure 3. Obtaining motion trajectories.

estimated motion parameters[19]. Both the shape and the intensities of the face model are modified using linear interpolation. Thus an approximation to the original video sequence can be synthesized.

EXPERIMENT FOR VIDEO COMPRESSION/INTERPOLATION

The human face modeling system was applied to the interpolation of video sequences. The 3D face model of the person appearing in the scene was built in advance. The model was then mapped to the person's face in the video and the movements of feature vertices of the model were tracked and the motion vectors obtained.

We collected a real smiling sequence of 20 frames as shown in Fig. 4(a). Each frame had size of 352×288 pixels with 8 bits per pixel, so the total data volume of the original sequence was 1.98 megabytes. The facial feature tracking algorithm was used to track all 22 feature vertices as shown in Fig. 3(a) through the whole sequence.

By analyzing the motion parameters obtained from the tracking algorithm, we found that the facial motions during the period of this sequence was relatively smooth. Therefore an interpolation procedure was applied to create all 18 in-between frames using the two end frames(frame 1 and frame 20). A sequence shown in Fig. 4(b) was synthesized by interpolating both the motion and intensity changes for each vertex.

The two end frames were encoded using JPEG which yielded $5,551$ bytes and $5,622$ bytes respectively. Among all 353 vertices on the face model, only 33 of them were chosen as key vertices for the interpolation. Each element of the vertex was specified by 2 bytes which required $33 \times 6 = 198$ bytes of storage. The motion vectors were actually represented in 2D and each of them had magnitude less than 16, so 1 byte was used to store the motion vector for each feature vertex which contributed other 22 bytes to the total data volume. The final amount of storage needed to reconstruct this sequence using our method was $11,393$ bytes which reflected a compression factor of 178. Assuming standard video rate, *i.e.*, 30frames/second, this method gives a bit rate of 136.7 kbps.

As a comparison, we also encoded this sequence using CCITT H.261 videoconferencing coding standards, as shown in Fig. 4(c), which gave a total compressed data volume of $24,591$ bytes. Thus the compression factor of H.261 was 82.4. The bit rate of H.261 codec was 288.2 kbps based on the same video rate. The compression factor of the model-based coding system was 116% larger than that of H.261's.

To see how accurately the original sequence was reconstructed, we calculated mean-squared errors between the coded and the original sequences. The MSE's between corresponding frames were computed and shown in Fig. 5. We emphasize that although

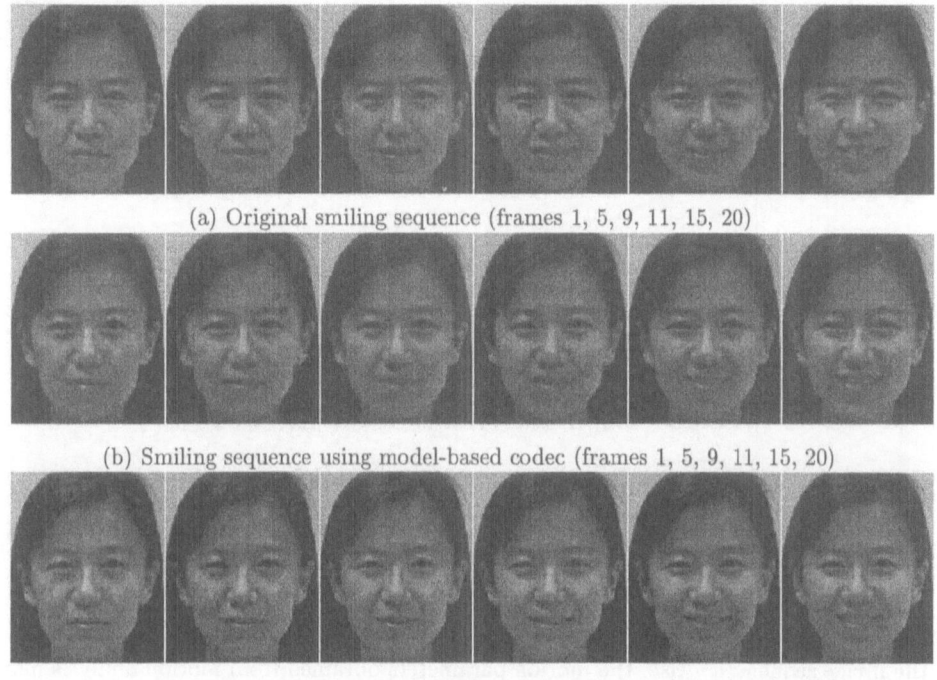

(a) Original smiling sequence (frames 1, 5, 9, 11, 15, 20)

(b) Smiling sequence using model-based codec (frames 1, 5, 9, 11, 15, 20)

(c) Smiling sequence using H.261 codec (frames 1, 5, 9, 11, 15, 20)

Figure 4. The original sequence vs. coded sequences.

the model-based results had higher MSE than the H.261 results, subjectively the visual qualities of the two sequences were virtually identical, as we could see from Fig. 4.

CONCLUSION

We have described a human face modeling system and its application to very low bit-rate video compression. The results show that this approaching is promising. However, a number of challenging problems still remain.

Better face models.   3D modeling of human face is perhaps the most important issue. Existing models are either too complex for facial motion analysis or too simple for expression synthesis. A hierarchical face model may be needed to better cope with the face modeling problem. Also, models for different facial components are needed for better results, for example, the tongue model and the hair model.

Automated face model fitting.   This is an essential requirement for many applications. It will be interesting to see how automatically a generic face model can be fitted to a specific person's face images if certain constraints are maintained. Further effort can be made to remove these constraints while preserving the automation.

Accuracy of face modeling.   From the experiments we have made so far, large errors occur in areas where no feature points can be located from the given views, *e.g.*, the cheek areas. One possible solution to this is to use multiple views. The shape of the cheeks can be modified using some intermediate views.

3D motion estimation and motion decomposition.   It is difficult to find robust algorithms to estimate 3D head motion from a set of 2D corresponding points in

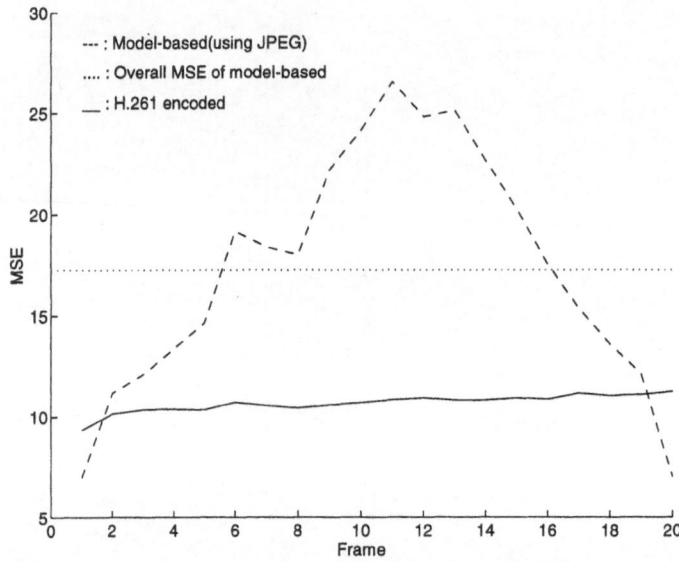

Figure 5. MSE per frame between the original sequence and encoded sequences.

the image sequences. Also, the motion parameters obtained from motion analysis may consist of both global head motions and local facial expressions. In many applications, we must deal with the motions separately, which means the motion parameters have to be decomposed.

Lighting effects on the face model. The original face model is completed by mapping the 2D image intensities to the 3D wire-frame face model. The texture of the face model does not change when it moves, which might be unnatural. A shading model should be added to the face model for more natural syntheses.

ACKNOWLEDGEMENTS

This work was supported by National Science Foundation Grant IRI-8908255 and by a grant from Texas Instruments.

REFERENCE

[1] K. Aizawa and T. S. Huang, Model-based image coding: advanced video coding techniques for very low bit-rate applications, *Proceedings of the IEEE*, Vol. 83, No. 2, February 1995, pp. 259-271.

[2] H. Li, P. Roivainen and R. Forchheimer, 3-D motion estimation in model-based facial image coding, *IEEE Transactions on Pattern Analysis and Machine Intelligence*, Vol. 15, No. 6, June 1993, pp. 545-555.

[3] D. Terzopoulos and K. Waters, Analysis and synthesis of facial image sequences using physical and anatomical models, *IEEE Transactions on Pattern Analysis and Machine Intelligence*, Vol. 15, No. 6, June 1993, pp. 569-579.

[4] W. J. Welsh, *Model-based Coding of Images*, Ph.D. dissertation, British Telecom Research Laboratories, January 1991.

[5] F. I. Parke, Parameterized models for facial animation, *IEEE Computer Graphics and Applications,* Vol. 12, November 1982, pp. 61-68.

[6] S. M. Platt and N. I. Badler, Animating facial expressions, *Computer Graphics,* Vol. 15, No. 3, 1981, pp. 245-252.

[7] K. Waters, A muscle model for animating three-dimensional facial expression, *Computer Graphics,* Vol. 21, No. 4, July 1987, pp. 17-24.

[8] P. Ekman and W. V. Friesen, *Facial Action Coding System,* Consulting Psychologists Press, Inc., Palo Alto, California, 1978.

[9] M. Rydfalk, *CANDIDE: A Parameterized Face,* Technical Report LiTH-ISY-I-0866, Department of Electrical Engineering, Linköping University, Sweden, October 1987.

[10] T. Akimoto and Y. Suenaga, 3D facial model creation using generic model and front and side views of face, *IEICE Transactions on Information & System,* Vol. E75-D, No. 2, March 1992, pp. 191-197.

[11] J. K. Aggarwal and N. Nandhakumar, On the computation of motion from sequences of images – a review, in *Proceedings of the IEEE,* August, 1988, pp. 917-935.

[12] R. Koch, Dynamic 3-D scene analysis through synthesis feedback control, *IEEE Transactions on Pattern Analysis and Machine Intelligence,* Vol. 15, No. 6, June 1993, pp. 556-568.

[13] K. Mase, Recognition of facial expression from optical flow, *IEICE Transactions,* Vol. E74, No. 10, October 1991, pp. 3474-3482.

[14] M. Kass, A. Witkin and D. Terzopoulos, Snakes: active contour models, in *Proceedings of International Conference on Computer Vision,* 1987, pp. 259-269.

[15] H. Agawa, G. Xu, etc. Image analysis for face modelling and facial image reconstruction, *SPIE: Visual Communications and image Processing,* Vol. 1360, 1990, pp. 1184-1197.

[16] F. I. Parke, Techniques for Facial Animation, in *New Trends in Animation and Visualization,* Eds. N. M. Thalmann and D. Thalmann, John Wiley & Sons Inc., 1991, pp. 229-241.

[17] K. Aizawa and H. Harashima, Model-based analysis synthesis image coding system for a person's face, *Signal Processing: Image Communication,* Vol. 1, No. 2, 1989, pp. 139-152.

[18] L. Tang, L. S. Chen, Y. Kong, T. S. Huang and C. R. Lansing, Performance evaluation of a facial feature tracking algorithm, in *Proceedings of NSF/ARPA Workshop: Performance vs. Methodology in Computer Vision,* Seattle, Washington, June 1994, pp. 218-226.

[19] L. Tang and T. S. Huang, Analysis-based facial expression synthesis, in *Proceedings of IEEE International Conference on Image Processing,* Vol. III, Austin, Texas, November 1994, pp. 98-102.

EXPLOITATION OF SPATIO-TEMPORAL INTER-CORRELATION AMONG MOTION, SEGMENTATION AND INTENSITY FIELDS FOR VERY LOW BIT RATE CODING OF VIDEO[1]

Aggelos K. Katsaggelos[†] and Taner Özçelik[‡]

[†]Northwestern University
Department of EECS
Evanston, IL 60208

[‡]SONY Electronics, Inc.
3300 Zanker Road
San Jose, CA 95134

INTRODUCTION

During the last decade there has been a dramatic increase in the number of the applications requiring video compression techniques. These applications, which range from high definition television (HDTV) and digital cable to video conferencing and picture phones, have varying bandwidth requirements. The challenging general problem is the representation of full motion video at very low bit rates (VLBR), possibly as low as 8 kbits/sec. The research that proposes workable solutions to the problem will directly impact the future developments and possible applications in the area, as well as, one of the functionalities of a new standard (MPEG-4). Potential applications include videophone, multi-media electronic mail, remote sensing, electronic newspapers, interactive multi-media databases, multi-media videotex, video games, interactive computer imagery, multi-media annotation, surveillance, telemedicine and communication aids for the hearing impaired.

There has been quite a few approaches proposed for the VLBR coding of video over the past few years. Although these approaches are quite different in nature, they can be grouped into: A) Traditional, B) Model-Based, and C) Object-Oriented. Traditional approaches are represented by the block-based approaches, where the scene is divided into blocks without paying attention to the content of the scene. Some of the already developed video coding standards such as H.261, H.263, MPEG-1 and MPEG-2 fall into this category. The traditional approaches decouple the video coding problem into 2 stages. The temporal decorrelation is achieved in the first stage through motion compensation, and the spatial decorrelation is achieved in the second stage using a transformation. Decoupled decorrelation of the video source forms one of the disadvantages of these techniques. On the other hand, since both the motion compensation and spatial transformation are block-based, objectionable artifacts occur in the reconstructed video. For instance, block-based motion models fail in cases of two or more differently moving objects in the same block. Thus, although the block-based approach minimizes the overhead information transmitted over the channel, it fails to provide high quality reconstructed video at very low bit rates.

Model-based techniques, on the other hand, rely on modeling certain objects of interest in the scene by structural models such as wireframe models[1]. Depending on the specific tech-

---

[1]This work was partially supported by Motorola, Inc.

nique that is used, this approach may be heavily dependent on the "type" of the sequences, i.e., the models assumed for the proper performance of the algorithm. This restriction represents a severe limitation to the wide use of model-based techniques.

Object-oriented approaches[2] analyze the sequence into several objects, followed by the synthesis of the objects to determine what needs to be transmitted across the channel. Object-oriented techniques are sequence independent and offer several advantages over traditional techniques. Since the scene is analyzed into objects, artifacts exhibit themselves as geometrical distortions, which are less objectionable than blocking or mosquito artifacts, exhibited by traditional techniques. Additionally, since object-based analysis is performed, more accurate motion compensation is obtained. The reported results in this area suffer, however, from the sizeable overhead necessary to represent the objects, which needs to be transmitted[3].

In this paper, we propose a video coding scheme which addresses some of the shortcomings of existing VLBR video coding schemes. We propose an approach to exploit the existing inter-correlations among the various components into which we decompose an image sequence[4]. A reduced overhead bit-rate results. We therefore address the shortcoming of existing approaches that of providing decoupled solution steps to problems which are inherently coupled. The proposed codec follows the *estimator encoder-decoder* paradigm we have been advocating[4,5,6], according to which additional intelligence is given to both encoder and decoder for solving estimation problems.

FRAMEWORK OF THE PROPOSED CODING APPROACH

In object-oriented coding images are first analyzed and separated into differently moving objects. The information pertaining to each object in terms of its motion, shape and intensity is then coded and transmitted through the channel. With such an approach a scene-optimized coding technique is generated. In this paper, we present a hybrid object-oriented approach to VLBR video coding. Our approach also aims at analyzing the image into separate objects according to their motion, shape and intensity prior to coding. It makes use, however, of the inter-correlations between the three parameter sets in order to reduce the amount of information to be transmitted over the channel. In other words, we make use of the fact that the three fields, the displacement vector field (DVF), the segmentation field, and the intensity field, are inter-correlated both spatially and temporally. In order to exploit these correlations we propose the coupled estimation and coding of the three fields instead of separating them into independent sets[2]. Thus, we divide the information that is necessary to reconstruct a scene into two parts: the information that is predictable by the decoder and the information that needs to be transmitted over the channel. The decoder is given therefore additional intelligence by which it estimates the part of the information which is not transmitted. On the other hand, the encoder dynamically decides at each frame which part of the information is predictable and which is not. The exploitation of the inter-correlations between the three fields not only assures more accurate estimation of the fields, but guarantees more efficient transmission of the video information. This is simply achieved by eliminating the need to transmit certain overhead information when it can be predicted accurately by the decoder.

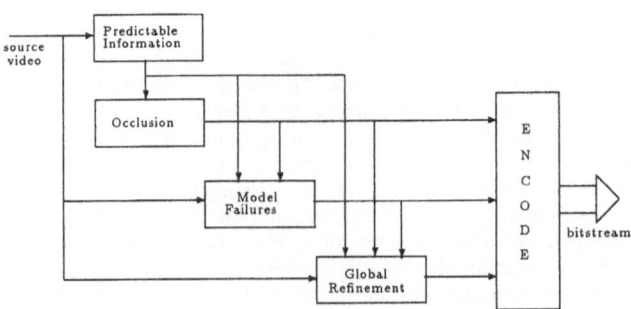

Figure 1: Hierarchical Coding Model

Let $I_k^p(\vec{r})$ denote the predictable information pertaining to the current frame k of an image sequence, at location $\vec{r}$, and $I_k^u(\vec{r})$ the unpredictable information, which is transmitted separately over the channel. The union of the two pieces of information, i.e.,

$$I_k(\vec{r}) = I_k^p(\vec{r}) \cup I_k^u(\vec{r}).$$ (1)

results in the information used to reconstruct the current frame. $I_k^p(\vec{r})$ refers to, for example, the predictable part of the DVF, the segmentation field or the intensity field. On the other hand unpredictable information refers to failures in motion, segmentation or intensity models. Thus, while the decoder regenerates the predictable part of the information, the unpredictable part needs to be transmitted. More specifically, the regeneration of the predictable information consists of the prediction of the motion and segmentation fields and the recovery of the intensity field based on assumed priori models. The unpredictable part of the information is described in a hierarchical fashion. This gives the encoder the flexibility to control the bit rate, as well as, the reconstruction quality efficiently. In the proposed coding scheme the unpredictable area information is divided into 3 hierarchical levels, as shown in Fig. 1. Based on the motion and segmentation models occluded areas are first detected. Due to the specific motion prediction model that is employed, the DVF in the covered areas is transmitted. Areas where the motion prediction model fails are detected next. Intensity information pertaining to these areas is transmitted. Finally, depending on the bits spent and the reconstruction quality, global refinement information can be transmitted as well.

MOTION AND SEGMENTATION ESTIMATION AND PREDICTION

The temporal redundancy in image sequences is removed with the use of motion estimation and compensation. Block-matching motion estimation techniques, which assign one vector to each square block, are widely used and are part of the video coding standards. Spatio-temporal gradient motion estimation techniques assign one vector to each pixel. They are capable of better describing the motion of objects, due to the use of a dense DVF. Their disadvantage is that the dense DVF can consume a large part of the bit budget if it is to be transmitted and is not efficiently compressed. In the proposed codec we utilize a recursive spatio-temporal gradient technique to estimate the dense motion field, as well as, its segmentation[7]. An object-based motion field is then obtained by a post-processing technique applied to the dense vector field.

Let the dense DVF of the current frame be denoted by $\hat{D}_k = \{\hat{d}_k(m,n)\}$, where $\hat{d}_k(m,n)$ denotes the vector at location (m,n), the object-based DVF by $\hat{D}_k^o = \{\hat{d}_k^o(m,n)\}$, and the segmentation field by $\hat{S}_k = \{\hat{s}_{d_k}(m,n)\}$ (henceforth, $(\hat{})$ denotes an estimated field). The motion and segmentation fields are estimated simultaneously in order to exploit their correlation. It is clear that the segmentation and the motion information together can be a large overhead to the bit rate if they are not coded efficiently. In order to reduce the amount of information that needs to be transmitted over the channel, we exploit the temporal correlation between the current and the past DVFs, as well as, the correlation between the current and the past segmentations. The current segmentation field, $S_k(\vec{r})$ is first predicted according to

$$\tilde{S}_k = \beta(\hat{S}_{k-1}, \hat{D}_{k-1}^o),$$ (2)

where β denotes a prediction operator (henceforth $(\tilde{})$ denotes a predicted field). We have investigated various forms of β[4,8]. The current DVF, D_k, is then predicted based on $\tilde{S}_k$ and $\hat{D}_{k-1}$ according to

$$\tilde{D}_k = \gamma(\tilde{S}_k, \hat{D}_{k-1}),$$ (3)

where γ denotes a prediction operator. More specifically, the following Gauss-Markov model is used[4,7,8]

$$\tilde{d}_k(\vec{r}) = \sum_{i,j \in \mathcal{R}_k} a_{i,j,0}^{\hat{s}_{d_k}(\vec{r})} \tilde{d}_k(m-i, n-j)$$

$$+ \sum_{i,j \in \mathcal{R}_{k-1}} a_{i,j,1}^{\hat{s}_{d_k}(\vec{r})} \cdot \left[\hat{d}_{k-1}(m-i-\tilde{d}_x(m,n), n-j-\tilde{d}_y(m,n)) \right]$$ (4)

where $\vec{r} = (m,n)$, $a_{i,j,s}|_{s=0,1}$ are 2×2 matrices containing the spatial ($s = 0$) and temporal ($s = 1$) prediction coefficients which are transmitted over the channel, $\hat{s}_{d_k}(\vec{r})$ represents the binary line field which provides the segmentation of the DVF, and

$$\begin{bmatrix} \tilde{d}_x(\vec{r}) \\ \tilde{d}_y(\vec{r}) \end{bmatrix} = \hat{d}_{k-1}(\vec{r} - \hat{d}_{k-1}(\vec{r})). \tag{5}$$

According to this method, the need to transmit the DVF and the line field for the entire frame is eliminated; therefore, only the motion and segmentation information which are unpredictable by the decoder are transmitted at each frame. In other words, the prediction of D_k and S_k are performed only in those areas of the scene where the motion is similar from frame to frame while the information for the rest of the scene is transmitted over the channel in a compressed form. General 3-D models of the motion trajectory can also be used for obtaining a prediction of the current DVF. For example, given the past 3 or more frames, complex motion parameters such as acceleration, rotation and zoom information can be estimated from the motion trajectory and used in predicting the current DVF.

Assuming that the two previous frames are available both at the transmitter and the receiver the DVF corresponding to the previous frame is estimated. Then, based on this estimate a spatio-temporal prediction of the DVF at the current frame is obtained. Once a good prediction of $\tilde{D}_k$ is obtained a prediction of the current frame can be obtained by motion compensation of the previous frame according to

$$\tilde{F}_k = \lambda(\hat{F}_{k-1}, \tilde{D}_k), \tag{6}$$

where $F_k = \{f_k(m,n)\}$ denotes the intensity field of the current frame, and λ the motion compensation operation on each pixel of the current frame, i.e.,

$$\lambda(\tilde{f}_k(\vec{r})) = \lambda(\hat{f}_{k-1}(\vec{r} - \tilde{d}_k(\vec{r})). \tag{7}$$

According to the proposed coding scheme the decoder needs to perform motion estimation between the two previous frames, which obviously increases the storage and complexity requirements for the decoder. However, the proposed technique reduces the bit rate by using the fact that a part of the source information is predictable from the past information which is available at the decoder. Thus this method saves the transmission of the overhead segmentation information, as well as, the predictable motion information.

OCCLUSIONS

One of the important existing challenges of any type of motion compensation is the accurate representation and compensation of information in the occluded areas of the image. Occluded areas are defined as the areas where part of the scene is covered or uncovered by a moving object. Most of the existing codecs fail in these areas. For instance, if the motion field model does not directly incorporate such occlusion effects, failures occur as a result of motion estimation in and around these areas. Similarly motion models fail around the covered background areas. Therefore, more often than not, information pertaining to these areas is transmitted over the channel with low efficiency in terms of the bit rate and the reconstructed image quality. The inefficiency stems mainly from the fact that the position information of the occluded areas is transmitted in addition to the actual motion or intensity information in these areas. In this paper, a method to detect and transmit both the covered and the uncovered areas of the scene with no position overhead is described. The method utilizes the temporal continuity of motion fields in order to achieve very low bit rates.

In the past there have been few approaches for the detection and transmission of the occluded area information in the context of very low bit rate coding. Most of the existing techniques take the approach of segmenting the scene into four regions; stationary background, moving object, covered background and uncovered background. According to these approaches, first a change detection mask is obtained based on two consecutive frames. Then motion estimation is performed resulting in a motion model for each object that is identified. Based on the change detection mask and the motion model the uncovered and covered regions are separated from the moving objects[2]. However, these approaches have several disadvantages. First, either the position information of the detected occluded areas or the change detection mask along with the motion vectors must be transmitted over the channel. The position information clearly can be very costly depending on the amount of motion from one frame to another. Second, these algorithms fail in detecting the uncovered areas in cases of

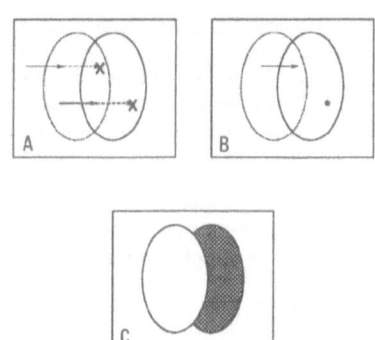

Figure 2: An Example for Occlusion Detection

sub-pixel motion since the reverse motion is considered in order to detect these areas[2]. Third, the temporal continuity of motion between successive frames is not incorporated into these algorithms. Fourth, such techniques tend to break down considerably in the presence of noise. Finally, these approaches make use of the intensity field instead of the DVF. This in turn can be the limiting factor, since the characteristics of the occluded areas are captured more efficiently with the DVF than the intensity field. This is due to the fact that occluded areas are inherently caused by inefficiencies in motion estimation and compensation.

Detection of Occluded Areas

In general, most encoder/decoder models proposed in the past are simply formulated to perform inverse tasks of each other. The following proposed technique differs from its counterparts in the sense that the decoder is given additional intelligence. Thus, like the encoder, the proposed decoder also attempts to take advantage of the correlations present in the image sequence. Let us assume that the decoder has reconstructed the frames up to frame $k - 1$, and about to decode frame k. Given a good prediction of the current DVF, both the encoder and the decoder can perform similar operations to detect the occluded areas. Thus the need to transmit the position information of the occluded areas is avoided. Consider the following difference,

$$\Delta(\vec{r} + \tilde{d}_k(\vec{r})) = \| \hat{d}_{k-1}(\vec{r}) - \hat{d}_{k-1}(\vec{r} + \tilde{d}_k(\vec{r})) \|, \tag{8}$$

where $\hat{d}_{k-1}$ represents the estimated DVF for the previous frame, f_{k-1}, and $\tilde{d}_k$ represents the predicted DVF for the current frame, f_k, described by Eq. (4). Equation (8) compares the displacement vectors of the previous frame at the current location, $\vec{r} + \tilde{d}_k(\vec{r})$, and the previous location $\vec{r}$. The comparison can be based on any norm (the ℓ_2 norm is used in our work). Δ is compared to a threshold; if greater than the threshold pixel $\vec{r} + \tilde{d}_k(\vec{r})$ belongs to a newly covered area. An approach to obtain an optimum value for the threshold is described in Ref. 4. We consider a simple example of one object translating in a stationary background, as shown in Fig. 2. Figures 2 (a), (b) depict an object at two time instances, $k - 1$ (dotted line) and k (solid line). Figure 2(a) shows two representative displacement vectors for the frame $k - 1$, $\hat{d}_{k-1}(\vec{r})$. It also shows the locations $\vec{r} + \hat{d}_{k-1}(\vec{r})$, denoted by an x. In Fig. 2(b) the vectors $\hat{d}_{k-1}(\vec{r} + \tilde{d}_{k-1}(\vec{r})) = \hat{d}_{k-1}(\vec{r} + \hat{d}_{k-1}(\vec{r}))$ are shown (dot represents a zero length vector). It is clear that by using Eq. (8), Fig. 2(c) results, where the shaded area denotes the covered area. Similarly, the uncovered background is determined by comparing to a threshold

$$\Delta(\vec{r} - \tilde{d}_k(\vec{r})) = \| \hat{d}_{k-1}(\vec{r}) - \hat{d}_{k-1}(\vec{r} - \tilde{d}_k(\vec{r})) \| . \tag{9}$$

Encoding of Occlusion Information

Once the occluded areas are determined, the remaining task is to encode the occluded area information in an efficient way. Since the proposed DVF prediction technique is likely to fail in covered areas, we propose the encoding of the DVF in these areas. On the other hand, since accurate motion compensation is generally unavailable in the uncovered areas, the intensity field information is encoded using the technique described in the next section.

In order to encode the displacement vectors of the occluded areas efficiently, Differential Pulse Code Modulation (DPCM) is employed. Thus, by this method, inspite of the increase in the dynamic range of the displacement vectors, the correlation between the neighboring vectors is exploited. It should be emphasized once again that since the decoder performs the same operations in order to detect the locations of the occluded pixels, there is no need to transmit the location information of these pixels. Thus, any scan order that is available to both the encoder and the decoder is sufficient in determining the position information of the transmitted displacement vectors. In our implementations a raster scan order is used.

MOTION PREDICTION FAILURES

In typical image sequences, complete motion compensation of the scene is not possible (consider, for example, cases when objects move into the scene or when the camera undergoes a global motion, exposing new areas to the scene). However, the most common type of imperfect motion compensation is the failure of the assumed motion models. For instance, in the case of motion estimation by block-matching (BM), the motion model fails to accommodate such motion as rotation, zoom, etc. This is due to the fact that BM assumes that a block of pixels undergoes a similar translational motion. A similar problem exists with object-based and pixel-based motion estimation algorithms. Due to the implicit assumptions in the models, these motion estimation algorithms may break down in areas of the scene where complex (very fast or unpredictable) motion occurs or the illumination changes, etc. Such areas of the scene where unsatisfactory motion compensation is achieved are in general referred to as *model failures*. One example of such instances is the motion of the eyes and the mouth in a typical video-conferencing scene. These areas usually constitute the most important areas of the scene since the human eye tends to track the moving areas more accurately than the slowly changing areas. Thus algorithms for the efficient detection and encoding of these model failure areas must be developed in order to achieve the targeted very low bit rates.

The proposed algorithms in the literature for the detection and encoding of the model failure area information are closely tied to the specific algorithm which is implemented. For instance, in the object-oriented encoding of video approach[2], the detection of the model failures is usually carried out during the motion estimation process. Thus the performance of the detection algorithm is clearly dependent on the performance of the motion estimator. Therefore the detection of the model failures can only be carried out in the context of a similar codec. However, as stated previously, model failures are not only a problem of object-oriented approaches, but depend on the specific characteristics of the scene. In this paper a general method to detect and encode model failures with no dependency on the type of the motion estimator used is introduced.

Detection of Model Failures

As described above, the unpredictable areas are defined as the areas where the motion prediction model fails. We demonstrate the use of our method in detecting the motion model failures based on the codec described above. In the proposed codec, the dense DVF is utilized to fully capture the motion of objects. However, since the dense DVF constitutes a large overhead if they are to be transmitted over the channel, a regeneration of the current DVF is performed both at the encoder and the decoder based on the previously transmitted information. In order to eliminate the errors due to the regeneration of the DVF in detecting the model failures, both the estimated and the predicted DVF are utilized in the proposed method. More specifically, model failures are detected by comparing the predicted DVF to the estimated DVF. In order to perform such a comparison we compute the motion compensated prediction error for the current frame based on the estimated and the predicted DVF. These two error signals are then compared and processed in order to detect the unpredictable areas where a considerable failure of the model occurs. Thus, the technique proposed here attempts to eliminate the false detection of the *unpredictable* regions where a failure is inevitable due to motion compensation even if the estimated motion vectors are used. For instance, such errors generally occur around the moving edges due to inaccurate description of the discontinuities in the motion field. However, since this type of failure occurs in general with most of the existing motion estimation/compensation algorithms, false detection of areas around moving edges is eliminated. Instead objects with complex motion such as the eyes and the mouth are detected as unpredictable areas by this approach. In other words, more emphasis is given to areas with complex or unpredictable motion than the geometrical distortions caused around the moving objects.

Encoding of the Model Failure Area Information

In order for the decoder to be able to reconstruct the unpredictable region information, both the position and the intensity field information of these regions have to be encoded. Several methods have been proposed in the literature for encoding the boundary information[9]. Of these techniques two major categories are known as chain coding and spline approximation. Although spline approximation methods have been very efficient in coding the segmentation boundaries in object-oriented codecs[2], their efficiency depends heavily on the size of the object to be encoded (they can loose their efficiency if the size of the object is small). As described above, the unpredictable regions are likely to be small regions. Thus we exploit the second category of boundary coding techniques, chain coding.

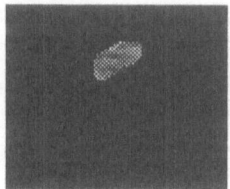 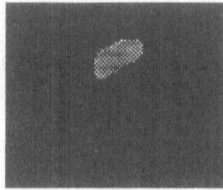

Figure 3: An example of motion prediction failure

In our specific implementation we exploit the other extreme case where we limit the number of possible directions. We begin by limiting the alphabet size to 3 symbols $A_c \equiv \{L, R, S\}$, where A_c denotes the alphabet, L denotes a left turn, R denotes a right turn and S denotes a straight move, based on the previously encoded symbol. Through a preprocessing step [4], the boundary is guaranteed to be represented with the available alphabet A_c. The generated string of symbols is encoded using Huffman encoding. Finally, in order to achieve efficient compression of the segmented intensity information, an algorithm based on Projection Onto Convex Sets (POCS)[10] is implemented with some modifications. Figure 3 shows an example of motion prediction failure. In this particular example the complex motion around the eye of a person is detected in the analysis stage at the encoder and encoded. The left figure shows the detected area of the original frame while the right figure shows the same area in the reconstructed frame. In this example, the overhead of transmitting the boundary and the intensity information of the detected area was approximately 0.8 bits per area pixel.

GLOBAL REFINEMENT

Once the occluded and model failure area information is encoded, a global refinement procedure is employed. According to this the difference between the intermediary reconstructed frame, synthesized using the predictable and the transmitted unpredictable information and the original current frame is analyzed. If this difference is significant it is encoded and transmitted. The global refinement encoder again takes advantage of the correlations in the spatial and temporal directions present in these difference fields[4]. This process ensures a constant quality of reconstructed video at the decoder.

INFORMATION	BITS
Occlusion	400
Motion Prediction Failure	900
Global Refinement	1400

Table 1: Bit distribution among different information channels

EXPERIMENTAL RESULTS

In this section, first the performance of the proposed DVF prediction is compared to a standard coding algorithm which is similar to H.261, i.e., a motion vector is computed for each 16×16 block of pixels at each frame and transmitted over the channel. The algorithm is tested on the standard video-conference image sequence "Miss America". According to our notation, $BR(k)$ is the bit-rate achieved at each frame, whereas, the average bit-rate is denoted by $\overline{BR}$.

As a measure of the reconstruction quality we use the reconstructed *signal-to-noise ratio* ($RSNR$). Figure 4 show the values of $RSNR$ at each frame for approximately the same bit-rate or 7.5 frames/sec of the *Miss America* sequence. The proposed coding approach shows a very significant improvement in quality when compared to the standard coding method.

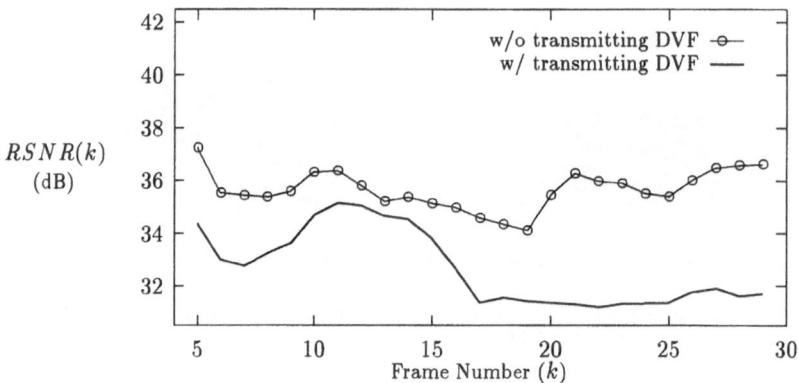

Figure 4: Comparison based on the $RSNR$ for approximately the same bit-rate ($\overline{BR} \approx$ 64 kbps) (*Miss America*).

When the proposed coding scheme is tested at very low bit rates, an overall reconstructed video quality of ≈ 33 dB was achieved at 20 kbps. The distribution of an average number of 2700 bits per frame with respect to different information segments transmitted across the channel is shown in Table 1.

CONCLUSIONS

In this paper, we propose a hybrid object-oriented algorithm for very low bit rate encoding of video. In order to achieve efficient compression, we exploit the inter-correlations between the intensity, segmentation and motion fields. More specifically, first the video information is separated into predictable and unpredictable parts. While the predictable portion requires no information transmission, the information pertaining to the unpredictable areas is transmitted hierarchically. In other words, assuming an intelligent decoder model, first the occluded areas of the scene are detected. While no position information for these areas needs to be transmitted, the relevant motion or intensity information in the corresponding areas is efficiently encoded. In order to compensate for the failures in motion prediction additional intensity information needs to be transmitted. We propose a technique to cope with these situations. Finally, if necessary, a global refinement procedure is employed in which extra global correction is transmitted. The proposed codec consists of a number of novel stages or tools which can be applied within the context of other codecs.

REFERENCES

1. K. Aizawa, H. Harashima and Saito, Model-based synthesis image coding (MBASIC) system for a person's face, *Signal Processing : Image Communication,* vol. 1, no. 2, pp. 139-152, Oct. 1989.

2. H G. Mussmann, M. Hotter and J. Ostermann, Object-oriented analysis-synthesis coding of moving images, *Signal Processing: Image Communication*, pp. 117-138, vol. 1, no. 2, 1989.

3. M. Hotter, Object-oriented analysis-synthesis coder based on the model of flexible 2-D objects, *Signal Processing : Image Communication*, vol. 2, no. 4, pp. 409-428, Dec. 1990.

4. T. Özçelik, A very low bit rate codec, *Ph.D. Dissertation*, Northwestern University, Dept. of EECS, Dec. 1994.

5. T. Özçelik, J.C. Brailean, and A.K. Katsaggelos, Image and video coding algorithms based on recovery techniques using mean field annealing, *IEEE Proceedings*, vol. 83, no. 2, pp. 304-316, Feb. 1995.

6. M.R. Banham, J.C. Brailean, C. Chan and A.K. Katsaggelos, Low bit rate video coding using robust motion vector regeneration in the decoder, *IEEE Trans. Image Processing*, vol. 3, no. 5, pp. 652-665, Sept. 1994.

7. J.C. Brailean and A.K. Katsaggelos, A recursive nonstationary MAP displacement vector field estimation algorithm, *IEEE Trans. Image Processing*, vol. 4, no. 4, pp. 416-429, Apr. 1995.

8. T. Özçelik and A. K. Katsaggelos, Robust displacement vector field prediction algorithms with application to very low bit rate video coding, *Proc. SPIE Visual Communications and Image Processing*, Cambridge, MA, vol. 2094, pp. 1378-1389, Nov. 1993.

9. M. Hotter, Predictive contour coding for an object-oriented analysis-synthesis coder, *Proc. IEEE International Symposium on Information Theory*, pp. 75, San Diego, CA, Jan. 1990.

10. H.H. Chen, M.R. Civanlar and B.G. Haskell, A block transform coder for arbitrarily shaped image segments, *Proc. Very Low Bit Rate Video Workshop*, paper no. 1.1, Essex, UK, 1994.

MORPHOLOGICAL MOVING OBJECT SEGMENTATION AND TRACKING FOR CONTENT-BASED VIDEO CODING

Chuang Gu, Touradj Ebrahimi, and Murat Kunt

Signal Processing Laboratory
Swiss Federal Institute of Technology
CH-1015, Lausanne, Switzerland

1. INTRODUCTION

Besides contour and texture as two visual primitives [1, 2], motion is another important visual primitive. The understanding of motion in a dynamic scene represents a principal research topic for many applications such as video coding, video indexing and annotation, matching and recognition. In fact, motion information contains very rich cue to describe how a dynamic scene is constructed. Such description about the independent movement of each region or object is of primary interest to all those applications. Obviously, this compact description can be utilized to improve the coding efficiency as well as to provide the foundation for all the new functionalities which are under intensive investigation in the framework of next international standard MPEG4 [3].

Our goal in this work is to develop a generic motion segmentation algorithm which tries to produce a description about moving objects with homogeneous motions in the dynamic scene. This algorithm should be generic in the sense that no priori knowledge about the dynamic scene as well as the motion of camera are assumed. Consequently, this task becomes a challenge one in the field of image sequence analysis. It is well known that generic motion segmentation is a difficult problem as it is related to two things: motion estimation and motion segmentation. In fact, there is a strong relation between these two processes. At a first glance, it seems clear that motion estimation produces a description about the movement of each pixels in the scene and motion segmentation tries to provide a partition of this description. However, these two steps depend tightly on each other. On the one hand, a precise motion estimation relies on the regions of support. This means that a good segmentation is required to provide this support. On the other hand, a precise motion segmentation needs a good motion estimation to describe coherent motions. To overcome this chicken-egg problem, a universal solution should be found to break this connected chain.

Motion segmentation has received high attention in the research community [4, 5, 6]. In this paper, a novel motion segmentation method will be introduced. We find that generic motion segmentation can be achieved if multiple sources, e.g. color and motion, are taken into account simultaneously. Using the strategy of divide-and-conquer, we can solve a big and difficult problem by dividing it into smaller and easier ones. For this reason, spatial-temporal segmentation techniques are investigated. The result of a hierarchical motion segmentation is combined with the result of a spatial segmentation in order to provide a precise partition of the motion field. Using this distributed system, the difficult motion segmentation problem can be solved by two independent easier ones where the constraints are less complex. Furthermore, the problem of motion tracking is solved to avoid the random fluctuation of the motion segmentation and establish the links of moving objects in the successive frames.

As a non-linear approach for signal processing, multivalued morphology [7] can deal with various multivalued images such as color images as well as motion vectors. Therefore, the problem of motion segmentation can be universally solved in the framework of multivalued morphology. Using a set-theoretical methodology for image sequence analysis, multivalued morphology can estimate many features of the geometrical structure in the signals where the partition of motion is just a particular case. We will show that multivalued morphological filters and segmentation tools are particularly suitable for solving the problems in moving object segmentation and tracking.

The structure of this paper is as follows. Based on the analysis of the requirements for motion segmentation and tracking, Section 2 gives a brief description of the proposed motion segmentation and tracking algorithm to satisfy these requirements. Section 3 introduces a hierarchical joint spatial-temporal motion segmentation method. Section 4 addresses the problem of moving object tracking. Section 5 demonstrates the overall performance of the proposed motion segmentation and tracking system and provides some conclusions.

2. OVERVIEW OF THE PROPOSED SYSTEM

Although the design of a good motion segmentation and tracking system may depend on the specific application under consideration, there exists common points regardless the environment and the classification. These common points construct the requirements as well as the goals for the design of our motion segmentation and tracking system. They are listed as follows.

- Generosity: The motion segmentation and tracking system should be generic without any specific prior knowledge on the types of motion.

- Flexibility: The motion segmentation system should be flexible to provide various levels of segmentation.

- Precision: The motion segmentation and tracking system should provide precise multiple motion boundaries. These boundaries should correspond to the real moving entities.

- Interactivity: The motion tracking system should avoid large delays which prevent the interactivity.

- Content-based accessibility: The system should know the evolution of current moving objects in the future. This is the foundation to provide the universal content-based accessibility.

- Complexity: The system should be simple to implement in software and/or hardware.

Taking these requirements into account, the proposed motion segmentation and tracking system is designed as follows. Our goal is to identify homogeneous moving objects with arbitrary shapes in the sequence. Therefore, the feature for the segmentation and tracking algorithm should be the motion information. A multi-grid motion estimation algorithm [8] is used to obtain a dense motion vector field. This dense motion vector field indicates the displacement of each pixel in the dynamic scene. The motion segmentation and tracking system deals with the resulting motion vector field which could be considered as a 3D multivalued image M:

$$M(x, y, t) = (V_x, V_y) \tag{1}$$

where x, y represent the spatial axis, t denotes the temporal axis and V_x, V_y are horizontal and vertical motion components.

We could then assume that the problem of motion segmentation and tracking is nothing else but a multivalued segmentation problem which can be solved by using multivalued morphological segmentation tools. In order to satisfy the requirement of interactivity, the motion vector sequence is divided into groups of pictures. This structure is the same as that of MPEG2. Each group is composed of two types of two dimensional images: one intraframe and the rest interframes. Hence, the whole segmentation task is also divided into two sub-tasks: intraframe motion segmentation and interframe motion segmentation which is also called tracking.

The intraframe motion segmentation utilizes a morphological hierarchical segmentation tool. The elementary structure of this hierarchical motion segmentation algorithm is composed of four steps: motion simplification, marker extraction for motion field, watershed for motion field and motion modeling. Using this hierarchical approach, different levels of motion segmentation could be provided to meet the requirement of flexibility.

However, in reality, a dense motion vector field could be very noisy due to the inaccuracy of motion estimation algorithm. Therefore, the result of motion segmentation relying only on motion information (called temporal segmentation) normally does not correspond to the real moving objects. The segmented motion boundaries could be distorted comparing to the real objective entities. In this case, the result of a spatial segmentation could be used to improve the accuracy of the temporal segmentation. The spatial segmentation also deals with a multivalued image: i.e. color image. It could provide a partition image satisfying homogeneous color criterion in the spatial domain. This spatial segmentation algorithm uses a similar hierarchical structure. The inaccurate temporal segmentation result could be refined using the result of this spatial segmentation. By taking into account the statistical distribution of spatial homogeneous region inside motion homogeneous regions, the small spatial regions could be clustered to form bigger regions with homogeneous motions. Because of this spatial support, the final joint spatial-temporal segmentation could result in very precise motion boundaries.

The interframe motion segmentation is considered as a tracking problem. A good tracking algorithm should extract real moving objects with homogeneous motions corresponding to the perception of human visual system. These objects should be temporally coherent, i.e. it should be possible to follow these regions along the temporal axis. Hence, the temporal relationship between successive frames should be constructed. Obviously, simple two dimensional motion segmentation can not fulfill this goal. Generally, it is very difficult to link two individual partitions in the temporal domain. For this reason, a 3D (2D plus time) segmentation algorithm for the motion vector field should be employed. As a result, the requirement for content-based accessibility could be satisfied because we know where the current objects will go in the next frames. In order to prevent large delay and to decrease the memory requirement, the 3D motion segmentation uses only two motion vector images: i.e. the past motion vector image and the current motion vector image. Obviously, the computation of these two motion vector images involves three consecutive frames in the image sequence.

The whole tracking procedure follows a recursive process. Since the motion partition of the previous frame is already known, the motion partition of the current frame can be considered as the evolution of the past motion partition. Therefore, the past motion partition is projected into the current motion vector field according to a homogeneous motion criterion. The projection for the first interframe is based on the motion partition of the intraframe. As a geometrical approach for signal processing, multivalued morphology is particularly suitable for this task. A 3D multivalued watershed algorithm [7] could be used. Since this projection only uses the motion information, the same problem of inaccurately tracked motion boundaries will occur. Therefore, a spatial segmentation for the current interframe should also be combined with this tracking result in order to obtain a more precise motion partition.

Concerning the complexity of the whole system, the simple max/min operations involved in the multivalued morphology allow the efficient implementation of motion segmentation and tracking in both software and hardware. The structure of the whole system is illustrated in Fig. 1.

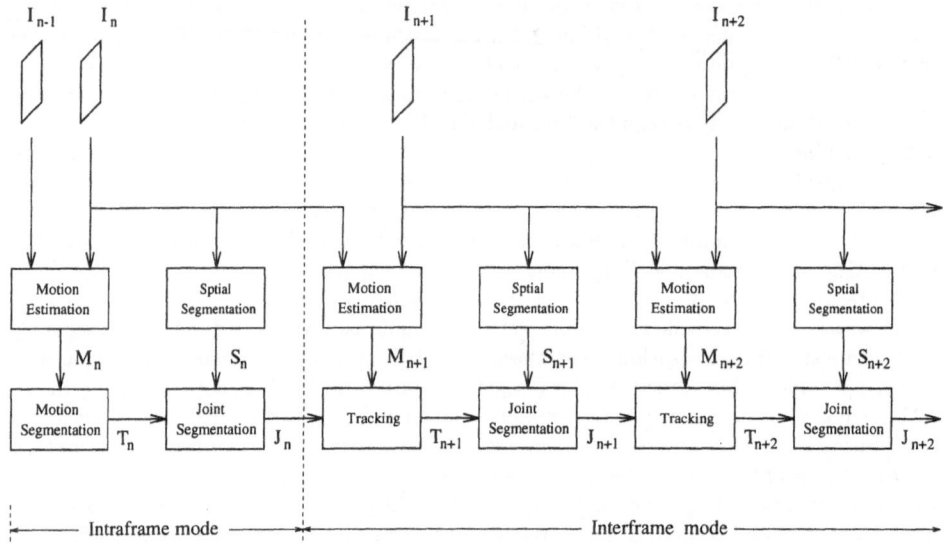

Figure 1: Diagram of the motion segmentation and tracking system. I_i: original input frame i; M_i: motion vector field of frame i; T_i: partition image of motion vector field for the frame i; S_i: partition image of spatial domain for frame i; J_i: partition image of joint spatial-temporal segmentation for the frame i.

3. INTRAFRAME MODE: HIERARCHICAL MOTION SEGMENTATION

The proposed motion segmentation method aims at segmenting a motion vector image into homogeneous regions with coherent motion using the characteristics of human visual system (HVS): what we see and how we see [1]. It is believed that HVS uses semantic unities to identify and recognize various moving objects in a scene. Therefore, the extracted regions should correspond as much as possible to the real semantic moving unities.

Unlike the traditional motion segmentation methods which try to give the result in a single level, the proposed method employs a multilevel segmentation which uses a hierarchical structure to provide the final result from coarse to fine. At the beginning, a coarse partition of motion field is computed. Then, the motion field residue which is the difference between the original motion vector field and the modeled motion vector field is evaluated. A big residue signal means that the motion segmentation is not sufficiently well estimated and needs further refinement. In this case, a fine motion segmentation is carried out in order to extract smaller regions with homogeneous motions. This process continues iteratively until a satisfactory result is obtained.

The basic motion segmentation steps involved in each level are motion simplification, motion marker extraction, motion watershed and motion modeling. The motion simplification achieves the simplification of the motion vector image. Motion marker extraction is carried out as the feature extraction to identify seeds with homogeneous motions. Motion watershed conducts the decision step. The motion modeling evaluates the quality of motion partitioning in the current level and generates the modeling error for the next level of the hierarchy. Finally, the result of hierarchical motion segmentation is combined with a spatial segmentation to produce a precise joint spatial-temporal motion segmentation.

3.1. Motion Simplification

The goal of motion simplification is to control the complexity of the motion information treated in the current level. The input signal for the motion simplification is the remaining motion data which is not appropriately modeled in the previous level. This motion data is called motion residue which is the difference between the original motion vector image and the modeled one in the previous level. Obviously, a large value in the motion residue means the poor modeling quality in this point and a small value indicates a better approximation. The motion simplification is an important stage in the basic hierarchical structure. It controls the type and the amount of data which is withdrawn from the motion vector image before the motion marker extraction step. This step makes the identification of homogeneously moving regions much easier. It also determines the notion of hierarchy in the motion segmentation algorithm.

Several alternative filters can be used for the motion simplification purpose, e.g. multivalued linear low-pass filter, multivalued opening and closing, multivalued opening and closing by reconstruction and multivalued area opening and closing [7]. Multivalued area filters can provide a precise size-oriented decomposition of multivalued signals. They are particularly suitable to simplify a motion vector image because a dynamic scene is composed of moving objects with different size. The small moving objects which most probably correspond to noise will be removed while the boundaries of big moving objects are precisely preserved.

3.2. Motion Marker Extraction

After motion simplification step, the resulting motion vector image is smoother than before which makes the identification of homogeneous motion regions easier. The feature extraction in the hierarchical motion segmentation algorithm is called motion marker extraction. The motion marker image is a special partition image where zero value represents uncertain areas. The homogeneous motion criterion (or similarity) used in this structure is a flatness one. By using the city block distance function, we are sure that the extracted markers really represent homogeneous motions in horizontal and vertical directions. Actually, the motion marker extraction procedure utilizes a region growing process. Conservatively, the threshold is settled to a small value in order to ensure the motion homogeneity. Accordingly, this leads to the robustness of the results. Once the motion marker extraction is accomplished, the region number in the current level of segmentation is found. Each marker represents a homogeneous motion region. In most cases, uncertain areas will appear in the motion marker image. These uncertain areas should be classified into existing regions (or markers) in the next step of the hierarchical motion segmentation structure: decision.

3.3. Decision: Motion Watershed

The goal of the watershed step is to determine the remaining uncertain areas after the motion marker extraction procedure. The pixels in these remaining uncertain areas should be classified into existing motion markers which represent the seeds of several homogeneous motion regions in the current level. The classification procedure is performed by using a motion watershed algorithm. It is implemented as a special version of the multivalued watershed algorithm [7]. Here, the multivalued input signal is the two motion components in horizontal and vertical direction. The distance function involved in this motion watershed algorithm employs the Euclidean distance measurement.

An important fact which should be taken into account in the motion watershed algorithm is the constraint from the previous level. Any marker may not cross the boundary of the partition image from the previous level. Similarly, the region growing process of motion watershed obeys the same rule. It is strictly constrained within the territory of the previous motion partition. During the region growing process, this constraint is checked whenever a new pixel is concerned. The growing procedure is interrupted if it causes the crossing. Gradually, uncertain areas are classified into the known regions (or markers). This procedure stops when there is no uncertain area left in the image.

3.4. Motion Modeling

Once the motion segmentation in the current level is done, a motion modeling process is carried out to evaluate the segmentation result. Since the segmentation procedure has already extracted a set of regions with homogeneous motions, the motion distribution inside each region is expected to be the low frequency components. Ideally, these low frequency components could be modeled using smooth functions such as polynomial ones. In our motion segmentation algorithm, an affine motion model is used in order to deal with complex scenes. This model is nothing else but a simple first order polynomial approximation of the motion field. Finally, the motion residue which is the error of the motion modeling is delivered to the next level of hierarchical motion segmentation.

3.5. Joint Spatial-Temporal Segmentation

It is well known that any given motion estimation algorithm may provide inaccurate results. This may happen due to various noises, incorrect motion model, motion distortion, occlusion, etc. Consequently, a motion segmentation algorithm which is based only on estimated motion information might as well introduce inaccuracies. This phenomena is particularly obvious in the motion boundaries.

Here, the result of a spatial segmentation could be used to improve the accuracy of the temporal segmentation which is only based on motion information. In fact, a spatial segmentation produces a partition to indicate the homogeneous spatial regions. In the real world, it has been observed that the real motion boundaries are normally located in the position of spatial boundaries. For this reason, a spatial segmentation is utilized to improve the precision of temporal segmentation. A simple yet comprehensive approach is developed to combine the result of the spatial and temporal segmentations into a precise segmentation for moving objects. For each spatial region, if the majority part of this spatial region is located in a temporal region, this spatial region is assigned to that temporal region. In this way, the spatial regions are clustered according to the constraint of the temporal regions which roughly describe the areas with homogeneous motions. Consequently, the temporal segmentation is refined which produces a precise motion segmentation fitting to the real moving objects.

4. INTERFRAME MODE: MOVING OBJECT TRACKING

Interframe motion segmentation can be considered as a tracking process. Since we want to find coherent moving objects through the time axis, the feature for tracking is the region with homogeneous motion. By taking into account the past motion partition in the current motion segmentation procedure, motion tracking can solve the possible random fluctuation problem in the motion segmentation algorithm. This step could also establish the links between the moving objects in the successive frames and therefore tackle the correspondence problem for moving objects.

Actually, tracking can be regarded as the detection of evolution for all the regions with coherent motion over time. In this case, 3D multivalued morphological segmentation tool is particularly suitable to fulfill this task. Assume the original previous and current motion vector frames are $M(x, y, t)$ $M(x, y, t+1)$. The previous motion partition image is $J(x, y, t)$.

The interframe motion segmentation should find the current segmentation $J(x, y, t+1)$, see Fig. 1.

In the first step, we want to find the evolution of previous motion partition in the current frame $T(x, y, t+1)$. No new region is allowed to appear. It constructs the correspondence between the previous motion partition and the current motion partition. Basically, it can be considered as a projection procedure: to project the previous motion partition to the current frame. Therefore, a 3D segmentation should be used instead of individual 2D ones. Multivalued watershed can exactly fulfill this task. Actually, the previous motion partition $J(x, y, t)$ can be considered as the markers to indicate homogeneous motion regions. Hence, marker extraction has already been accomplished. As a result, the decision of region-growing multivalued watershed extends the previous motion partition into the current uncertain area. The pixels in the current motion vector frame are classified to the correspondent regions (markers) in the previous motion vector frame. This gives the tracking result $T(x, y, t+1)$.

However, as we explained in the intraframe motion segmentation, $T(x, y, t+1)$ may contain inaccurate motion boundaries because the tracking process relies only on motion information. Therefore, a spatial segmentation $S(x, y, t+1)$ for the current frame $I(x, y, t+1)$ is used to improve the precision of $T(x, y, t+1)$. After the joint spatial-temporal segmentation described previously, the final precise motion partition for the current frame $J(x, y, t+1)$ could be obtained.

Because of the 3D multivalued segmentation, the link between the previous motion partition $J(x, y, t)$ and current motion partition $J(x, y, t+1)$ are naturally established.

Figure 2: Result of joint spatial-temporal motion segmentation.

Figure 3: Result of moving object tracking.

5. Results and conclusion

In this section, the performance of the proposed motion segmentation and tracking algorithm will be demonstrated for a real image sequence "Foreman". The original sequence is a color image sequence in QCIF format. For the sake of convenience, only luminance component is shown in all experimental results. The original frame rate has been considered as 25 Hz. A frame rate of 12.5 Hz is selected for motion segmentation and tracking which means one frame is skipped between every two frames. This also means that the motion estimation is done at 12.5 Hz.

The result of first motion image segmentation, i.e. intraframe motion segmentation, is shown in Fig. 2. In this figure, the five images presented are original image, computed motion vector image, temporal segmentation for motion vector image, spatial segmentation and joint spatial-temporal segmentation. The boundaries of all the segmentation are superposed on the original image to see where they are located. We can find that the joint spatial-temporal segmentation combines the two partitions (spatial partition and temporal partition) into a precise motion partition to describe the boundaries of real moving objects. The result of interframe motion segmentation which has been considered as a tracking procedure is illustrated in Fig. 3. By using the 3D multivalued watershed and joint spatial-temporal segmentation, the boundaries of moving objects are precisely tracked. This step avoids the random fluctuation of the motion segmentation and naturally establishes the links between moving objects in the successive frames.

As a conclusion, the proposed generic motion segmentation and tracking algorithm can identify real semantic moving objects efficiently. Using multivalued morphology and joint spatial-temporal segmentation, the boundaries of moving objects can be precisely allocated. This automatic algorithm can be used to improve the video coding efficiency as well as to provide the foundation of all the new functionalities for various new content-based video applications.

References

[1] M. Kunt, A. Ikonomopoulos, and M. Kocher. Second generation image coding techniques. *Proceedings of the IEEE*, Vol. 73, No. 4, pp. 549–575, April 1985.

[2] P. Salembier, L. Torres, F. Meyer, and C. Gu. Region-based video coding using mathematical morphology. *Proceedings of the IEEE*, Vol. 83, No. 6, pp. 843–857, June 1995.

[3] Ad Hoc Group on MPEG4 Test Procedures. MPEG4 Proposal Package Description. In *ISO/IEC JTC1/SC29/WG11/N998*, Tokyo, July 1995.

[4] P. Bouthemy and F. Meyer. Motion segmentation and qualitative dynamic scene analysis from an image sequence. *Journal of Computer Vision*, Vol. 10, No. 2, pp. 157–182, 1993.

[5] T.S. Huang and A.N. Netravali. Motion and structure from feature correspondences: A review. *Proceedings of the IEEE*, Vol. 82, No. 2, pp. 252–268, February 1994.

[6] J.Y.A. Wang and E.H. Adelson. Representing moving images with layers. *IEEE Tran. on Image Processing*, Vol. 3, No. 5, pp. 625–638, Sept. 1994.

[7] C. Gu. Multivalued morphology: its theoretical framework. In *Submitted to International Symposium on Mathematical Morphology*, Atlanta, USA, 1996.

[8] F. Dufaux and F. Moscheni. Motion estimation techniques for digital TV: a review and a new contribution. *Proceedings of the IEEE*, Vol. 83, No. 6, pp. 858–879, June 1995.

SEGMENTATION OF IMAGE AREAS CHANGED DUE TO OBJECT MOTION CONSIDERING SHADOWS

Jörn Ostermann[1]

AT&T Bell Laboratories
Room 4E518
101 Crawfords Corner Rd
Holmdel NJ 07733–3030
Phone (908) 949 6683
Email: ostermann@big.att.com

ABSTRACT

In this paper, an algorithm for detecting moving objects in front of a static background is presented. Moving shadows are not considered to be part of the moving objects. The algorithm computes for each pel of two consecutive images three criteria based on the luminance difference, the temporal contrast and edges. The joint evaluation of these criteria enables the change detector to detect the moving objects. Knowledge about the scene illumination is not required.

1 INTRODUCTION

For low bit–rate video coding, several segmentation–based coders have been proposed which describe a moving object by its shape, motion and its surface texture. Proposals based on region–based techniques define a region with homogenous texture in a still frame and track the motion of a region over time [5][9]. Regions with similar motion might be combined into objects. In object–based analysis–synthesis coding, an object is defined as having homogenous motion [7][3][11]. Each object is tracked over time. These two classes of segmentation–based coders apply motion–compensation in order to predict the current frame from the previous frames. The motion–compensation relies on the assumption that changes between consecutive frames are due to motion of opaque objects. As far as illumination is concerned, they assume constant local illumination and diffuse reflecting surfaces. Hence, an algorithm is required which estimates the image regions changed due to object motion. Moving shadows should not be identified as moving objects, since motion compensation does not work on shadows.

Assuming a stationary camera as in applications like desktop video–conferencing, this task is frequently tackled using a temporal change detector [2][6][4][1]. These algorithms cannot distinguish between changes due to moving objects or moving shadows. In order to tackle this problem, the evaluation of highpass–filtered images was suggested in [10]. In [12],

---

[1]Research was carried out at the Institut für Theoretische Nachrichtentechnik and Informationsverarbeitung, Universität Hannover, Germany.

Multimedia Communications and Video Coding
Edited by Y. Wang *et al.*, Plenum Press, New York, 1996

an algorithm using a shading model has been proposed to detect moving objects. However, the shading model only allowed an evaluation for blocks of size 5x5 pels and larger. It does not allow the precise estimation of object boundaries. Furthermore, this criterion is not able to detect moving objects with a smooth surface texture [12].

Here, an algorithm is proposed that takes two frames of an image sequence, computes three criteria based on these images and evaluates the results in order to generate the change detection mask. Section 2 describes the three criteria and how they are combined for the new change detector. Section 3 gives experimental results.

2   DETECTION OF IMAGE AREAS CHANGED BY OBJECT MOTION

The proposed algorithm takes as input the two original luminance images l_k and l_{k+1}. Three criteria are computed for each pel of these images. Temporal changes are detected using a temporal difference criterion (DC) and moving object boundaries are computed using an edge criterion (EC) [8][10] (Figure 1). The output of the three criteria are evaluated such that the image areas changed due to object motion, i.e. the moving object and the area uncovered due to object motion, are estimated as changed areas. The results of the change detector will be demonstrated on the two test sequences *Claire* and *Erik* (Fig. 2). The well known test sequence Claire shows a talking person in front of a blue screen. The scene is illuminated using diffuse and direct light. Furthermore, the sequence is contaminated with a noise bar moving slowly from the top of the image to the bottom. Test sequence Eric is illuminated using direct and diffuse illumination. The shadow of Eric is visible on the picture in the background.

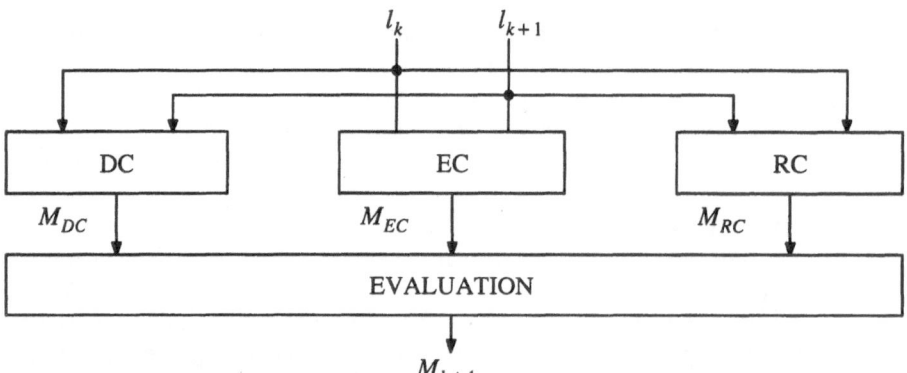

Figure 1.   Block diagram of the detector of the image regions changed due to object motion (luminance signal l_k, difference (DC), edge (EC) and reflection (RC) criterion).

Figure 2.   Frame 1 of the test sequences (a) *Claire* and (b) *Erik*.

2.1 Difference Criterion (DC)

It is assumed that moving objects generate significant temporal changes in consecutive images. The purpose of the DC is to distinguish those parts of an image with significant temporal changes due to object motion from those parts that are unchanged or only changed due to noise. The DC binarizes the difference image between the luminance images l_k and l_{k+1} using a noise adaptive threshold. A median filter and morphological operators are applied to the binary image, in order to smooth boundaries and eliminate small regions [4]. Figure 3 shows the mask M_{DC} for the two test sequences.

2.2 Edge Criterion (EC)

The purpose of the edge criterion is to find potential object boundaries. The assumption is that moving objects have luminance edges around themselves. The EC computes for both images l_k and l_{k+1} an edge image. The edges have a width of 1 pel. A pel in the mask M_{EC} is set to *static* if there is an edge in both edge images. A pel is set to *changed* if there is an edge only in one of the two images. The label *unknown* is assigned to the remaining pels of the mask M_{EC}. This label *unknown* indicates that the criterion is not able to decide about the status of the pel. Figure 4 shows mask M_{EC} with the *changed* (white) and *static* (black) edges for the two test sequences.

2.3 Reflection Criterion (RC)

The RC should provide an indication of which parts of an image are changed due to a moving object itself and which parts are unchanged (background and moving shadows). The RC assumes that the image signal $l(x)$ depends on the illumination $E(x)$ and the bidirectional reflection function $R(x)$. $R(x)$ accounts for the wavelength of the illumination, surface material and the geometrical arrangement of illumination, camera and surface. The illumination $E(x)$ depends on ambient and direct light. Assuming diffuse illumination, parallel projection and a constant k_b, the image signal is given by the reflection model

$$l(x) = k_B \cdot E(x) \cdot R(x). \tag{2.1}$$

Assuming this reflection model, a temporal contrast can be computed:

$$K(x) = \frac{l_k(x)}{l_{k+1}(x)} = \frac{E_k(x)}{E_{k+1}(x)} \cdot \frac{R_k(x)}{R_{k+1}(x)}. \tag{2.2}$$

If neighboring pels of an image do not change over time, $K(x)$ is 1 for these pels. If the illumination of these pels changes due to a change of lighting or a moving shadow, $E_{k+1}(x)$ changes for the neighboring pels by the same amount and $R(x)$ stays constant. Hence, $K(x)$ is constant for neighboring pels. The neighborhood is defined by a window of size 5x5 pels. On the other hand, if an object moves, $E(x)$ stays constant but $R(x)$ at a location x changes arbitrarily for each x depending on the surface texture of the object. Therefore, $K(x)$ changes arbitrarily for neighboring pels due to object motion. Hence, this reflection model allows to distinguish between temporal changes caused by object motion and moving shadows. However, certain restrictions apply: If an object with a smooth surface texture like a grey scale ramp moves, $K(x)$ stays constant. Furthermore, this criterion is fairly noise sensitive. Therefore, pels with $K(x)$ close to 1 are declared as *static*, pels with $K(x)$ significantly different from 1 are declared *changed*. The label *unknown* is assigned to the remaining pels. Figure 5 shows the masks M_{RC} for the two test sequences. The areas of the moving shadow are correctly classified as *static*. However, facial parts of the persons are also declared as *static*.

2.4 Evaluation

The output of the criteria DC, EC and RC are combined in order to generate the final pel–wise mask M_{k+1} which marks the image areas changed due to object motion. The evalua-

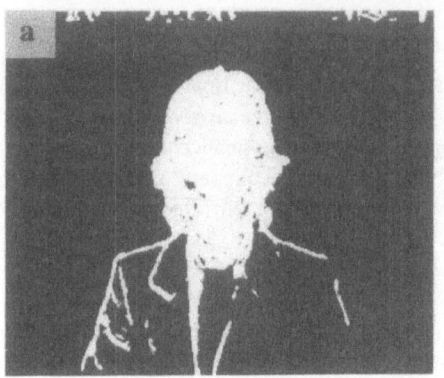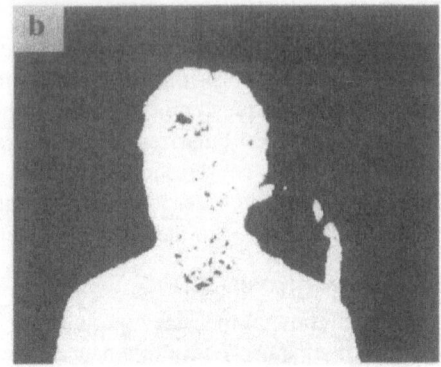

Figure 3. Mask M_{DC} as computed by the difference criterion (DC) for the frames 1 and 2 of the test sequences (a) *Claire* und (b) *Erik* (black = *static*, white = *changed*).

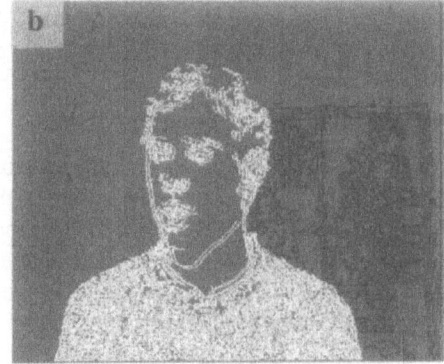

Figure 4. Mask M_{EC} as computed by the edge criterion (EC) for the frames 1 and 2 of the test sequences (a) *Claire* und (b) *Erik* (black = *static*, white = *changed*, gray = *unknown*).

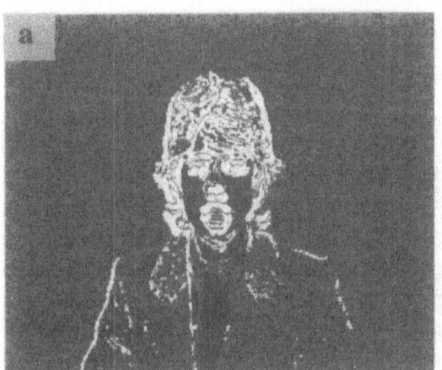

Figure 5. Mask M_{RC} as computed by the reflection criterion (RC) for the frames 1 and 2 of the test sequences (a) *Claire* und (b) *Erik* (black = *static*, white = *changed*, gray = *unknown*).

tion procedure is based on heuristic experiments. In a first step, the three masks M_{DC}, M_{EC} and M_{RC} are superimposed. Neighboring pels with the same labels in all three masks are considered one region. To each region the label *static* or *changed* is assigned based on its own labels and the labels of its neighboring regions. I.e., a region with all labels being *changed* is declared *changed*. If the DC-label is *static*, the region is declared *static*. Of special interest are the regions of moving shadows. For those, the DC-label is *changed* and the other two labels are *static*. This region will only be declared as *changed* if it is surrounded by moving edges (regions with the EC–label *changed*). This rule allows the correct classification of the face and the moving shadow. Figure 7 shows the results after evaluating several frames of the test sequences.

3 EXPERIMENTAL RESULTS

Experiments have been carried out on several natural and artificial image sequences. In comparison to a change detector evaluating the difference signal only, it was found that the edge criterion allows for a more precise estimation of the object boundaries and that the reflection criteria in conjunction with the DC and RC criteria is able to distinguish between moving objects and moving shadows. Moving shadows are erroneously declared as moving objects if the boundary of the shadow is picked up by the edge detector. For synthetic images with very sharp shadow boundaries shadows are classified correctly as long as the shadow does not change the image intensities by more than 20%. For natural images where the shadow boundary is softer, changes up to 50% of the image intensity are acceptable for correct classification of shadows.

The change detectors have been tested in an object–based analysis–synthesis coder (OBASC). Since an OBASC has knowledge of the moving objects it uses a background memory. The background memory is initialized with the first frame of the image sequence and updated with image areas uncovered due to object motion. Input to the change detector are the

Figure 6. Masks M_{k+1} according to [4] for the test sequences (a) *Claire* and (b) *Erik* after 30 and 10 frames respectiveley. In order to ease the evaluation, the changed areas are taken from the original images.

 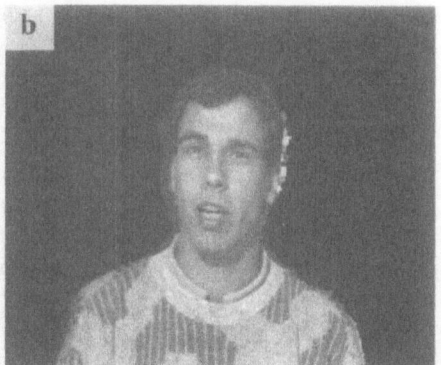

Figure 7. Masks M_{k+1} as computed by the new detector for the test sequences (a) *Claire* und (b) *Erik* after 30 and 10 frames respectiveley. In order to ease the evaluation, the changed areas are taken from the original images.

current image I_{k+1} and the previously coded image I'_k which partially contains the background memory. Figure 6 shows the moving objects for the test sequences Claire and Eric after 30 and 10 frames, respectively. Figure 7 shows the results using the new change detector. Apparently, the new detector is not as sensitive to noise or shadows as the algorithm based on evaluating the luminance difference only.

4   CONCLUSIONS

In this paper, a change detector is proposed that evaluates three criteria in order to determine image areas changed due to object motion. The temporal difference criterion as used in other publications is augmented by a reflection criterion that is based on a simple illumination model for detecting moving shadows. The third criterion is an edge criterion that detects static and moving edges. The joint evaluation of these three criteria results in a change detector that is able to detect moving objects in front of a static background and that classifies moving shadows correctly as static as long as the direct illumination of the scene accounts for less than 50% of the total illumination.

5   LITERATURE

[1]   T. Aach, A. Kaup, R. Mester, "Statistical model–based change detection in moving video", Signal Processing, Vol. 31, No. 2, pp. 165–180, March 1993.

[2]   J.C. Candy, M.A. Franke, B.J. Haskell, F.W. Mounts, "Transmitting Television as Clusters of Frame–to–Frame Differences", Bell System Technical Journal, Vol. 50, pp. 1877–1888, 1971.

[3]   P. Gerken, "Object–based analysis–synthesis–coding of image sequences at very low bitrates", COST 211ter European workshop on new techniques for coding of video signals at very low bitrates, Hannover, Germany, December 1993.

[4]   M. Hötter, R. Thoma, "Image segmentation based on object oriented mapping parameter estimation", Signal Processing, Vol. 15, No. 3, pp. 315–334, October 1988.

[5]   M. Kunt, "Second–generation image–coding techniques", Proceedings of the IEEE, Vol. 73, No. 4, pp. 549–574, April 1985.

[6]   F.W. Mounts, "A video encoding system employing conditional Picture–Element Replnishment", Bell System Technical Journal, Vol. 48, pp. 2545–2554, September 1969.

[7]   H.G. Musmann, M. Hötter, J. Ostermann, "Object–oriented analysis–synthesis coding of moving images", Signal Processing: Image Communication, Vol. 1, No. 2, pp. 117–138, November 1989.

[8]   C. Lettera, L. Masera, "Foreground/background segmentation in videotelephony", Signal Processing: Image Communication, Vol. 1, No. 2, pp. 181–189, Oktober 1989.

[9]   W. Li, M.Kunt, "Morphological segmentation applied to displaced difference coding", Signal Processing, Vol. 38, No. 1, pp. 45–56, July 1994,

[10] J. Ostermann, "Modelling of 3D moving objects for an analysis–synthesis coder", SPIE/SPSE Symposium on Sensing and Reconstruction of 3D Objects and Scenes, B. Girod Hrsg., Proc. SPIE 1260, Santa Clara, California, U.S.A., February 1990.

[11] J. Ostermann, "Object–based analysis–synthesis Coding based on the source model of moving rigid 3D objects", Signal Processing: Image Communication, No. 6. pp. 143–161, 1994.

[12] K. Skifstad, R. Jain, "Illumination Independent Change Detection for Real World Image Sequences", Computer Vision, Graphics, and Image Processing, vol. 46, no. 3, pp. 387–399, June 1989.

OBJECT-SCALABLE CONTENT-BASED 2-D MESH DESIGN FOR OBJECT-BASED VIDEO CODING[1]

Yucel Altunbasak and A. Murat Tekalp

Department of Electrical Engineering and Center for Electronic Imaging Systems
University of Rochester, Rochester, NY 14627

1 INTRODUCTION

It is generally accepted that the object-based coding (OBC) technology offers the potential to deliver object-scalable digital video at a quality higher than currently available through the ITU-T H.263 standard [1] at similar bitrates. An object is a significant feature of the scene, such as for example, the head of a participant in a video phone conversation. Recent advances in low bitrate, object-based video compression have been summarized in [2, 3] and the references therein. These techniques range from sophisticated 3-D object based algorithms, including those using customized wireframe models, to more simple and general 2-D object based methods.

3-D OBC methods may not be suitable for real-time general purpose video communications, because they often require customized 3-D wireframe models based on prior knowledge of the scene content. 2-D OBC methods, such as those using triangular/quadrilateral meshes and spatial transformations, offer more readily implementable solutions which are superior to standard block-based motion compensation. Existing international standards such as H.263 and MPEG 1-2 all use block motion compensation (in order to limit the number of bits reserved for the motion vectors) which suffers from blocking artifacts at low bitrates [4]. A promising approach to avoiding blocking artifacts but still using a coarsely sampled motion field is motion compensation by means of 2-D mesh models (which may be viewed as irregular sampling of the dense motion field). Brusewitz [5] was among the first who proposed triangle-based motion compensation, where a triangular mesh is overlaid on the image. Sullivan and Baker [6] used quadrilateral meshes for motion compensation under the name control grid interpolation. Motion-compensation within each mesh element (patch) is accomplished by means of a spatial transformation (affine, bilinear, etc.) whose parameters can be computed from node-point motion vectors. Typically, affine and bilinear mappings are used with triangular and quadrilateral patches, respectively [7, 8, 9]. Recently, Nakaya et al. [7] proposed a hexagonal search procedure for motion estimation and compensation based on a regular mesh; while Wang et al. [9] advanced an optimization framework for motion compensation based on an active mesh. However, the former method may suffer from visual artifacts resulting from multiple motions within a patch and the latter does not address occlusion-adaptive temporal tracking (linking) of the resulting nonuniform 2-D mesh.

The success of temporal tracking (linking mesh elements from frame to frame) strongly

---

[1]This work is supported in part by a National Science Foundation IUCRC grant and a New York State Science and Technology Foundation grant to the Center for Electronic Imaging Systems at the University of Rochester, and a grant by Eastman Kodak Company.

depends on how well the 2-D mesh model fits the scene content, and how well we can detect occlusion boundaries and estimate the motion field in the vicinity of these boundaries. Occlusion regions, classified as background to be covered (BTBC) and uncovered background (UB) may appear at the object boundaries (due to global object motion) or within the objects (due to local motion or deformations). The latter is generally referred to as self-occlusion. If the connectivity of the entire mesh is preserved at all times, then a 2-D mesh model cannot properly handle occlusion regions. In this paper, we propose occlusion-adaptive mesh models, where we suppress mesh-connectivity constraints at occlusion boundaries. In this scheme, no node points are allowed to be placed in the BTBC region(s), and patches in the previous frame corresponding to the BTBC region(s) are forced to disappear in the current frame. Another desired property of 2-D meshes is object scalability, which allows various objects in the scene to be treated differently. For example, more bits may be spent on an object of interest. To this effect, this paper describes methods to design object-scalable, content-based, occlusion-adaptive 2-D mesh models for object-based coding, where accurate occlusion detection is addressed in Section 2 and methods for content-based and object-scalable 2-D mesh model design are covered Section 3.

2 ACCURATE OCCLUSION DETECTION

Accurate estimation of occlusion boundaries is a challenging task, since motion estimates around occlusion regions are generally inaccurate [10, 11, 12]. In the following, we first describe a fast method to obtain an initial estimate of the occlusion boundaries. We, then, elaborate on how inaccuracies in the initial occlusion boundaries lead to significant performance degradation in motion tracking and compensation. Next, we propose an algorithm for improved occlusion detection by edge alignment and segmentation-based search. Finally, the improved occlusion boundaries are approximated by polygons.

2.1 Initial Occlusion Detection

Initial boundaries of the occlusion regions can be estimated by thresholding the displaced frame difference (which is a function of a dense motion field). Clearly, the accuracy of the initial occlusion boundaries will depend on the accuracy of dense motion estimates, especially in the vicinity of the occlusion regions. 2-D dense motion estimation methods include block-based, optical-flow equation based, pel-recursive, and Bayesian methods [4]. In this work, we have used optic-flow equation based methods, such as the method of Lucas-Kanade [14] and Horn-Schunck [15], since they yield smoother (hence closer to the actual) motion fields than, for example, block-matching (which has been adopted in the international standards, such as H.261 and MPEG 1-2). It should be noted that values of the smoothness and/or blurring parameters in these methods are important, since if the motion field is oversmoothed, motion tracking over small regions becomes impossible. For the sake of minimizing the computational burden, 2-D dense motion estimation and occlusion detection can be performed only over the changed regions (which can be easily detected from the frame difference). The resulting occlusion map is usually post-processed by appropriate median and/or morphological filtering to form contigious regions.

A summary of the initial occlusion (BTBC) detection algorithm is as follows:

1. Calculate the frame difference (FD) and find the change detection mask (CDM).

2. Estimate the dense motion field from frame $k - 1$ to frame k within the CDM.

3. Motion compensate frame $k - 1$ from frame k to compute the predicted estimate $\tilde{I}_{k-1}$.

4. If the displaced frame difference, $DFD(x, y) = I_{k-1}(x, y) - \tilde{I}_{k-1}(x, y)$, where $I_{k-1}(x, y)$ is
   the observed intensity, is greater than a threshold T_1, then label (x, y) as an occlusion pixel.

5. Perform post-processing to form smooth BTBC regions as follows:

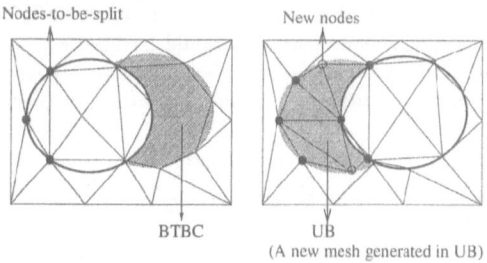

Figure 1: Occlusion-adaptive mesh design and tracking.

(a) Apply median filtering with a 5×5 kernel.
(b) Apply three successive morphological closing operations with a 3×3 kernel, followed by three morphological opening operations with the same kernel.
(c) Eliminate regions whose area is smaller than N_1 (predefined) pixels.

Because most simple dense motion estimation algorithms are not sufficiently accurate near occlusion regions, the initial occlusion boundaries computed as above tend to overestimate the actual boundaries [10, 11, 12]. This leads to severe problems in motion tracking and compensation in the vicinity of occlusion regions. For example, our ability to track lip motion [20] is directly related to how well the boundaries of the UB (when the mouth is opening) and BTBC (when it is closing) can be estimated.

2.2 Improved Occlusion Detection

Accurate occlusion detection plays an important role in 2-D mesh-based motion tracking and compensation. In an occlusion-adaptive 2-D mesh, all patches within the BTBC region in frame $k - 1$ should vanish in frame k and no pixel from frame $k - 1$ should map into the UB region in frame k, provided that occlusion regions can be perfectly detected and motion vectors at the occlusion boundaries can be accurately estimated. The concept of an occlusion-adaptive mesh is illustrated in Fig. 1.

Let's investigate the effects of inaccurate BTBC detection by an example. Suppose a person is speaking in a videophone scene; and his/her mouth is open in the previous frame, and closed in the current one. Then, the area between the lips in the previous frame is backgorund to be covered (BTBC). Two cases, corresponding to overestimation and underestimation (threshold T_1 too high) of the actual BTBC boundaries, are depicted in Figures 2 (a) and (b), where the solid and dotted lines show the actual and the estimated BTBC boundaries. In the former case, motion vectors along the estimated BTBC boundary (dotted) may not represent the actual motion of the mouth (motion of the actual boundary). In the latter case, although the BTBC region should vanish, the region between the dotted and solid lines will not be forced to vanish. It should be clear from the above discussion that accurate detection of occlusion regions plays an important role in minimizing the area of model failure (MF) regions after motion compensation.

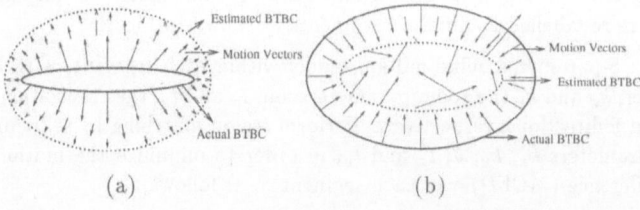

Figure 2: Effects of a) overestimating and b) underestimating the occlusion region.

Motion models and/or assumptions used in dense motion estimation fail in certain areas of the image (where proper motion compensation would have been possible using more complicated models) resulting in overestimation of occlusion regions. There are three main reasons for inaccurate dense motion estimation: i) multiple motions or partial occlusion within the local motion estimation window, ii) local deformations, and iii) intensity variation due to shading. In the following, we describe a two-step algorithm composed of edge alignment and segmentation-based local deformation checking to refine the initial occlusion map to address i) and ii) above, respectively.

2.2.1 Edge Alignment

Motion segmentation and occlusion detection are closely related, since occlusion regions are most likely located near boundaries of independently moving objects. However, search for possible occlusion regions and/or occlusion refinement cannot be limited to object boundaries (given in case of MPEG-4 test sequences) since there is also self-occlusion which can happen anywhere within objects. To this effect, we first impose the constraint that occlusion boundaries must align with spatial edges. Because, in practice, there may be several edges within the initial BTBC map, here we perform edge alignment with the outermost edge locations identified by a sequence of morphological operations. The outline of the proposed edge-alignment algorithm is as follows:

1. Estimate image intensity gradients I_x and I_y in the x and y directions, respectively, within the initial BTBC mask. Then, all pixels for which the magnitude of the image intensity gradient is above a threshold T_2 are labelled as edge pixels (edgels).

2. Apply morphological closing operation on the binary edge image as many times needed to close all holes inside the edge image.

3. Apply morphological opening operation as many times as in the previous step.

4. A pixel belongs to the refined BTBC region only if it is both in the initial BTBC region and the postprocessed edge image.

2.2.2 Segmentation-based Refinement

The edge alignment step eliminates uniform image areas outside the outermost edge locations within the initial occlusion map. Next, we address eliminating regions which may have multiple motion, partial occlusion, and/or local deformations. This is done by segmenting the BTBC map (after edge-alignment) into regions of uniform color (gray-scale) and performing region matching (for each region) in a two or six dimensional motion parameter space. Regions for which a match can be established are eliminated from the BTBC map. This procedure guarantees alignment of occlusion boundaries with spatial edges, since intensity boundaries are a superset of actual motion boundaries.

A summary of the complete procedure is given below:

1. Perform intensity/color segmentation within the refined BTBC region in frame I_{k-1} using the adaptive K-means algorithm (starting with an initial number of segments and merging those with close mean intensity/color). Spatially disconnected regions are treated as different segments even if they have the same label. To this effect, all distinct regions are re-labelled as different segments as shown in Fig. 3.

2. Assume a 5-parameter affine motion model within each segment, with θ, the rotation parameter, k_x and k_y the zoom parameters and t_x and t_y, the translation parameters in the x and y directions, respectively. Perform region matching by searching for the best set of parameters k_x, k_y, θ, t_x and t_y, in order to minimize the motion compensated frame difference (MCFD) over each segment S_i as follows:

   (a) Find the center of gravity (x_c, y_c) of the segment S_i.
   (b) Employ a logarithmic search for the parameter set which minimizes the MCFD as follows:

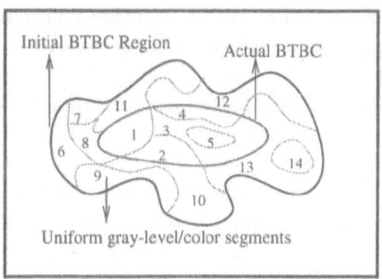

Figure 3: Segmentation for accurate occlusion detection.

   i. Calculate the range and step size for each of the five parameters about the best solution from the previous iteration.

   ii. Assuming the center of rotation of the segment S_i is (x_c, y_c), calculate the MCFD for each set of search parameters and find the best set.

   iii. Iterate a predetermined number of times.

3. If MCFD is below a predefined threshold, then take that segment out from refined BTBC map. Otherwise, label that segment as the "occlusion segment".

The proposed improved occlusion detection method can be used as a preprocessing step with any motion estimation and/or segmentation algorithm. After having achieved an accurate occlusion detection, motion vectors in the vicinity of the improved occlusion boundaries should be revised.

2.3 Polygon Approximation

The boundaries of the occlusion (BTBC) regions need to be approximated by a shape that can be represented with a few parameters. Most commonly used shape models are polygonal and B-spline approximations. Here, we employ a polygonal approximation method, similar to that proposed by the COST 211ter group [16], because of its simplicity and robustness. Furthermore, a polygonal boundary naturally fits with the boundaries of the proposed occlusion-adaptive mesh model. That is, the nodes of the polygon approximation will also serve as nodal points of the 2-D mesh to be able to accurately track the motion in the vicinity of occlusion regions by means of motion vectors at these node points.

3 OBJECT-SCALABLE CONTENT-BASED MESH DESIGN

Here, we propose two algorithms for designing 2-D content-based mesh models with triangular patches, such that patch boundaries conform with the boundaries of objects in the scene, BTBC and UB regions, and multiple motions. Object-scalable extensions of the proposed design procedures are given in Section 3.2.

3.1 Content-Based Mesh Design

A content-based mesh refers to a nonuniform mesh structure, where patch boundaries are designed to fit the scene content and motion edges; whereas a regular (uniform) mesh has equal size patch elements whose boundaries are determined only by the node density. Because regular meshes can be compactly described by horizontal and vertical node densities, which can be transmitted at every frame, regular meshes can be designed independently for each pair of frames alleviating the mesh tracking problem [8]. On the other hand, nonuniform meshes provide superior motion compensation since they can be designed such that each patch contains a single motion which can be accurately described by a spatial transformation. However, transmission of mesh information for each frame as overhead is costly, which necessitates mesh tracking.

There have been two approaches reported for nonuniform mesh design: optimization and split-and-merge methods. Designing an optimal mesh structure requires global optimization of a suitable cost function. However, most practical optimization methods converge to a local minimum which is "close" to a uniform mesh (assuming the initial mesh is a uniform one) [9]. Split and merge methods successively divide, starting possibly with a uniform mesh, patches which do not satisfy a predetermined criterion in an attempt to find locally optimum solutions [8, 19, 20]. Some of the inserted nodes may need to be treated specially to preserve the connectivity (geometry) of the mesh and/or children-parent patch relationships in hierarchical mesh refinement procedures [18, 8]. In the following, we propose a split-and-merge method which is a variation of the method given in [19] (we ensure that no patches cross over the BTBC polygon) and a novel spatio-temporal gradient based approach, which consists of node-point selection followed by triangulation.

3.1.1 Split and Merge Method

The split-and-merge method aims to design a locally optimal mesh structure in the sense of minimizing the displaced frame difference (synthesis error). The algorithm is composed of a split stage followed by a merge stage, which are described below.

Split-stage:

1. For image regions outside the BTBC polygon(s), design a uniform triangular mesh with prespecified horizontal and vertical node densities. Triangular patches in the vicinity of the BTBC polygon, formed by connecting nearby node points with the corner points of the polygon, may be nonuniform. Order all patches sequentially.

2. For each patch,

   (a) Find motion vectors at the vertices (see [21]), and compute an affine mapping.

   (b) Compute the average and the maximum DFD within the patch.

   (c) If the average and/or maximum DFD are larger than some threshold(s), then insert a new node at the point of the largest DFD. Connect this new node with the existing three vertices. Then, go to step 2a and repeat this procedure for each of the three new patches.

   (d) If the average and/or maximum DFD are smaller than some thresholds, go to the next patch (step 2).

Merge-stage:

For each node point, except those on the boundary of the BTBC polygon(s), consider the smallest polygon (composed of triangular patches) enclosing that node point. If the average and/or maximum DFD within the polygon are smaller than some thresholds, kill the node point and perform a new triangulation for the polygon under consideration. Note that this triangulation is not unique.

The split and merge stages are repeated successively, until no further splitting or merging are required.

3.1.2 Spatio-Temporal Gradients Method

The algorithm consists of a nonuniformly spaced node-point selection procedure, followed by triangulation using the selected node points. The procedure aims to partition the image into triangles in such a way that a predefined function of the displaced frame difference (DFD) within each patch attains approximately the same value. An outline of the algorithm is as follows:

1. Estimate a 2-D dense motion field between the present and reference frames. Label all pixels, except those in the BTBC polygon(s), as "unmarked." Include all corner points of the BTBC polygon(s) in the list of selected node points.

2. Compute the average displaced frame difference DFD_{avg} given by

$$DFD_{avg} = \frac{\sum_{(x,y)}(DFD(x,y))^p}{K} \tag{1}$$

where the summation is over all unmarked points, K is the number of unmarked points, and p is a positive number.

3. Compute a cost function $C(x,y)$ associated with each unmarked pixel as a predefined function of spatio-temporal intensity gradients. The cost function includes terms that are functions both spatial and temporal intensity gradients so that selected node points, hence the boundaries of the patches, coincide with spatial edges and motion edges.

4. Find the unmarked pixel with the highest $C(x,y)$ which is not closer to any other previously selected node point than a prespecified distance. Label this point as a node point.

5. Grow a region about this node point until $\sum(DFD(x,y))^p$ in this region is greater than DFD_{avg}. Label all pixels within this region (depicted in Figure 4) as "marked."

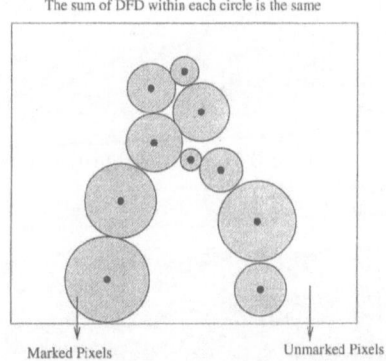

Figure 4: Demonstration of proximity constraints in the mesh design algorithm.

6. Go to 2 until a desired number of node points, N, are selected.

7. Given the selected node points, apply a triangulation procedure (e.g., Delauney triangulation [22]) to obtain a content-based mesh.

3.2 Object Scalability

This section addresses object-scalable content-based mesh design assuming the boundaries of individual objects are known. Object-scalability provides object-dependent quality scalability as well as ability to extract specified object layers from the encoded bitstream. Object-scalable mesh design introduces the additional consideration that no patch should cross over the object boundaries in any frame (which also imposes a constraint on the mesh tracking algorithm to be discussed in the next section).

Object scalable mesh design is realized in two steps: i) Approximate the boundary of each individual object by a polygon, and ii) apply constrained Delauney triangulation, where line segments representing the boundary of the object polygons are passed as constraints. When two objects have a common boundary segment, polygon approximation of the boundaries requires special attention. In this case, intersection points (corners) of these boundaries are first computed and entered as node points in the polygon approximation algorithm.

4   RESULTS

Experimental results are provided to compare the performance of the proposed 2-D mesh design algorithms, with and without occlusion adaptivity, versus a knowledge based 2-D mesh (projection of the CANDIDE wireframe model) using frames 1 and 4 of the sequence Miss America. In particular, the initial and final BTBC maps (obtained as described in Sections 2.1 and 2.2, respectively) are shown in Fig. 5 (a) and (b). Fig. 5 (c) shows a projection of the 3-D wireframe model CANDIDE after it is scaled to match frame 1. Next, we designed two 2-D mesh models (without taking the BTBC regions into account) using the split method and the spatio-temporal gradients method, respectively, where each mesh has 155 nodes (which matches that of the CANDIDE model). They are shown in Fig. 6 (a) and (b). Finally, Fig. 6 (c) shows an occlusion-adaptive 2-D mesh designed by using the spatio-temporal gradients method, where the BTBC regions (painted black) are approximated by polygons. Node points are selected at the boundary (corner points) of these polygons but not inside them.

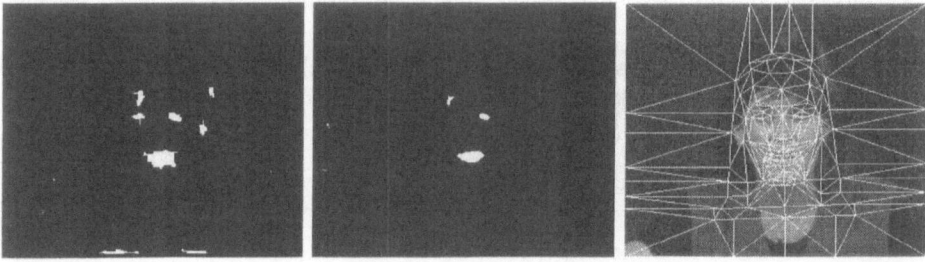

Figure 5: a) Initial BTBC map, b) final BTBC map, c) projection of the CANDIDE wireframe model.

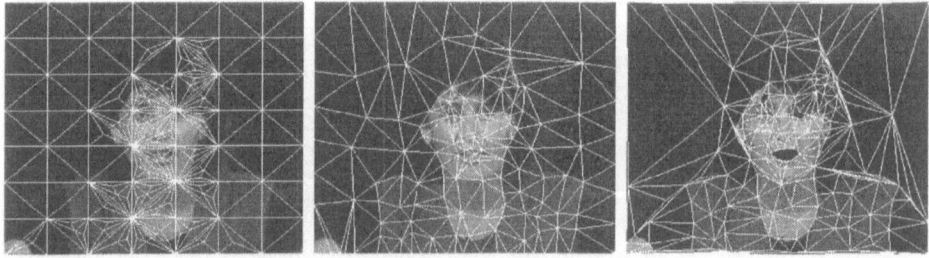

Figure 6: a) Split-merge method, b) Spatio-temporal gradient method, c) object-scalable design.

Motion compensation (MC) PSNR achieved by these mesh models, as well as meshes with 200 and 250 node points, are listed in Table 4. It can be seen that all of the proposed methods outperform the MC PSNR that can be obtained by the knowledge-based 2-D mesh model.

	CANDIDE	Adaptive	Content	Scalable	Scalable w/occ.
155 node	36.44	36.69	37.24	37.48	37.53
200 node	-	36.78	37.44	-	-
250 node	-	-	37.64	37.54	-

Table 1: MC PSNR between frames 1 and 4 using various 2-D meshes.

References

[1] ITU-T Expert Group on Very Low Bitrate Video Coding, Simulation test model TMN3, Technical Report, July 1994.

[2] H. Li, A. Lundmark, and R. Forchheimer, "Image sequence coding at very low bitrates: A Review," *IEEE Trans. Image Proc.*, vol. 3, pp. 589-609, Sept. 1994.

[3] K. Aizawa and T. S. Huang, "Model-based image coding: Advanced video coding techniques for very low bit-rate applications," *Proc. IEEE*, vol. 83, pp. 259-271, Feb. 1995.

[4] A. M. Tekalp, *Digital Video Processing*, Prentice Hall, 1995.

[5] H. Brusewitz, "Motion compensation with triangles," *Proc. 3rd Int. Conf. on 64-kbit coding of moving video*, Rotterdam, Netherlands, Sept. 1990.

[6] G. J. Sullivan and R. L. Baker, "Motion compensation for video compression using control grid interpolation," *Proc. ICASSP'91*, pp. 2713-2716, Toronto, Canada, 1991.

[7] Y. Nakaya and H. Harashima, "Motion compensation based on spatial transformations," *IEEE Trans. Circ. and Syst.: Video Tech.*, vol. 4, pp. 339-356, June 1994.

[8] C. L. Huang and C. Y. Hsu, "A new motion compensation method for image sequence coding using hierarchical grid interpolation," *IEEE Trans. Circ. and Syst. Video tech.*, vol. 4, no. 1, pp. 72-85, 1994.

[9] Y. Wang and O. Lee, "Active mesh - A feature seeking and tracking image sequence representation scheme," *IEEE Trans. Image Proc.*, vol. 3, pp. 610-624, Sept. 1994.

[10] W. B. Thompson, K. M. Mutch, and V. A. Berzins, "Dynamic occlusion analysis in optical flow fields," *IEEE Trans. Patt. Anal. Mach. Intel.*, vol. 7, pp. 374-383, 1985.

[11] M. Hoetter and R. Thoma, "Image segmentation based on object-oriented mapping parameter estimation," *Signal. Proc.*, vol. 15, pp. 315-334, 1988.

[12] M. Irani, B. Rousso, and S. Peleg, "Computing occluding and transparent motions," *Int. J. Comp. Vision*, vol. 12:1, pp. 5-16, 1994.

[13] Y. Wang, R. S. Wang, O. Lee, T. Chen, H. H. Chen, and B. G. Haskell, "Mouth shape detection and tracking using an active mesh," *Vis. Comm. and Image Processing Conf.*, SPIE vol. 2501, pp. 1141-1152, Taiwan, May 1995.

[14] B. D. Lucas and T. Kanade, "An iterative image registration technique with an application to stereo vision," *Proc. DARPA Image Understanding Workshop*, pp. 121-130, 1981.

[15] B. K. P. Horn and B. G. Schunck, "Determining optical flow," *Artif. Intel.*, vol. 17, pp. 185-203, 1981.

[16] "Simulation model for object-based coding," COST 211ter, Simulation Subgroup, SIM(93) 50, Oct. 1993.

[17] D. Terzopoulos and M. Vasilescu, "Sampling and reconstruction using adaptive meshes," *Proc. IEEE Int. Conf. Computer Vision and Pattern Recognition (CVPR'91)*, pp. 70-75, June 1991.

[18] J. Flusser, "An adaptive method for image registration," *Patt. Recog.*, vol. 25(1), pp. 45-54, 1992.

[19] W.-F. Lee and C.-K. Chan, "Two-dimensional split and merge algorithm for image coding," *Vis. Comm. and Image Proc.'95*, SPIE vol. 2501, pp. 694-704, May 1995.

[20] O. Lee and Y. Wang, "Nonuniform image sampling and interpolation over deformed meshes and its hierarchical extension," *Vis. Comm. and Image Proc.*, SPIE 2501, pp. 389-400, May 1995.

[21] Y. Altunbasak, A. M. Tekalp, and G. Bozdagi, "Two-dimensional object-based coding using a content-based mesh and affine motion parameterization," *Proc. IEEE Int. Conf. Image Proc.*, Washington, DC, Oct. 1995.

[22] D. T. Lee and B. J. Schachter, "Two algorithms for constructing a Delauney triangulation," *Int. J. Comp. and Info. Sci.*, vol. 9, no. 3, 1980.

A REGION-BASED VIDEO CODER
USING A FORWARD TRACKING ACTIVE MESH

Yao Wang[1], Anthony Vetro[1], and Ouseb Lee[2]

[1]Polytechnic University, Brooklyn, NY 11201
[2]Research and Development Center, POSDATA LTD., Seoul, Korea

INTRODUCTION

In an earlier study, we have reported the use of an active mesh representation for describing a video sequence [1]. In this representation, a *scene adaptive* mesh is generated for an initial frame, such that, the nodes fall on distinct features in the image and each element covers a 2D patch corresponding to a single object. The resulting nodes are then tracked in successive frames, so that the color patterns over corresponding elements in these frames match one another. The motion field between every two frames is interpolated from the displacements of the control nodes. In this paper, we present a coder that employs the active mesh structure. With this coder, an image is partitioned into regions of distinct motions. Each region is described by an ensemble of connected quadrilateral elements embedded in a mesh structure. The shape and color of each region are described by the nodal positions and color functions of the elements in this region in an initial frame; while its motion (including shape deformation) is described by the nodal trajectories in the following frames, which are in turn specified by a few motion parameters. Each motion class in this coder corresponds to a region (not necessarily connected), which may or may not correspond to a separate physical object. For this reason, we call this coder a *region–based* coder. This coder has been applied to a typical CIF resolution, head-and-shoulder type of sequence. At 50 Kbits/sec (for the luminance component only), the algorithm can render the motion more accurately and naturally than the H263-TMN4 algorithm.

In the following, we first give an overview how the coder is structured. Then we describe the coding algorithm for each parameter. This will be followed by a report of the simulation results and concluding remarks.

OVERVIEW OF THE CODER STRUCTURE

The proposed coder divides a given sequence into groups of pictures (GOPs) so that the scene content in each GOP does not change dramatically. For each GOP, a scene adaptive mesh is first generated for the starting frame. This frame is called an *I–frame*. The nodes in this mesh are then tracked in the following frames successively. The image pattern in each new frame is predicted by warping the previous frame based on nodal correspondences. The elements in the mesh are grouped into different regions, so that the elements in the same region have similar motion characteristics. Rather than specifying the nodal positions at every frame, we assume that the nodal trajectory within each region is piecewise linear. For each region, frames at which the region experiences motion transition are identified as *MT-frames* of this region.

The coder specifies i) nodal positions and color patterns of all elements in the I-frame; ii) initial element classification map and update information; iii) nodal displacements between each MT-frame and its previous frame for each region; and iv) prediction errors in MT–frames and possibly ME-frames. The quantization parameters for the color pattern in the I-frame and prediction errors in MT-frames and ME-frames can differ among separate regions. In order to further reduce bit rates and/or computation time, the coder could also skip frames. In the decoder, the skipped frames can be interpolated from the previous and following coded frames, based on the decoded nodal trajectories. The simulation results presented here however were obtained by coding every frame.

The mesh generation for the starting frame and forward nodal tracking in the following frames are accomplished by the energy minimization approaches presented in [1]. For motion based element classification, we are investigating two alternatives [2]: a parallel approach using the K-means clustering algorithm, and a layered approach using a robust estimation method. However, simulation results reported in this paper are generated based on manual segmentation results. One segmentation map is generated for each GOP, without sequential update.

The detection of the MT-frames is illustrated in Fig. 1. For each node, we linearly extrapolate its position in the current frame based on its positions in previous two frames. This extrapolated nodal position is then compared to the estimated position produced by the nodal tracking algorithm. Finally, the nodal extrapolation error for region l is calculated according to:

$$E_l = \frac{1}{N_l} \sum_{\mathbf{p}_n \in R_l} w_n \|\mathbf{p}_{n,\mathrm{tr}} - \mathbf{p}_{n,\mathrm{ex}}\|^2,$$

where R_l represents region l, N_l the number of nodes in R_l, $\mathbf{p}_{n,\mathrm{tr}}$ the tracked position of the n-th node, and $\mathbf{p}_{n,\mathrm{ex}}$ the extrapolated position of the n-th node. If $e_l > T$, then the region is treated as an MT-frame. The threshold T is selected based on the allowable amount of nodal error over a region. The coefficient w_n is dependent on the edge magnitude f_n at the node n through a sigmoidal relation:

$$w_n = \frac{1}{1 + \exp\{-\lambda(f_n - f_c)\}},$$

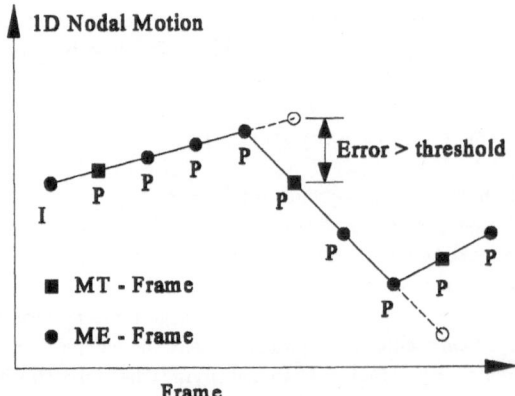

Figure 1: Illustration of MT-Frame Detection.

where λ and f_c are user selectable constants. The rationale behind the use of the edge based weighting function are two-fold. First, the tracked nodal position may not be accurate in a smooth region; Second, nodal errors in smooth regions are not critical. By applying higher weights to nodes in edge regions, we essentially discount the effect of nodal errors in smooth regions.

CODING OF DIFFERENT PARAMETERS

The following information is to be coded in each GOP: the nodal positions in the I-frame and the initial and updated segmentation map, which specify the *shape information*; the image patterns in individual elements in the I-frame and the prediction errors in P-frames, which constitute the *color information*; and finally the motion parameters at every MT-frame for each region, which comprise the *motion information*. In general, the coding accuracies for the color and motion parameters can vary among different regions.

For the color information, the coding is done element by element. In general, it is hard to represent a 2D function over an arbitrary quadrilateral. We developed a *Warped DCT* method. It first warps each element into a square block by using a forward bilinear mapping. It then codes the warped square block by a block-DCT based coding method. In the decoder, the warped block is first obtained by inverse DCT coding. The original element (with quantization artifacts) is then recovered by using the inverse bilinear mapping. In our computer simulation, we have adopted the coding schemes used in the MPEG1 coder for the intra– and inter–modes for elements in the I-frame and P-frames, respectively.

For nodal positions in the I-frame, a second order predictive coding scheme is used, by which the position of each node is predicted from those nodes to its left and top. The prediction error in each coordinate is quantized to quarter-pel accuracy and then entropy coded. The number of bits required for each node, B_n, is estimated by the the sum of the entropies of the prediction errors in the vertical and horizontal positions.

To specify the element classification map, we use a simple prediction scheme. If the class index of an element is the same as that of its previous element, a *No-Change* symbol is coded. Otherwise, the actual index is coded. The number of bits required for specifying the index of each element, B_r, is estimated by the entropy of all possible symbols (the No-Change symbol and all class indices).

Finally, for each class, to specify the nodal displacements at MT-frames, these displacements are converted into a set of bilinear motion parameters. Rather than specifying the $N_m = 8$ bilinear coefficients directly, they are first converted to 8 displacement values at four selected positions in the region. These values are then quantized to quarter-pel accuracy and entropy coded. The number of bits required to code each motion parameter, B_m, is estimated from the entropy of the quantized displacement values.

SIMULATION RESULTS

The proposed coding method has been used to code 50 frames of a test sequence *Miss America* in the CIF resolution (30 frames/sec, 352×288 pels/frame, YCbCr). Only the luminance component is processed. All 50 frames are grouped into one GOP, with a single I–frame in the beginning. The number of elements is $N_e = 44 \times 36$, with an average element size of 8×8, and the number of active nodes (those not on the frame boundary) is $N_n = 43 \times 35$.

As described previously, in this experiment, element classification is performed manually. One classification map is generated manually for the first frame, which is then used in all the following frames in this GOP, without sequential update. The classification map for the first frame is shown in Fig. 2(a). A total of 5 regions are used, which correspond to the eyes, the mouth, the remainder of the head, the shoulder, and the background. The prediction error is coded only for elements in the eye, mouth, and head regions, and is invoked at both

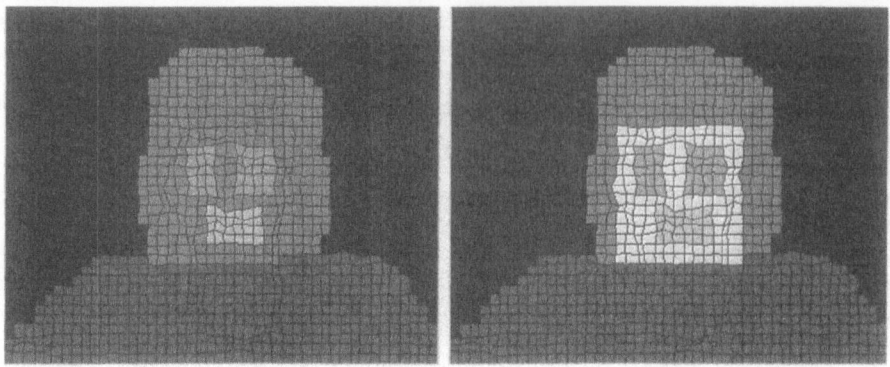

Figure 2: The element classification map used in the coder: (a) The motion class map obtained manually; There are a total of five motion classes (background, shoulder, head, eye, and mouth) and each class is denoted by a distinct gray level. (b) The motion class map with an added interior face region, represented by the brightest level.

MT- and ME-frames. In order to render the face region more accurately, an interior face region is specified in the head region, as shown in Fig. 2(b). This region belongs to the same motion class as the rest of the head, but is coded using a smaller QP. In the shoulder and background regions, each element is warped from its corresponding element in the previous frame recursively without error compensation. The QP for the I-frame is 15 for all classes. For the P-frames, the QPs for elements in the eye, mouth, face, and the rest of the head region are set to 16, 16, and 21, and 31, respectively. The total number of bits for this GOP of 50 frames is estimated to be $B = 84,716$ bits, which translates into a bit rate of 50.8 Kbits/sec (Kbps).

For comparison, we also applied the TMN4 test model of the ITU-T low-bit-rate coding standard H.263 to the same test sequence [3]. The coder automatically adjusts the frame rate for a given bit rate. The total number of coded frames is 20, with an average frame distance of 2.5. The average QPs used for the I-frame and P-frames are both 15. The total number of bits used for these 50 frames is 83,400, which translates into a bit rate of 50 Kbps. Three types of reconstruction schemes have been simulated for the uncoded frames: i) repeating from the previous coded frame; ii) linear interpolation between coded frames; and iii) motion-compensated linear interpolation based on the motion field generated by the block matching method. The last method yielded some unacceptable artifacts because of the error in block-based motion interpolation. Therefore, we only show the results obtained by the first two methods, which will be referred to as TMN4 and TMN4i, respectively.

Figs. 3(a) and 3(b) show the original image and the generated mesh for the first I-frame. Figs. 3(c) and 3(d) show the decompressed images by the proposed coder and the TMN4i coder, respectively, for this frame. Figs. 4(a) and 4(b) present the original images and meshes for selected P-frames (only a sub-region is shown with magnification). Figs. 4(c) and 4(d) compare the decompressed images by the proposed coder and the TMN4i coder. These frames are selected to include a variety of facial motions and a good mix of MT-frames and ME-frames. It can be seen that even though the compressed images by the proposed method are blurred compared to the original ones, there are no annoying visual artifacts. With the TMN4i method, the coded frames have noticeable blocking and quantization artifacts, while the interpolated frames suffer from motion blurring. For example, frame 4 appears to have a transparent eye. When observed in real time at 30 frames/sec, the motions of the eyes, the mouth, the face, as well as the head-and-shoulder part are all rendered smoothly and naturally in the compressed sequence using our method. These motions appear very jerky in the compressed sequence by the TMN4 method. With the sequence produced by the TMN4i

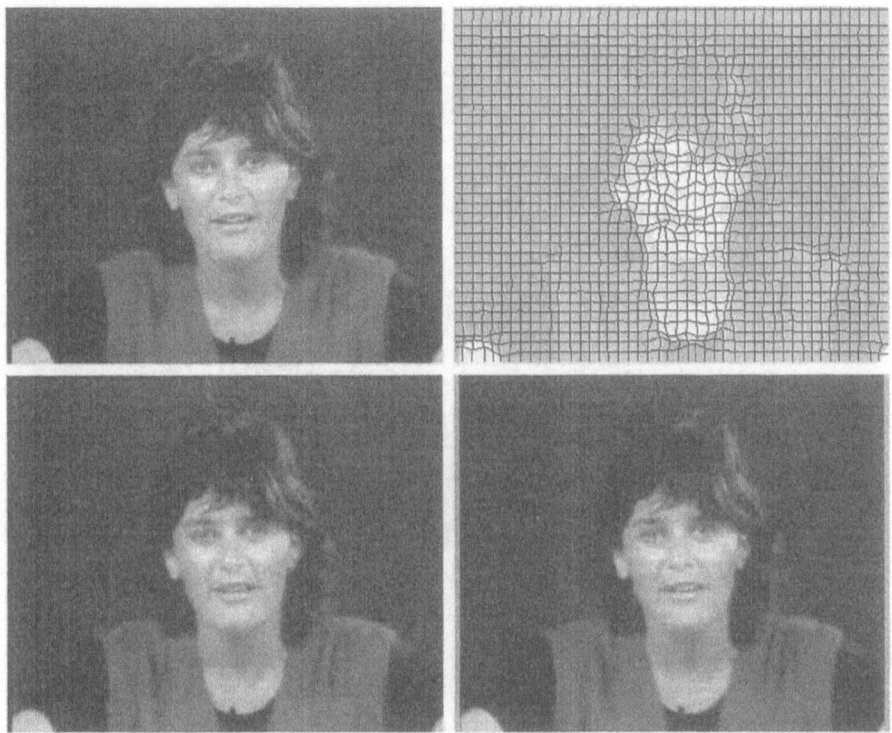

Figure 3: Simulation results for the I-frame (frame 0) of *Miss America*: (a) The original image; (b) The generated mesh; (c) The decompressed image using the proposed coder (0.20 bpp, PSNR= 36.59 dB); (d) The decompressed image using the TMN4i coder (0.24 bpp, PSNR = 36.99 dB.

method, these motions are rendered smoothly, but some eye movements are not reproduced accurately.

For a quantitative comparison of the proposed method with the TMN4 method, the PSNR curves for the coded sequences by the proposed region based coder (RBC), the TMN4 and the TMN4i methods are given in Fig. 5 (a). It can be seen that the PSNR performance of the TMN4 coder is on an average higher than that of the proposed method, especially towards the end of the sequence. This is because the proposed coder does not apply error compensation over the majority of the image frame, as explained before. We would like to point out that in spite of the lower PSNR values, the visual quality (mainly in terms of motion rendition) of the proposed method is significantly better. To give a more meaningful quantitative comparison, Fig. 5(b) shows the PSNR values calculated over the pixels in the sub-region included in Fig. 4, which is the region that attracts the most attention from the viewer. As expected, the PSNR values of the proposed coder are more comparable to TMN4i coder in this region.

CONCLUDING REMARKS

One unique feature of the reported coder is that it predicts the nodal positions in a present frame from their positions in previous frames by piecewise linear extrapolation. In addition, the majority of each frame in the scene is predicted (by mesh-based warping) from its previous frame recursively without error compensation. This type of motion extrapolation and forward prediction is possible only with forward tracking mesh-based motion representation.

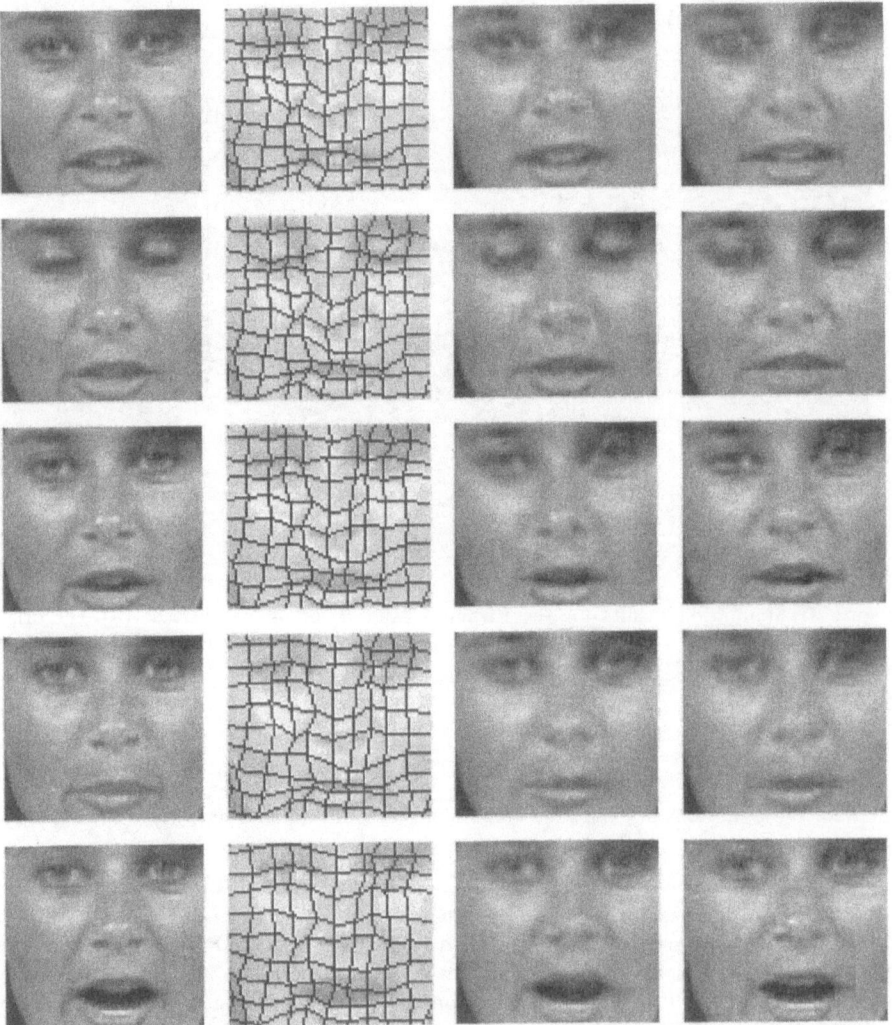

Figure 4: Simulation results for a sub-region in several P-frames of *Miss America*: (a) Original images; From top to bottom are frames 4, 7, 12, 16, and 21, which are selected to illustrate typical facial motions in the sequence. (b) The meshes generated for images in (a) using nodal tracking, starting from the mesh shown in Fig. 1(b). (c) Decompressed images using the reported coder. There is a good mix of MT- and ME-frames in these 5 frames for each class. (d) Decompressed images using the TMN4i coder, in which frames 12 and 21 are coded, while frames 4, 7, and 16 are linearly interpolated from surrounding coded frames.

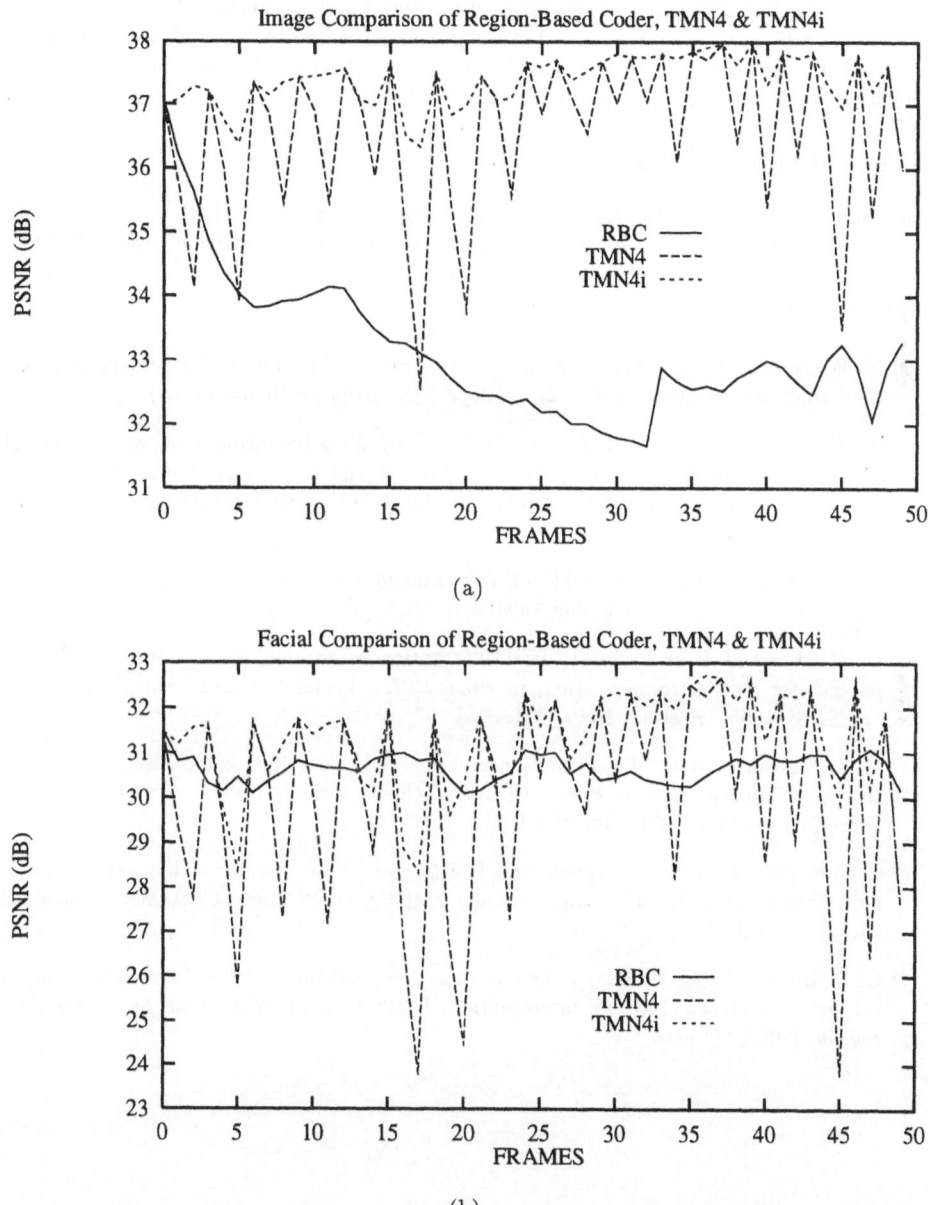

Figure 5: The PSNR curves for the decompressed images of *Miss America* by the proposed region based coder (RBC), the H.263-TMN4 coder (TMN4), and the H.263-TMN4 coder with frame interpolation (TMN4i). The PSNR values in (a) are calculated over the entire image, and those in (b) are calculated over the sub-region included in Fig. 5.

Another important merit of using the mesh structure for region (or object) description is that the motion and shape of each region can be jointly and efficiently described by the nodal trajectories. In the past, the mesh structure has been predominantly incorporated in block–based video codecs to improve the accuracy of backward motion compensated prediction and interpolation [4, 5, 6, 7]. We believe that the greater potential of the mesh structure lies in its capability for object tracking over multiple frames, which can be more fully exploited in region-based, object–oriented, and model–based video coder.

ACKNOWLEDGEMENT

The simulation results for the H.263-TMN4 algorithm were generated with a code developed at AT&T Bell Laboratories, Visual Communications Research Department, Holmdel, NJ.

REFERENCES

[1] Y. Wang and O. Lee, Active mesh — a feature seeking and tracking image sequence representation scheme, *IEEE Trans. Image Processing*, 3:610–624 (1994).

[2] Y. Wang, X.-M. Hsieh, J.-H. Hu, and O. Lee, Region segmentation based on active mesh representation of motion: comparison of parallel and sequential approaches, in *Proc. Second IEEE International Conference on Image Processing*, (Washington DC), Oct. 1995.

[3] ITU-T Study Group 15, Draft ITU-T Recommendation H.263 — Video Coding for Low Bitrate Communication, October 1994.

[4] Y. Nakaya and H. Harashima, Iterative motion estimation method using triangular patches for motion compensation, in *Proc. SPIE: Visual Commun. Image Processing*, vol. SPIE–1605, (Boston), Nov. 1991.

[5] G. J. Sullivan and R. Baker, Motion compensation for video compression using control grid interpolation, in *Proc. IEEE Int. Conf. Acoust., Speech, Signal Processing*, (Toronto), pp. 2713–2716, July 1991.

[6] J. Nieweglowski, T. G. Campbell, and P. Haavisto, A novel video coding scheme based on temporal prediction using digital image warping, *IEEE Trans. Consumer Electronics*, 39:141–150 (1993).

[7] C. L. Huang and C. Y. Hsu, A new motion compensation method for image sequence coding using hierarchical grid interpolation, *IEEE Trans. Circuits Syst. for Video Technology*, 4:42–52 (1994).

AUTOMATIC LIPREADING RESEARCH:
HISTORIC OVERVIEW AND CURRENT WORK

Eric Petajan and Hans Peter Graf

AT&T Bell Labs
Murray Hill, NJ 07974
edp@research.att.com

INTRODUCTION

Acoustic automatic speech recognition (ASR) systems tend to perform poorly with noisy speech. Unfortunately, most application environments contain noise from machines, vehicles, others talking, typing, television, sound systems, etc. In addition, system performance is highly dependent on the particular microphone type and its placement, but most people find head-mounted microphones uncomfortable for extended use and they are impractical in many situations. Fortunately, the use of visual speech (lipreading or, more properly, speechreading) information has been shown to improve the performance of acoustic ASR systems especially in noise. This paper outlines the history of automatic lipreading research and describes the authors current efforts.

SPEECHREADING SYSTEMS OVERVIEW

The first speechreading system was completed in 1984 by Petajan[1] for his PhD in electrical engineering. As shown in Tables 1a,b,c, automatic speechreading has been an increasingly active research area which has paralleled acoustic speech recognition research. Petajan[1,2,3,4], Finn[5], and Mase[6] used template matching of visual speech parameters during the mid 1980s. Petajan's first system used 4 mouth image parameters derived from binary mouth images which were automatically registered by tracking the nostrils. Petajan's later system used vector quantization of binary mouth images before template matching with dynamic time warping. Finn placed dots on the speakers face and hand traced the dot positions before the automated recognition process. Brooke[7] applied a radial measure to the inner lip contour for vowel recognition experiments. Nishida[8] focused on detecting word boundaries from visual speech and Mase applied optical flow analysis to the mouth area before template matching. In both Petajan systems, the speechreading results were used to correct the output of commercial acoustic speech recognizers. Finn, Nishida and Mase registered the mouth location manually.

Neural networks were applied to a mixed pattern of visual and acoustic information by Yuhas[9], Wu[10] and Stork[11]. Stork tracked facial dots automatically while Yuhas and Wu manually registered mouth images. Goldschen[12,13] Bregler[14,15,16,17,18], and Silsbee[19,20,21,22,23] used Hidden Markov Models (HMM) for connected and continuous speech recognition. Goldschen used the 450 TIMIT[24] sentences collected by Petajan's system to perform visual sentence recognition. Bregler combined the acoustic signal with mouth dot positions for spelling tasks; and Silsbee used post-recognition integration (acoustic and visual) on a 500 word corpus with manually registered mouth images. All systems which incorporated

Table 1a. Speechreading systems (template matching)

Table 1a. Speechreading systems (template matching)

AUTHOR	DATA SOURCE	PROCESSING METHOD	VOCAB-ULARY	ACOUSTIC/ VISUAL INTEGRATION
Petajan 1984	Nostril tracked oral cavity parameters	Template matching	Digits, letters, assorted words	Post-recognition integration
Finn 1986	Mouth dots (hand traced)	Template matching	23 consonants in VCV	None
Brooke/Petajan 1986	Nostril tracked binary mouth images with blackened teeth	Radial function	CVC	None
Nishida 1986	Video disk	Image differences (for word boundaries)	5 words	None
Petajan 1987	Nostril tracked binary mouth images	Vector quantization and dynamic time warping	Digits, letters	Post-recognition integration
Mase/Pentland 1989	Mouth images	Optical Flow	Connected digits	None
Smith 1989	Mouth images	Area and height derivative	4 words	None

Table 1b. Speechreading systems (neural nets and HMMs)

AUTHOR	DATA SOURCE	PROCESSING METHOD	VOCAB-ULARY	ACOUSTIC/ VISUAL INTEGRATION
Yuhas 1989	Mouth images	Neural network	Vowels	Pre-recognition integration
Wu 1991	Mouth images	Neural network	Vowels	Pre-recognition integration
Stork 1992	Mouth dots	Time delay neural network	Consonant Phonemes	Pre-recognition integration
Goldschen 1993	Nostril tracked oral cavity parameters	HMM	450 TIMIT sentences	None
Bregler 1993	Mouth dots	HMM	Connected letters	Pre-recognition integration
Silsbee 1993	Mouth images	HMM	500 words	Post-recognition integration

acoustic speech recognition were able to demonstrate accuracy improvements using visual speech, especially with acoustic noise.

Recently, many automatic speechreading efforts have been started using a variety of techniques as shown in Table 1c. Most of these systems are focused on improved mouth parameterization. Hennecke[25], Rao[26], Adjoudani[27] Coianiz[28], Vogt[29], and Luettin[30], and Petajan[31] are investigating lip models with constraints for more robust mouth feature analysis.

Table 1c Speechreading systems (recent efforts)

AUTHOR	DATA SOURCE	PROCESSING METHOD	VOCAB-ULARY	ACOUSTIC/ VISUAL INTEGRATION
Hennecke 1994	Mouth images	Deformable templates	Not yet	Not yet
Cosi 1994[32]	Multiple cameras	Neural network	VCV	Pre-recognition integration
Rao 1994	Mouth images	Lip modeling	Wow and mom	None
Adjoudani/ Benoit 1995	Painted lips	HMM	54 VCVCV	Pre and post-recognition integration
Movellan 1995[33]	Mouth images	Neural network	Digits	Pre-recognition integration
Coianiz 1995	Mouth images	Lip modeling	Not yet	Not yet
Vogt 1995	Mouth images	Model matching	Not yet	Not yet
Dalton 1995[34]	Mouth images	Template matching with dynamic time warping	40 words	Pre-recognition integration
Luettin 1995	Mouth images	Active shape models	Not yet	Not yet
Lu 1995	Mouth images	Neural network		
Petajan/Graf 1995	Nostril tracked color mouth images	Color thresholds with constrained inner lip contour	Not yet	Not yet

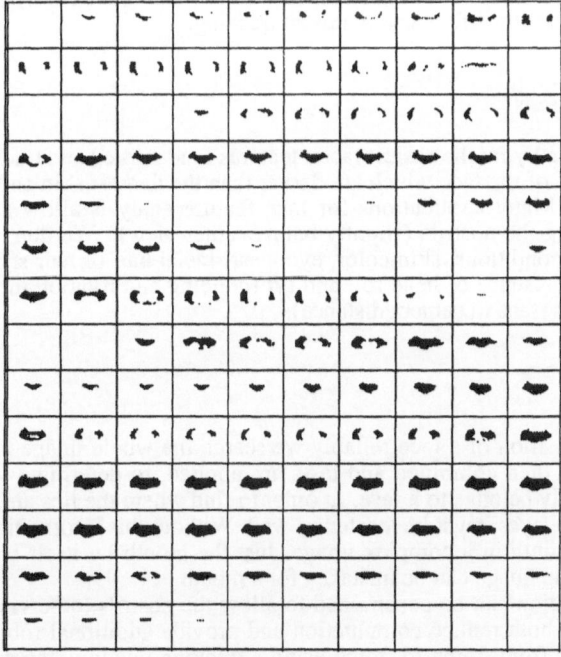

Figure 1. TIMIT sentence "Jane may earn more money by working hard. "

PREVIOUS MOUTH IMAGE ANALYSIS DESCRIPTION

Several systems use various lip markers (dots or lipstick) to aid the machine vision system. These approaches are aimed at capturing visual speech information for scientific investigation of articulation without regard for practical applications. Our previous automatic speechreading systems were designed to demonstrate that visual speech information could improve the performance of acoustic speech recognition systems. Therefore, the lighting and viewing parameters were fixed, and the pixel intensity thresholds were manually set in order to collect the most consistent visual speech data possible. Efficient image processing algorithms were implemented in real-time hardware to enable hundreds of processed contiguous video frames to be stored in the limited computer memory available at the time. The previous systems and our current system uses nostril tracking for robustness. The previous systems processed monochrome images from progressive scan CCD cameras using intensity thresholding followed by contour coding. Various parameterizations were successfully applied to the sequences of small binary mouth images. A typical two second sequence is shown in Figure 1 representing one of the TIMIT sentences (the frames are in raster scan order). This sequence contains four bilabial closures (rows 3,5,6,8), examples of tongue visibility (rows 22,3,7,10), and upper teeth visibility (rows 4,11). The side of the dark patterns are always the inner lip contour (mouth corners).

ROBUST FACE FEATURE ANALYSIS

Speechreading is readily used by humans in a large variety of adverse viewing conditions. This contributes greatly to our ability to recognize speech in even the most visually difficult and acoustically noisy situations. In fact, most people have little difficulty conversing in dark noisy nightclubs or across large distances.

A practical speechreading machine must also perform well under a variety of viewing conditions. Our current system adapts to lighting variations, head motion (tilt and range), and variations between speakers (glasses, facial hair, and skin color). The system also achieves computationally efficiency by nostril tracking.

Practical assumptions

The nostrils are by far the easiest facial features to identified and tracked. They are two holes in the middle of the face which are darker than the darkest skin and are almost never obscured by hair. Many applications for face feature analysis allow camera placement suitable for viewing the nostrils (slightly below center view). We don't assume anything about the lighting conditions, skin color, eye wear, facial hair or hair style The system is also robust in the presence of head roll and tilt (nodding and rotation in the image plane), and scale variations (face to camera distance).

System overview

To find the position of a face reliably we search the whole image for facial features. These features are then combined and tests are applied, to determine whether any such combination actually belongs to a face. In order to find where the lips are, other features of the face, such as the eyes, must be located as well. Without this information it is difficult to reliably find the mouth in a complex image. Just the mouth by itself is easily missed or other elements in the image can be mistaken for a mouth.

If camera position can be constrained to allow the nostrils to be viewed, then nostril tracking is used to both reduce computation and provide additional robustness. Once the nostrils are tracked from frame to frame using a tracking window, the mouth area can be isolated and normalized for scale and rotation. A mouth detail analysis procedure is then used to estimate the inner lip contour, and teeth and tongue regions. The inner lip contour and head movements are then mapped to synthetic face parameters to generate a graphical talking head synchronized with the original human voice. This information can also be used as the basis for visual speech features in an automatic speechreading system. Similar features were used in our previous automatic speechreading systems.

The face recognition algorithm periodically finds faces in isolated frames using morphological filtering and the relative positions of eye, nose and mouth blobs. This information is used to start and then verify the nostril tracking system. A mouth window is formed using the nostril positions. A closed mouth line (shown in Figure 2) is combined with the resulting open mouth inner lip contours to give the final inner lip contour estimate.

Fast and Robust Nostril Tracking

The size of the face is determined from the eye and eye-nose-mouth distance and a nostril tracking window is formed approximately around the nose with a size small enough to not include eyes or mouth yet large enough to contain the nostrils in the next frame after the fastest possible head movement.. A set of red, green and blue intensity thresholds (YUV are alternatively used) are adjusted up or down until the nostrils are found in succeeding steps. First ,the pixels in the nostril tracking window (NTW) are tested for nostril area and skin color area. Nostril area is simply the number of pixels which are below color nostril threshold and skin color area is the number of pixels which are labeled as skin color in a pre-determined skin color lookup table using the most significant 6 bits of the RGB values as table indices. If either test fails then the nostril thresholds are adjusted and the tests are applied again. If the nostril area and skin color area are within acceptable range then nostril size, separation, and contiguity tests are applied. This is done using horizontal and vertical projections or histograms of nostril pixels as shown in Figure 3. A pair of nostril size values is determined for each nostril by thresholding the projections for each nostril and computing the run length of projection values which exceed the threshold. All threshold values are tested and the value which produces the best match (absolute value distance) from a scaled template is used. Each of nostril height, width, and separation are weighted and summed after differencing with the template values. The template values are scaled based on the previous frame nostril separation to provide viewing distance normalization. Head tilt around the camera view direction is also estimated from the previous frame nostril position and compensated in the current frame. One or two vertical nostril pixel projection runs are expected in case of uncompensated head tilt. This algorithm is insensitive to noise, illumination variations, head tilt and scale variations, and nostril shape.

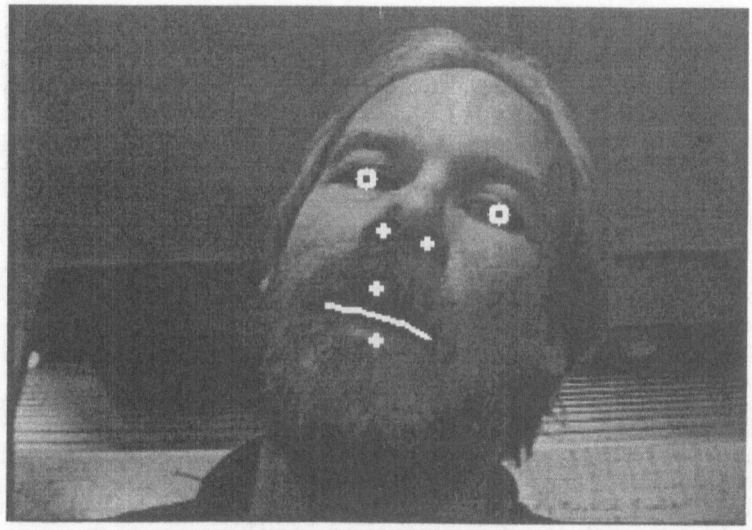

Figure 2. Isolated face image analysis (non-tracking)

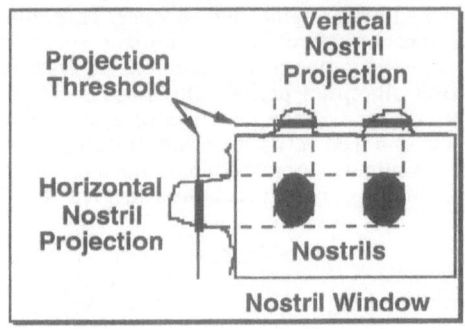

Figure 3. Nostril analysis and tracking

Mouth Detail Analysis

Given the head position, scale and tilt estimation from the nostril tracking system, a window is formed around the mouth area which is fixed distance from the nostrils. After compensating the mouth image for rotation and scale, a set of inner lip color thresholds are trained whenever the mouth is closed. Mouth closure is detected by the face recognition system which looks for a single vertical valley in the mouth region. The inner lips thresholds are taken to be 90% of the minimum color intensity in each column of pixels in the mouth window resulting in a color threshold for each horizontal position. The inner lip contour threshold array is then applied to the mouth window. Each pixel which is below threshold is labeled an inner mouth pixel. Isolated inner mouth pixels are removed (unlabeled). The teeth are then detected by forming a bounding box around the inner mouth pixels and testing all non-inner mouth pixels for teeth color given as a fixed set of red, green and blue ranges (or YUV ranges). Then a closed contour is formed around the inner mouth and teeth pixels starting from the mouth corners. The upper contour is constrained to have increasing or constant height as approaching the face centerline. The lower contour is have increasing or constant distance from the upper contour. These constraints were derived from anatomical considerations. Figure 4 shows the analysis results overlaid on a video frame.

Mapping Inner Lip Contour to Synthetic Face

The head position and inner lip contours are used to control a synthetic talking head model[35,36,37,38]. The head position is taken as the nostril position and controls the horizontal and vertical head model position relative to the camera model in the graphics rendering system. The head tilt is controlled by either the angle between the eye line or nostril line and horizontal but the eye line is less noisy. The final tilt angle value is a running average over a few frames for increased smoothness. The inner lip contour is compared to the inner lip contour of the head model directly. The screen image of the model mouth is analyzed and the inner lip contour is computed by coloring each lip with a unique color so that the vertical transitions between lips and mouth opening are easily detected. The teeth and tongue are removed before analysis. In general, a purely geometric computation of inner lip contour is possible given a 2D or 3D model but a more expedient approach was taken. The distance measure between the real and the model inner lip contour is the summed vertical distance between the contours. An additional weighting factor is applied if the real contour lies outside the model contour since the real contour tends to err on the side of smallness or concavities. The head model mouth parameters (jaw rotation, upper and lower lip position, mouth corner position and mouth scale) are adjusted in combination for each frame until a sufficiently close match is achieved. For greater computational efficiency, the model parameters values from the previous frame are used to constrain the search for the optimal parameter values in the current frame. Since the jaw rotation is not directly estimated from the real face, it is estimated as a small fraction of the vertical mouth opening. Figure 5 was generated using the head tilt and inner lip contours from Figure 4.

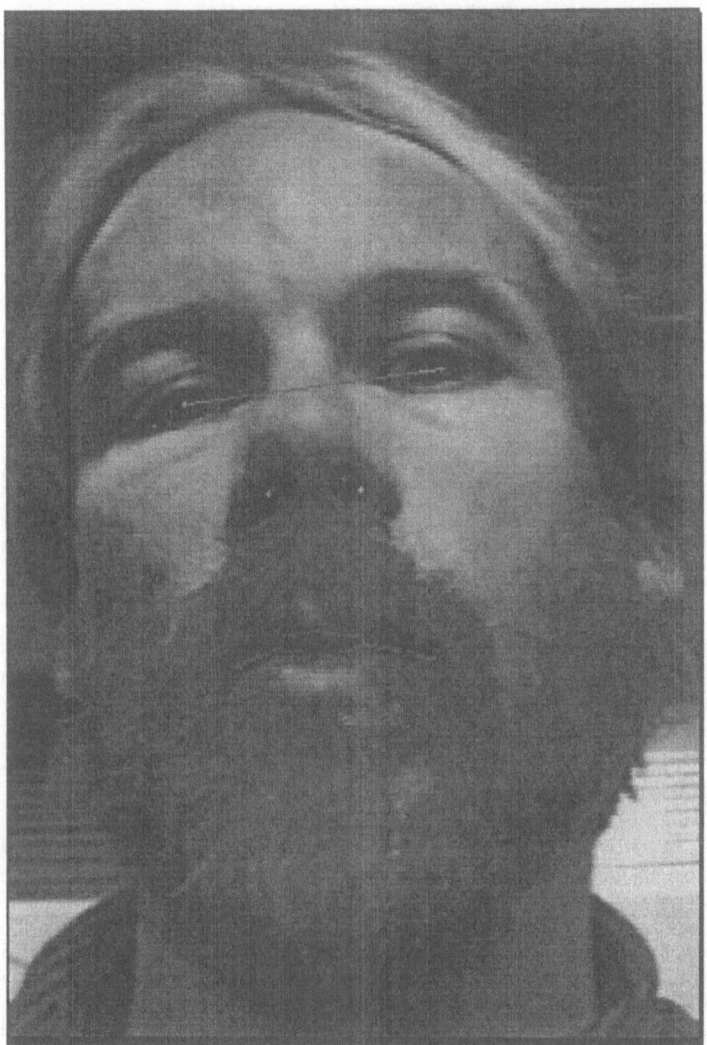

Figure 4. Video frame with overlaid face features

APPLICATIONS

Small color video cameras are inexpensive and easily positioned for ideal face viewing in many applications. For example, a camera placed just below a computer monitor provides a good view of all facial features (including nostrils) throughout all normal head motions, e.g., looking down at the keyboard and slouching. Other application environments which allow ideal camera position include bank machines, cars, kiosks, point of sale terminals (cash registers), laptop or notebook computers, copying machines, access control stations, aircraft cockpits, and personal digital assistants.

The minimum set of face parameters needed to drive a synthetic talking face is quite small. All of these parameters can be encoded into a very low bitrate signal for transmission over ordinary phone line using either voice/data modem or data modem with digital audio compression. The applications for this technology include video conferencing, model-based coding (MPEG-4), networked interactive games, televideo marketing, enhanced public address in noisy environments, entertainment, speech recognition, and enhanced computer/human interaction. If the system is simultaneously

Figure 5. Synthetic face driven from Figure 4.

used for video coding and speech recognition, then the cost of implementation is more easily justified. Since the face features are easily transmitted using the telephone network, acoustic/visual speech recognition in the network is feasible.

CONCLUSIONS

Since 1984, 23 different machine speechreading efforts have been initiated. Some researchers focused on speech science and some focused on face feature analysis. Recently, we have developed a system for acquisition of facial features which only assumes that the camera position enables viewing the nostrils. Details of the mouth are obtained by the system including the inner lip contours, teeth location and some tongue information. The algorithms are efficient and are designed for real-time implementation. The visual speech information yielded by this system is of sufficiently high quality to control the lip movements of a synthetic face in a convincing manner. More importantly, this information should be ideal for automatic speechreading.

The ability of personal computers to process video data is increasing rapidly with the advent of video conferencing, desktop video editors, and video compression. Simultaneously, high quality video cameras are becoming small and cheap enough to be placed almost anywhere in the application area. These developments provide much of the infrastructure needed for automatic speechreading on personal computers.

REFERENCES

1. E. D. Petajan, *Automatic Lipreading to Enhance Speech Recognition,* PhD Thesis, University of Illinois at Urbana-Champagne, (1984).

2. E. D. Petajan, Automatic lipreading to enhance speech recognition, *Proceedings Globecom Telecommunications Conference*, pp. 265-272, IEEE, Atlanta(1984).

3. E.D. Petajan, Automatic lipreading to enhance speech recognition", *Proceeding of the IEEE Conference on Computer Vision and Pattern Recognition*, pp 40-47, IEEE, San Francisco(1985).

4. E. D. Petajan, N.M. Brooke, G.J. Bischoff, and D.A. Bodoff, An improved automatic lipreading system to enhance speech recognition, *Proc. Human Factors in Computing Systems*, pp.19-25, ACM, Washington, DC(1988).

5. K. Finn, *An Investigation of Visible Lip Information to be Used in Automatic Speech Recognition*, PhD Thesis, Georgetown University, Washington, DC (1986).

6. K Mase and A. Pentland, Automatic lip-reading by optical flow analysis, *Media Lab Report 117,* MIT, Cambridge (1991).

7. N.M. Brooke and E.D. Petajan, Seeing Speech: Investigations into the synthesis and recognition of visible speech movements using automatic image processing and computer graphics, *Proceedings IEE Conference on Speech Processing*, IEE, London (1986).

8. Nishida, Speech recognition enhancement by lip information, *ACM SIGCHI Bulletin*, 17, no. 4, pp 198-204 (1986).

9. B.P. Yuhas, M.H. Goldstein, T.J. Sejnowski, R.E. Jenkins, Neural network models of sensory integration for improved vowel recognition, *Proceedings IEEE*. vol.78, no.10, pp1658-1668 (Oct 1990).

10. J. Wu, S. Tamura, H. Mitsumato H. Kawai, K Kurosu and K. Okazaki, Neural network vowel recognition jointly using voice features and mouth shape image, *Pattern Recognition*, Vol 24, No. 10, pp 921-927, (1991).

11. D.G. Stork, G.J. Wolff, E.P. Levine, Neural network lipreading system for improved speech recognition, *Proceedings International Joint Conference on Neural Networks*, Vol 2, pp 289-295, (1992).

12. A. Goldschen, *Continuous Automatic Speech Recognition by Lipreading*, PhD, George Washington University, (1993).

13. A. Goldschen, O. Garcia, E. Petajan, Continuous optical automatic speech recognition, *Proceedings of the 28th Asilomar Conference on Signals, Systems, and Computers*, pp. 572-577, IEEE, (1994).

14. C. Bregler, H. Hild, S. Manke and A. Waibel, Improving connected letter recognition by lipreading,*ICASSP*, pp 557-560, Minneapolis (1993).

15. C. Bregler, S. Omohundro, and Y. Konig, A hybrid approach to bimodal speech recognition, *28th Annual Asilomar Conference on Signals, Systems, and Computers*, pp 556-560, Pacific Grove (1994).

16. C. Bregler, S. Manke, and A. Waibel, Bimodal sensor integration on the example of speech--reading, *Proc. of IEEE Int. Conf. on Neural Networks* (1993).

17. C. Bregler and S.M. Omohundro, Surface learning with applications to lipreading, *Advances in Neural Information Processing Systems*, Vol 6, ed. Cowan, Tesauro, and Alspector, pub Morgan Kaufmann (1994).

18. C. Bregler and Y Konig, Eigenlips for robust speech recognition, *ICASSP*, pp 669-672, Adelaide (1994).

19. P. L. Silsbee, Sensory integration in audiovisual automatic speech recognition, *28th Annual Asilomar Conference on Signals, Systems, and Computers*, pp 561-565, Pacific Grove (1994).

20. P. L. Silsbee, Motion in deformable templates, *First IEEE Intl. Conf. on Image Processing*, IEEE (1994).

21. P. L. Silsbee and A. C. Bovik, Medium vocabulary audiovisual speech recognition, *New Advances and Trends in Speech Recognition and Coding*, pp 13-16, NATO ASI (1993).

22. P. L. Silsbee and A. C. Bovik, Audio-visual speech recognition for a vowel discrimination task, *Proc. Visual Commun. and Image Process. '93*, pp 84-85, SPIE (1993).

23. P. L. Silsbee, *Computer Lipreading for Improved Accuracy in Automatic Speech Recognition*, PhD Thesis, University of Texas (1993).

24. *DARPA TIMIT CD-ROM*, 1988, NIST, Gaithersburg, MD 20899.

25. M. E. Hennecke, K.V. Prasad and D.G. Stork, Using deformable templates to infer visual speech dynamics, *28th Annual Asilomar Conference on Signals, Systems, and Computers*,, Pacific Grove (1994).

26. R.R. Rao and R.M. Mersereau, Lip modelling for visual speech recognition, *28th Annual Asilomar Conference on Signals, Systems, and Computers*, Pacific Grove (1994).

27. A. Adjoudani and C. Benoit, Audio-visual speech recognition compared across two architectures, *Proc. Eurospeech '95*, Madrid, (Sept. 1995).

28. T. Coianiz, L. Torresani, B. Caprile, 2D deformable models for visual speech analysis, *NATO ASI 940584 Speechreading by Man and Machine: Models, Systems and Applications* (1995).

29. M. Vogt, Fast matching of a dynamic lip model to color video sequences under regular illlumination conditions, *NATO ASI 940584 Speechreading by Man and Machine: Models, Systems and Applications* (1995).

30. J. Luettin, J.A. Thacker, S. W. Beet, Active shape models for visual speech feature extraction, *NATO ASI 940584 Speechreading by Man and Machine: Models, Systems and Applications* (1995).

31. E. Petajan and H. Graf, Robust face feature analysis for automatic lipreading and character animation, *NATO ASI 940584 Speechreading by Man and Machine: Models, Systems and Applications* (1995).

32. E. Magno and P. Cosi, Spatio-temporal characteristics of lips and jaw movements in relation to speechreading: experimental data and their use for bimodal phonetic recognition application with neural networks, *NATO ASI 940584 Speechreading by Man and Machine: Models, Systems and Applications* (1995).

33. J. Movellan and G. Chadder, Channel separability in the audio-visual integration of speech: implications for engineers and cognitive neuroscientists, *NATO ASI 940584 Speechreading by Man and Machine: Models, Systems and Applications* (1995).

34. B. Dalton, R. Kaucic, and A. Blake, Automatic speechreading using dynamic contours, *NATO ASI 940584 Speechreading by Man and Machine: Models, Systems and Applications* (1995).

35. M. Cohen and D. Massaro, *Modeling Coarticulation in Synthetic Visual Speech"*, *Models and Techniques in Computer Animation*, Springer-Verlag, 1993.

36. F. Parke, *A Parametric Model for Human Faces*, PhD Thesis, University of Utah, (1974).

37. F. Parke, A model for human faces that allows speech synchronized animation, *Journal of Computers and Graphics*, (1975).

38. F. Parke, A parameterized model for facial animation, *IEEE Computer Graphics and Applications*, (1982).

ON THE PRODUCTION AND THE PERCEPTION OF AUDIO-VISUAL SPEECH BY MAN AND MACHINE

C. Benoît

Institut de la Communication Parlée
Unité de Recherche Associée au CNRS N° 368
INPG/ENSERG - Université STENDHAL
BP 25X - F38040 Grenoble, France
(benoit@icp.grenet.fr)

INTRODUCTION

Since the Fifties, several experiments have been run to evaluate the "benefit of lip-reading" on speech intelligibility, all presenting a natural face speaking at different levels of background noise: Sumby and Pollack, 1954; Neely, 1956; Erber, 1969; Binnie et al., 1974; Erber, 1975. We here present a similar experiment run with French stimuli.

Experiments run by McGrath (1985) and then by Summerfield et al. (1989) showed that human lips alone carry more than half the visual information provided by the whole natural face of an English speaker. They also showed that vision of the teeth somewhat increases the intelligibility of a message: it disambiguates sounds differing in jaw position like /bib/ vs. /bab/. A series of experiments have been recently carried out at the ICP to confirm and extend this findings to French language. In addition, we compared the overall performance of normal hearers in audio-visual intelligibility tests where the visual displays were made of a natural face (Benoît et al., 1994) natural lips alone (Le Goff et al., 1995), and a bunch of 3D parametric models of the main components of a speaker's face: the lips, the jaw, and the skin (Guiard-Marigny et al., 1995).

The same parameters as those used to animate our synthetic models of the face have been measured on nine repetitions of an extended version of the same corpus as above in order to evaluate the performances of an HMM classifier in an identification process analogous to the task performed by our human subjects (Adjoudani & Benoît, 1995). In addition, we compared the visual recognition scores of subsets of those parameters in order to identify which parameters transmit the most information on visible gestures in speech.

THE AUDIO-VISUAL CORPUS

The same corpus was used all through the perceptual and automatic recognition tests reported below. Similarly, the same parameters measured on this very corpus have been used to animate our synthetic models and to train and test an HMM classifier. The reason is that this reference corpus and the same characteristic parameters had primarily been widely studied through multidimensional and clustering analyses, so that the effect of coarticulation within each test-word is well-know: Benoît et al. (1991); Benoît et al. (1992). These geometrical studies finally allowed us to make predictions on visual confusions between coarticulated segments.

This corpus was made of transitions between French phonemes including most of the labial actions (lip protrusion, rounding, spreading, closure) and jaw positions in French. It

Multimedia Communications and Video Coding
Edited by Y. Wang *et al.*, Plenum Press, New York, 1996

involved ten repetitions of three vowels ($V_i \in \{i, a, y\}$) and six consonants ($C_j \in \{b, v, z, 3, 1, r\}$) embedded in various vocalic combinations to create 54 sentences of the type: *"C'est pas [$V_iC_jV_kC_jV_iz$]. ?"*. The last vowel was stressed because of the interrogative form and lengthened due to the following /z/. Each sentence was repeated ten times in a pseudo-random order, modified to avoid consecutive repetitions of the same stimulus.

A male French native-speaker was pre-selected on the basis of sufficiently high dynamics and good symmetry in his lip displacements. He pronounced all stimuli in a single session (plus others not used in the studies reported below.) The corpus was audio-visually recorded: acoustic signal, plus front and profile views at 50 ips.

3300 video fields were labeled on the center of the acoustic wave forms of each constituting segment.

We used a subset of our audio-visual corpus to serve as reference for further perceptual tests. In order to test human perceivers, we selected one representative utterance from each of the first nine repetitions of the 18 different stimuli of the form /VCVCVz/, each presenting the same consonant (among six) repeated two times and embedded in-between the same vowel (among three) repeated three times. Those stimuli were also selected as reference for further animation of the various facial models. Finally, all utterances of the 54 different stimuli [$V_iC_jV_kC_jV_iz$] were used as training and test sets for automatic recognition of visible speech.

THE OPTICAL PARAMETERS

Several parameters have been measured on the speaker's lips and chin, both carefully made-up so that the front and profile views of the vermilion area were easy to extract. Several parameters have then been automatically measured from the lip contours, e.g., the width, height, and area of the internal and external lip contours, the upper and lower lip protrusion, the lip contact protrusion, the chin lowering, etc. (See Figure 1.)

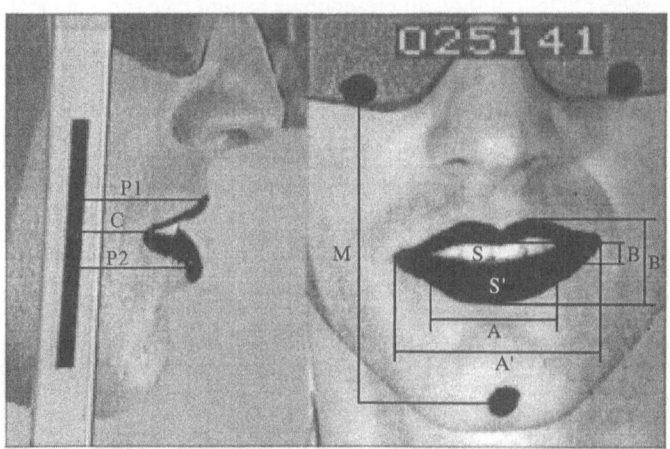

Figure 1. Schematic of the front (right) and profile (left) parameters measured on the speaker's face.

GEOMETRICAL ANALYSIS OF LIP/JAW GESTURES

We performed a multidimensional analysis of the above mentioned parameters measured at the center of the acoustic realization of the allophones from our reference corpus (Benoît et al., 1992). Several observations came out of this earlier study. Five main categories of labial shapes were identified: closed lips (e.g., /b/); spread lips (e.g., /i, e, ε, a/); open rounded lips (e.g., /ɜ/); closing rounded lips (e.g., /y, u/), and neutral lips (e.g., /ɔ, ã/). The entire space of coarticulated lip shapes is much wider than the subspace made of the vowels alone. There are remarkably high correlations between some parameters, e.g., the internal area and the product of the internal width by the internal height (r = .999). A discrete sampling of the lip/jaw space could be obtained with a score of "visemic classes", as seen on Figure 2. And finally, the parameters best discriminating those classes were identified: internal width and height, lip contact protrusion.

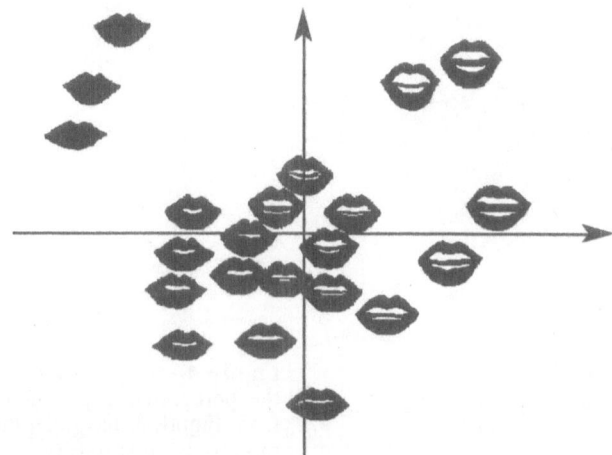

Figure 2. Projection in a factorial plane of "basic" lip shapes identified from the labial gestures of a French speaker without expressions. Coarticulation is taken into account.

PARAMETRIC MODELS OF THE FACE

The lip model

The twenty or so representative shapes of the "French lips" identified in the above mentioned study (Benoît et al., 1992) served as a reference to work out a high-resolution 3D model of the lip contours (Guiard-Marigny et al., to appear): See Figure 3. The lip model is ultimately controlled by means of five parameters: internal lip height and width, protrusion of the upper lip, lower lip and lip contact.

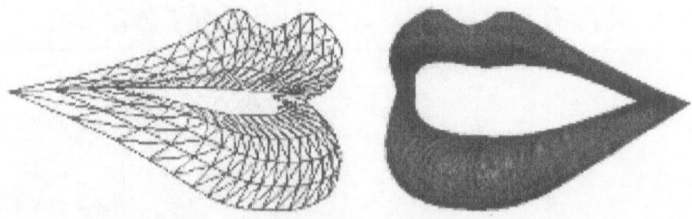

Figure 3. The 3D model of the lip contours developed at the ICP, in its underlying wireframe structure (left) and rendered in Gouraud-shading and texture-mapping (right).

The face model

A model of the face has been worked out too, from the face model originally developed by Parke (1974) and enhanced for speech production by Cohen and Massaro (1993). Its low-resolution lip model has been replaced by our lip model above mentioned (Le Goff et al., 1995). Only the motion of the skin around the lips is controlled, by the same parameters that control the lip model. In addition, jaw (or better chin) rotation is controlled by an extra-parameter. This face model is displayed on Figure 4.

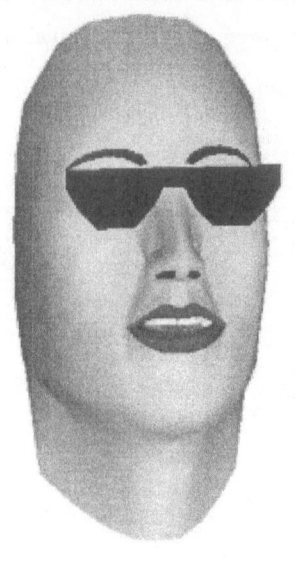

Figure 4. Image of the facial model used in the perceptual experiments. The lip model from Figure 3. has been integrated. The face model is ultimately controlled by six parameters.

The jaw/skull model

A 3D model of the jaw has been developed to account for the six degrees of freedom in its natural motion relative to the skull in speech and mastication (Guiard-Marigny & Ostry, 1995). A simplified set of control commands was then used to fairly approximate jaw motion in speech by means of only one parameter, namely chin vertical position. Finally, it is possible to control with good coherence in time and space the jaw/skull model and the lip model (Guiard-Marigny et al., 1995). Figure 5 shows the two models superimposed as well as their control parameters.

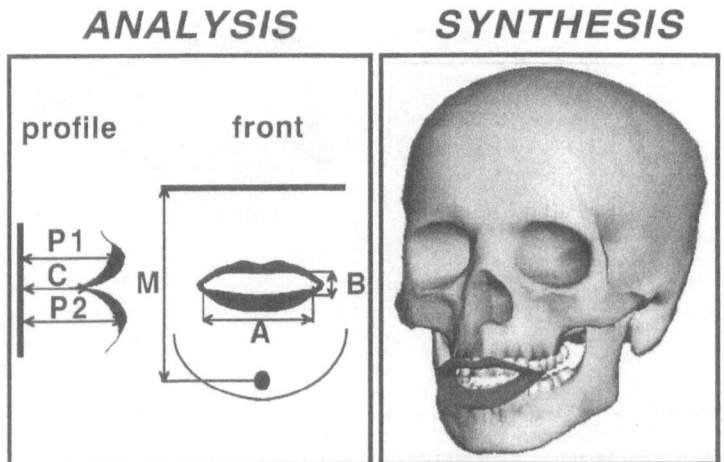

Figure 5. Image of the jaw/skull model and of the superimposed lip model used in the perceptual tests. Right: Schematic of the control parameters which animate the whole model.

INTELLIGIBILITY TESTS WITH NATURAL SPEECH

In a pilot experiment (Benoît et al., 1994), we have evaluated the relative intelligibility of the 18 stimuli made of VCVCV, as uttered by our speaker, against five conditions of background noise in two modes of presentation: audio alone and audio-visually with his whole face displayed from front in black and white. In a follow-up (Le Goff et al., 1995), we have tested the intelligibility of the speaker's lips alone, displayed in a binary mode, with the same speech material, to another group of listeners in the same test conditions. Auditory alone scores were in close agreement to those from the first group. We see from Figure 6. that the "audio + lip" scores stand in-between the auditory alone and the "audio + face" scores. We can conclude from these experiments that lips alone carry on average the two-thirds of the speech intelligibility carried out by the whole natural face. This finding thus confirms those by McGrath (1985).

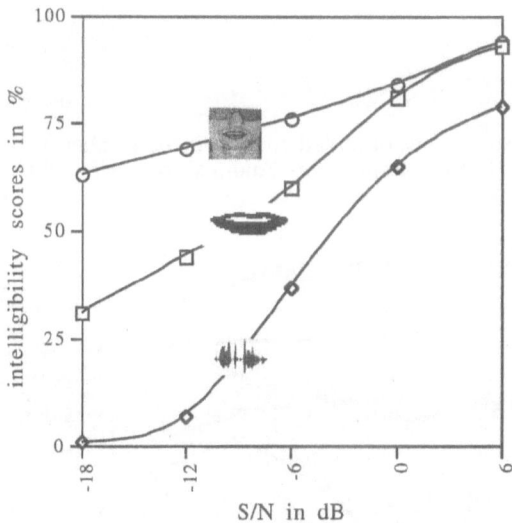

Figure 6. Audio-visual scores observed with natural speech under different conditions of background noise. Bottom curve: Audio alone condition. Middle curve: Audio + natural lips. Top curve: Audio + natural face.

INTELLIGIBILITY TESTS WITH SYNTHETIC DISPLAYS

Later on, tests have been run to evaluate the intelligibility of synthetic displays synchronized with the same acoustic material as in the above experiments. The intelligibility scores obtained with the audio alone in this experiment were comparable to those reported above. Figure 7. shows the scores obtained by the facial model, the jaw/lip model, and the lip model in a bimodal mode of presentation. The lip model restores approximately a third of the missing information when the acoustic signal is degraded, whatever the level of degradation. Moreover, a noticeable gain in speech intelligibility is observed when the synthetic jaw is added to the synthetic lips. However, the jaw/lip model does poorer than the facial model in term of audio-visual intelligibility although both are controlled by the same six parameters. This is obviously due to the lack of "naturalness" of the jaw/lip display. Subjects take better advantage of a "coherent" display where they can see the skin of the chin lowers around the jaw than when they have to deal directly with the view of the jaw bone motion.

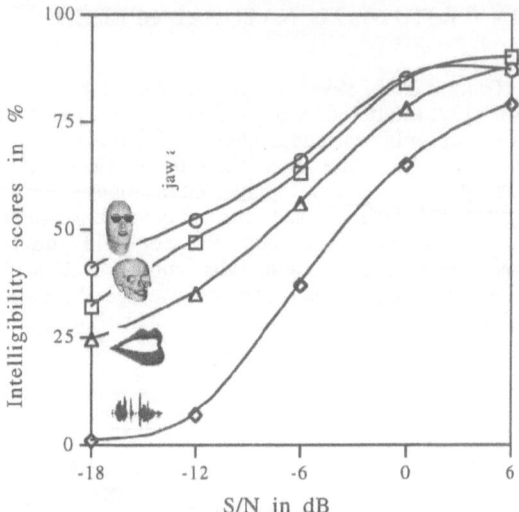

Figure 7. Audio-visual scores observed with synthetic displays under different conditions of background noise. From bottom to top: Audio alone; audio + lip model; audio + lip/jaw models; audio + face model.

RECOGNITION SCORES WITH HMMs

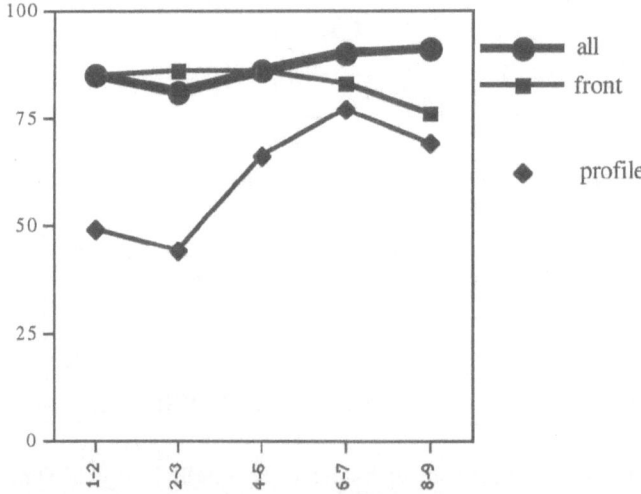

Figure 8. Recognition scores obtained with a HMM classifier trained with 7 repetitions and tested with 2 repetitions of 54 different non-sense words in a jack-knife technique. Abscissa gives the test-word numbers in the pair according to the time of recording. Three set of optical parameters have been used: All, front and profile. See text.

Our reference database and our set of optical parameters have been used elsewhere to assess the best architecture for automatic recognition of audio-visual speech with HMMs (Adjoudani and Benoît, to appear). In addition, the visual classifier has been used to compare the recognition scores of 54 different non-sense words $V_i C_j V_k C_j V_i z$. Training was performed with seven repetitions and test with two repetitions of each word. A jack-knife technique was used to evaluate the performance of the visual HMM along various training-test sets of words. Results are presented in Figure 7. Three main sets of parameters have been tested to evaluate the respective quantity of information associated with the angle of view: front, profile, and both. As expected, we see that the best recognition scores are obtained with all the parameters. Then front parameters alone give much better scores than profile parameters. In addition, it doesn't look like the profile information adds a lot to the front information since differences between scores obtained with all parameters and only the front parameters are pretty small.

CONCLUSION

Visible speech is located everywhere on the talker's face. Whether natural or synthetic, the most visual cues are given to perceivers, the better they decode visible speech, and the better they integrate audio-visual speech. It is not surprising that visible speech is better decoded with a whole natural face than with the natural lips alone, since vision of the teeth, tongue, chin and cheeks is important too. However, lips play the most important role in speechreading (hence the term lipreading?).

When dealing with synthetic displays of components of the face, subjects better decode visible speech when those displays are "coherent" or "ecological", in the sense that the articulators should be presented with natural links. Otherwise, the perceiver has to make an effort to integrate different pieces of information. This is why a facial model covered with a synthetic skin is more intelligible than its underlying structure onto which the same lips are superimposed, even though both objects are animated in an identical manner, with exactly the same number of parameters.

Machines recognize visible speech at its best when they are given the largest quantity of information. However, parameters measured on the front view of the speaker's face seem to be more informative than parameters measured on the profile view.

Recognition scores by man and machine are in agreement with each other. However, if "good" parameters are certainly needed by machines, one may wonder which kind of visual cues humans actually deal with in speechreading. More holistic approaches to "face processing" should be investigated in the future.

ACKNOWLEDGEMENTS

This research was partly supported by the French CNRS and by the ESPRIT project n° 8579 "MIAMI".

REFERENCES

Adjoudani, A., and Benoît, C., to appear, On the integration of auditory and visual parameters in an HMM-based ASR, in: *Speechreading by Man and Machine*, D. Stork, Ed., NATO-ASI series, Springer-Verlag (1996).

Benoît, C., Boë, L.J., and Abry, C., 1991, The effect of context on labiality in French, *Proceedings of the 2nd Eurospeech Conference*, Vol. 1, 153-156, Genoa, Italy.

Benoît, C., Lallouache, M.T., Mohamadi, T.M., and Abry, C., 1992, A set of French visemes for visual speech synthesis, in: *Talking Machines: Theories, Models, and Designs*, G. Bailly & C. Benoît, Eds, Elsevier Science Publishers, North-Holland, Amsterdam, 485-503.

Benoît, C., Mohamadi, T., and Kandel, S., 1994, Efefct of phonetic context on audio-visual intelligibility of French, *Journal of Speech and Hearing Research*, 37, 1195-1203.

Binnie, C.A., Montgomery, A.A., and Jackson, P.L., 1974, Auditory and visual contributions to the perception of consonants, *Journal of Speech and Hearing Research*, 17, 619-630.

Cohen, M.M., & Massaro, D.W., 1993, Modeling coarticulation in synthetic visual speech, *Computer Animation'93*, N. Magnenat-Thalmann & D. Thalmann, Eds, Springer-Verlag.

Erber, N.P., 1969, Interaction of audition and vision in the recognition of oral speech stimuli. *Journal of Speech & Hearing Research*, 12, 423-425.

Erber, N.P., 1975, Auditory-visual perception of speech. *Journal of Speech & Hearing Disorders*, 40, 481-492.

Guiard-Marigny, T. and Ostry, D.J., 1995, Three-dimensional visualization of human jaw motion in speech, *Meeting of the Acoustical Society of America*, Washington.

Guiard-Marigny, T., Benoît, C. and Ostry, D.J., 1995, Speech intelligibility of synthetic lips and jaw, *Proc. of the 13th Int. Congress of Phonetic Sciences*, Vol. 3, 222-226, Stockholm, Sweden.

Le Goff, B., Guiard-Marigny, T., and Benoît, C., 1994, Real-time analysis-synthesis and intelligibility of talking faces, *Proc. of the 2nd International Workshop on Speech Synthesis*, 53-56, New Paltz (NY), USA.

Le Goff, B., Guiard-Marigny, T., and Benoît, C., 1995, Read my lips... and my jaw! How intelligible are the components of a speaker's face?, *Proceedings of the 4thEurospeech Conference*, Vol. 1, 291-294, Madrid, Spain.

McGrath, M., 1985, *An examination of cues for visual and auso-visual speech perception using natural and computer-generated faces*, Ph.D Thesis, University of Nottingham, UK.

Neely, K.K., 1956, Effect of visual factors on the intelligibility of speech, *Journal of the Acoustical Society of America*, 28, 1275-1277.

Sumby, W.H., & Pollack, I., 1954, Visual contribution to speech intelligibility in noise. *Journal of the Acoustical Society of America*, 26, 212-215.

Summerfield, Q., MacLeod, A., McGrath, M., & Brooke, M., 1989, Lips, teeth, and the benefit of lipreading, in Handbook of Research on Face Processing, A.W. Young & H.D. Ellis, Eds, Elsevier Science Publishers, North-Holland, Amsterdam, 223-233.

PERCEPTUAL QUALITY EVALUATION OF LOW-BIT RATE MODEL-ASSISTED H.261-COMPATIBLE VIDEO

Arnaud Jacquin

AT&T Bell Laboratories
Signal Processing Research Department
Murray Hill, NJ 07974

ABSTRACT

We describe an experiment conducted so as to evaluate the perceptual quality of low-bit rate H.261-compatible video. The video coding systems tested are based on CCITT Recommendation H.261, operating at the bit rates of 48 and 24 kbps, at the constant frame rates of 7.5, 5 and 2 fps, and are followed by a postprocessing module. One system uses the concept of *model-assisted coding* which results in a better rendition of facial features—producing better eye contact and lip synchronization—at the expense of a slightly lower quality in the rest of the image.

The experiment shows that a majority of non-expert subjects was sensitive to the improved facial features resulting from model-assistance. This was reflected qualitatively by the fact that 70% of the subjects indicated a marked preference for model-assisted H.261 six or more times out of eight test sequences. This was also reflected quantitatively by non-negligible MOS increments obtained by the model-assisted coders versus their non-assisted counterparts. At 48 kbps, model-assistance increases the MOS by 0.6, elevating the subjective audiovisual quality from *less-than-fair* to *fair-to-good*. The experiment also shows a slight preference for the frame rate of 5 fps at 48 kbps, and a strong dislike of images coded at 2 fps at 24 kbps, where the image quality is fair but the lip synchronization between video and audio is completely lost.

INTRODUCTION

It is usually assumed that in very low bit rate video teleconferencing situations, viewers typically focus their attention to the *face(s)* of the person(s) on the screen, rather than to areas such as clothing or background [1] . Although fast motion is known to mask coding artifacts, the human visual system has the ability to *lock on* and *track* particular moving objects, such as a person's face.

Multimedia Communications and Video Coding
Edited by Y. Wang *et al.*, Plenum Press, New York, 1996

The goal of the perceptual evaluation experiment of low-bit rate video described in this Memorandum was twofold. First, to try to determine a number of preferred combinations of coding rate and frame rate for a video coding system based on CCITT Recommendation H.261. Second, to try to evaluate and measure the sensitivity of viewers to the rendition of facial detail which affects *eye contact* and *lip synchronization* in low bit rate coding situations. The latter was made possible by the availability of a flexible low bit rate video coding software platform based on an algorithm referred to as *model-assisted H.261* described in [2, 3] . In effect, this coder operates a transfer of bits from the region outside the face location to a rectangular area which surrounds the eyes, nose and mouth of the person in the image, thereby increasing coding quality in the facial area at the expense of the outside region.

Importantly, the video sequences in the experiment all had a synchronized audio track, since lip-sync or the lack thereof can only be perceived in the presence of the speech signal. The subjects consisted of a group of twenty video coding non-experts, i.e. not used to the coding artifacts which arise with low-bit rate digital video coding systems.

METHODOLOGY AND EXPERIMENTAL SET-UP

Video coding systems tested

The two video coding systems used are based on CCITT Recommendation H.261 [4, 5] which describes an algorithm for video coding at the rates of $p \times 64$ kbps, where p is an integer in the range $1, 2, \ldots, 30$, and which is used in most commercial video tele-conferencing systems to date. The first coder, which we refer to simply as H.261, is consistent with Reference Model 8 (RM8), the "reference implementation" of an H.261 encoder developed and used internally by the H.261 working group [6] . It can operate at various bit rates and various constant frame rates. For the experiment described here, the two bit rates considered were 48 and 24 kbps, and the three frames rates were 7.5, 5 and 2 fps.

The second coder, which we refer to as *model-assisted H.261* (MA H.261), is also fully compatible with Recommendation H.261 but makes use of the concept of *model-assisted coding* introduced in [7, 8] . It includes a preprocessing module which performs automatic face location detection and accurately tracks eyes-nose-mouth regions in head-and-shoulders images. This makes it possible for the encoder to use a region-selective quantization strategy which results in coded images where facial features appear sharper at the expense of the rest of the image. The details of the implementation which uses a *virtual modulated buffer* to achieve the desired rate control are fully described in [2] .

The input color sequences to the coders were in QCIF format, i.e. with luminance images of size 180×120. The very low bit rates of interest for this experiment make that choice the most reasonable one. The images produced by both coders were then postprocessed by a postfiltering module described in [9, 10] , which removes much of the coding artifacts produced by the coders. Some coding artifacts typically remain in the postfiltered images however, especially in regions of uncovered background, whenever motion content is high, such as when arms move and/or hands appear in the image.

Description of experimental set up

Tape generation Two original digital color video sequences were used. The first one, about 30-seconds long, referred to as 'jelena', shows a woman wearing a dark plaid jacket discussing her thesis work. The amount of motion in the sequence is low but by no means nonexistent, since her shoulders rotate throughout the sequence. During most of the sequence, jelena does not directly look at the camera but rather slightly downwards or to one side. The second sequence, about 20-seconds long, referred to as 'roberto,' shows a man facing directly the viewer, telling a story about a trip to a city in Thailand involving a bridge. The motion content is high throughout the sequence, with a lot of hands and arms motion, which makes this sequence very challenging to code at low bit rates.

The coded digital video sequences in QCIF format were interpolated to CIF resolution for display. They were then digitally recorded on D1 tape, with two sequences displayed side by side. A synchronized uncoded speech signal was then overlaid on the tape. The audio track was therefore the same throughout the tape.

The sequence-by-sequence layout of the tape indicating which coder was used for which sequence and at which combination of bit rate and frame rate is shown in Table 1. Throughout the tape, a coded 'jelena' sequence was systematically followed by a coded 'roberto' sequence with the same encoder configuration. Sequences 1 through 4 consist of material coded at 48 kbps at the two frame rates of 7.5 and 5 fps, respectively for Sequences 1, 2 and Sequences 3, 4. Sequences 5 through 8 consist of material coded at 24 kbps at the two frame rates of 5 and 2 fps, respectively for Sequences 5, 6 and Sequences 7, 8.

Experimental set up The pool of subjects consisted of twenty Lab members—eight females, twelve males—randomly selected among video coding non-experts, who

Table 1: Tape layout.

	Image A	Image B
Sequences 1, 2	MA H.261	H.261
	48 kbps, 7.5 fps	48 kbps, 7.5 fps
Sequence 3, 4	H.261	MA H.261
	48 kbps, 5 fps	48 kbps, 5 fps
Sequence 5, 6	H.261	MA H.261
	24 kbps, 5 fps	24 kbps, 5 fps
Sequence 7, 8	MA H.261	H.261
	24 kbps, 2 fps	24 kbps, 2 fps

Table 2: Subject preferences at 48 kbps.

Seq. no.	Seq. 1	Seq. 2	Seq. 3	Seq. 4	Avg.
MA H.261 preferred	17	12	16	11	70%
H.261 preferred	1	2	1	5	11%
no preference	2	6	3	4	19%

individually took part in the experiment. Coded video sequences were shown in side-by-side pairs on a 13-inch video monitor[1], with the image on the left labelled "A", and the image on the right labelled "B". The subjects were seated two to three feet away from the monitor. Eight sequences corresponding to either different input video or different coder configurations were shown in succession—each sequence shown twice with a three-second intermission in between. The subjects were encouraged to focus on the image on one side of the screen during the first showing, then on the other side during the second showing, rather than try to compare the two images "on-the-fly", even though that particular option was also possible. This was suggested in order that the viewers average out their perception of a particular coded sequence over the entire length of the sequence (approximately half a minute) rather than base their ratings on different clips[2].

Video quality rating The subjects were asked to evaluate the quality of the audio-visual signals on a five-point scale from 1 to 5, with the correspondence:

1: bad, **2**: poor, **3**: fair, **4**: good, **5**: excellent.

They were also asked to indicate whether they preferred one image versus the other, and if this was the case, to indicate what motivated their selection. This was designed in order to detect slight preferences for one coder or the other, which might not otherwise be reflected by the numerical ratings.

ANALYSIS OF EXPERIMENTAL DATA

The tabulation of subject preferences for either model-assisted or unassisted H.261 for the various input sequences and coder configurations is shown in Tables 2 and 3. At both coding rates of 48 kbps and 24 kbps, 70% of the subjects indicated a preference for model-assisted coding. It is of interest to note the following.

- A preference for model-assisted coding **six or more times** (out of a total of eight sequences) was indicated by 70% of the subjects—88% of the female subjects, and 42% of the male subjects, which indicates a very good consistency of individual

---

[1]The approximate size of the images on the screen was therefore 5in × 3.5in.

[2]With each of the coders tested, which produce decoded images at a constant frame rate, the image quality can vary significantly, according mostly to varying motion content.

Table 3: Subject preferences at 24 kbps.

Seq. no.	Seq. 5	Seq. 6	Seq. 7	Seq. 8	Avg.
MA H.261 preferred	17	9	16	13	69%
H.261 preferred	2	3	0	1	7%
no preference	1	8	4	6	24%

Table 4: Mean opinion scores and MOS increments for model–assisted H.261.

	H.261 MOS score and std dev.	**MA H.261** MOS increment
Sequence 1	2.60, $\sigma = .73$	$\Delta = .91$
Sequence 2	2.58, $\sigma = .55$	$\Delta = .35$
Sequence 3	2.50, $\sigma = .80$	$\Delta = .85$
Sequence 4	2.70, $\sigma = .64$	$\Delta = .31$
Sequence 5	2.08, $\sigma = .46$	$\Delta = .55$
Sequence 6	1.75, $\sigma = .54$	$\Delta = .31$
Sequence 7	1.50, $\sigma = .73$	$\Delta = .43$
Sequence 8	1.63, $\sigma = .78$	$\Delta = .36$

subjective evaluations. All of these subjects indicated that a motivation for their preference for images coded with model-assisted H.261 was either: "better lip synchronization," "better facial features," or "clearer, less blocky faces."

• When considering the female population only (eight subjects), 81% indicated a marked preference for model-assisted coding at 48 kbps, and 78% indicated the same preference at 24 kbps. For the male population these percentages drop to 65% at 48 kbps, and 56% at 24 kbps. The female subjects were therefore clearly more sensitive to better-rendered facial features than the males.

Mean Opinion Scores The mean ratings, referred to as Mean Opinion Scores (MOS), as well as standard deviations are given is Table 4. Table 4 also shows the average MOS increments obtained by model-assisted coders versus their non-assisted counterparts. Model-assistance results in average MOS increments of .6 at 48 kbps and .4 at 24 kbps—corresponding to a 25% increase. Qualitatively, this increment elevates the the subjective audiovisual quality of sequences coded at 48 kbps from somewhere halfway between 'poor' and 'fair' to slightly better than 'fair' when model-assisted H.261 is used. At 24 kbps, the rating of model-assisted H.261 is only slightly better than 'poor'. The MOS increments for 'jelena' were consistently higher than those for

Table 5: 90% confidence intervals.

	H.261	**MA H.261**
48 kbps, 7.5 fps	±.24	±.26
48 kbps, 5 fps	±.27	±.28
24 kbps, 5 fps	±.18	±.27
24 kbps, 2 fps	±.28	±.31

'roberto', a result which can be explained by the much more noticeable coding artifacts on 'roberto', due to high motion content, accross all the coders tested. Too many artifacts can *mask* improvements in facial details.

A large majority of subjects showed little sensitivity to the drop of frame rate from 7.5 fps to 5 fps at 48 kbps[3]. The slight increase in MOS for the sequence 'roberto' at 5 fps versus the one at 7.5 fps seems to denote the subjects' sensitivity to reduced coding artifacts at the lower of the two frame rates. However, at 24 kbps, subjects reacted strongly on average against the lower frame rate of 2 fps, for which the motion is extremely choppy and synchronization between lip mouvements and the audio track has completely disappeared. Interestingly, an informal comparison of still frames from sequences coded at 48 kbps, 5 fps and at 24 kbps, 2 fps shows a very similar image quality. The great difference in MOS scores (of one point) for these two coding configurations must be therefore attributed to motion rendition alone—in particular to the presence or absence of *lip synchronization*[4].

Confidence intervals Table 5 gives the 90% confidence intervals for the coders and coder configurations considered. These intervals indicate, assuming a normal distribution of the ratings, how future subjects would rate these coders. These intervals were computed according to the formula (see [11]):

$$I = \overline{m} \pm 1.645 \times \frac{\overline{\sigma}}{\sqrt{n}},$$

where $\overline{m}$ and $\overline{\sigma}$ respectively denote the average MOS and average standard deviation over the two sequences evaluated, for a particular coding system, and n denotes the number of subjects.

CONCLUSION

A large majority of subjects, none of them expert in very low bit rate video coding, was found to be sensitive to the improvement of facial features which is provided by model-assistance in the context of a standard low bit rate coding algorithm such as H.261. This was reflected qualitatively by the fact that 70% of the subjects indicated a

---

[3]Only four subjects commented that they were annoyed by the low frame rate/poorer lip-sync at 5 fps.

[4]This should illustrate the danger in trying to evaluate coded video quality in the absence of the accompanying speech signal for video teleconferencing and video telephony.

marked preference for model-assisted H.261 six or more times out of eight test sequences, and quantitatively by non-negligible MOS increments obtained by model-assisted coders versus their non-assisted counterparts of .6 at 48 kbps, and .4 at 24 kbps.

For video coded at very low bit rates, i.e. 48 kbps and lower, most viewers found that a constant frame rate of 5 fps is acceptable. Most viewers expressed a preference for video coded at 48 kbps at the lower frame rate of 5 fps, for which image quality is better than at 7.5 fps, especially for challeging sequences. However, this situation is reversed below 5 fps. Viewers expressed a strong dislike of the very low frame rate of 2 fps, even though the spatial image quality at this rate is quite acceptable.

Acknowledgements

The author would like to thank Nikil Jayant and Larry Rabiner who provided the inspiration for this work by emphasizing the importance of metrics of performance, James Pawlyk for his assistance with the taping phase, and the twenty subjects who graciously took part in the experiment.

REFERENCES

[1] J. S. Angiolillo, H. E. Blanchard, E. W. Israelski, "Video telephony," *AT&T Technical Journal*, vol. 72, no. 3, May/June 1993.

[2] A. Eleftheriadis, A. Jacquin, "Low bit rate model-assisted H.261-compatible coding of video," *Proceedings ICIP '95*, October 1994.

[3] A. Eleftheriadis, A. Jacquin, "Automatic face location detection for model-assisted rate control in H.261-compatible coding of video," *Signal Processing: Image Communication*, To appear.

[4] M. Liou, "Overview of the px64 kbit/s video coding standard," *Communications of the ACM*, vol. 34, no. 4, April 1991.

[5] "Draft revision of recommendation H.261: video codec for audiovisual services at $p \times 64$ kbit/s," *Signal Processing: Image Communication*, vol. 2, no. 2, pp. 221–239, August 1990.

[6] "Description of reference model 8 (RM8)," CCITT SGXV WG4, Specialists Group on Coding for Visual Telephony, Doc. 525, June, 1989.

[7] A. Eleftheriadis, A. Jacquin, "Model-assisted coding of video teleconferencing sequences at low bit rates," *Proc. ISCAS '94*, May–June 1994.

[8] A. Eleftheriadis, A. Jacquin, "Automatic face location detection and tracking for model-assisted coding of video teleconferencing sequences at low bit rates," *Signal Processing: Image Communication*, To appear.

[9] T-S Liu and N. Jayant, "Adaptive post-processing algorithms for low bit rate video signals," *Proceedings ICASSP '94*, 1994.

[10] J. Apostolopoulos, N.S. Jayant, Private communication.

[11] A. Papoulis, Probability, Random Variables, and Stochastic Processes, Third Edition, McGraw–Hill, Inc., 1991.

SPEECH ASSISTED MOTION COMPENSATION
IN VIDEOPHONE COMMUNICATIONS

Fabio Lavagetto

DIST
University of Genova
Via Opera Pia 13
Genova 16145 Italy

INTRODUCTION

The basic objective addressed by this work is that of approaching the problem of videophone coding from an audio-video joint point of view, both for the analysis and for the synthesis. The motivating idea is that interpersonal audio-video communications represents an easily modelable information source, characterized in audio by a human speaker's voice and in video by the same speaker's face. In the very large majority of cases a videophone communication consists exactly of these two strongly correlated items: a talking face together with its synchronous speech.

The objective is the better exploitation of the video bandwith in low bitrate videophone communication where, as an example, an average video frequency of a few Hz is expected: this means that images can be separated in time even by hundreds milliseconds. In interframe image refresh the encoder computes the motion field MF between the current frame $I_d(t)$ (the same available at the decoder, where the notation d means decoded) and the new incoming frame $I(t+1)$. Since these two frames, as said before, can be separated by a long time interval, their differences (expecially in correspondence of the mouth region) are significant and the smoothness of the resulting motion field is low. This implies that both MF and the error signal $e(t+1) = I(t+1) - I_{mc}(t+1)$ (where the notation mc stands for motion compensated) will require many bits for being encoded. It is just because of these facts that the interframe refresh frequency is so low.

The proposed solution is to process the speech interval which separate $I_d(t)$ from $I(t+1)$ and be able to transform $I_d(t)$ into a new image $I_p(t+\frac{1}{2})$ more similar to $I(t+1)$. At this point, the motion field can be computed between the so got $I_p(t+\frac{1}{2})$ and $I(t+1)$ thus obtaining a smoother motion field and an error signal with lower entropy. All these facts allow to reduce the bitrate and the video bandwidth left can be used to send further information. The goal is to raise the interframe refresh frequency without requiring further bandwidth. This result can be reached since the generation of

$I_p(t + \frac{1}{2})$ depends only on speech analysis and can be computed at the decoder exactly in the same way it has been done previously at the encoder.

The key element is the capability to derive visual information from the analysis of speech. This information is the one which must be used to apply the suitable modifications for passing from $I_d(t)$ to $I_p(t+\frac{1}{2})$. Both the encoder and the decoder must analyze speech after encoding and decoding in order to be coupled. These operations can be done with some major constraints being a good quality of the speech signal and acceptable visibility of the mouth in frame $I_d(t)$. The encoder can therefore operate as follows:

1 process a speech segment s;

2 estimate the articulatory vector v;

3 apply v onto $I_d(t)$ and generate $I_p(t + \frac{1}{2})$;

4 compute the motion field MF between $I_p(t + \frac{1}{2})$ and $I(t + 1)$;

5 apply MF on $I_p(t + \frac{1}{2})$ and generate $I_{mc}(t + 1)$;

6 compute the error $e = I(t + 1) - I_{mc}(t + 1)$;

7 encode and transmit (MF, e).

The desired effect of motion field smoothing, obtained through speech analysis, obviously affects only the image region in correspondence to the mouth which can be identified therefore as Region Of Interest (ROI). A reasonable solution can be therefore that of interlacing each full-screen refresh with a few ROI refreshes: in this way the mouth region is updated at higher frequency than the other image regions, subjectively less important. The ROI refresh frequency can be varied dynamically depending on the motion in the scene for maintaining a target constant bitrate. Considering that the high frequency refresh of the mouth region relies heavily on the speech-assisted prediction $I_p(t + \frac{1}{2})$, the decoded audio must guarantee a minimum quality to allow a reliable articulatory estimation. In case of too noisy channels the articulatory vector v can be trasmitted directly to the decoder avoiding its estimation from corrupted speech: this asks for some side information to be transmitted but definitely represents a handy back-door for overcoming these impairments. Similar considerations also apply to the ROI position and size encoding: they can be estimated at the decoder from frame $I_d(t)$ by means of the same algorithms applied at the encoder to track the mouth shape or, alternatively, can be trasmitted directly as side information.

SPEECH-ASSISTED FRAME PREDICTION

Articulatory estimation is performed by means of a bank of 5 Time-Delay neural networks[1] each of them trained to associate a 12-dimensional cepstrum vector to a specific mouth visible parameter: external width (W) and height (H), upper lip offset (Lup), lip contact segment (dw) and nose-to-chin distance (LM). From these 5 almost uncorrelated parameters, other 4 correlated parameters are estimated (lup, w, h and LC) as shown in Figure 1.

The networks configuration has been based on the experimental results reported in[2] using a training procedure based on the maximization of the pattern-target cross-correlation[3]. The position and the shape of the mouth in frame $I_d(t)$ has been extracted

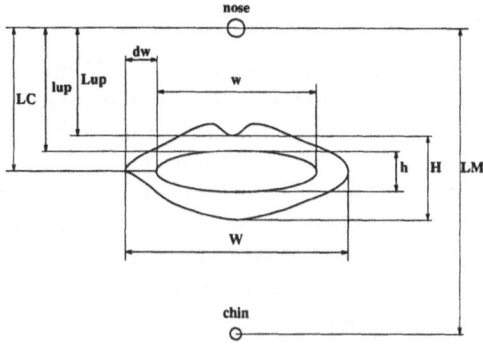

Figure 1: Wire-frame model of the mouth driven by speech parameters

by means of symmetry-based and edge-based algorithms[4] which have provided a line description of the external lip contour, while the internal contour is approximated by an ellipse. The mouth model of Figure 1 is first scaled and adapted to the extracted lip descriptors before being deformed according to the parameters estimated from speech. As shown in Figure 2, the lip contour extracted from frame $I_d(t)$ (drawn in solid line) is put in correspondence to the lip contour predicted from speech analysis (drawn in dashed line) by subdividing the rectangular region of interest into sectors of $\frac{\pi}{8}$ each, with center in correspondence to the center of the predicted contour. The intersection points with the old and the new contours define 72 triangles, 36 on each side with respect to the mouth symmetry axis, 16 of them belonging to the internal mouth region (modeled as an ellipse), 32 to the lip region and 24 to the surrounding skin. Each triangle of the mouth model undergoes an affine transformation which can be determined and applied to the texture, triangle by triangle, thus providing the predicted frame $I_p(t + \frac{1}{2})$.

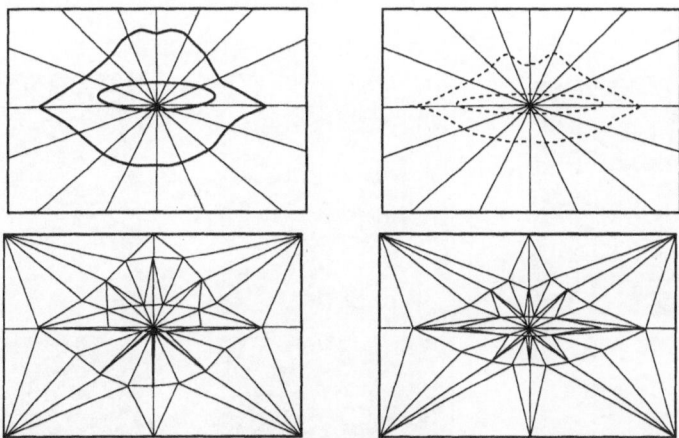

Figure 2: The mouth contour extracted from $I_d(t)$ (top-left) is put in correspondence to the mouth contour predicted from speech analysis (top-right). The wire-frame model is adapted to the extracted mouth contour (bottom-left) and to that predicted from the speech (bottom-right).

Any algorithm for the estimation of the motion field between $I_p(t + \frac{1}{2})$ and $I(t+1)$ can be then applied with reference only to the region of interest.

3

EXPERIMENTAL RESULTS

The work which is here reported has approached the problem from a very basic point of view simplifying as much as possible the experimentation framework. The elementary goal which was addressed was that of evaluating the effectiveness of speech-assisted prediction in terms of smoothness of the motion field and entropy of the prediction error. Experimentations have been done with gray-level images, 8 bpp, with reference to a region of interest of 128x128 pixels. The algorithms employed for the extraction of the lips and the identification of the region of interest were those previously employed in the work reported in [4]. In order to facilitate the lip contour extraction lipstick was used to enhance the contrast with respect to the surrounding skin. The position of the mouth was frontal to the camera without appreciable rotations or scale changes. In a first simulation the affine transformation was computed for each of the 72 triangles of the mouth model while, in a second simulation it has been limited to the 36 triangles on the right half of the model constraining triangles on the left hand-side to follow symmetrical deformations.

In Figure 3 three consecutive frames $(I(t), I(t+\frac{1}{2}), I(t+1))$ of a closing mouth, extracted from the original 25 Hz sequence, are shown (top) together with the synthesized 12.5 Hz sequence (bottom) with the $I(t+\frac{1}{2})$ frame predicted from speech analysis. The inverse sequence, showing an opening mouth, is presented in Figure 4. The adapted and affine deformed triangle wire-frames are shown in Figure 5.

A 3-step block-matching algorithm has been used to compute the motion field between $I_d(t)$ and $I(t+1)$ from the original 25 Hz sequence, leading to the motion-compensated and residual frames shown at the top of Figure 6 in case of a closing mouth. The motion field has been then computed between $I_p(t+\frac{1}{2})$ and $I(t+1)$ leading to the motion-compensated and residual frames shown at the bottom of Figure 6. Similar results are shown for the sequence of an opening mouth in Figure 7.

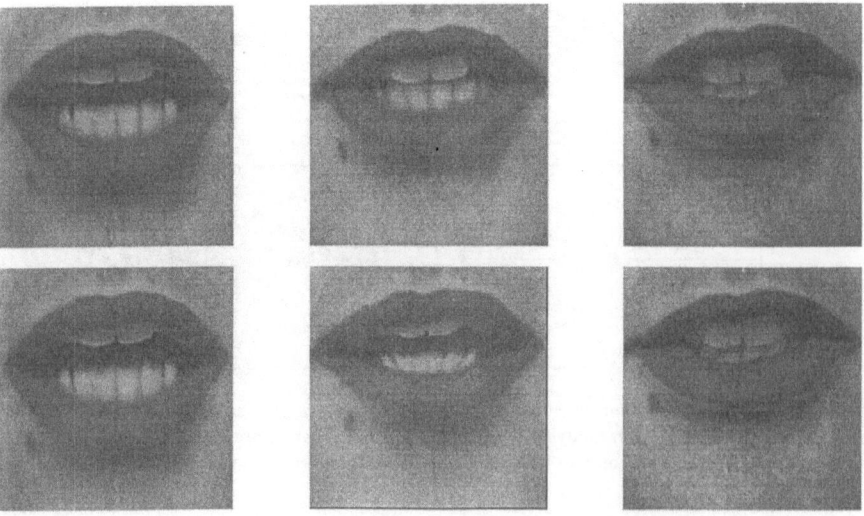

Figure 3: Examples of a closing mouth from a 25 Hz sequence. (Top) Original 25 Hz frames $I(t)$, $I(t+\frac{1}{2})$ and $I(t+1)$. (Bottom) Corresponding 12.5 Hz sequence in which the in-between frame $I_p(t+\frac{1}{2})$ has been predicted from speech. In case of closing mouth the prediction of $I(t+\frac{1}{2})$ is good reaching a PSNR of 35 dB since all the texture information is present in $I(t)$.

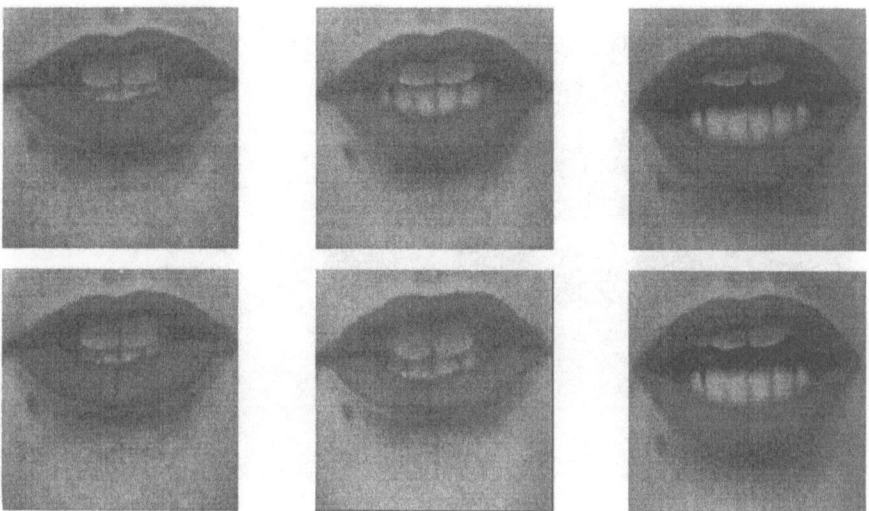

Figure 4: Examples of an opening mouth from a 25 Hz sequence. (Top) Original 25 Hz frames $I(t)$, $I(t + \frac{1}{2})$ and $I(t + 1)$. (Bottom) Corresponding 12.5 Hz sequence in which the in-between frame $I_p(t + \frac{1}{2})$ has been predicted from speech. In case of an opening mouth, less quality is achieved with a PSNR of 31 dB since poor texture information is typically available from $I_d(t)$ on the internal mouth region.

Figure 5: Adaptation and affine deformation of the wire-frame in case of a closing (top) and of an opening mouth (bottom).

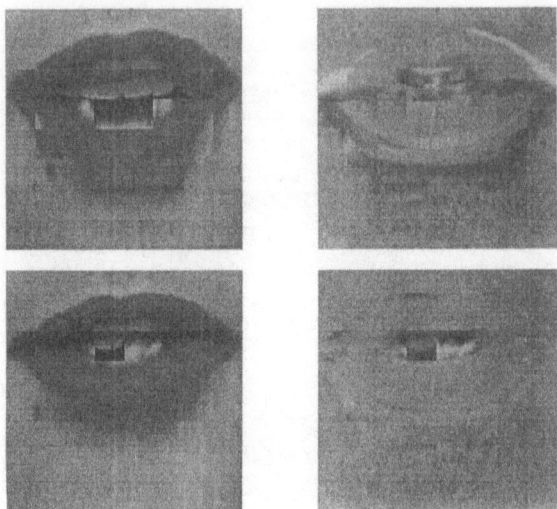

Figure 6: Closing Mouth: motion compensated frame $I_{mc}(t+1)$ (16x16 pel blocks) and residual image obtained by means of the motion field computed between $I_d(t)$ and $I(t+1)$ (top), and by means of the motion field between $I_p(t+\frac{1}{2})$ and $I(t+1)$ (bottom).

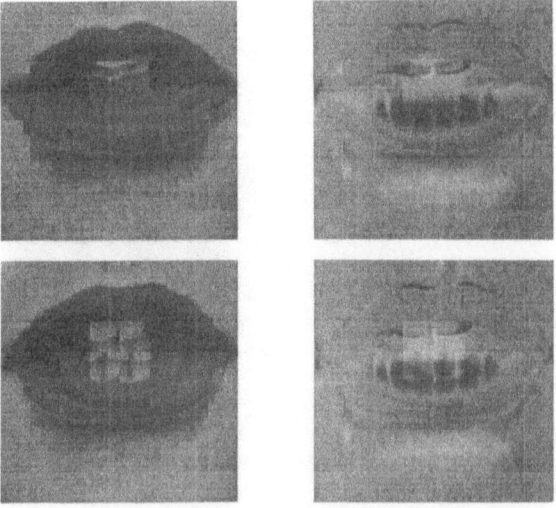

Figure 7: Opening mouth: motion compensated frame $I_{mc}(t+1)$ (16x16 pel blocks) and residual image obtained by means of the motion field computed between $I_d(t)$ and $I(t+1)$ (top), and by means of the motion field between $I_p(t+12)$ and $I(t+1)$ (bottom).

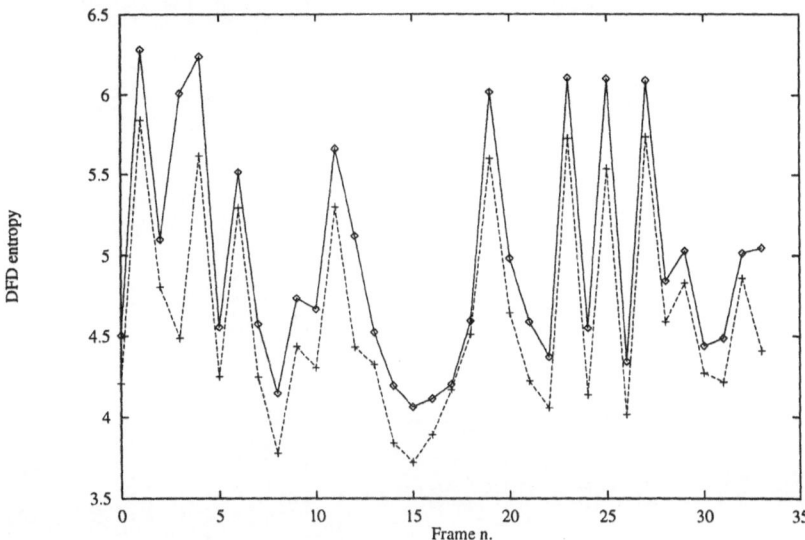

Figure 8: Entropy of the prediction error (expressed in bits/pel) computed on the motion compensated 12.5 Hz sequence with and without speech-driven prediction.

The entropy of the residual images (prediction error) has been evaluated on a sample sequence consisting of two italian words separated by a short silence interval with closed mouth, leading to the results shown in Figure 8.

REFERENCES

1. F. Lavagetto. Converting Speech into Lip Movements: A Multimedia Telephone for Hard of Hearing People, *IEEE Trans. on RE.* 3:1 (1995).
2. A. Waibel, T. Hanazawa, G. Hinton, K. Shikano and K.J. Lang. Phoneme recognition using time-delay neural networks, *IEEE Trans. on ASSP.* 37:3 (1989).
3. F. Lavagetto. Maximizing the target-pattern cross-correlation for training Time-Delay Neural Networks, *Proc. IEEE ICNN-95.* (1995).
4. F. Lavagetto and S. Curinga. Object-oriented scene modeling for interpersonal video communication at very low bitrate, *Signal Processing: Image Communication.* 6:1 (1994).

CROSS-MODAL PREDICTIVE CODING FOR TALKING HEAD SEQUENCES

Ram R. Rao[1] and Tsuhan Chen[2]

[1]Georgia Institute of Technology
Atlanta, GA 30332   rr@eedsp.gatech.edu
[2]AT&T Bell Laboratories
Holmdel, NJ 07733   tsuhan@research.att.com

ABSTRACT

Predictive coding of video has traditionally used information from previous video frames to help construct an estimate of the current frame. The difference between the original and estimated signal can then be transmitted to allow the receiver to fully reconstruct the original video frame. In this paper, we explore a new algorithm for use in coding the shape of a person's lips in a head-and-shoulder video sequence. This algorithm uses the same predictive coding loop, but instead of forming an estimate of the lip image using motion compensation and previous video frames, it forms an estimate from the associated acoustic data. Since the acoustic data is also transmitted, the receiver is able to reconstruct the video with very little side information. In this paper, we will describe our predictive coding system and analyze methods for converting from the acoustic data to visual estimates.

INTRODUCTION

Traditional video coding algorithms use motion-compensated prediction to form an estimate of the current video frame from past and/or future video frames. The difference between the original frame and the estimate can be coded and transmitted to the receiver. The receiver would use its own reconstructed copy of past frames to predict the current frame. It would then add the incoming difference signal to reconstruct the original frame. This method has been proven extremely useful for removing the temporal redundancy in video.

We want to explore methods for removing cross-modal redundancy (redundancy between the acoustic and video signals). The basic premise is there is information in the acoustic signal which can be used to help predict the video signal. This is particularly

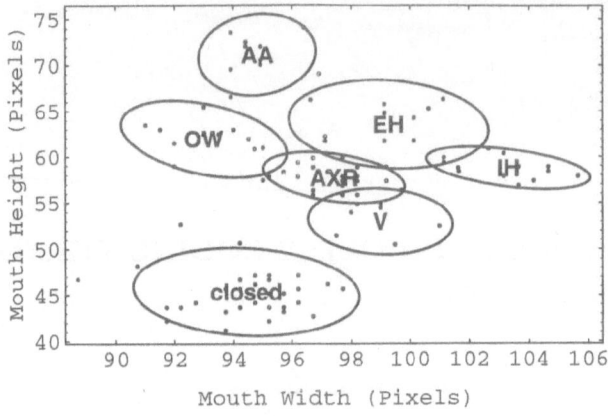

Figure 1: Height vs. width for certain vowel sounds.

true of talking head sequences. For example, if the audio indicates that a vowel is being said, one could predict that the person's mouth is open. Likewise, if the audio indicated that a /p/, /b/, or /m/ were being spoken, one could predict that the person's mouth is closed. Moreover, as shown in Figure 1, the height and width of the lips tend to cluster in different areas for differing vowel sounds. Therefore, if we had knowledge of what sound was being spoken, we could form a good estimate of the lip height and width.

Our cross-modal predictive coding system is shown in Figure 2. The input image is analyzed, and a set of visual parameters is extracted. For example, one might measure the height between the lips, and the width between the corners of the mouth. In a model-based coder[1], this parameter set would be quantized and transmitted to the receiver which would then animate the lips accordingly. Since both the transmitter and receiver have access to the acoustics, we would like to remove the redundancy that exists between the two streams of information, namely the audio and video. This is accomplished by predicting the visual parameters from the acoustics. If this prediction works well, no side information needs to be sent, if it is slightly off, an error signal can be sent, and if the prediction is poor, the actual parameter set would be sent.

More precisely the decision of what information is sent can be made based on the "rate-distortion criterion." Let X denote the real value of the parameter set, and $\widehat{X}$ denote the result of the A-to-V prediction. Also, suppose it requires B_0 bits to transmit the parameter set X, and B_1 bits to send the difference between X and $\widehat{X}$. Therefore, if the decision module decides to send nothing, the information rate (number of bits

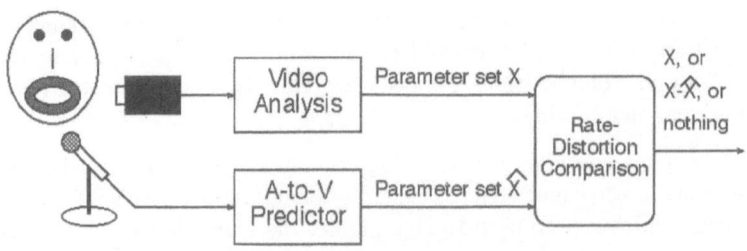

Figure 2: Cross-modal predictive coding system

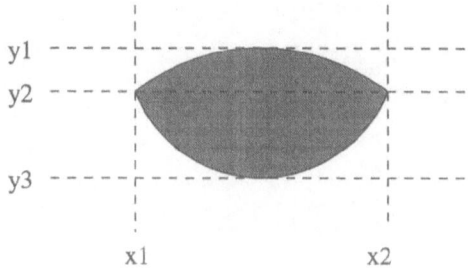

Figure 3: Template. $\lambda = [x1, x2, y1, y2, y3]$.

required) equals 0, and the distortion (parameter error at receiver) equals $|x - y|$. If the decision module decides to send the difference, $x - y$, then the rate equals B_1, and the distortion equals 0 (because the receiver can reconstruct X based on the prediction $\widehat{X}$ from the A-to-V prediction, and the difference received). Finally, if the decision module decides to send the real value of X, the rate would equal B_0, and the distortion would equal 0.

To decide which of the above three cases is the final decision, the cost function, $C = R + wD$, is computed for the three cases, and the one that gives the minimum C is chosen as the final decision. Here, the value w is a parameter that can be set by the user. It determines the trade-off between the bit rate and the parameter error. Typically, one chooses a large w when small distortion is considered more important than a low bitrate, and a small w when the bit rate is considered more important than the error.

VISUAL ANALYSIS

The first problem which must be addressed is extracting parameters from the video sequence. To accomplish this, we use a variant of deformable templates to track the shape of the lips through successive video frames [2]. A few assumptions are made concerning the structure of the input images. First, it is assumed that the image can be divided into foreground (pixels within the outer contour of the lips) and background (pixels which are part of the face) regions. Next, we assume that the shape of the foreground can be modeled by two parabolas as shown in Figure 3. This is our template, and it is completely specified by the parameter, λ, which is five dimensional ($\lambda = [x1, x2, y1, y2, y3]$). When the value of λ changes, the shape and position of the template changes. This template is used to divide the image into foreground and background regions. Finally, we assume that there are distinct probability density functions (pdf) which govern the distribution of pixel colors in the foreground and background. Since the lips and face have different colors, the assumption is valid.

If we have estimates for the foreground (pixels within the lips) pdf and background (pixels of the face) pdf, we can evaluate the joint probability of all pixels in the image. This joint probability is given by:

$$P[I|\lambda] = \prod_{(x,y)\epsilon fg} b_{fg}(I(x,y)) \prod_{(x,y)\epsilon bg} b_{bg}(I(x,y)) \tag{1}$$

where $P[I|\lambda]$ is the joint probability, $I(x,y)$ is the three dimensional pixel value at location (x,y), and b_{fg} and b_{bg} are the foreground and background pdf's, respectively.

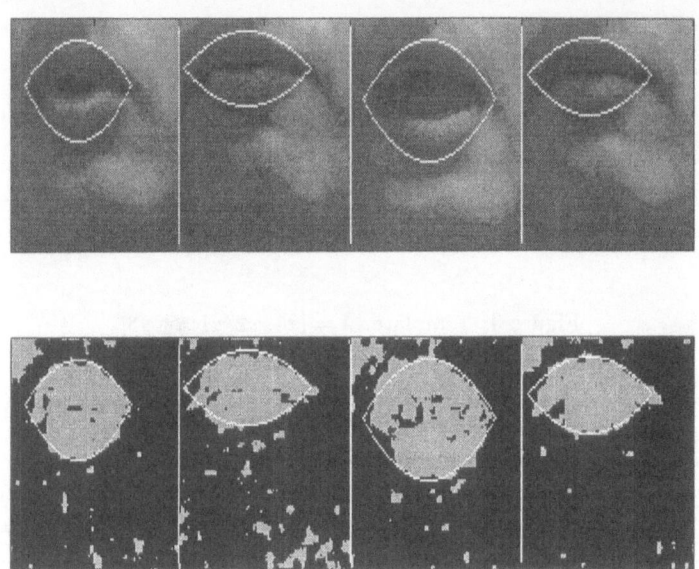

Figure 4: Tracking Results

Notice the dependence on λ: if λ is changed, different pixels become part of the foreground and background, thus changing the joint probability value. Our visual analysis sytem uses a maximization algorithm to find the parameter, λ which maximizes the joint probability of the pixels in the image.

Our analysis system works as follows. In the initial frame, the template is manually aligned with the subject's lips. The foreground and background probability density functions, b_{fg} and b_{bg} are modeled as Gaussian mixtures with two Gaussians per mixture. The initial segmentation of the data is used to estimate the parameters of the Gaussian mixtures. For each new image, we run a log search maximization algorithm to find the parameter, λ, which maximizes $P[I|\lambda]$. Results of our tracking algorithm are shown in Figure 4. Lip height and width can easily be determined from the parameter, λ.

ACOUSTIC TO VISUAL MAPPING

The most important part of our coding system is a reliable acoustic to visual conversion system. There has been some research in this area in recent years [3,4,5]. Chen et al. [3] showed that it is possible to use acoustic information to improve video quality when frames were lost. They used a speech recognizer to segment the speech into acoustic units, and a table lookup to convert from each acoustic unit to a visual parameter set. Lavagetto [4], on the other hand, directly converted from cepstral coefficients to visual parameters by using a time delay neural network.

In general, each of these three systems can be viewed as having a classification stage followed by a mapping as shown in Figure 5. The audio gets classified into one of many groups, and the visual centroid associated with the acoustic group is output by the mapping. There are many approaches to training the classifier and mapper. We will examine two approaches in particular. The first approach (clustering acoustics first)

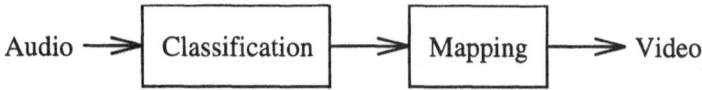

Audio → Classification → Mapping → Video

Figure 5: Classification and Mapping Strategy

was used by Morishima, et al. [5] They trained a vector quantizer on the acoustic data (cepstral coefficients). Then, for each acoustic group, the corresponding visual data is averaged to compute the visual centroid. Therefore, incoming acoustic signals are classified according into an acoustic group, and the visual centroid for that group is output.

The second training approach (clustering video first) starts by vector quantizing the visual features. A Gaussian density is then fit to the acoustics that correspond to each of the different visual groups. Therefore, an acoustic observation can be classified into a group using these probability estimates. The visual centroid can then be output.

Intuitively, there is a slight difference between the two training algorithms. This is shown in Figure 6. Consider training the acoustics first. The acoustic centroids would be at the location marked with asterisks. Everything to the left of the dotted line would be classified to one group, while everything to the right would be classified in another. Since there is overlap between the two groups, errors will be made. If the visual centroid for the data left of the dotted line were computed, it would likely be at the left 'x'. This would result because most of the visual tokens would come from the shaded group, but some would come from the other group. This would skew the centroid to the right. Likewise, the visual centroid for the second group would be skewed to the left.

Now, let's examine what happens if we clustered the visual features first. The visual centroids would be at the positions marked with an 'o', and averaging the corresponding acoustic tokens would place the acoustic centroids at the locations marked with '*'. This analysis would suggest that clustering the video first has some advantages over clustering the audio first.

How do the two approaches affect the acoustic to video conversion? There is a tradeoff between classification error and the quantization error. The visual centroids obtained by clustering the video first are optimal for faithfully reconstructing the signal, but the acoustic centroids are suboptimal for classifying the acoustic signal. Hence the quantization error is low, but the classification error is high. On the other hand, when the acoustics are clustered first, the centroids are optimal for classification, but suboptimal for reconstruction.

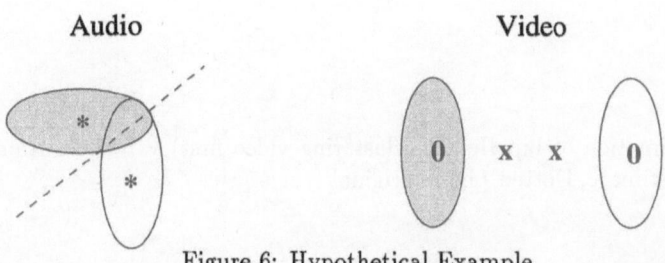

Figure 6: Hypothetical Example

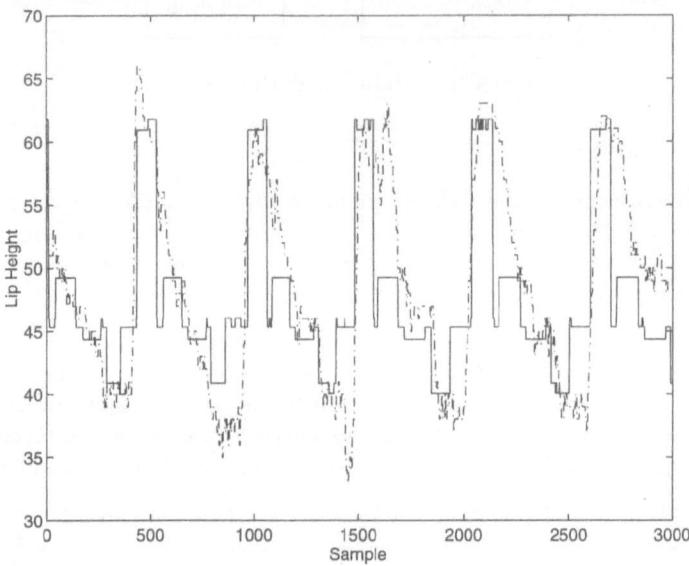

Figure 7: Estimation of Lip Height (clustering audio first) vs. audio frame number. Solid Line – estimate, Dotted Line – original.

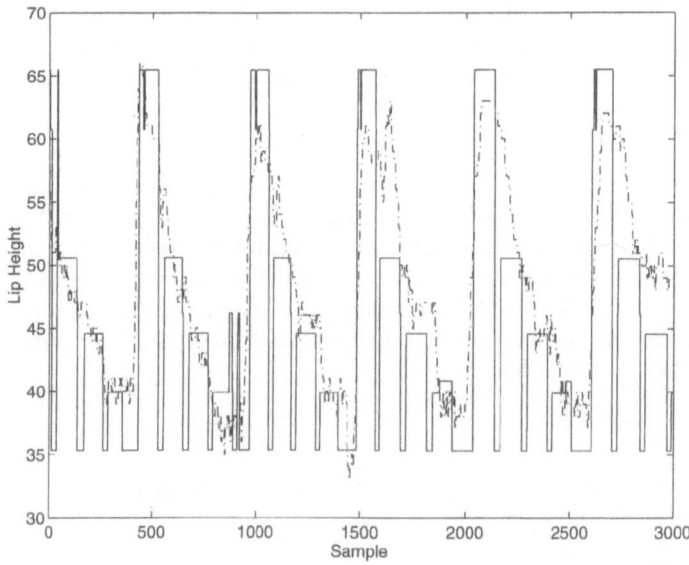

Figure 8: Estimation of Lip Height (clustering video first) vs. audio frame number. Solid Line – estimate, Dotted Line – original.

EXPERIMENTAL RESULTS

Our experimental data consisted of synchronized audio-visual data. The subject repeatedly cycled through four vowel sounds (/eh/, /iy/, /ah/, /axr/), and the audio and video was synchronously captured. There were approximately 10 to 12 repititions of each vowel sound. Both training and testing data was captured for three different sets of vowel sounds. The 8 kilohertz audio was converted to eight LPC-derived cepstral coefficients (frame size=240 samples, frame rate= 100 hertz). The visual parameters were captured as fast as possible with time stamps (between 15 and 30 hertz), and upsampled to 100 hertz (by repeating parameters). Once again, our data consisted of 8 LPC cepstral coefficients for the audio, and the associated lip height and width for the video.

The training data was used to train the audio to visual conversion systems as mentioned before. The audio-visual data was clustered into eight classes in both cases. In the first case, the audio was clustered into groups followed by the associated video. In the second case, the video was clustered into groups, followed by the associated audio. The audio in the testing data was input into the A-V mapping, and the visual output was passed through a median filter to remove impulsive noise. Results showing the actual visual data and reconstructed visual data are shown in Figure 7 and Figure 8.

One can see that when the audio is clustered first, the reconstructed signal is close to the original, but does not do a good job of modeling the peaks and valleys. When the video is clustered first, the peaks and valleys are modeled better, but there are a number of areas which have classification errors, and thus the reconstructed parameters are significantly off. In fact, the periodic valleys in the reconstructed visual signal correspond to silence in the acoustic signal, whereas the each plateau corresponds to a different vowel sound.

CONCLUSION

In this paper, we have developed a framework for removing redundancy between the acoustic and visual signals in a model-based coding system. Our data shows that cross-modal predictive coding is possible, and the practicality of such a system depends on the accuracy with which acoustic data can be converted to visual estimates. It was seen that there is a tradeoff between the classification error associated with classifying the audio, and the quantization error associated with reconstructing the video. The joint design of a classifier and mapper which allows control of the trade-off between classification error and quantization error is currently under investigation. Possible alternatives include use of neural networks, or clustering methods which jointly cluster the audio-visual data.

REFERENCES

1. Aizawa, K., Harashima, H., Saito, T., "Model-based synthesis image coding (MBASIC) system for a person's face," Signal Processing: Image Communication, volume 1, number 2, pages 139–152, Oct. 1989.

2. Rao, R. and Mersereau, R., "State-Embedded Deformable Templates," to appear in ICIP '95, Washington, D.C., 1995.

3. Chen, T., Graf, H. P., and Wang, K., "Speech-assisted video processing: Interpolation and low-bitrate coding," 28th Annual Asilomar Conference on Signals, Systems, and Computers, Pacific Grove, CA, October 1994.

4. Lavagetto, F., "Converting speech into lip movements: A multimedia telephone for hard of hearing people," IEEE Transactions on Rehabilitation Engineering, Vol. 3, No. 1, March 1995.

5. Morishima, S. , Aizawa, K., and Harashima, H., "An intelligent facial image coding driven by speech and phonemes," ICASSP '89, Glasgow, UK, 1989.

TIME-VARYING MOTION ESTIMATION ON A SEQUENCE OF IMAGES*

R. Leonardi and A. Iocco

Signals and Communications Lab., Dept. of Electronics for Automation
University of Brescia, Brescia (BS), I-25123, Italy
Ph: +(39-30) 371-5434, Fax: +(39-30) 380-014
e-mail: leon@bsing.ing.unibs.it

Abstract

This work is concerned with the estimation of time-varying motion field in a sequence of images assuming that the motion is translational. We consider only the apparent motion of any point in the scene, i.e. its projection over the image plane, rather than the three dimensional motion of points in the scene. Unlike many other approaches that use simply a motion model with constant velocity for the pixels in the image, a second-order time-varying motion model is considered.

We believe that this approach is more suited to describe the motion of the real-world objects, as their motion is obeying the fundamental laws of classical mechanics. We will show how this second-order motion model outperforms the classical constant-velocity model for predicting frames of an image sequence. This could suggest that, with very few parameters, the motion of portions of an image sequence can be correctly estimated with significant impact in a video compression scheme or in a frame rate conversion application.

INTRODUCTION

A precise motion analysis seems relevant for many applications, e.g. the study of medical or satellite images, the analysis of the human body, the surveillance of garages, subways, highways, the identification and tracking of objects. The correct knowledge of the motion field in a scene, allows also to reduce the image sequence redundancy and may be thus efficiently used to increase coding efficiency. Several are also the applications for robot motion control.

---

* This work was supported by the European Union, under the HCM program (MANADIX project).
Experimentation was carried out at the Signal Processing Laboratory of Swiss Federal Institute of Technology, Lausanne.

In the first section we introduce the second-order model. Section 2 describes how its motion parameters can be estimated. In section 3 we present some simulation results, while section 4 gives some concluding remarks.

1 A SECOND ORDER MOTION MODEL

In the literature[1,2] several techniques have been proposed to estimate apparent or 3D motion from an image sequence. We can group them in four main classes: Transform Domain Techniques, Gradient Based Techniques, Point Correspondence Techniques and Block Matching Techniques.

We have chosen to work within the framework of block matching techniques to estimate and predict motion information. Block matching techniques are easier to implement than others; they have a large diffusion and are computationally efficient; finally they do not use additional constraints in the estimation process, while others, such as the gradient based methods, do.

In a linear motion model we assume that objects in the scene are subject to a constant velocity within a reasonable period of time. If $\mathbf{x}$ is the reference to a block of a set of homogeneous moving pixels in the image, and $\mathbf{v_0}$ is its constant velocity, we have:

$$\mathbf{d}(\mathbf{x}, t) = \mathbf{v}_0 \cdot t \tag{1}$$

where $\mathbf{d}$ is assumed to be a translational displacement. Under such model, if we know the displacement for the given block of pixels between any two instants of time, we can estimate the velocity $\mathbf{v_0}$ through equation (1).

The displacement $\mathbf{d}$ can first be obtained through the block matching technique, according to the classical minimization of the absolute displaced frame difference signal between two given frames that occur at time r and n respectively:

$$\mathbf{d}(\mathbf{x}, n) = \arg \min_{\mathbf{d} \in [D_{min}, D_{max}]} \sum_{x \in N(x)} |I(\mathbf{x}, r) - I(\mathbf{x} + \mathbf{d}, n)| \tag{2}$$

Later we can predict the position of the block referred to by $\mathbf{x}$, at any other instant of time using the following relationship:

$$\mathbf{d}_{pr}(\mathbf{x}, t) = \mathbf{v}_0 \cdot t \tag{3}$$

The most important problem using this approach is that we assume a linear time variation as a model for motion and this is often questionable when modeling motion of real-world objects, especially over a significant amount of time. Therefore we propose a model that incorporates both an initial velocity $\mathbf{v_0}$ and a constant acceleration $\mathbf{a}$. More clearly the displacement equation is described by:

$$\mathbf{d}(\mathbf{x}, t) = \mathbf{a} \cdot \frac{t^2}{2} + \mathbf{v}_0 \cdot t \tag{4}$$

Certainly the model assumes that objects in the scene are subject to constant acceleration $\mathbf{a}$, during the entire time frame during which equation (4) holds.

2 ESTIMATION PROCESS

To estimate the parameters of the time varying model (acceleration **a** and initial velocity $\mathbf{v_0}$) we need at least two displacement estimates, $\mathbf{d_p}$ and $\mathbf{d_n}$, for any block of pixels referred to by **x** (figure 1). These estimates correspond to the location of the block in at least two instants in time, with respect to a current reference frame denoted by index r. This makes this approach quite different from the one proposed by Chahine and Konrad[3], where a gradient based technique was used to estimate the motion model with acceleration.

Using a block matching approach, two pairs of displacements are estimated using three different frames, p, r, n. The first displacement for a given block is identified by $\mathbf{d_p}$ and describes the motion occurred between frame p and frame r, the reference frame. Similarly, for the same block, a second displacement $\mathbf{d_n}$ is estimated between frame n and frame r. These two displacements can be combined to form the set of equations:

$$\mathbf{d}_p(\mathbf{x}, t_1) = \mathbf{a} \cdot \frac{t_1^2}{2} + \mathbf{v}_0 \cdot t_1 \tag{5}$$

$$\mathbf{d}_n(\mathbf{x}, t_2) = \mathbf{a} \cdot \frac{t_2^2}{2} + \mathbf{v}_0 \cdot t_2 \tag{6}$$

where **a** and $\mathbf{v_0}$ are the acceleration and the initial velocity of the block assuming that the second order model is at least valid from frame p to frame n, and can remain as such till the instant in time at which we want to predict the displacement of the current block.

From the system of two equations, (5) and (6), with two vector unknowns we obtain the acceleration **a** and initial velocity $\mathbf{v_0}$ vectors.

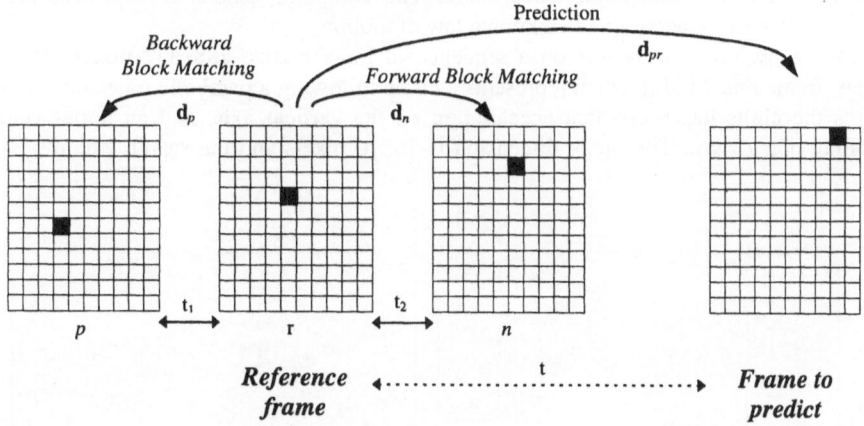

Figure 1. Model of motion estimation and prediction.

Solutions to (5) and (6) can be expressed as follows:

$$\begin{bmatrix} t_1 & t_1^2/2 \\ t_2 & t_2^2/2 \end{bmatrix} \cdot \begin{bmatrix} v_x \\ a_x \end{bmatrix} = \begin{bmatrix} d_{p_x} \\ d_{n_x} \end{bmatrix} \tag{7}$$

$$\begin{bmatrix} t_1 & t_1^2/2 \\ t_2 & t_2^2/2 \end{bmatrix} \cdot \begin{bmatrix} v_y \\ a_y \end{bmatrix} = \begin{bmatrix} d_{p_y} \\ d_{n_y} \end{bmatrix} \qquad (8)$$

It can be noted that the matrix to be inverted only depends upon time.

At any other instant in time during which the constant acceleration model holds a predicted displacement $\mathbf{d}_{pr}$ can be estimated using the following equation:

$$\mathbf{d}_{pr}(\mathbf{x}, t) = \mathbf{a} \cdot \frac{t^2}{2} + \mathbf{v}_0 \cdot t \qquad (9)$$

Clearly if such prediction is applied on a block by block basis this may lead to holes and overlap of the predicted blocks in the predicted frame. In our experimentation we have not handled this aspect of the problem, setting the value of holes to a fixed level.

We have worked with an integer pel accuracy. This means that if the resulting displacements are small, inaccuracy in measuring the correct displacements results in estimations of vectors $\mathbf{a}$ and $\mathbf{v}_0$ which are biased.

Let us assume each displacement $\mathbf{d}_{p/r}$ has been measured with a precision $\delta\mathbf{d}_{p/r}$. The inaccuracy in the prediction of a frame occurring at time t, with respect to the reference frame, will increase quadratically with respect to t, but will be inversely proportional to the time interval that separates frame n from frame p.

3 EXPERIMENTAL RESULTS

We have tested the performance of the algorithms presented in the section 1, on computer generated and natural sequences. The computer generated sequences contain objects which move according to a known law of motion.

We show now the results on a sequence of images (size 288x352 pixels, 256 gray levels, frame rate 25 Hz) which represents a circle following a parabolic trajectory. In other words the circle has a constant acceleration on the vertical axis, and an initial velocity pointing upper right. The block dimension is 16x16 pixels and the search window 64x64 pixels.

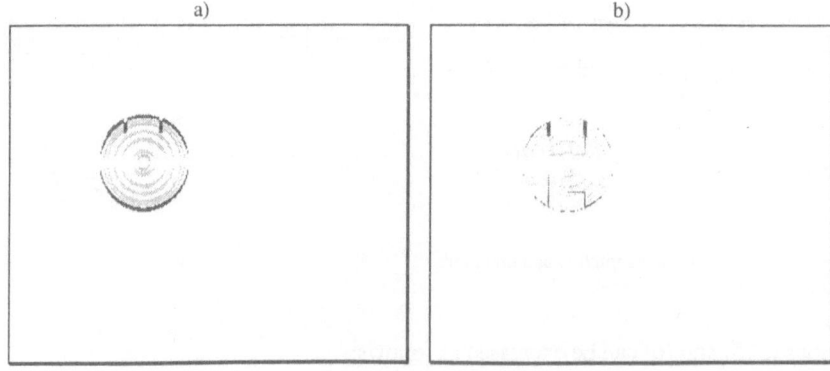

Figure 2. Displaced frame difference image; a) constant velocity method, b) constant acceleration method.

In figure 2 we present a comparison between two displaced frame difference images. These displaced frame difference images where obtained by estimating on a predicted frame three frames after the reference frame the position of each block and subtracting the resulting frame estimate from the actual data. Motion parameters were obtained using frames that were two frames apart from the reference frame. Figure 2 shows the results of such difference signal using both the constant velocity and constant acceleration method, respectively. The darker is the image the larger is the compensation error.

It can be observed that the constant acceleration model outperforms the constant velocity method in estimating the real motion of the object.

Figure 3 shows the tracking of acceleration on the sequence. We have predict every image of sequence using the approach presented in section 2.

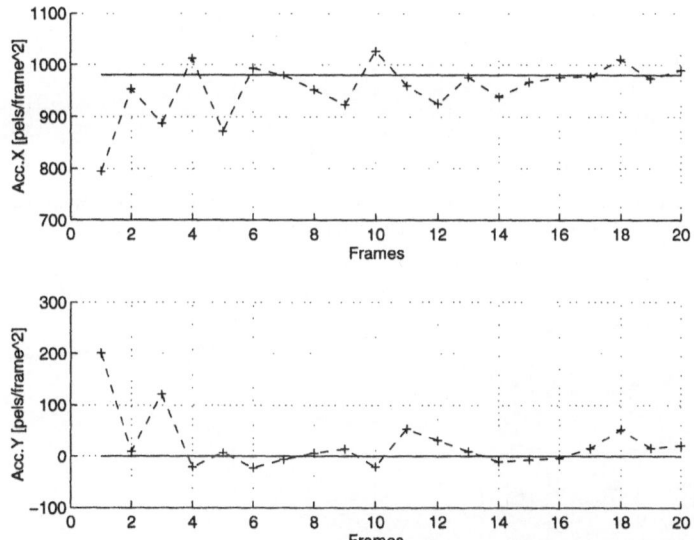

Figure 3. Acceleration tracking- continuous line: real acceleration; dotted line: acceleration estimated with the constant acceleration method.

The predicted acceleration performs well the real acceleration of the object, except for the first frames where the object is too near at the left side of the image which leads to improper displacement estimations.

Figure 4 presents the tracking of the velocity, tracking obtained using either the constant velocity method or the constant acceleration method. The constant acceleration method estimates very accurately the real velocity.

Figures 3 and 4 simply indicate the mean of the acceleration and velocity values for all the blocks which have a displacement estimate different from zero.

Figure 5 shows the displaced frame difference estimate for the sequence "Table Tennis" using either methods.

It can be noted that the ping pong ball, the only object which is less subject to the camera zooming, is must better identified from the constant acceleration method than the constant velocity method (in this case the error signal is larger at the ball location). The old estimation method identifies two balls, either in a wrong position.

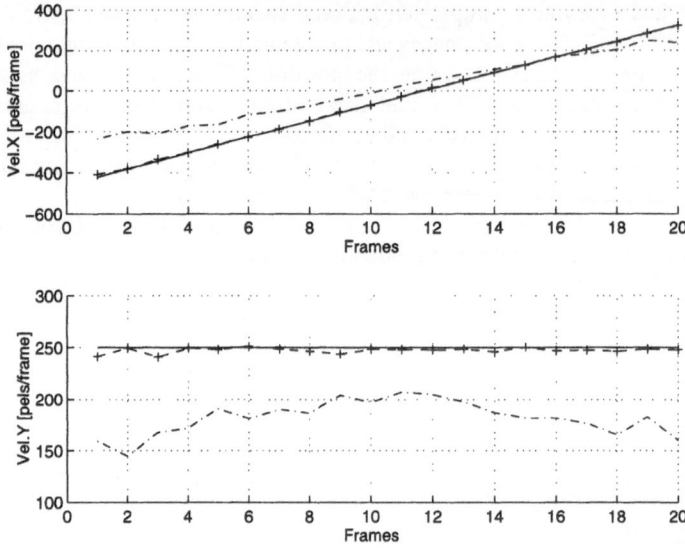

Figure 4. Velocity tracking- continuous line: real velocity; dotted line: velocity estimated with constant velocity method; point dotted line: velocity estimated with constant velocity method.

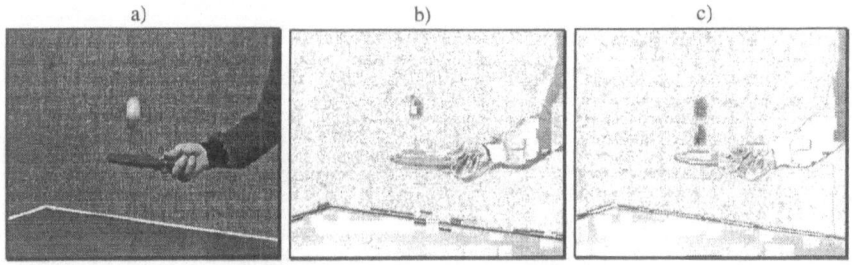

Figure 5. a) real image to predict; b) difference image with constant acceleration method; c) difference image with constant velocity method.

4 CONCLUSIONS

In this paper we have demonstrated that, in a sequence where some objects are subject to a constant acceleration, the knowledge of this parameter leads to substantial reductions of the compensation error with respect to linear methods.

The principal results are concerned with a good estimation of displacement vectors and a great and careful tracking of the velocity and positions of objects.

To improve this method we plan to regularize the acceleration and velocity estimates; we are also investigating how a fractional pel accuracy leads to better estimates of the motion models.

An important issue to analyze remains the estimation of the time interval during which a given constant acceleration model holds for a given block of pixels. This will enable to design proper motion predictive schemes that could be used in a video coding context.

We are also currently investigating how to handle the holes and overlaps problem, by estimating the motion of arbitrarily shaped regions rather than square blocks of pixels.

ACKNOWLEDGMENT

We thank Prof. Murat Kunt to let us use the facilities of the Signal Processing Laboratory at the Swiss Federal Institute of Technology, Lausanne, for performing some simulations presented in this work. We would like also to thank Mr. Markus Shutz and Dr. Touradj Ebrahimi for fruitful discussions regarding this work.

REFERENCES

1. J.K. Aggarwal and N. Nandhakumar, On the computation of motion from sequences of images-a review, *Proc. of the IEEE*, (1988).
2. M. Orchard, A comparison of techniques for estimating block motion in image sequence coding, *SPIE Visual Communications and Image Processing IV*, (1989).
3. M. Chahine and J. Konrad, Motion compensated interpolation using trajectories with acceleration, *Proc. of the Digital Video Compression: Algorithms and Technologies 1995 Conference*, 2419:152-163 (1995).

A LOGARITHMIC-TIME ADAPTIVE BLOCK-MATCHING ALGORITHM FOR ESTIMATING LARGE-DISPLACEMENT MOTION VECTORS*

Michael C. Chen and Alan N. Willson, Jr.

Department of Electrical Engineering
University of California, Los Angeles
Los Angeles, CA 90095-1600

INTRODUCTION

Since its inception in the early 1970s [1]-[3], the concept that image quality of temporal prediction can be enhanced with the help of motion information has had a substantial influence on the progress of image sequence coding. The use of motion estimation/compensation, which attempts to obtain knowledge of the displacements of moving objects in successive frames and apply it to remove temporal redundancy, has been suggested as one of the important reasons that current video coding schemes have achieved a level of performance that was widely considered impossible thirty years ago.

In general, block-matching algorithms, which estimate the amount of motion on a block-by-block basis, are widely adopted by most video systems and are recommended by several standards committees (i.e., ITU-T H.261(p×64), H.263, and ISO MPEG-1, MPEG-2) because of their simplicity and high efficiency. Among the various block-matching algorithms, the full search (FS) algorithm, which exhaustively searches every possible displaced candidate within a search area, is normally considered the optimal method[†]. However, the FS method demands a tremendous amount of computation, which may not be practical for real-time applications such as HDTV where the sample-rates are high and the search ranges are large, and for portable communication systems where low power consumption is an essential concern. For these reasons, there has been considerable interest in the development of efficient fast block-matching motion estimation techniques in order to relieve the tremendous computational overhead imposed by the FS algorithm. Most of the current fast algorithms can be broadly classified as the *pel-subsampling* (PS) methods and the *search candidate reduction* (SCR) methods. PS reduces computational complexity by limiting the number of pixels within each block in estimating motion vectors (MVs), whereas SCR achieves the goal by selectively checking only a small number of positions, under the assumption that the distortion measure decreases monotonically towards the best matching position.

Typically, the number of computations for the PS algorithms is proportional to the square of the search range, as opposed to the linear or logarithmic complexity of the SCR algorithms. Notice that the computational requirements are comparable when a small to medium search range

---

*This work was supported by the Office of Naval Research under Grant N00014-95-1-0231.

[†]In [4], we have shown that the FS method provides only sub-optimal performance for the motion-compensated transform coding system. The optimal solution can only be obtained by taking into account rate and distortion during the optimization process.

TABLE I. Computational complexity per 16×16 block for various algorithms.

Algo.	Order	$N = 8$ Op.	$N = 8$ α	$N = 16$ Op.	$N = 16$ α	$N = 32$ Op.	$N = 32$ α
FS	$O(N^2)$	16384	1.0	65536	1.0	262144	1.0
M41SUB	$O(N^2)$	3072	5.3	9216	7.1	33792	7.8
1DFS	$O(N)$	5120	3.2	11264	5.8	23552	11.1
LOG	$O(\log N)$	4352	3.8	6400	10.2	8448	31.0
Proposed	$O(\log N)$	6656	2.5	8704	7.5	10752	24.4

is desired. For the PS methods, Koga *et al.* [5] first introduced a subsampling scheme. Later, Liu and Zaccarin [6] proposed the modified 4-to-1 subsampling technique (M41SUB) with an $O(N^2)$ complexity (N is the number of search candidates in the horizontal or vertical dimension). For the SCR methods, Koga *et al.* [5] suggested the $O(\log N)$-time logarithmic search method (LOG) (also known as the three-step hierarchical search), and Chen *et al.* [7] developed the $O(N)$-complex one-dimensional full search algorithm (1DFS).

Traditionally, these fast algorithms were developed for slowly-varying video applications where a small to medium search range ($N = 8 \sim 16$) is adequate. However, as is evident by the fact that a higher interframe coding efficiency is obtainable with the use of large-size search windows, industrial standards (H.263, for instance) specify a search range of $[-16, 15]$ ($N = 32$) in order to ensure effective motion compensation for applications having many large-displacement MVs—those with fast-moving objects, quick camera motions, or low sampling rates.

To increase the search area without incurring an extensive computational effort, the LOG algorithm possesses a distinct advantage, as shown in Table I (α is defined here as the ratio between the number of operations of the FS and that of the various algorithms). However, the frequent failure of the monotonically increasing distortion (MID) assumption severely impairs the SCR method's capability to locate the best matching MVs. Thus, our efforts attempt to reduce the local minimum occurrences associated with the LOG algorithm by adaptively adjusting the search window size, and its search scheme, based on the temporal and spatial interblock motion correlation (IBMC). Although a $[-15, 15]$ search range is used in our discussion, the same principles can also be applied to a larger search range.

THE CONVENTIONAL LOGARITHMIC-TIME ALGORITHMS

Concerning block-matching algorithms, the current frame X_t is divided into blocks of size $k \times k$. For each block in the current frame, the MV $V_t(i, j)$ is obtained by finding the displaced block within a predefined search area Ω in the previous frame X_{t-1} such that

$$V_t(i, j) = (p', q') \quad : \quad \Phi(i, j, p', q') \leq \Phi(i, j, p, q) \quad \text{for all } (p, q) \in \Omega \tag{1}$$

$$\text{where} \quad \Phi(i, j, p, q) = \frac{1}{k^2} \sum_{m=0}^{k-1} \sum_{n=0}^{k-1} | X_t(i \times k + m, j \times k + n)$$
$$- X_{t-1}(i \times k + m + p, j \times k + n + q) | \tag{2}$$

$$\text{or} \quad \Phi(i, j, p, q) = \frac{1}{k^2} \sum_{m=0}^{k-1} \sum_{n=0}^{k-1} | X_t(i \times k + m, j \times k + n)$$
$$- X_{t-1}(i \times k + m + p, j \times k + n + q) |^2 \tag{3}$$

The LOG algorithm achieves significant computational savings by employing a coarse-to-fine search strategy, as shown in Fig. 1. In a square-shaped tracking window, the initial nine search positions are uniformly displaced with a stepsize of d, where $d = (w + 1)/4$ and w is the verti-

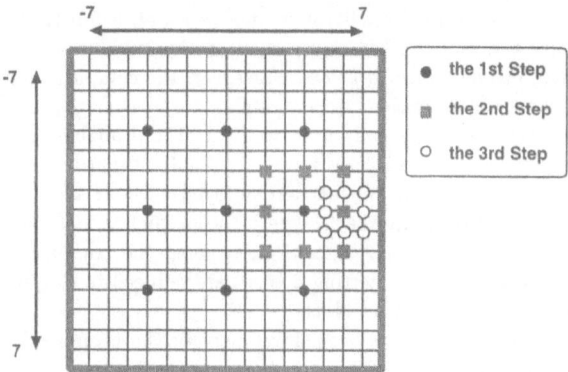

Fig. 1. An example of the LOG block-matching algorithm.

cal/horizontal search size. Among these initial positions, the best matching position is searched for and it becomes the search center for the next stage as the stepsize and the search window are halved. The process continues and the motion vector is finally determined when the search area converges to a single search candidate.

When a larger search range is employed, the FS method experiences a PSNR performance gain due to an increase of the search space. However, this performance improvement is not guaranteed for the LOG algorithm. Depending on the image sequence, the increasing local minimum problems that originate from the further displaced initial search positions sometimes dominate, causing a severe performance degradation. One way to alleviate this problem is to identify the region of interest based on the extracted motion characteristics, thereby reducing the effective search window size and its displacement accordingly. Adopting temporal interblock correlation is suggested by Chen *et al.* [8]. Unfortunately, the success of that approach relies heavily on a very high frame rate, which significantly limits its potential applications. Seferidis and Ghanbari [9] propose a texture analysis approach to derive the velocity of moving objects. Despite a satisfactory performance, the analysis is rather complex, involving histogram and normalization manipulations. It is obvious that a simpler yet comparably powerful technique is necessary.

INTERBLOCK MOTION CORRELATION

The spatial IBMC was found very useful in improving the performance of fast block-matching algorithms [6], [10]-[11]. Basically, the use of spatial IBMC can be viewed as a linear prediction process that manages to locate the current block's initial MV with the help of the adjacent blocks' MVs . Let $\tilde{V}_t(i,j)$, $\hat{V}_t(i,j)$, and $V_t(i,j)$ denote the initially predicted MV, the final estimated MV, and the mean-square-error-optimal MV of the i-th row and j-th column block at frame t, respectively. Let $a_t(p,q)$ represent the predictor coefficient of the adjacent MV $\hat{V}_t(i+p,j+q)$ and let $\hat{S}$ represent the prediction region of support. For a given set of final estimated MVs $\hat{\mathcal{V}}_t = \{\hat{V}_t(i+p,j+q) \mid (p,q) \in \hat{S}\}$, the initially predicted MV $\tilde{V}_t(i,j)$ can be formulated as

$$\tilde{V}_t(i,j) = \sum_{(p,q)\in\hat{S}} a_t(p,q)\hat{V}_t(i+p,j+q) \tag{4}$$

$$\text{subject to} \qquad \sum_{(p,q)\in\hat{S}} a_t(p,q) = 1. \tag{5}$$

Hence, the objective of the linear prediction is to find the coefficient set $\hat{A}_t = \{\hat{a}_t(p,q) \mid (p,q) \in \hat{S}\}$ such that the expected mean square difference between $V_t(i,j)$ and $\tilde{V}_t(i,j)$ is minimized. That is,

$$\hat{A}_t = \arg\min_{A_t} E\{|\, V_t(i,j) - \tilde{V}_t(i,j)\,|^2\}. \tag{6}$$

Clearly, obtaining the MAE-optimal predictor coefficient set $\mathcal{A}_t$ is not trivial. This soft-decision solution requires not only the motion field is assumed stationary but also the cross correlation between $\tilde{V}_t(i,j)$ and $V_t(i,j)$ to be known *a priori*. Moreover, the computational burden can easily make this approach an unattractive choice.

Since the linear prediction process is only used to locate the probable best-matching positions for the LOG search stage, it does not directly affect the final estimation result. Consequently, a less accurate but simpler prediction scheme can be employed without sacrificing much estimation accuracy. The hard-decision solution approach, where the predictor coefficients $a_t(p,q)_{(p,q)\in \hat{S}}$ are forced as either 0 or 1, is adopted in our algorithm, to keep the computational cost as low as possible. Therefore, the tasks involved in getting $a_t(p,q)$ are reduced to simply identifying the best motion-correlated block among neighboring blocks.

An efficient spatial IBMC measure has been developed for the proposed algorithm. The basic idea of this measure is to classify all blocks into nine categories, according to their directions of motion. Then the quantitative IBMC between each adjacent block pair can be obtained by comparing their motion. Since iterative procedures are avoided in our method, the linear prediction is restricted to causal adjacent blocks. In addition, only the four nearest blocks are chosen in our algorithm because the degree of IBMC usually decreases drastically with an increase of distance.

THE PROPOSED ALGORITHM

Provided that the best matching position can be appropriately predicted, narrowing the search range can alleviate the local minimum problems encountered in the LOG method. On the other hand, this could cause a severe performance degradation if the best matching position is excluded from the newly defined search area. Therefore, much caution is needed when the spatial IBMC is used to determine the initially predicted MVs and the corresponding search scheme.

In most video sequences the real motion field is smooth and slowly-varying. However, with occlusion regions and illumination changes present, the motion vectors estimated from block-matching algorithms can become very noisy. In addition, the sole use of causal blocks and their final estimated MVs is not sufficient to extract the existing spatial IBMC [12]. As a result, the motion vectors cannot always be predicted correctly both vertically and horizontally. To best use the information from the linear prediction process, the vertical and horizontal search regions, as well as search ranges, are determined *independently*.

Quantitative criteria incorporated with the temporal and spatial IBMC are developed to indicate the propriety in using the initially predicted MVs for each block. Our criteria contain three condition tests, namely: the spatial correlation test (SCT), the spatial MV test (SMT), and the temporal MV test (TMT). The SCT evaluates the degree of spatial correlation among the surrounding blocks, whereas the SMT and the TMT detect the spatial and temporal variation of motion characteristics, respectively. The vertical and horizontal dimensions are evaluated independently. Only if all three conditions are met is the search range in the corresponding dimension reduced.

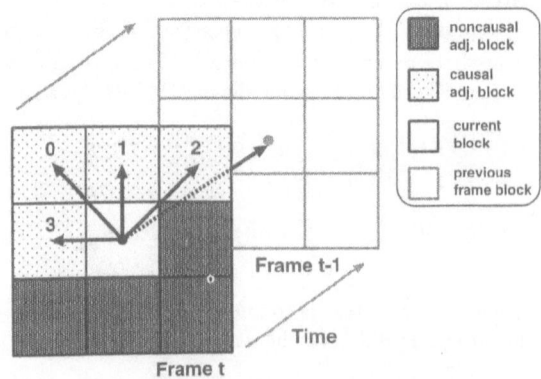

Fig. 2. An illustration of the spatial and temporal interblock correlation.

Spatial Correlation Test

As shown in Fig. 2, each of the four causal blocks is assigned an individual number to ease procedure descriptions. For a given pair $(i,j) \in \{(0,1), (1,2), (0,3), (1,3)\}$, the SCT estimates the spatial IBMC of all pairs, based on the spatial MV distance measure (SMDM), defined as

$$SMDM_{i,j} = (m_i - m_j)^2 \tag{7}$$

where m_i and m_j are the x-components (or y-components) of the i-th block's and j-th block's final estimated MV, respectively.

If $SMDM_{i,j}$ is smaller than or equal to T_o, two blocks are considered as motion correlated in the testing dimension[‡]. Based on experimental evidence we have set the SCT criterion to require that two or more pairs are found to be correlated. It is then very likely that the blocks belong to the same rigid body.

Spatial MV Test

The motion of each block is classified into one of nine directions [12]. To evaluate the accuracy of the initially predicted MVs, the SMT compares the directions of motion of the current block and the best correlated block using horizontal and vertical SMT measures. They are given by

$$HSMT = (x_{cur} - x_{best})^2 \tag{8}$$
$$VSMT = (y_{cur} - y_{best})^2 \tag{9}$$

where (x_{cur}, y_{cur}) and (x_{best}, y_{best}) represent the directions of motion of the current block and the best correlated block, respectively. If $HSMT = 0$ $(VSMT = 0)$, the initially predicted MV is likely to correctly reflect correlated horizontal (vertical) motion. Thus, the test is passed in the horizontal (vertical) direction.

Temporal MV Test

Based on the assumption that most MVs change smoothly with time, the TMT determines the appropriateness of the initially predicted MV $\tilde{V}_t(i,j)$ by comparing it with the final estimated MV $\hat{V}_{t-1}(i,j)$.

The temporal measure (TDM) is given by

$$TDM = (\tilde{m}_t - \hat{m}_{t-1})^2 \tag{10}$$

where $\tilde{m}_t$ and $\hat{m}_{t-1}$ are the x-components/y-components of $\tilde{V}_t(i,j)$ and $\hat{V}_{t-1}(i,j)$, respectively. The test is passed if TDM is smaller than or equal to T_t (T_t is usually set to 16 to allow a small-scale temporal motion variation).

Adaptive Window Size Search Scheme

To maintain the simplicity of the search procedures, the newly defined search range for each dimension is chosen as either [-15,15] or [-7, 7] (with an associated displacement of 8 and 4 pixels, respectively). Regardless of the search range decisions, the number of search steps is fixed at four, which is sufficient to permit all search-plane sizes to converge to a single position. Since only three steps are required for a [-7, 7] search range to converge, an additional step is available if a reduced search range, i.e., [-7, 7], is chosen. The questions here are how and when to utilize this search step such that the estimation performance can be further enhanced without greatly affecting search regularity. Statistically, the best matching positions tend to be close to the initially predicted MVs. Moreover, due to the hierarchical search characteristics of the LOG, the early-stage search decisions usually influence the final search performance the most. Based on these observations, the additional search step is allocated to the second step and the positions adjacent to the initially predicted MVs are checked. As a variation from the conventional LOG method, the search center of

---

[‡]According to our simulations, the algorithm performs best when T_o is set to 4, which takes into account noise effects from estimation inaccuracy.

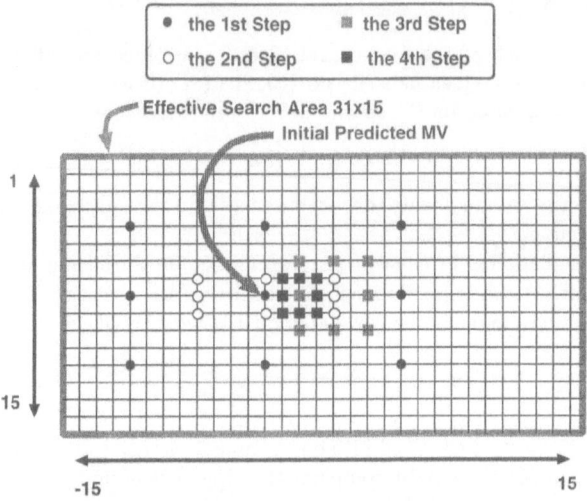

Fig. 3. An example of the adaptive window size LOG algorithm.

the third step is selected by comparing the best matching positions from the first and second steps. An example of the proposed search scheme is illustrated in Fig. 3, where the initial horizontal and vertical displacements are 8 and 4 pixels, respectively. As can be seen, in the second step the search procedure takes advantage of available vertical IBMC by assigning more checking positions close to the initially predicted MV vertically while the search procedures in the horizontal dimension are kept the same as in the conventional LOG.

Proposed Algorithm Summary

Our proposed algorithm consists of three main stages, namely: linear prediction, window size adjustment, and hierarchical search. It can be summarized as follows:

- **Linear Prediction** For each block, the spatial IBMC extraction process determines the direction of motion by finding the best matching position among nine uniformly displaced search positions with a search center at $(0,0)$ and a displacement size of 4 pixels. Then the linear prediction is performed by comparing the direction of motion of the current block with those of the four causal adjacent blocks. The one with the closest direction of motion is chosen as the best correlated block and its final estimated MV is used as the initially predicted MV for the current block.

- **Window Size Adjustment** The SCT, SMT and TMT criteria are used to determine the search window size. If a block meets all three criteria, its initially predicted MV is considered reliable. The search range, therefore, is reduced by half and the search procedure is conducted in a more condensed manner.

- **Hierarchical Search** With the search range determined in the window size adjustment stage, the search procedure, centered at the initially predicted MV, proceeds in a coarse-to-fine manner as described above. The final estimated MV is determined as the search area is reduced to a single candidate.

PERFORMANCE COMPARISONS AND EXPERIMENTAL RESULTS

To test the performance of the proposed method, simulations were run on four test video sequences: the Susie, Trevor, and football sequences, with a format of 256×256 pixel/frame, and the flower garden sequence with a frame size of 720×480. A sampling rate of 10 frame/sec was

TABLE II. Comparisons of the PSNR performance for various algorithms.

Algo.	Search Range	Susie PSNR	Trevor PSNR	garden PSNR	football PSNR
FS	$[-15, 15]$	31.99	31.74	20.82	20.24
FS	$[-7, 7]$	31.04	31.54	17.12	19.01
Proposed	$[-15, 15]$	31.68	31.51	20.41	19.68
LOG	$[-15, 15]$	29.98	30.83	18.72	19.14
LOG	$[-7, 7]$	30.12	31.09	16.55	18.53

chosen. A 16×16 block size and a maximum displacement of 15 pixels in both the horizontal and vertical directions were employed.

Table II compares our approach with the FS and LOG algorithms, in terms of average PSNR. To facilitate our discussion, the experimental results for a 15×15 search range are also included. In contrast to the FS, the LOG algorithm fails to improve consistently with a larger search range. It, in fact, has a performance degradation for the two videophone sequences (Susie and Trevor), which contain a standard slow-moving talking head-and-shoulder scene with a still background. The proposed method, however, is able to deliver an excellent estimation, which substantially outperforms the LOG method both in the case of the $[-7, 7]$ and the $[-15, 15]$ search range.

Fig. 4 shows the PSNR performance of various algorithms on a frame-by-frame basis for the 150-frame Susie sequence. As can be seen, the $[-7, 7]$ LOG method experiences a huge degradation as an object undergoes a large motion (from frame 35 to frame 80), whereas the performance of the $[-15, 15]$ LOG approach declines when the motion is relatively small (from frame 100 to frame 140). Unlike LOG, the performance of the proposed method closely follows the FS. It constantly possesses a significant margin over the two LOGs for almost all frames. This indicates that our approach is capable of automatically adjusting its search range and search scheme as the motion characteristics in the video sequence vary. Hence, it is less sensitive to the image characteristics in the video sequence.

CONCLUSION

In this paper, we have shown that a large search range, which has rarely been effectively tackled by most of the current fast algorithms, is necessary to achieve a better coding efficiency for MPEG or videophone sequences. Despite its advantage of computational simplicity, the conventional LOG

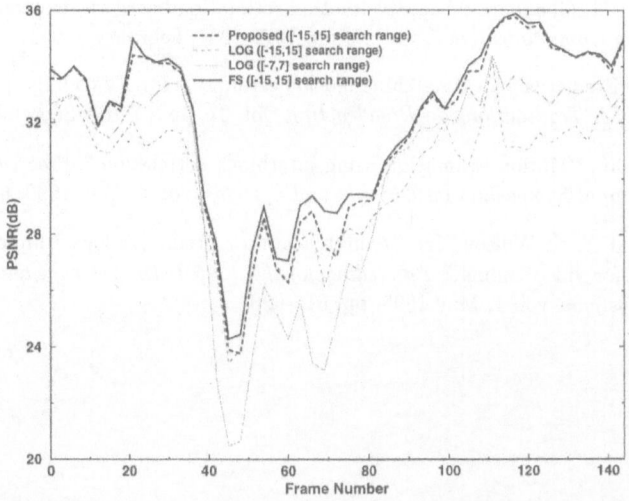

Fig. 4. The PSNR performance of the motion-compensated frames for the Susie sequence.

algorithm is not sufficiently effective to provide performance gains in estimating the MVs, due to the deteriorating condition of the MID hypothesis. Without resorting to a linear, or even higher, computational complexity, the proposed algorithm exhibits a performance relatively close to the FS method, while only requiring a regular search scheme and a level of computations comparable to the LOG algorithm.

REFERENCES

[1] F. Rocca and S. Zanoletti, "Bandwidth reduction via movement compensation on a model of the random video process," *IEEE Transactions on Communications,* vol. COM-20, pp. 960-965, October 1972.

[2] J. O. Limb and J. A. Murphy, "Measuring the speed of moving objects from television signals," *IEEE Transactions on Communications,* vol. COM-26, pp. 573-578, April 1975.

[3] A. N. Netravali and J. D. Robbins, "Motion-compensated television coding: Part I," *Bell System Technical Journal,* vol. 58, pp. 631-670, March 1979.

[4] M. C. Chen and A. N. Willson, Jr., "Rate-distortion optimal motion estimation algorithm for video coding," to appear in *Proceedings of the 1996 IEEE International Conference on Acoustics, Speech and Signal Processing,* May 1996.

[5] T. Koga *et al.,* "Motion compensated interframe coding for video conferencing," *Proceedings of the 1981 IEEE National Telecommunications Conference,* vol. 4, November 1981, pp. G5.3.1-5.3.5.

[6] B. Liu and A. Zaccarin, "New fast algorithms for the estimation of block motion vectors," *IEEE Transactions on Circuits and Systems for Video Technology,* vol. 3, pp. 148-157, April 1993.

[7] M.-J. Chen and L.-G. Chen, "One-dimensional full search motion estimation algorithm for video coding," *IEEE Transactions on Circuits and Systems for Video Technology,* vol. 4, pp. 504-509, October 1994.

[8] L.-G. Chen *et al.,* "A predictive parallel motion estimation algorithm for digital image processing," *Proceedings of the 1991 IEEE International Conference on Computer Design,* October 1991, pp. 617-620.

[9] V. Seferidis and M. Ghanbari, "Adaptive motion estimation based on texture analysis," *IEEE Transactions on Communications.,* vol. 42, pp. 1277-1287, February 1994.

[10] S. Zafar *et al.,* "Predictive block-matching motion estimation for TV coding - part I: inter-block prediction," *IEEE Transactions on Broadcasting,* vol. 37, pp. 97-101, September 1991.

[11] C. H. Hsieh *et al.,* "Motion estimation using interblock correlation," *Proceedings of the 1990 IEEE International Symposium on Circuits and Systems,* vol. 2, May 1990, pp. 995-998.

[12] M. C. Chen and A. N. Willson, Jr., "A high accuracy predictive logarithmic motion estimation algorithm for video coding," *Proceedings of the 1995 IEEE International Symposium on Circuits and Systems,* vol. 1, May 1995, pp. 617-620.

PRUNED TREE-STRUCTURED VECTOR QUANTIZATION IN THE HIGHER FREQUENCY VIDEO SUBBANDS

Chin-Hua Hsu[1] and Robert E. Van Dyck[2]

[1]Center for Computer Aids for Industrial Productivity
Rutgers University, Piscataway, NJ 08855
[2]GEC-Marconi Electronic Systems Co.
Wayne, NJ 07474

Abstract

Entropy-Constrained Vector Quantization (ECVQ) has been used as an effective method for coding the higher spatial frequency subbands in a wavelet video coding system. To achieve good distortion-rate performance, some of the subbands require a bit rate over one bit/pixel. For four-dimensional, full-searched ECVQ, this translates into over 800 codevectors/subband. In this paper, the full-searched ECVQ is replaced by Entropy-Pruned Tree-Structured Vector Quantization (EPTSVQ), and performance evaluations are made. The number of operations required for the quantization of the higher frequency subbands (HFS) is reduced by factors of 25 to 50, depending on the subband. Performance of error concealment methods for ATM transmission are also examined.

1   Present System Overview

For our present wavelet video coding system [1], the luminance component is decomposed into seven subbands and each chrominance component is decomposed into four subbands, as shown in Figure 1. A slightly modified version of the H.261 algorithm is run on the lowest frequency luminance and chrominance subbands. The other subbands are ECVQ [2] coded. The motion vectors are only computed for the lowest frequency luminance subband, $Ybll$, and these vectors are then used by all subbands that are predictively coded ($Ybll$, $Crll$, $Cbll$, and $Yblh$).

The system provides good quality video for moderate motion, almost CIF sized images at bit rates under 500 kb/s, and for teleconferencing scenes at bit rates from 128 to 180 kb/s. Frames can be either intraframe or interframe coded. For intraframes, all subbands are coded without regard to any previous frames. For interframes, four subbands are predictively coded (as mentioned above) using the previous frame, and the remaining 11 subbands are intraframe coded. Again, the quantization of the HFS is ECVQ, but subband $Yblh$ can also be ECTCQ [3] coded to give increased performance.

2   Pruned Tree-Structured VQ

Despite the good distortion-rate performance of full-search ECVQ, the computational complexity makes this method unattractive for real-time implementation. Further studies were undertaken to determine the bit rate increase necessary to yield similar quality video using EPTSVQ [4].

Fifteen Subband System

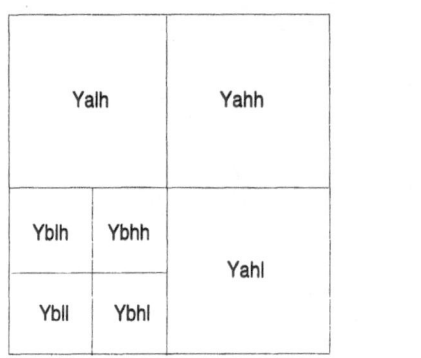
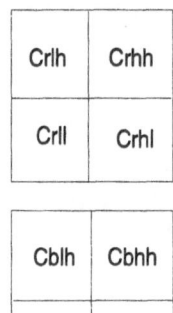
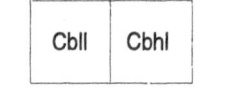

Figure 1: Frequency Domain Partition for the Fifteen Subband Decomposition.

Separate EPTSVQ codebooks were designed for all of the higher frequency luminance and chrominance subbands. A training sequence, consisting of 61 frames from each of the four image sequences "Flower Garden," "Ferris Wheel," "Table Tennis," and "Dave," was decomposed into its fifteen subbands. The smaller subbands had 1,093,120 pixels and the larger subbands had 4,372,480 pixels. The pixels in each subband were grouped into 2×2 blocks, creating 4-dimensional vectors, which became the training sequence for that subband. Different training sequences were used to design codebooks for the motion-compensated (M.C.) predicted subbands. For these bands, the M.C. frame differences were computed, and the differences between the actual and predicted pixel values were used for the training sequences.

2-1 TSVQ Codebook Design

The actual design of the codebooks was accomplished in two steps. The first step was the design of a Tree-Structured Vector Quantizer (TSVQ). After some initial experimental designs, it was decided that a TSVQ with eleven levels (2048 codevectors) was sufficient. This is mainly due to the fact that eleven levels gets the distortion down to near the same values as in the ECVQ codebooks designed previously. Also, the distortion-rate curve for a thirteen level TSVQ was not any better. Although there was a large number of samples in the training sequence, the thirteen level quantizer had a number of leaf nodes with no training vectors.

For each higher frequency subband, given a training set T_1 with K-dimensional vectors, a binary tree-structured VQ codebook C of depth L was generated by the algorithm given in [5, p. 410]. The subsets of the training set reaching each node of the tree are denoted by $T_2, T_3, \cdots, T_{2^{L+1}-1}$, where the nodes are labeled as in Figure 2. The number of training vectors in each node t is given by $|T_t|$, and $|T_1| = N$. For a given depth, the codewords are $C(1), C(2), \cdots, C(2^{L+1} - 1)$.

Instead of growing the tree one node at a time as advocated in [6], we grew the tree one level at a time. Starting with the centroid of the training sequence, splitting using $\pm\epsilon$ was done to obtain two codevectors. Each vector in the training sequence was then mapped to the closest codevector giving training sequences T_2 and T_3. The centroids of each of these sequences were computed to get a level 1 TSVQ. The process of splitting, mapping the training sequence to the closest codevector, and computing the centroids, was repeated until the desired number of levels was reached.

2-2 Entropy Pruning

The second step in the codebook design was pruning the TSVQ using the entropy at each node. Starting with the complete tree, the number of training vectors, $n(t)$, and the distortion, $d(t)$,

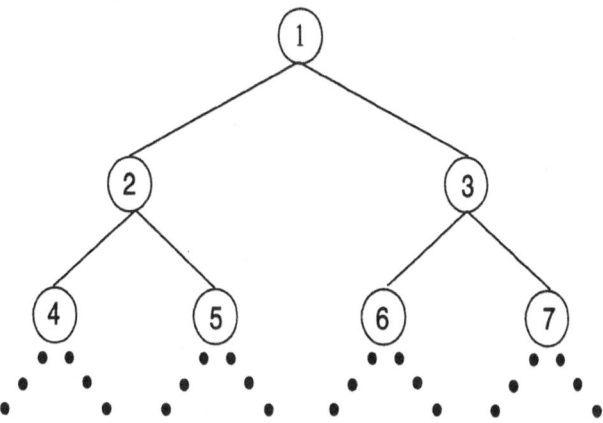

Figure 2: Node Labeling in the Binary Tree.

for each node t are calculated according to

$$n(t) = |T_t| \tag{1}$$

$$d(t) = \frac{1}{n(t)} \sum_{i=0}^{n(t)-1} \|\mathbf{x}_i - \mathbf{c}(t)\|^2. \tag{2}$$

The $\mathbf{x}_i$ are the training vectors and $\mathbf{c}(t)$ is the centroid (codevector) for the node, and $\|\cdot\|$ is the Euclidean norm divided by the vector dimension K. The parameters $\lambda(t) \leftarrow 0$ for all $2^L - 1$ interior nodes in the tree.

The entropy pruning algorithm [4] has the following steps:

Step 1. (Calculation of λ): For all interior nodes $t = 1, \cdots, 2^L - 1$

$$\Delta d(t) = (d(2t)n(2t) + d(2t+1)n(2t+1) - d(t)n(t))/N \tag{3}$$

$$\Delta r(t) = n(t)/N \tag{4}$$

$$\lambda(t) = -\frac{\Delta d(t)}{\Delta r(t)}. \tag{5}$$

Step 2. (Pruning): Set $\lambda_{min} = \infty$.
    while $((2t)$ and $(2t+1)$ are leaves)
      if $\lambda(t) < \lambda_{min}$
        $\lambda_{min} = \lambda(t)$
        $i = t$
    delete $C(2i)$ and $C(2i+1)$.

Step 3: Repeat Step 2. until the desired number of codevectors is reached. The distortion and rate of the pruned tree is stored as well as the codevectors.

Figure 3 shows the distortion-rate curves for subband $Yblh$ (intraframe coded). Similar distortion-rate curves were designed for the remaining subbands. One can see that the distortion-rate curve for the EPTSVQ falls between the curve for the full-searched VQ designed using the Generalized Lloyd algorithm and the Entropy-Constrained VQ. At high rates, the TSVQ and the EPTSVQ have similar performance, and meet at the distortion-rate point of the fully populated eleven level tree. However, the latter has better performance as the bit rate is reduced. Since this subband is typically allocated 1 to 1.3 bits/pixel in our application, the difference is significant.

The generalized BFOS algorithm is employed to solve the bit allocation problem [7]. For the MPEG "Bicycles" sequence coded at 10 frames/sec and 450 kb/s, approximately 1 bit/pixel

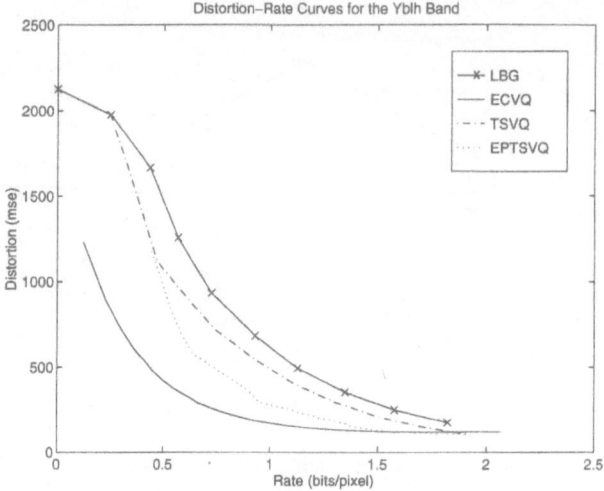

Figure 3: Distortion-Rate Curves for subband Yblh.

is needed for ECVQ coding of subbands $Yblh$ and $Ybhl$. The EPTSVQ bit allocation leads to less than a 50 percent increase in distortion for these perceptually important subbands, as well as greater distortion in the HFS.

3 Simulation Results

3-1 Comparison with ECVQ

Figure 4 shows the bit rate and peak signal-to-noise ratio (PSNR) for EPTSVQ and ECVQ, with H.261 coding of the lowest frequency subband (LFS); twenty frames of the "Bicycles" sequence from the MPEG test sequences are used. The results for ECVQ coding are taken from [1]. The image size is 352×224 pixels, and the frame rate is 10/sec. The lower frame rate is achieved by coding every third frame. The coded frames are divided up into blocks of ten. The first frame in each block is intraframe coded, and the following nine frames are coded using M.C. The average bit rate is approximately 470 kb/s.

These figures show that for similar bit rates, EPTSVQ can come within 1 to 1.5 dB in PSNR. Since the perceptual quality is mainly determined by the LFS which is H.261 coded, the perceptual quality is similar (although slightly degraded). The "Miss America" sequence was coded by both systems, and the relative difference in quality also holds. The bit rates and PSNRs are shown in 5. The image size of this sequence is 352×288 pixels, and the bit rate average about 150 kb/s.

3-2 Higher Frame Rate Video

Of particular interest to us is the performance of the algorithm for both infra-red and color video at 15 frames/sec. The higher frame rate should lead to better motion-compensated prediction for the LFS and subband $Yblh$. The infra-red imagery is compressed by treating it as a luminance image and using the seven luminance subbands for the decomposition. The same EPTSVQ codevectors, designed using the color test sequence discussed above, are used.

Figure 6 shows the bit rates and PSNRs for the infra-red sequence, "IR 1". This sequence contains a moving train in a railroad yard. Four different simulations were run, with average bit rates of 254 kb/s, 269 kb/s, 315 kb/s, and 427 kb/s. The image size is 256×192 pixels. The images were created by grabbing NTSC video at 15 frames/sec with a TARGA board to get 512×400 pixel frames. Each frame was de-interlaced by taking the even field, and horizontally

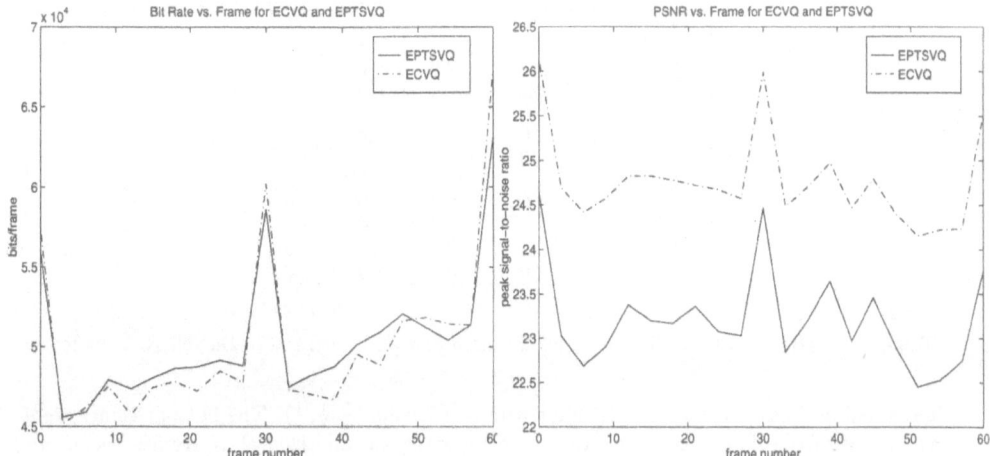

Figure 4: Bit rate and PSNR as a function of frame number for the H.261/ECVQ and H.261/EPTSVQ coders operating at 10 frames/sec. "Bicycles" sequence.

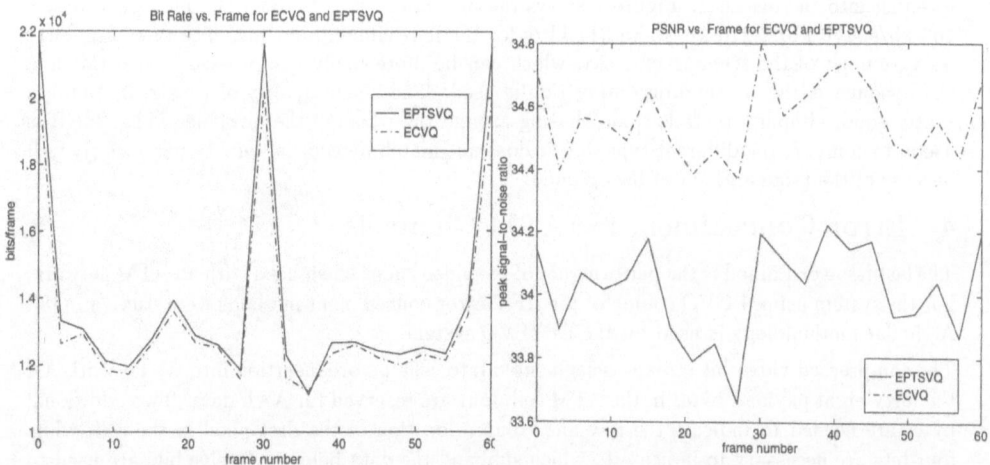

Figure 5: Bit rate and PSNR as a function of frame number for the H.261/ECVQ and H.261/EPTSVQ coders operating at 10 frames/sec. "Miss America" sequence.

	Frame 0		Frame 3	
Subband	Bits	Cells	Bits	Cells
$Ybll$	23,173	73	19,103	60
$Crll$	1,295	5	1,156	4
$Cbll$	1,828	6	1,817	6
$Yblh$	8,015	26	7,889	25
$Ybhl$	7,870	25	6,753	22
$Ybhh$	3,843	13	0	0
$Yalh$	9,922	32	8,364	27
$Yahl$	0	0	0	0
$Yahh$	0	0	0	0
$Crlh$	0	0	0	0
$Crhl$	0	0	0	0
$Crhh$	0	0	0	0
$Cblh$	0	0	0	0
$Cbhl$	0	0	0	0
$Cbhh$	0	0	0	0

Table 1: Bits/frame and Cells/frame for an intra and an interframe. "Bicycles" sequence.

filtered and decimated using the FIR filter with coefficients $(\frac{1}{4}, \frac{1}{2}, \frac{1}{4})$. The bottom 8 lines were then cropped from the image. Since the input sequence was digitized at 15 frames/sec, every frame is coded.

Notice that there is a scene change between frames 0 and 1. Despite this change occurring during a predicted frame, the quality of the frame is still good. The algorithm allocates more bits to this frame than to other predicted frames. The subjective quality of simulation runs 2, 3, and 4 is quite good. Run one, having the lowest bit rate, suffers from some blocking distortion. Run 2 changes the H.261 quantization parameter to allocate more bits to the LFS, especially during intraframes. The total bit rate is similar, but the perceptual quality is significantly better.

The last sequence tested is taken from the movie "Top Gun." It was created using the same procedure used for the "IR 1" sequence. The sequence shows two airplanes making a turn and receding into the distance. Figure 7 shows the bits rates and PSNRs. The bit rates average 163 kb/s, 176 kb/s, 210 kb/s and 311 kb/s for the four simulation runs. The rates are lower because most of the scene is blue sky, which can be more easily compressed. Since this is a color sequence, the 15 band decomposition is used. The image quality of runs 2, 3, and 4 is pretty good, although there is some aliasing around the wings of the airplanes. The PSNR is useful to compare the different runs, but it does not match perceptual quality particularly well because of the large amount of background.

4 Error Concealment for ATM Channels

The last issue examined is the performance of the video coder when used with an ATM network. For the system using ECVQ coding of the HFS, error concealment methods were studied in [8]. A similar methodology is used for the EPTSVQ system.

The compressed video bit stream is segmented into cells before insertion into the network. Of the forty-eight payload bytes in the ATM cell, four are reserved for AAL data. Two additional bytes are needed to indicate the row and column location of the first pixel in the cell, while four bits are necessary to indicate to which subband the data belong. Twelve bits are used to provide a sequence number for the cells, leaving 40 bytes for the video data. Table 1 shows the number of bits and ATM cells necessary to transfer two typical frames from the "Bicycles" sequence.

Cell losses usually occur in the network due to a lack of buffer space in the switches or congestion on a particular link. These problems manifest themselves to a video application as bursts of lost cells, rather than as random cell losses. When cell loss has been detected by the receiver,

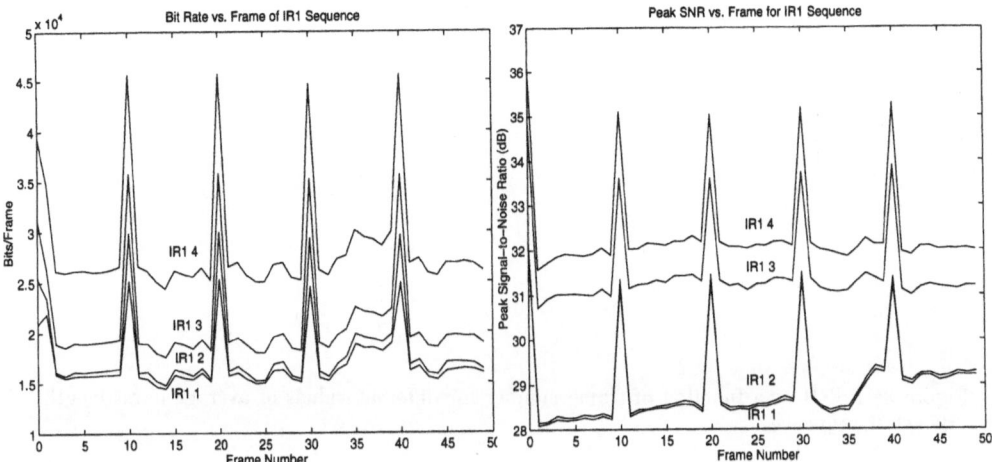

Figure 6: Bit rate and PSNR as a function of frame number for the H.261/EPTSVQ coder operating at 15 frames/sec on the "IR 1" sequence.

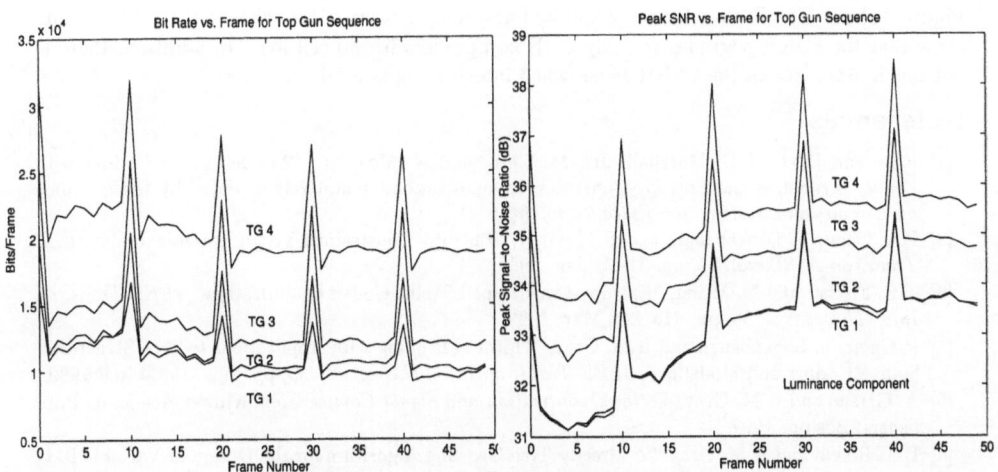

Figure 7: Bit rate and PSNR as a function of frame number for the H.261/EPTSVQ coder operating at 15 frames/s on the "Top Gun" sequence.

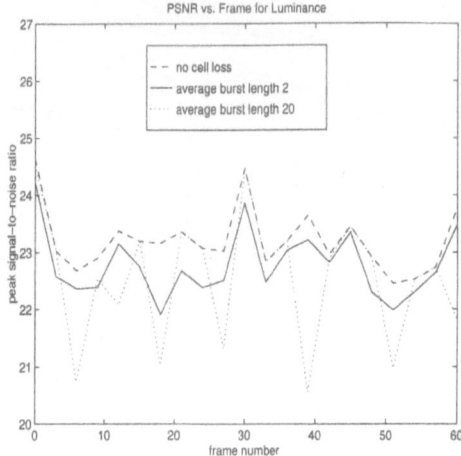

Figure 8: PSNR as a function of frame number for different values of average burst length. "Bicycles" sequence.

concealment techniques can be used to replace the missing cells with other data in order to minimize the degradation of the video.

In our scenario, the LFS and the motion vectors have high priority, and they are assumed to have no ATM cell loss. The HFS have low priority and may suffer from cell losses as high as ten percent. Gilbert's model [9] is used to simulate these channel losses. The version of the model used is a two state Markov chain where each transition corresponds to the transmission (or loss) of a single cell. By properly choosing the transition probabilities, the desired burst length and average cell loss rate are specified.

Since the HFS exhibit little interframe correlation, the missing cells are set to zero. Simulations with average burst lengths of 2, 5, 10, and 20 cells were run for a 10 percent cell loss rate. Figure 8 shows PSNR plots for the 2 and 20 burst length cases, along with no loss. The results show that the PSNR degrades roughly 2 dB compared with no cell loss. In addition, there is not much difference in the PSNR sense when interleaving is used.

References

[1] R.E. Van Dyck, T.G. Marshall, Jr., M. Chin, and N. Moayeri, "Wavelet Video Coding with Ladder Structures and Entropy-Constrained Quantization," tentatively accepted by *IEEE Trans. on Circuits and Systems for Video Technology*.

[2] P.A. Chou, T. Lookabaugh, and R.M. Gray, "Entropy-Constrained Vector Quantization," *IEEE Trans. on ASSP*, vol. 37, pp. 31-42, Jan. 1989.

[3] T.R. Fischer and M. Wang, "Entropy-Constrained Trellis Coded Quantization," *IEEE Trans. on Info. Theory*, vol. 38, pp. 415-426, Mar. 1992.

[4] P. Chou, T. Lookabaugh and R.M. Gray, "Optimal Pruning with Applications to Tree-Structured Source Coding and Modeling," *IEEE Trans. on Info. Theory*, vol. 35, pp. 299-315, March 1989.

[5] A. Gersho and R.M. Gray, *Vector Quantization and Signal Compression*, Kluwer Academic Publishers, Boston, 1992.

[6] E.A. Riskin and R.M. Gray, "A Greedy Tree Growing Algorithm for the Design of Variable Rate Vector Quantizers," *IEEE Trans. on Signal Processing*, vol. 39, pp. 2500-2507, Nov. 1991.

[7] E.A. Riskin, "Optimal Bit Allocation via the Generalized BFOS Algorithm," *IEEE Trans. on Info. Theory*, vol. 37, pp. 400-402, March 1991.

[8] R.E. Van Dyck, C.-H. Hsu, and C.-F. Hsu, "A Two-Layer Subband/H.261 Video Coder for ATM Networks," *Proc. Sixth International Workshop on Packet Video*, pp. A3.1-A3.4, Sept. 1994.

[9] E.N. Gilbert, "Capacity of a Burst-Noise Channel," *Bell Syst. Tech. J.*, vol. 39, pp. 1253-1266, Sept. 1960.

A CELLULAR CONNECTIONIST ARCHITECTURE FOR CLUSTERING-BASED ADAPTIVE QUANTIZATION WITH APPLICATION TO VIDEO CODING*

Lulin Chen, Chang Wen Chen and Jiebo Luo

Department of Electrical Engineering
University of Rochester
Rochester, NY 14627-0231

ABSTRACT

This paper presents a novel cellular connectionist model for a clustering-based adaptive quantization for video coding applications. The adaptive quantization is designed for a wavelet-based video coding system with the desired scene adaptive and signal adaptive quantization because of its ability to differentiate whether a specific coefficient is part of the scene structure through Gibbs random field constraints. The adaptive quantization is accomplished through an MAP estimation based clustering process whose massive computation of neighborhood constraints makes it difficult for a real-time implementation of video coding applications. The proposed cellular connectionist model aims at designing an architecture for the real-time implementation of the clustering-based adaptive quantization. With a cellular neural network architecture mapping onto the image domain, the powerful Gibbs spatial constraints are realized through interactions among neurons connected with their neighbors. In addition, the computation of coefficient distribution is designed as an external input to each component of a neuron or processing element (PE). We prove that the proposed cellular neural network does converge to the desired steady state with proposed update scheme. This model also provides a general architecture for image processing tasks with Gibbs spatial constraint-based computations.

INTRODUCTION

We describe in this paper a novel cellular connectionist model for the clustering-based adaptive quantization for video coding applications. The adaptive quantization is an innovative and integral component in a wavelet-based video coding system[1, 2]. The most important feature of this adaptive quantization scheme is its ability to quantize a given wavelet coefficient according to not only its own magnitude, but also the magnitudes of its

* This research is supported by NSF Grant EEC-92–09615 and a New York State Science and Technology Foundation Grant to the Center for Electronic Imaging Systems at the University of Rochester.

neighbors. Such constrained quantization is accomplished through the incorporation of the Gibbs random field model and is capable of achieving the scene adaptive quantization. That is, the available bits can be allocated to the prominent scene structures rather than some isolated coefficients whose perceptual importance is negligible. Such adaptive quantization has been shown to perform better than traditional quantization schemes in low bit rate image and video coding because of its simultaneous signal adaptive and scene adaptive capability[1, 2]. The success of the adaptive quantization is due largely to the incorporation of the Gibbs random field neighborhood constraints. However, the massive computation of the neighborhood constraints involved in the MAP estimation would generally need tremendous computational power, which, in general, limits the real-time implementation of the adaptive quantization algorithm to video coding applications.

There have been many attempts to solve the problems of image processing using connectionist architectures to achieve the real-time implementation potential. Usually, the problems are mapped onto one of the well-known models, such as Hopfield full-connection model[3], Kohonen self-organization model[4] or multi-layered perceptron with error-backpropagation algorithm[5]. The cellular connectionist model or cellular neural network[6] has recently been widely used because of its easy hardware implementation. The cellular connectionist architecture proposed here is intended to relief the extensive computational requirement needed for the proposed constrained clustering-based adaptive quantization in video coding application. With a cellular neural network structure mapping onto the image domain, the powerful Gibbs spatial constraints are realized through instantaneous interactions among neurons connected with their neighbors. Such network model is structurally similar to the cellular network models proposed by Chua and Yang[6], although the processing elements designed in this paper are functionally more versatile that those used by Chua and Yang[6]. In addition, the conditional density computation is designed as an external input to each component of a neuron or processing element (PE). We prove that the proposed cellular neural network does converge to the desired steady state with suggested update scheme. This model gives rise to a general architecture for image processing with spatial constraint-based computations.

THE ADAPTIVE QUANTIZATION ALGORITHM

The adaptive quantization that we intend to implement with cellular connectionist architecture is an integral part of a low bit rate wavelet video coding scheme proposed recently by Luo et al.[1, 2]. The overall video coding scheme is shown in Figure 1. The input video signal is first taken two frames at a time and passed through the two-tap Haar filterbanks which will result in two temporal subbands: HPT (high-pass temporal) and LPT (low-pass temporal). The LPT subband will further be decomposed into seven subbands (two levels) and the HPT subband will be decomposed into four subbands (one level) using the Doubechie's 9/7 filterbanks. These decomposed subbands will be then be quantized, coded and transmitted over a communication channel. At the receiving end, these subbands will then be decoded and reconstructed using the corresponding filterbanks to obtain the reconstructed video frames. This general scheme of video coding can be applied to many image communication applications, including multimedia communication, HDTV, videoconferencing, and videophone. In many such applications, the communication bandwidth is usually limited, therefore, low bit rate video coding is required for a successful transmission. To meet the low bit rate requirement posed by the communication system, coarse quantization becomes inevitable.

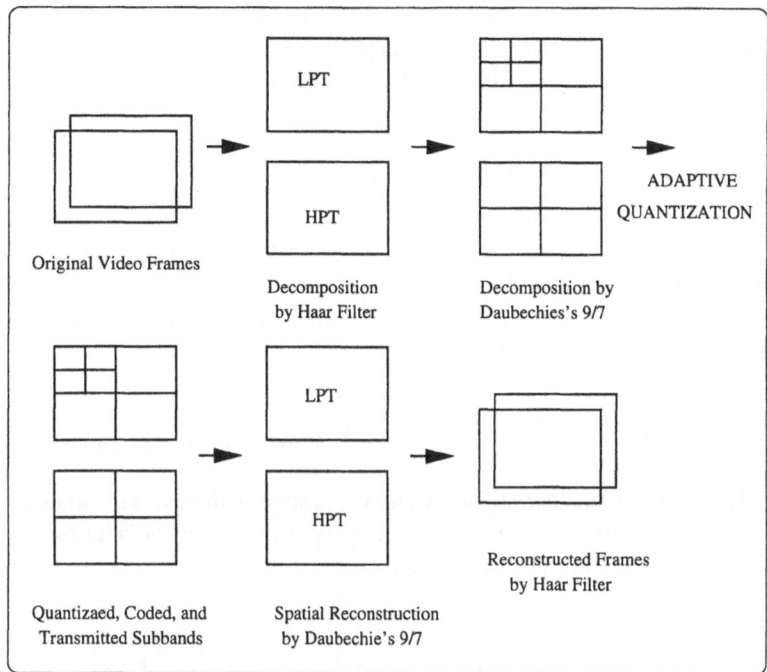

Figure 1 The overall video coding scheme based on wavelet transform and adaptive quantization.

Since the quantization is often the sole source of information loss, the design of quantization scheme is critical for video coding, especially at low bit rate. With same bit rate, the minimization of perceivable information loss caused by quantization is an important issue in low bit rate video coding. It is generally agreed that the distortion of scene structures is more perceivable than the missing of isolated details. Therefore, it is ideal for a quantization scheme to be able to differentiate coefficients related to scene structures from isolated coefficients and then discard the isolated coefficients if there is not enough bit to code. This desired differentiation capability can be achieved when wavelet coefficients are quantized through constrained clustering as described in the following. Upon wavelet decomposition of video signals, the high frequency subbands would generally exhibit "edges" and "impulses" corresponding to intensity discontinuities in either spatial or temporal domains. Strong and clustered "edges" and "impulses" are often of visual importance while weak and non-structural "impulses" are generally of negligible visual importance but would need considerable amount of bit to code. At low bit rate, traditional quantization schemes often destroy the inherent scene structure, since it is difficult for either scalar quantization or vector quantization to incorporate scene related constraints into their implementations.

The proposed adaptive quantization based on clustering with spatial constraints is able to incorporate scene structures through Gibbs random field model. If we denote a given subband by y and a quantization of this subband by x, according to Bayes' rule, the *a posteriori* probability can be expressed as:

$$p(x|y) \propto p(y|x)\, p(x)\,, \tag{1}$$

where $p(x)$ is the *a priori* probability of the quantization, and $p(y|x)$ represents the conditional probability of the subband distribution given the clustering. The Gibbs random field can be characterized by a neighborhood system and a potential function. A Gibbs random field constrained clustering is accomplished by assigning labels to each pixel in the given image. A label $x_s = i$ implies that the pixel s belongs to the i-th class of the K classes. Therefore, we have:

$$p(x_s \mid x_q, \ \forall q \neq s) = p(x_s \mid x_q, \ q \in N_s), \tag{2}$$

where N_s represents the defined neighborhood for pixel s. Associated with each neighborhood system are cliques and their potentials. A clique C is a set of sites where all elements are neighbors. If we consider that a 2D image is defined on the Cartesian grid and the neighborhood of a pixel is represented by its 8 nearest pixels[7], the Gibbs distribution with two-point clique potentials can be defined as :

$$p(x) \propto \exp\left\{-\sum_C V_C(x)\right\}, \qquad V_C(x) = \begin{cases} \beta, & \text{if } x_s = x_t \text{ and } s, t \in C \\ -\beta, & \text{if } x_s \neq x_t \text{ and } s, t \in C \end{cases} \tag{3}$$

If we model the conditional density as a Gaussian process with mean an variance at a pixel location s, then it can be written as spatially varying density function with respect to pixel location s, then, the overall probability function will be:

$$p(x|y) \propto \left\{\sum_s \left[-\frac{1}{2\sigma_s^2}(y_s - \mu_s)^2\right] - \sum_C V_C(x)\right\}. \tag{4}$$

There are two components in the overall probability function. One corresponds to the adaptive capability that force the quantization to be consistent with local intensity distribution. The other corresponds to the spatial continuity constraint characterized by the clique potentials within a given 2D lattice.

THE CELLULAR CONNECTIONIST ARCHITECTURE

The proposed cellular connectionist architecture is shown in Figure 2. The layout of the proposed network model is similar to those proposed by Chua and Yang[6] and Basak et al.[8, 9] with each neuron connected to its eight nearest neighbors as shown in Figure 2. However, the proposed network model is different from conventional models[6] in that each neuron or PE in the present model has a number of independent state components which are updated by the signals from neighboring neurons and external inputs. Each neuron in this model represents a pixel in image domain which is connected to its eight neighboring pixel as defined by the Gibbs random field neighborhood system. Within each PE, a number of independent state components, as shown in Figure 3, are updated by the inputs from neighboring neurons as a mean of enforcing the Gibbs spatial constraints. MATCH part of the PE accomplishes the equivalent conditional density computation as an External Input to each PE components. Suppose that the state vector of a PE is written as:

$$u_s = \left(u_s^1, u_s^2, \ldots u_s^i\right), \tag{5}$$

and the transfer function maps the state vector to the output by:

$$x_s = g_m(u_s^m) \qquad if \qquad u_s^m \geq u_s^n, \qquad 1 \leq n \leq i. \tag{6}$$

To ensure the convergence of the network, the transfer function $g(.)$ can be chosen as an increasing function such as sigmoid function and asymptotically almost saturating. The update of the PE state can be derived from original Bayesian formulation by expanding the overall probability function for the MAP estimation shown in Equation (4), and leave only the terms that are related to the iteration. Such update can be written as:

$$\frac{du_s^i}{dt} = c_s \left(y_s - \mu_s^i \right)^2 + \sum_{q \in N_s} \left\{ w_{sq} f_s^i (x_q) \right\} , \qquad (7)$$

where w_{sq} is the weight computed from the contribution of all cliques and,

$$f_s^i (x_q) = 2\delta \left(x_s^i - x_q \right) - 1 . \qquad (8)$$

Notice that there are two terms in the update formula (7). The first term corresponds to the conditional density iterative estimation and the second term corresponds to the Gibbs spatial constraints. When $x_s^i = x_q$, the function $f_s^i(x_q) = 1$. This means that the clustering is consistent with its neighbors. When $x_s^i \neq x_q$, the function $f_s^i(x_q) = -1$. This means that the clustering is different from its neighbors. All neighbors of a given pixel within the predefined neighborhood system will be evaluated in order to enforce the spatial constraints characterized by the Gibbs random field.

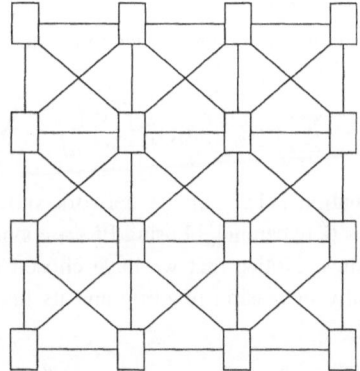

Figure 2 2D connection of cellular neural network.

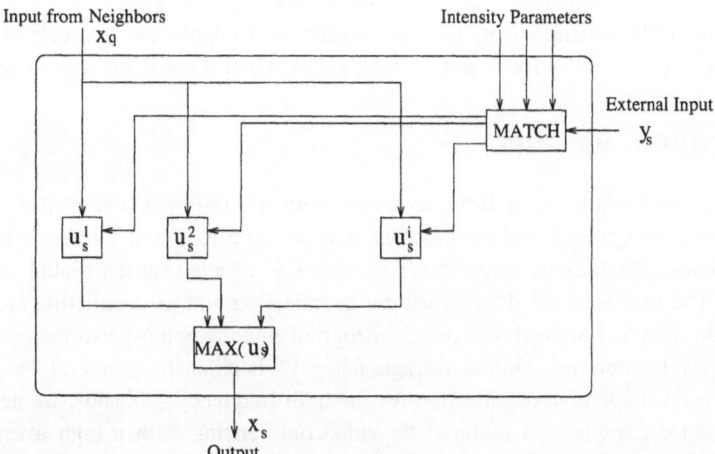

Figure 3 Diagram of a typical Processing Element (PE) in the connectionist model.

PROOF OF NETWORK CONVERGENCE

Fro any given neural network architecture, it is necessary to ensure that the network converges with respect to a specific update scheme. With the proposed transfer function, update scheme, and weight computation formula, we will prove, in the following, the proposed cellular network converges to the desired steady status.

Notice that the following equations have been used to describe the proposed network:

$$
\begin{cases}
\frac{du_s^i}{dt} & = c_s\left(y_s - \mu_s^i\right)^2 + \sum_{q \in N_s} \left\{w_{sq} f_s^i(x_q)\right\} \\
f_s^i(x_q) & = 2\delta\left(x_s^i - x_q\right) - 1 \\
w_{sq} & = w_{qs} \\
x_s & = g_m\left(\max_{1 \le m \le n}\left(u_s^i\right)\right)
\end{cases}
\tag{9}
$$

We expect the network described above to stabilize as the time passes. Let us define the energy function of the network as:

$$
E = -\frac{1}{2} \sum_s \sum_{q \in N_s} w_{sq} f_s^i(x_q) x_q - c_s\left(y_s - \mu_s^i\right)^2 x_s .
\tag{10}
$$

After some algebraic manipulation, the energy change of the network with respect to time can be written as:

$$
\frac{dE}{dt} = \sum_s \frac{\partial E}{\partial x_s} \frac{dx_s}{dt} = \frac{1}{2} \sum_s \sum_q (w_{sq} - w_{qs}) f_s^i(x_q) \frac{dx_s}{dt} - \sum_s \frac{dg^{-1}(x_s)}{dx_s}\left(\frac{dx_s}{dt}\right)^2
\tag{11}
$$

There are two terms in Equation (11). For the network shown in Figures 2 and 3, and Equation (9), the connections between neighboring PEs are symmetric. Therefore, the first term in Equation (11) vanishes. Notice that we have chosen g to be a sigmoid function. Hence, g^{-1} is a monotonically increasing function and its first derivative is positive. We can therefore conclude that,

$$
\frac{dE}{dt} = -\sum_s \frac{dg^{-1}(x_s)}{dx_s}\left(\frac{dx_s}{dt}\right)^2 \le 0 .
\tag{12}
$$

When $\frac{dx_s}{dt} = 0$,, we will have, $\frac{dE}{dt} = 0$, for all s .

Equation (12) indicates that the state update will enable the network to reach the minimum energy status and will stay at such status when there is no change for x_s.

VIDEO CODING RESULTS

Experimental results have been obtained using the CIF video sequence "salesman" with complex background and complicated motion. The temporal filterbank is the 2-tap Haar filterbank. Daubechies' wavelet 9/7 filterbank is selected for the spatial analysis and synthesis. The results of the decomposition, quantization and reconstruction are shown in Figures 4 through 6. For display purpose, histogram equalization is performed on Figures 4 and 5 to boost the contrast. On the average, using 11–band spatio-temporal decomposition and up to 7–level adaptive quantization for the high frequency subbands, we achieved the 40:1 luminance compression required by videoconferencing with a high average PSNR of 32.97 dB. The high perceptual quality of the reconstructed images suggests that this adaptive quantization scheme is promising for various low bit rate image and video coding applications.

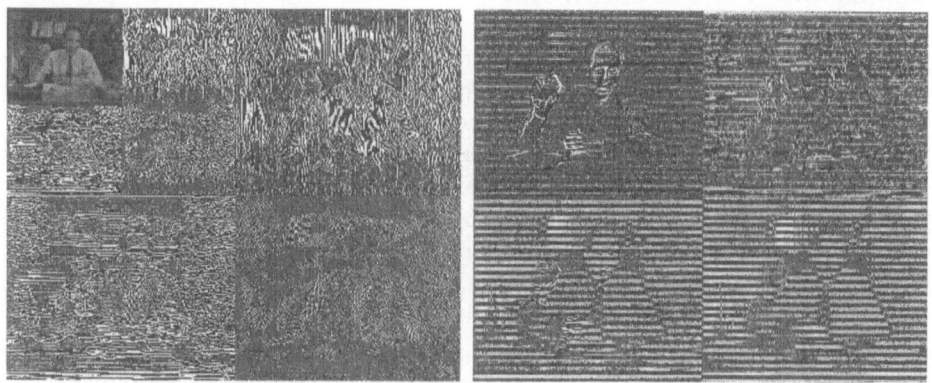

Figure 4 11–band decomposition of "salesman" sequence. Left:
Low-pass temporal subbands; Right: High-pass temporal subbands.

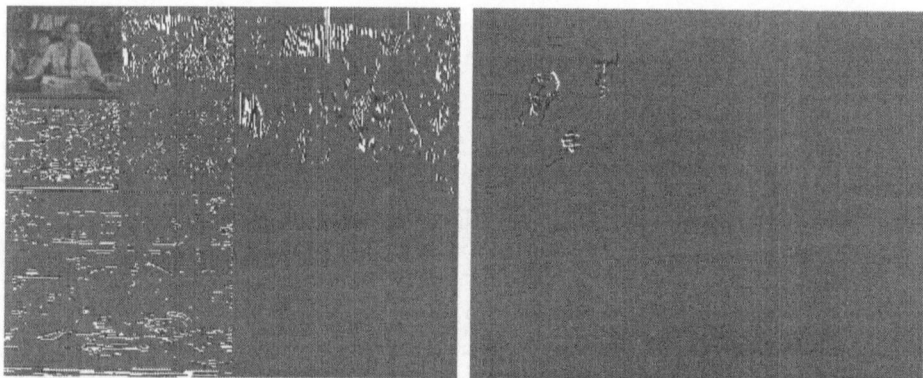

Figure 5 Quantized subbands. Left: Low-pass temporal subbands; Right: High-pass temporal subbands.

Figure 6 Left: Original frame; Right: Reconstructed frame after adaptive quantization.

DISCUSSION AND FUTURE WORK

The proposed cellular connectionist architecture described in this paper for the clustering-based adaptive quantization can be generalized and extended to other image processing tasks with spatial constraint-based computations. A spatial constraint can be realized by the basic cellular structure and additional considerations can be implemented as inputs to the PEs of the network. Within the architecture, we may also add the feedback part to each PE, i.e. modifying Equation (9) with additional term $w_s x_s$ to introduce directly the impact of current PE states to the next states. The weight w_{sq} in Equation (9) can be modified to differentiate the horizontal or vertical priority of the clique strengths for some visual communication applications.

We are currently investigating the modification of MATCH part of each PE to handle different types of conditional density distribution for a wider range of applications. For example, it has been shown that[10] Laplacian densities are more suitable for the modeling of each cluster in the high frequency subbands. With more suitable distribution modeling by Laplacian density, we expect to achieve better coding results in terms of both compression ratio and reconstructed image quality.

REFERENCES

1. J. Luo, C. W. Chen, K. J. Parker, and T. S. Huang, "Three dimensional subband video coding with segmentation in high frequency subbands," in *Proc. International Conf. Image Processing*, (Austin, TX), pp. III–255–259, November 1994.

2. J. Luo, C. W. Chen, K. J. Parker, and T. S. Huang, "Signal adaptive and scene adaptive quantization for subband image and video compression using wavelet," *IEEE Trans. Circuit and System for Video Technology*, May 1995. (In Press).

3. J. J. Hopfield and D. W. Tank, "Neural computation of decisions in optimizing problems," *Biological Cybern.*, vol. 52, pp. 141–152, 1985.

4. T. Kohonen, *Self-Organization and Associative Memory.* Berlin: Spring-Verlag, 1988.

5. D. E. Rumelhart, G. E. Hinton, and R. J. Williams, "Learning internal representations by error propagation," in *Parallel Distributed Processing* (D. E. Rumelhart, J. L. McClelland, and the PDP Research Group, eds.), vol. 1, Cambridge, MA: MIT Press, 1986.

6. L. O. Chua and L. Yang, "Cellular neural networks: Applications," *IEEE Trans. Circuits and Systems*, vol. 35, pp. 1273–1290, 1988.

7. J. Besag, "Spatial interaction and the statistical analysis of lattice system," *Journal of the Royal Statistical Society*, vol. 36, pp. 192–326, 1974.

8. J. Basak, B. Chanda, and D. Dutta Majumder, "On edge and line linking with connectionist models," *IEEE Trans. Syst. Man, Cybern.*, vol. 24, pp. 413–428, 1994.

9. J. Basak, C. A. Murthy, S. Chaudhury, and D. Dutta Majumder, "A connectionist model for category perception: theory and implementation," *IEEE Trans. Neural Networks*, vol. 4, pp. 257–269, 1993.

10. J. Luo, C. W. Chen, K. J. Parker, and T. S. Huang, "Adaptive quantization with spatial constraints in subband video compression using wavelets," in *International Conf. Image Processing '95*, (Washington DC), pp. II–583–586, October 1995.

A COMPARITIVE STUDY OF EXISTING APPROACHES FOR MOVING PICTURE FRACTAL CODING USING I.F.S.

J.-L. Dugelay, J.-M. Sadoul, and M. Barakat

Institut EURECOM, Multimedia Communications Dept.
2229 route des crêtes, B.P. 193, F-06904 Sophia Antipolis Cedex
fax. +33 93 00 26 27
e-mail. dugelay@eurecom.fr
url. http://www.cica.fr/~image

INTRODUCTION

The I.F.S. (Iterated Functions System) technique was been invented by the mathematician J. Hutchinson in the early eighties. It defines iterative processes which converge towards a fixed point, independently of their starting point. The fixed point is called the attractor of the IFS. This notion is part of a more general theory developed by the mathematician B. Mandelbrot known as Fractal theory. Barnsley has developed a general formulation for the use of I.F.S. for still image coding applications. The current reference algorithm has been proposed by A. Jacquin, who introduced the notion of Local-IFS[1]. This algorithm is briefly reviewed in section "A review of still image fractal compression using IFS". Since then, several relevant papers have put forward IFS as a promising technique for still image coding and proposed some improvements on Jacquin's algorithm[2]. Nevertheless, multimedia applications often need moving picture rather than still image compression in order to, for example, store video databases on CD-ROM. An overview of fractal video compression using IFS[3], yields that articles in this field can be divided into two categories (if one does not consider algorithms using a frame by frame approach): The first one is a mixed -or joint- approach and is based on a combinaison between an inter-frame coding using DPCM and, an intra-frame coding using IFS This approach is similar to the MPEG video coding scheme in that it uses IFS instead of DCT. The second one is a cubic -or volumetric- approach and is based on the extension of blocks to cubes (the third dimension is the temporal dimension). These two approaches are tested, compared and discussed, in the section "Comparison between the mixed and cubic approaches", in terms of compression rate versus quality, complexity and computational cost, possibilities of zoom. In the last section, several improvements are considered in order to optimize each of them.

Multimedia Communications and Video Coding
Edited by Y. Wang *et al.*, Plenum Press, New York, 1996

A REVIEW OF STILL IMAGE FRACTAL COMPRESSION USING I.F.S.

Jacquin's Algorithm

Coding Stage. Let x_c be the image to be compressed, x_0 an initial image, x_a the attractor and, x and y two generic images; Collage Theorem states:

$$if\ \exists \omega:\ d(x_c, \omega(x_c)) \leq \varepsilon\ and,\ d(\omega(x), \omega(y)) \leq \sigma.d(x, y)\ with\ \sigma < 1\ (\omega\ contractive)$$

$$then\quad d(x_c, x_a) \leq \varepsilon / (1 - \sigma)\quad with:\ x_a = \lim_{n \to \infty} \omega^{0,n}(x_0)$$

The reference algorithm in still image coding applications is Jacquin's algorithm: image "x_c", to be coded, is partitioned twice at two levels of resolution; for instance, this may be into squared-blocks of size *BxB* and *2Bx 2B* (generally *B* is fixed at 8). The former are called range blocks (R) and the latter are called domain blocks (D). For each range block, the algorithm searches for the best matching with a domain block according to the following criterion,

$$err_k = \sum_{pixels \in R_k} (W_k(D) - R_k)^2$$

where,

R_k designates the range block number k,
$W_k(D)$, the transformed domain block associated with R_k.

Before the matching stage, domain blocks are transformed as follows:
- sub-sampling by a factor two (in each direction),
- geometric transformations (eight isometries are considered)
- scale and shift of luminance values.

Finally, each area of the image (i.e. each range block) has an associated affine transformation,

$$W_k \begin{pmatrix} i \\ j \\ l \end{pmatrix} = \begin{pmatrix} a_k & b_k & 0 \\ c_k & d_k & 0 \\ 0 & 0 & s_k \end{pmatrix} . \begin{pmatrix} i \\ j \\ l \end{pmatrix} + \begin{pmatrix} e_k \\ f_k \\ o_k \end{pmatrix}$$

where,

$a_k, b_k, c_k, d_k, e_k, f_k$ represent the geometric transformation,
s_k, o_k represent the grey level transformation,
i,j the pixel coordinates,
l the grey-level.

The contractive transformation constraint is respected firstly by sub-sampling domain blocks by of factor 2, and secondly by making the factor s inferior to 1.

Decoding Stage. In order to decode image "xc", from its IFS code and from any image "x_0", the algorithm proceeds as follows: Image "x_0" is partitionned into a set of square-blocks. Each area of the image is computed by taking the associated block in image "x_0" and applying an associated contractive affine transformation defined during the coding stage. Then, image "x_1" is obtained. The algorithm iterates this process to obtain "x_2" from "x_1", ... , until it reaches "x_a". In practice, less than ten iterations are needed for convergence.

Current Work in Still Image Coding and Processing using I.F.S.

Current works in still image fractal compression using I.F.S. can be classified into four main categories, depending upon the problem(s) considered:
. **statement**, formulation and basic aspects of the algorithm;
This category mainly covers works linked to iteration contractivity constraint notions.
. **implementation**, reference algorithm computation and optimization;
This category covers the majority of studies. The addressed problems are segmentation, domain block classification, reduction of computational complexity, ...
. **functionality**, image processing related to an application or a service;
This category covers articles about zoom, progressive decoding,...
. **extension**, generalization of the reference algorithm for moving/color images.
The following sections deal only with the extension of the coding algorithm from still images to moving pictures.

MOVING PICTURE FRACTAL COMPRESSION USING I.F.S.

The first author who considered IFS technique in a moving picture coding scheme was Beaumont in 1991[4]. Nevertheless, most papers dealing with the subject mainly dated from 1994 (Table 1). Currently, two approaches are proposed in the literature (plus some hybrid versions based on these two approaches) for a possible extension of Jacquin's algorithm for moving picture coding:

. An "intra" coding mode using IFS associated with an "inter" coding mode using block-matching. This method is close to the MPEG standard using the D.C.T. (Discrete Cosine Transform) and the block-matching technique (Figure 1).

. The sequence to be coded is partitioned into several GoP (Group of Picture). Instead of segmenting an image into a set of square blocks, each GoP is segmented into a set of cubes. The third dimension is associated to the temporal dimension (Figure 2).

Table 1. Number of publications/year on moving picture fractal compression using I.F.S.

Year	1990	1991	1992	1993	1994
publications	0	1	2	5	14

The Mixed Approach: I.F.S and Block-Matching

Hürtgen and Büttgen's algorithm[5] uses DPCM for inter-frame coding and IFS for intra-frame coding. The basic idea included in this paper consists of considering two kinds of range blocks: moving range blocks which will be coded using IFS and still range blocks which will be reused from the previous frame. The authors point out that it is more efficient to encode the image itself rather than inter-frame differences. Simulations confirm this but only if the video sequence includes a lot of motion. As a perspective in order to improve this scheme, it is suggested to substitute DPCM by a procedure of motion estimation-compensation using a block-matching technique. ·
Other authors propose schemes which are very similar to Hurtgen's algorithm: Fisher, Rogovin and Shen[6] suggest encoding all range blocks of each picture from domain blocks

included in the previous frame. Hence, no iterations are needed during the decoding stage. This approach could be assimilated to a modified block-matching procedure.

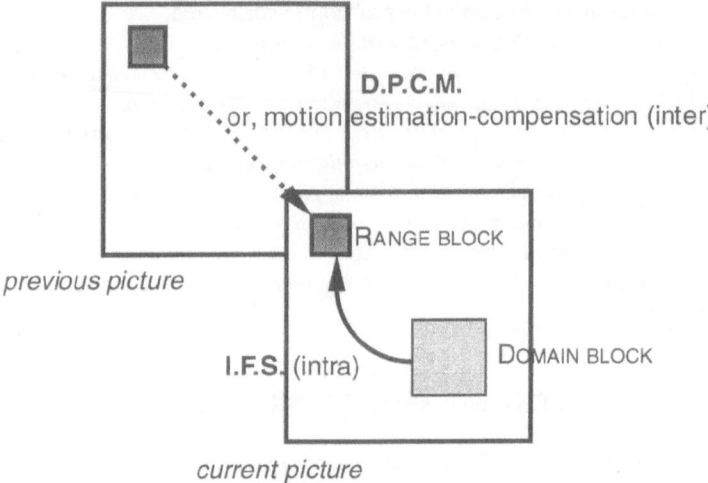

Figure 1. In the mixed approach, images of video sequence are considered two by two.

The Cubic Approach: Extension from Block to Cube Primitives

Lazar and Bruton's algorithm[7] considers GoP instead of still images and uses a segmentation into cubes instead of blocks. Domain blocks are sub-sampled by averaging a 2x2x2 pixels group into one pixel. Cubes are considered as a set of blocks (a block per frame). Hence, the temporal dimension is not considered in the same way as spatial dimensions. The usual transformations defined for blocks in still image coding can be successively applied to each block of a cube. Nevertheless, an additional transformation (i.e. temporal reverse) is used which consists of inverting the order of blocks inside a domain cube.

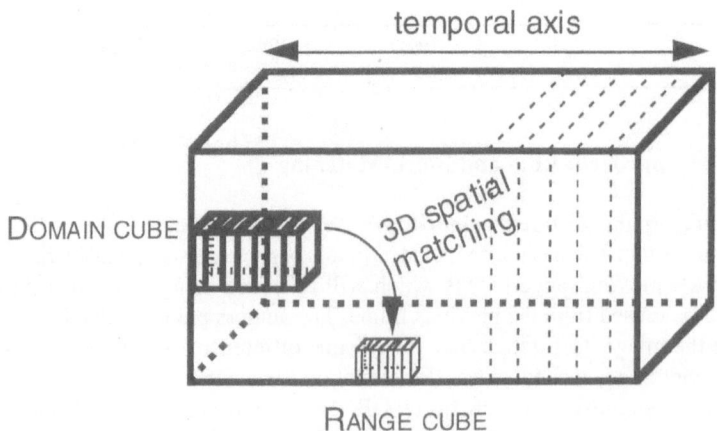

Figure 2. In the cubic approach, images of video sequence are considered GoP by GoP.

In an application other than coding (i.e. interpolating intermediate view in a multi-view system), Naemura and Harashima introduce some 3-D isometries in order to match domain cubes with range cubes[8].

COMPARISON BETWEEN THE MIXED AND CUBIC APPROACHES

Comparative tests have been realized on several sequences: "Miss America", "Salesman" (16 consecutives images of size 256x256) on each approach, described in the previous section.

Compression Rate Versus Quality

In the case of the cubic approach, the code associated to each range cube is always the same:
- a luminance shift (o_c bits),
- a contrast scale (s_c bits),
- an index associated to the selectionned predefined geometric transformation (i_c bits),
- and the address of the associated domain cube (respectively, x_c, y_c and z_c bits).

That is to say, by using a regular segmentation of the sequence into n_c range cubes, the total amount raw information required for each GoP of sequence is,

$$n_c.(o_c + s_c + i_c + x_c + y_c + z_c)$$

In contrast to the cubic approach, the mixed scheme has a variable bit rate: the number of bad predicted blocks (n'_m) depends on the temporal activity of the sequence. The code should include a binary table (n_m) in order to indicate for each block if it is a block coded by using DPCM (inter-mode) or by using I.F.S. (intra-mode). The code associated to each range block is,
- a luminance shift (o_m bits),
- a luminance scale (s_m bits),
- a geometric transformation index (i_m bits)
- and the address of the associated domain block (respectively, x_m and y_m bits).

That is to say, by using a regular segmentation of each image of the sequence into n_m blocks, and by assuming that n'_m blocks are coded by IFS, the total amount raw information required for each image is,

$$n_m + n'_m.(o_m + s_m + i_m + x_m + y_m)$$

Several sets of parameters have been defined in order to obtain either, the same image quality (i.e. PSNR), or the same compression rate.
An acceptable quality (that is to say, degradations are barely visible) is obtained for the cubic approach using a size of range cubes of 4x4x4 for the cubic approach and using a size of range blocks of 8x8 for the mixed approach. With these sets of parameters, at equal PSNR, the compression rate of the mixed approach is twice that of the cubic one. In two cases, the search strategy for the matching between range and domain blocks or cubes is a full search strategy; shift and scale parameters are quantized using the same number of bits.
Of course, if the set of parameters associated to the mixed approach is modified in order to obtain the same compression rate as the cubic approach (that is to say, using smaller range blocks of size 4x4 and using a new threshold which allows to consider about only the third of blocks to be coded using intra mode), the PSNR associated to the mixed approach

increases. Then, for equal compression rates, the mixed approach provides a better PSNR than the cubic one.

Complexity and Computational Cost

It is clear that cubic approach is harder to implement than mixed approach. The first one requires memory for only two frames, whereas the second one which needs to store all the GoP during the process of coding and decoding. Moreover, transformations on cubes are very expensive in term of computational cost.

Zoom

Several authors who work in the field of still image coding, assert that IFS may be used to realize a zoom in addition to compression. This is due to the property of invariance by scale included in fractals. An IFS code is theoretically independent of the size of the original image, and then this code can be used to reconstruct an image at any level of resolution. This aspect is very interesting in some multimedia applications which need a reduced (i.e. icon) as well as an enlarged (i.e. zoom) version of a given image[9] (Figure 3) In moving picture coding, the mixed approach, which uses IFS only in intra-mode, keeps only the possibility of a spatial zooming whereas the cubic approach completely keeps this aspect and yields to zoom the sequence in all the dimensions, spatials as well as temporal ones.

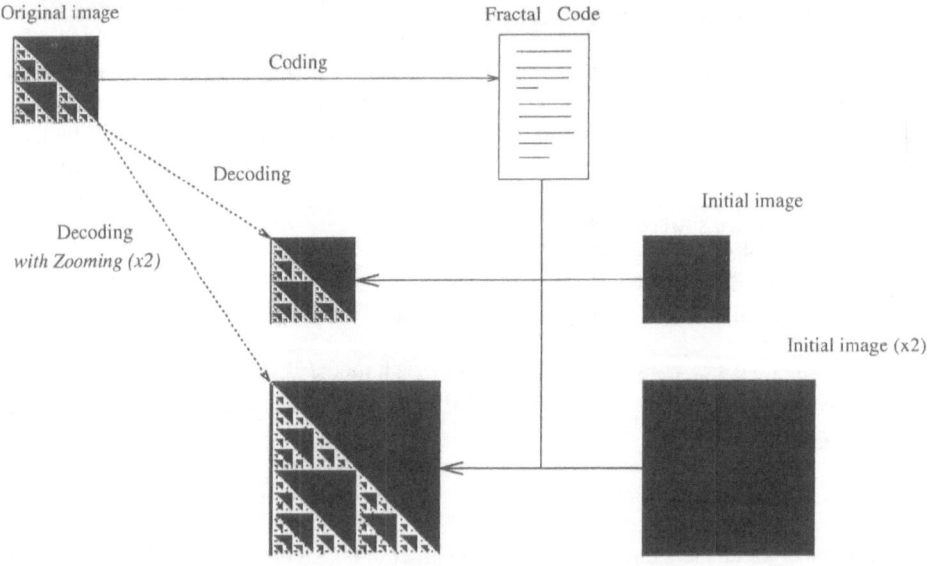

Figure 3. Basic illustration on how to zoom in addition to compress.

Discussion

After having studied the two approaches, several basic questions remain : The main question brought up by the mixed approach comes from the space search chosen for the Domain Block. With a mixed scheme, a frame can be divided into two parts: the background which has no or little temporal activity and therefore is predicted with the motion prediction scheme; and the foreground which has moved and which is coded using IFS. For this IFS coding process, should we use domain blocks for the entire picture, that is to say allow the matching of a block not coded by IFS ? or even restrict the search of the domain to the previous frame ? It surely insures an exact IFS code, because the matching is done with blocks that are known to the decoder. But, the iterative process is no longer needed, hence it is far more dubious to talk about convergence to an attractor. And the scale constraint is not any more needed. So the problem is to decide whether this coding scheme can still be named "fractal" or if it is not just a better block-matching procedure (i.e. using affine transforms). In the cubic approach, the extension using an additional dimension, that is to say, the temporal dimension, significantly changes the problem regarding contractivity constraint, which is still not perfectly controlled in still image coding. How to consider the temporal dimension versus the two spatial dimensions (horizontal and vertical) on one hand, and the luminance information on the other hand ? Which extension can be made to generalize the isometries applied to blocks in order to define geometric transformations to apply to cubes ?

CONCLUDING REMARKS

In this paper, two possible extension of the fractal compression algorithm defined for still image to moving picture are studied. The first one is a mixed approach based on a combination between inter-frame coding using DPCM, and intra-frame coding using IFS. The second one is a cubic approach which is based on the extension of block to cube primitives. Those two approaches are tested and then compared (Table 2).

Table 2. Summary table of advantages and disadvantages of each approach.

	compression versus quality	complexity	zoom
Mixed	+		+
Cubic		-	++

The mixed approach is more immediate to implement, compared to the cubic approach, for at least two reasons: the computation cost is lower and the scheme is closer to the existing video coding schemes such as MPEG. In any case, some improvements must be added to those current schemes. A possible way concerning the mixed approach could be the modification of the block-matching mode to a more complex procedure called super block-matching. The aim would consist in offering the same possibilities of transformations in the two modes, intra and inter frame. A possible improvement to augment the cubic approach could be the introduction of new possibilities in constructing domain cubes, in order to compensate a motion inside a GoP. The cubic approach is more complex than the mixed one. Nevertheless, this approach is closer than the other one to the basic algorithm proposed by A. Jacquin. For instance, it yields the complete possibility of zooming included in fractal theory.

ACKNOWLEDGMENTS

This work is in part supported by FRANCE TELECOM, CNET. The authors thank J. Signès, from C.C.E.T.T. (Rennes), for helpful discussions. We also thank F. Dubois and R. Knopp, from the EURECOM institute, for reviewing the contents of this paper.

REFERENCES

1. A.E. Jacquin, Image coding based on a fractal theory of iterated contractive image transformations, *IEEE Trans. on Image Processings*, 2,1 (1992), pp. 18-30.
2. Y. Fisher et al, *Fractal Image Compression: Theory and Application*, 1995 Springer-Verlag New York, Inc.
3. J.-L. Dugelay, J.-M. Sadoul and M. Barakat, Moving picture fractal coding using I.F.S.: *A review, available via URL. http://www.cica.fr/~image.*
4. J.-M. Beaumont, Image data compression using fractal techniques, British Telecom Technology Journal, Vol. 9, No. 4 (1991), pp. 93-109.
5. B. Hürtgen and P. Büttgen, Fractal approach to low rate video coding, *VCIP'93*, Boston, SPIE Vol. 2094, pp. 120-131.
6. Y. Fisher, D. Rogovin and T.P. Shen, Fractal (self-VQ) encoding of video sequences, *VCIP'94*, Chicago.
7. M.S. Lazar and L.T. Bruton, Fractal block coding of digital video, *IEEE Trans. on Circuits and Systems for Video Technology*, 4,3 (1994), pp. 297-308.
8. T. Naemura and H. Harashima, Fractal coding of a multi-view 3d image, *IEEE ICIP'94*, Austin, Texas, November 1994.
9. E. Polidori and J.-L. Dugelay, Zooming using I.F.S., *a NATO ASI*, July 8-17 (1995), Trondheim, Norway.

STATUS AND DIRECTION OF THE MPEG-4 STANDARD

Atul Puri

AT&T Bell Labs
101 Crawfords Corner Road
Holmdel, NJ 07733

INTRODUCTION

The increasing variety of applications involving video and the resulting large volume of generated data that needs to be stored or transmitted have created an enormous need for video compression. For many years, research has been ongoing in digital video coding with the primary goal of achieving high compression with good subjective quality. Often, from a practical standpoint, implementation complexity has been an additional important concern due to its adverse effect on cost effectiveness. Of recent, significant advances have been made in video compression research as well as what can be implemented cost effectively. Further, due to the increasing requirements of interchange of video data between applications, terminals and services, interoperability is necessary for seamless integration. The interoperability requirements, maturity of efficient coding schemes and potential for cost-effective implementations have all been key factors leading to the world-wide standardization effort. Towards this end, considerable progress has been made, in particular, the ISO Moving Picture Experts Group (MPEG), which after having completed the first phase standard (MPEG-1), has, now completed the second phase, MPEG-2, except for ongoing extensions. The MPEG-1 video standard started in late 1988 achieved a stable stage of Committe Draft (CD) by late 1990 and the MPEG-2 video standard which was started in late 1990 achieved the CD stage in late 1993. In early 1993, the next phase of MPEG, called MPEG-4 was started and is expected to reach the CD stage by late 1997. The final stage in ISO MPEG standards is called International Standard (IS), and is typically reached one year after the CD stage. Before we discuss goals and direction of MPEG-4, it is useful to understand the trend leading upto MPEG-4 and thus a brief review of MPEG-1 and MPEG-2 is in order.

The MPEG-1 standard although originally intended for audio-visual coding at about 1.5 Mbit/s for digital storage media (DSM) applications, was designed to be fairly flexible, efficient and thus useful in a number of applications and at a wide range of bit-rates. Besides, efficient video compression, MPEG-1 had another important requirement, that is, basic interactivity with the bitstream such as random access, fast forward, fast reverse etc. In terms of video compression method, MPEG-1 uses block motion compensated DCT coding structure, frequent Intra (I-) pictures to allow random access, and Bidirectional (B-) pictures for increased coding efficiency. Due to the bitrates of optimization which were related to immediate target applications of MPEG-1, it was intended for coding of noninterlaced (progressive) video of lower than TV resolution. The main goal of the second phase, MPEG-2, was to address the more demanding applications such as broadcast TV at bit-rates in range of 4 to 10 Mbit/s. The MPEG-2 standard [1,2,3] is thus optimized for efficient coding of interlaced video of TV and higher resolution and builds on the motion compensated DCT structure and picture organization of MPEG-1 and includes key adaptations for efficient coding of interlaced video. During development of MPEG-2 new important requirements were identified which required functionalities of backward compatibility with H.261/MPEG-1 and scalability. Scalability is the property that enables decodability of subsets of entire bitstream on decoders of less than full complexity to produce useful video from the same bitstream. Scalability in picture quality, spatial resolution and temporal resolution were considered important [2, 3] giving rise to SNR scalability, Spatial scalability and Temporal scalability. Scalable video coding in MPEG-2 was a major step forward in MPEG-2 in terms of increased functionality as compared to MPEG-1. The MPEG-2 standard also supported the standard functionalities of interactivity with the bitstream of MPEG-1. Overall, as is well recognized now, the MPEG-2 video standard is truly generic addressing a variety of applications efficiently by virtue of flexibility in coding methods, picture resolution/formats and bitrates. More specifically, it not only offers good quality video for broadcast applications, but also includes a useful set of scalability tools for applications in ATM, multipoint video, migration to high temporal resolution HDTV, multimedia retrieval and others. Both MPEG-1 and MPEG-2 standards only specify a bitstream syntax and decoding semantics, allowing considerable innovation in optimization of encoding techniques.

The next MPEG standard, MPEG-4, was started in 1993 with the goal of very high compression coding at very low bitrates of 64 kbit/s or under. Coincidentally, the ITU-T (formerly CCITT) also started two very low bit-rate video coding efforts: a short term effort to be completed by 1995 intended to improve H.261 for coding at 20 to 30 kbit/s, and a long term effort to be completed by 1998 intended to achieve higher compression coding at similar bitrates. Currently, the ITU-T short term standard is nearly complete, and it has been decided that the long term ITU-T effort will be based on MPEG-4. When MPEG-4 was started, it was anticipated that with the continuing advances in non-block based coding, for example, in object or model based coding, a scheme capable of very high compression, mature for standization would emerge. By mid 1994, two things became clear. First, improvements in video coding efficiency were likely to be moderate say upto a factor of 2 or so not as compared to the original goal of MPEG-4 within its expected time frame. Second, a new class of multimedia applications were emerging that required increasing levels of functionality [4,5] then that provided by any other video standard at bitrates in range of 10 kbit/s to 1000 kbit/s. This lead to the redefinition of the original scope of MPEG-4 in terms of important classes of functionalities [6] - Content Based Interactivity, High Compression, and Universal Accessibility.

FUNCTIONALITIES GOALS OF MPEG-4

Although the scope of MPEG-4 is now defined in terms of application classes, several functionalities are allowed in each class and are briefly explained in Table 1.

Table 1 Functionalities supported by MPEG-4 and their explanation

Content Based Interactivity
Content Based Multimedia Data Access Tools: The access of multimedia data with tools such as indexing, hyperlinking, browsing, uploading, downloading etc.

Content Based Manipulation and Bitstream Editing: The ability to provide manipulation of contents and editing of bitstreams without need for transcoding.

Hybrid Natural and Synthetic Data Coding: The ability to code and manipulate natural and synthetic image data including decoder controllable compositing of scenes.

Improved Temporal Random Access at Low Bitrates: Efficient methods for random access at very low bitrates.

Compression

Improved Coding Efficiency: The ability to provide subjectively better audio-visual quality at bitrates compared to existing or emerging video coding standards.

Coding of Multiple Concurrent Data Streams: The ability for efficient coding of multiple views of a scene by exploiting redundancies between various views of a scene.

Universal Access

Robustness in Error Prone Environments: The capability to allow robust access to applications over a variety of wireless and wired networks and storage media.

Content Based Scalability: The ability to achieve scalability with fine granularity in spatial, temporal or amplitude resolution, quality or complexity.

SYNTAX DIRECTION

Due to the diverse requirements of the various functionalities and the increasing flexibilities offered by software based processing, it is expected that syntax will play a major role in the development of the MPEG-4 standard. Unlike syntax of other standards, a more flexible and extensible syntactic framework is sought. This is expected to be achieved by defining a syntax descriptive language by employing the object oriented programming methodology. The general requirements of MPEG-4 syntax description language (MSDL) are listed in [8]. The extent of influence of MSDL is expected to be higher than channel coding and lower than presentation layer for audio-visual data; howowever, there may be some regions of overlap. The location of MSDL [8] in a vertical stack of protocols is shown in Figure 1.

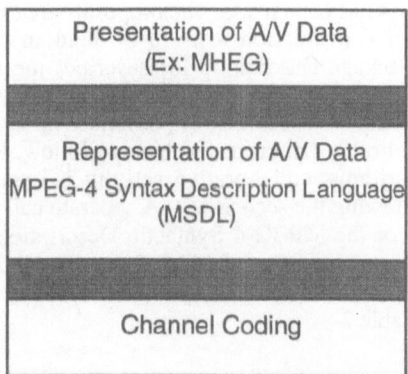

Figure 1 Position of MPEG-4 Syntax Description Language in vertical stack of protocols

It is expected that MPEG-4 will eventually not just standardize a single coding algorithm but will establish a set of *tools* which will be used to form coding *algorithms*, and further, coding algorithms would be customized for specific applications to make *profiles*. Thus, tools, algorithms and profiles are the coding objects of interest and can be thought of as consisting of an independent core and a standard interface [9]. The glue that binds the coding objects together is expected to be the MSDL.

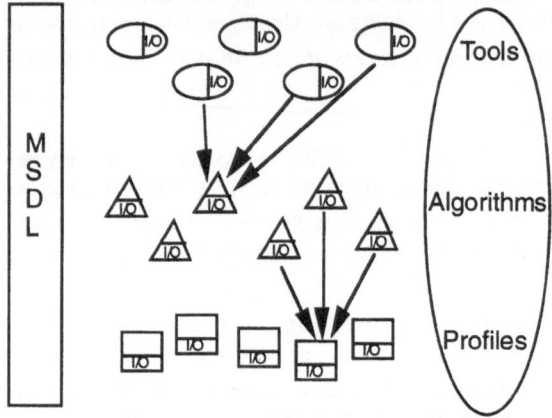

Figure 2 The function of MPEG-4 Syntax Description Language

In terms of MPEG-4 decoder complexity, both hardware as well as, software based decoders with a range of complexity and functionality are expected. It is anticipated that all MPEG-4 decoders will be able to support a *configuration phase* and a *transmission phase* . Some decoders may also support a *learning phase*. A configuration phase involves call set up and exchange of parameters such as availability of needed objects between coder and decoder. In a learning phase, coder either sends the missing objects or informs the decoder regarding where to find these objects. A transmission phase involves sending data in the format negotiated by other two phases.

STAGES OF DEVELOPMENT

A call for proposals requesting proposals for algorithms and tools for consideration for standardization for MPEG-4 has been made. The proposals are expected to be evaluated in a series of two competetive evaluation tests to be held in Nov. 95 and Nov. 96 respectively. The Proposal Package Description [6] describes the various functionalities that MPEG-4 is addressing in each test for which algorithm and tools proposals are sought. A separate Test/Evaluation Procedures document [7] describes how proposals will be tested or evaluated as well as general rules that proposers have to follow. Based on the submitted proposals for tools and algorithms, collaborative activity is expected to take place in between the two tests and following the second test. A separate call for proposals has been sent out to invite proposals for the MPEG-4 Syntactic Description Language. Although some collaborative work on syntax has been ongoing, the evaluation of submitted proposals will identify how work will proceed in future. A detailed development plan for the MPEG-4 standard [7] is outlined in Table 2.

Table 2 Development Plan for the MPEG-4 Standard

Nov. 94	Call for Proposals for First Evaluation
	Proposal Package Description version 1
March 95	Proposal Package Description version 2
	Initial Definition of First Evaluation
	Call for proposals for MPEG-4 Syntactic Description Language

July 95	Proposal Package Description version 3
	Evaluation of MPEG-4 Syntax Description Language begins
	Finalize Definition of First Evaluation
Nov. 95	First Evaluation begins
	Syntax Description Language Evaluation
	Call for proposals for Second Evaluation
Jan. 96	Verification Models version 1
	Definition of the Second Evaluation Process
March 96	Collaboration on Verification Models resulting in version 2
July 96	Verification Models Improvement
Nov. 96	Second Evaluation begins
	First Working Draft
July 97	Final Working Draft
Nov. 97	Committee Draft
March 98	Draft International Standard
Nov. 98	International Standard

FIRST EVALUATION

For the first evaluation tests, proposers were asked to submit algorithm proposals for formal subjective testing or/and tools/algorithms for functionalities not formally tested in first evaluation. Any tools/algorithms that are not formally tested are evaluated by a panel of experts and this type of informal assesment is referred to as *evaluation by experts*.

The framework of first evaluation has involved standardizing test material to be used in the first evaluation [7]. Towards this end, video scenes are classified from relatively simpler to complex by categorizing them into three classes, referred to as *Class A*, *Class B* and *Class C*. There are two other classes of scenes, Class D and Class E, where Class D contains Stereoscopic video scenes and Class E contains Hybrid of natural and synthetic scenes.

Since MPEG-4 addresses many different classes of scenes and many types of functionalities, it was necessary to devise 3 types of test methods. The first type of test method used is called Single Stimulus Method (SS) [7] and involves rating the quality of coded scene on a 11 point scale from 0-10. The second type of test method used is called Double Stimulus Impairment Scale (DSIS) [7] and involves presenting to assessors a reference scene (coded by a known standard) and after a 2 sec gap a scene coded by a candidate algorithm, with impairment of candidate algorithm to be compared to reference using a 5 level impairment scale. The third test mehod is called Double Stimulus Continuous Quality Scale (DSCQS) and involves presenting two sequences with a gap of 2 secs in between. One of the two sequences is coded reference and the other is the one coded by the candidate algorithm, however, the assessors are not told of the order of reference and the candidate coded scene. In DSCQS method a graphical continuous quality scale is used and is later mapped to discrete representation on a scale of 0 to 100.

Table 3 summarizes the list of formal tests, explanation of tests and type of test method employed.

Table 3 List of MPEG-4 First Evaluation Formal Tests and their explanation

Compression
Class A sequences at 10, 24 and 48 kbit/s: Coding to achieve the highest compression efficiency. Input video resolution is CCIR-601 and although any spatial and temporal resolution can be used for coding, display format is CIF on a windowed display. The test method employed is SS.
Class B sequences at 24, 48 and 112 kbit/s: Coding to achieve the highest compression efficiency. Input video resolution is CCIR-601 and although any spatial and temporal resolution can be used for coding, display format is CIF on a windowed display. The test method employed is SS.
Class C sequences at 320, 512 and 1024 kbit/s: Coding to achieve the highest compression efficiency. Input video resolution is CCIR-601 and although any spatial and temporal resolution can be used for coding, display format is CCIR-601 on a full display. The test method employed is DSCQS.
Scalability
Object Scalability at 48 kbit/s for Class A, 320 kbit/s for Class E and 1024 kbit/s for Class B/C sequences: Coding expected as permitting dropping of specified objects resulting in remaining scenes at lower than total bit-rate; these scenes are separately evaluated by experts. Display format for Class A is CIF on a windowed display and for Class B/C as well as Class E is CCIR-601 on full display. The test method employed for Class A is SS, for Class B/C is DSCQS and for Class E is DSIS.
Spatial Scalability at 48 kbit/s for Class A and 1024 kbit/s for Class B/C/E sequences: Coding expected as two spatial layers with each layer using half of the total bit-rate, however, full flexibility in choice of spatial resolution (of objects in each layer) is allowed. Display format for Class A is CIF on a windowed display and for Class B/C/E is CCIR-601 on full display. The test method employed for Class A is SS, for Class B/C/E is DSCQS.
Temporal Scalability at 48 kbit/s for Class A and 1024 kbit/s for Class B/C/E sequences: Coding expected as two temporal layers with each layer using half of the total bit-rate, however, full flexibility in choice of temporal resolution (of objects in each layer) is allowed. Display format for Class A is CIF on windowed display and for Class B/C/E is CCIR-601 on full display. The test method employed for Class A is SS, for Class B/C/E is DSCQS.
Error Robustness
Error Resilience at 24 kbit/s for Class A, 48 kbit/s for Class B and 512 kbit/s for Class C: Tested at high random bit error Rate (BER) of 10^{-3}, multiple burst errors with 3 bursts of errors and 50% BER within a burst, and a combination of high random bit errors and multiple burst errors. Display format for Class A and Class B is CIF on a windowed display and for Class C is CCIR-601 on full display. The test method employed for Class A and Class B is SS and for Class C is DSCQS.
Error Recovery at 24 kbit/s for Class A, 48 kbit/s for Class B and 512 kbit/s for Class C: Tested with long burst errors of 50% BER within a burst and a burst length of 1 to 2 secs. Display format for Class A and Class B is CIF on a windowed display and for Class C is CCIR-601 on full display. The test method employed for Class A and Class B is SS and for Class C is DSCQS.

In the first evaluation, algorithms for functionalities not tested formally as well as tool submissions are evaluated by a panel of experts. Proposers of these tools and algorithms have been asked to provide functional as well as performance specification of their tools/algorithms and what type of method is needed for testing their proposals by other MPEG-4 members. During 1996, core experiments are expected to be designed to test various tools/algorithms judged to be promising by the panel of experts.

At the time of writing of this paper, for first evaluation tests, proponents had already registered for algorithm and tool submissions and first evaluation was expected to be performed within the next 3 weeks. A total of 34 organizations or so have registered for MPEG-4 first evaluation, with a few organizations submitting separate proposals for seperate functionalities whereas a few others testing few particular classes of sequences in the various tests.

SUMMARY

To summarize, the MPEG has established international standards for audio-video coding by successful completion of MPEG-1 and MPEG-2 standards. MPEG-4 is the new activity by MPEG standards commitee and is intended to address a new range of bit-rate constrained multimedia applications which demand an increasing levels of interactivity, compression and accessibility. While there are some similarities in what will be standardized in MPEG-4 as compared to MPEG-1/MPEG-2. there are also major differences. One difference is the requirement of flexible syntax description language instead of a rigid syntax structure. Although MPEG-4 will standardize coding tools and algorithms for cost-effective implementation, it is also expected to allow considerable flexibility including ability to download new tools and coding algorithms. Currently MPEG-4 is about to conduct its first evaluation involving formal subjective tests and evaluation by experts. Test sequences, bit-rates, test conditions and the rules to be followed by proposers have all been finalized. A clear development plan for MPEG-4 exists and calls for completion of the standard by late 1998.

REFERENCES

1. Test Model Editing Committee, "MPEG-2 Video Test Model 5," ISO/IEC JTC1/SC29/WG11 Doc. N0400, April 1993.

2. Video Draft Editing Committee, "Generic Coding of Moving Pictures and Associated Audio," Recommendation H.262, ISO/IEC 13818-2, Draft International Standard for Video, Grimstadt, Norway, July 1994.

3. A. Puri, "Video Coding Using the MPEG-2 Compression Standard," Proceedings of SPIE Visual Communications and Image Processing, Boston, Mass., Nov. 1993, pp. 1701-1713.

4. A. Puri, "MPEG-4 Revised Scope and Functionality Proposal," ISO/IEC JTC1/SC29/WG11 Doc. MPEG94/362, Singapore, Nov. 1994.

5. MPEG-4 Requirements Ad hoc Group, "MPEG-4 Functionalities," ISO/IEC JTC1/SC29/WG11 Doc. MPEG94/399, Singapore, Nov. 1994.

6. AOE Sub Group, "MPEG-4 Proposal Package Description (PPD) - Revision 2.0," ISO/IEC JTC1/SC29/WG11 Doc. N937, Lausanne, March 1995.

7. MPEG-4 Test Procedures Ad hoc Group, "MPEG-4 Test/Evaluation Procedures Document - Revision 1.1," ISO/IEC JTC1/SC29/WG11 Doc. MPEG95/262, Tokyo, July. 1995.

8. MPEG-4 Syntax Description Language Ad hoc Group, "Requirements for the MPEG-4 Syntax Description Language - Revision 1.0," ISO/IEC JTC1/SC29/WG11 Doc. N999, Tokyo, July. 1995.

9. Cliff Reader, "MPEG Update, Where It's at, Where It's Going," 19th International TV Symposium, Montreux, June 1995, pp.117-127.

A RATE-CONSTRAINED ENCODING STRATEGY FOR H.263 VIDEO COMPRESSION

Thomas Wiegand[1], Michael Lightstone[2], T. George Campbell[3]
and Sanjit K. Mitra[1]

[1]Center for Information Processing Research
Department of Electrical and Computer Engineering
University of California, Santa Barbara, CA 93106

[2]Chromatic Research, Inc.
800A East Middlefield Road
Mountain View, CA 94043-4030

[3]Compression Labs, Inc.
2860 Junction Avenue
San Jose, CA 95134-1900

INTRODUCTION

In recent years numerous standards such as H.261 [1], MPEG-1 [2], and MPEG-2 [3] have been introduced to address the compression of video data for digital storage and communication services. Together, the applications for these standards span the gamut from low bit-rate video telephony to high quality HDTV with a new emerging standard, H.263 [4], targeting the low bit-rate end. More specifically, the primary mission for H.263 is traditionally regarded as the coding of digital video at rates suitable for transmission over public switched telephone network (PSTN) lines. Fast modems suited for this application typically run at 28.8 Kbits per second (Kb/s) within which video, audio, data, and overhead must be transmitted. This places a demanding rate constraint on the video coder which in most cases must operate at less than 24 Kb/s. In terms of wireless mobile networks whose capacities are often less than 19.2 Kb/s [5], this range of operation is also very conducive. Not surprisingly then, in addition to traditional telephony, there has been a significant and growing interest in the extension of the H.263 standard to mobile and wireless applications [6].

With regards to efficient realization, a key problem for H.263 as well as the other standards is the operational control of the encoder. Whereas most video standards uniquely stipulate the bit-stream syntax and, in effect, the decoder operation, the exact nature of the encoder is generally left open to user specification. Ideally, the encoder should balance the quality of the decoded images with channel capacity. This problem is compounded by the fact that typical video sequences contain widely varying content and motion that is often more effectively quantized if different strategies are permitted to code different regions. Currently, most video coding standards, including H.263, address this problem by utilizing several modes of operation which are selected on a block-by-block basis. The advantage of the multi-mode approach is that its inherent adaptability lays the foundation for better coding results. In this paper, the goal is to provide a practical, yet efficient, algorithm for exploiting this potential in a real H.263 video encoder. It is important to note that while H.263 shares many common elements with past standards, several unique attributes such as hierarchical motion estimation and overlapped motion compensation distinguish it from the others.

To improve overall encoder performance, past papers have successfully applied rate-distortion theory to optimize the frame type and/or the quantizer selection in an MPEG system [7], [8]. A potential drawback for these approaches, however, is that the problem of selecting the best encoding strategy for a frame is not considered at the macroblock level. Rather, the optimization is accomplished by assuming a fixed number of quantization choices for each frame. For a given number of frames, a diverging trellis is generated whose paths correspond to all possible combinations of quantization choices. The job of the encoder is to find the path with the lowest total cost in the rate-distortion sense. Unfortunately, the size of the tree grows exponentially with the tree depth, and only if the number of quantization choices is relatively small can the optimal solution be feasibly found. For systems like H.263 [4], and even MPEG [2], [3], this scenario is not very likely unless the multi-mode flexibility is restricted so as to lessen the number of possible quantization choices for each frame.

In this paper, we employ a new technique [9] that formalizes the problem of encoder optimization on a macroblock-by-macroblock basis using a rate-constrained product code framework [10]. An associated Lagrangian formulation leads to an unconstrained cost function and, in the special case of mode selection, a non-diverging trellis whose associated paths correspond to all possible operational rate-distortion points for the specified image region. The best path in the trellis can be efficiently located using a dynamic programming solution based on the Viterbi algorithm [11]. The final result is an H.263 encoder algorithm that, when applied to a macroblock slice, selects the optimum combination of macroblock modes and associated mode parameters so as to minimize the overall distortion for a given bit-rate budget. To this end, the paper is organized as follows. First, we formulate the mode selection problem as it pertains to a general block-based multi-mode video coding system, and then examine a solution for obtaining the best achievable performance in the rate-distortion sense. Next, the application of this technique to H.263 is described, and results are presented for various video phone sequences at data rates from 8 to 20 Kb/s.

EFFICIENT MODE SWITCHING

Consider an image region which is partitioned into a group of blocks (GOB) given by $\mathcal{X} = (\mathbf{X}_1, \ldots, \mathbf{X}_N)$. For a multi-mode video coder, each macroblock in $\mathcal{X}$ can be coded using only one of K possible modes given by the set $\mathcal{I} = \{I_1, \ldots, I_K\}$. Let $M_i \in \mathcal{I}$ be the mode selected to code block $\mathbf{X}_i$. Then for a given GOB, the modes assigned to the elements in $\mathcal{X}$ are given by the N-tuple, $\mathcal{M} = (M_1, \ldots, M_N) \in \mathcal{I}^N$. The problem of finding the combination of modes that minimizes the distortion for a given GOB and a given rate constraint R_c can be formulated as

$$\min_{\mathcal{M}} D(\mathcal{X}, \mathcal{M}) \text{ subject to } R(\mathcal{X}, \mathcal{M}) \leq R_c. \tag{1}$$

Here, $D(\mathcal{X}, \mathcal{M})$ and $R(\mathcal{X}, \mathcal{M})$ represent the total distortion and rate, respectively, resulting from the quantization of the GOB $\mathcal{X}$ with a particular mode combination $\mathcal{M}$.

To simplify this constrained optimization problem, we can employ a rate-constrained product code framework [10]. Assuming an additive distortion measure, the cost function and rate constraint can be simultaneously decomposed into a sum of terms over the elements in $\mathcal{X}$ and rewritten using an unconstrained Lagrangian formulation so that the objective function becomes

$$\min_{\mathcal{M}} \sum_{i=1}^{N} J(\mathbf{X}_i, \mathcal{M}), \qquad (2)$$

where $J(\mathbf{X}_i, \mathcal{M})$ is the Lagrangian cost function for block $\mathbf{X}_i$ and is given by

$$J(\mathbf{X}_i, \mathcal{M}) = D(\mathbf{X}_i, \mathcal{M}) + \lambda \cdot R(\mathbf{X}_i, \mathcal{M}). \qquad (3)$$

It is not difficult to show that each solution to (2) for a given value of the Lagrange multiplier λ corresponds to an optimal solution to (1) for a particular value of R_c [12], [13]. Unfortunately, even with the simplified Lagrangian formulation, the solution to (2) remains rather unwieldy due to the rate and distortion dependencies manifested in the $D(\mathbf{X}_i, \mathcal{M})$ and $R(\mathbf{X}_i, \mathcal{M})$ terms. Without further assumptions, the resulting distortion and rate associated with a particular block in the GOB is inextricably coupled to the chosen modes for every other block in $\mathcal{X}$. On the other hand, for many video coding systems, the bit-stream syntax imposes additional constraints that can further simplify the optimization problem.

For example, we can restrict the codec so that both the rate and distortion for a given image macroblock are impacted by only (i) the content of the current block and its respective operational mode, (ii) the content of the current and previous block, and (iii) the content of the current, previous, and ensuing block. These three cases correspond to simplified Lagrangians given by

$$\text{(i)} \quad J(\mathbf{X}_i, \mathcal{M}) = J(\mathbf{X}_i, M_i), \qquad (4)$$

$$\text{(ii)} \quad J(\mathbf{X}_i, \mathcal{M}) = J(\mathbf{X}_i, M_{i-1}, M_i), \quad \text{and} \qquad (5)$$

$$\text{(iii)} \quad J(\mathbf{X}_i, \mathcal{M}) = J(\mathbf{X}_i, M_{i-1}, M_i, M_{i+1})$$
$$= J'(\mathbf{X}_i, M_{i-1}, M_i) + J''(\mathbf{X}_i, M_i, M_{i+1}), \qquad (6)$$

respectively. For scenario (i), the optimization problem of (2) can be easily minimized by independently selecting the best mode for each macroblock in the GOB. For this particular case, the problem formulation is equivalent to the bit allocation problem for an arbitrary set of quantizers, proposed earlier by Shoham and Gersho in [13], and specifically for video coding by Wu and Gersho in [14]. The drawback is that this structural constraint is rather restrictive and does not correspond to the way macroblocks are coded in most video coding standards such as H.261, MPEG-1, MPEG-2, and especially H.263.

For instance, a block-to-block dependency typically exists due to the differential encoding of motion vectors such that the rate term for a given macroblock is dependent not only on the current mode but on the modes of previous blocks as well, leading to case (ii) and the Lagrangian of (5). For overlapped motion compensation (as found in H.263), the dependency also manifests itself in the distortion terms, introducing dependencies on past and future blocks. This behavior corresponds to case (iii) and the Lagrangian of (6)[1].

---

[1]Here we assume that the influence of the previous block can be separated from the influence of

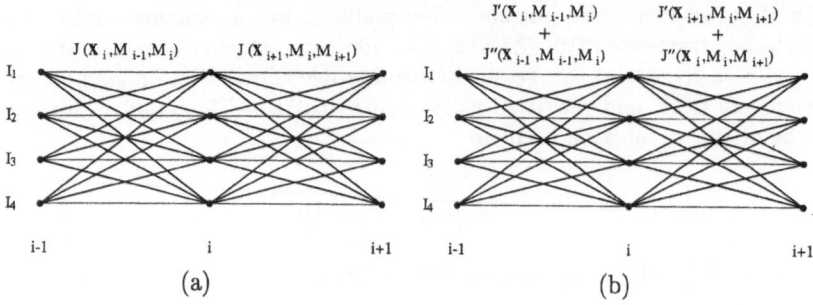

Figure 1. Resulting multi-mode trellis for the cases when the rate and distortion dependencies are (a) on past macroblocks and (b) on past and future macroblocks.

For both (5) and (6), we can obtain the optimal solution by viewing the search for the best combination of N modes in the GOB as an equivalent search for the best path in a trellis of length N. For case (ii), the nodes in the trellis for $i = 1, \ldots, N$, are given by the elements in $\mathcal{I}$, and the transitional costs from node M_{i-1} to node M_i are given by Lagrangian cost terms specified in (5). This trellis, shown in Fig. 1(a) for $K = 4$, can be efficiently searched using the Viterbi algorithm to obtain the optimal solution to (2). Similarly, the Viterbi algorithm can be implemented to obtain an optimal solution for case (iii) by searching the trellis described in Fig. 1(b). Further details on this mode selection technique, including the joint optimization of the mode selection with the selection of macroblock parameters such as the quantization step size, can be found in [9].

A final critical consideration with regards to mode selection is the determination of the Lagrange multiplier λ. Recall that while the solution to the unconstrained Lagrangian cost function for any value of λ results in minimum distortion for some rate, the final rate cannot be specified a priori. Often it is desirable to find a particular value for λ so that upon optimization of (2), the resulting rate closely matches a given rate constraint R_c. Because of the monotonic relationship between λ and rate, a possible solution is the bisection search algorithm [15], [16]. However, the computation associated with the re-optimization of (2) for numerous values of λ may preclude such a search in a practical encoder. As an alternative, we have considered a variety of successful approaches including a frame-to-frame update of λ using least-mean-squares (LMS) adaptation [17]. In our experiments (provided at the end of the paper), we employ a method for determining the LMS step-size dynamically for each frame or GOB (indexed by k) of the video sequence [18]. The strategy effectively reduces the bursty behavior of adaptation and results in an update procedure for the Lagrange multiplier given by

$$
\begin{aligned}
\lambda_{k+1} &= \lambda_k + \frac{1}{R_k^2}(R_c - R_k)R_k \qquad (7) \\
&= \lambda_k + \left(\frac{R_c}{R_k} - 1\right).
\end{aligned}
$$

In summary, it is important to note that whether the bisection or LMS algorithm is uti-

---

the subsequent block—which is the case in H.263. For this situation, the transitional cost from node M_{i-1} to node M_i is given by the sum of two terms, $J'(\mathbf{X}_i, M_{i-1}, M_i)$ and $J''(\mathbf{X}_{i-1}, M_{i-1}, M_i)$ which constitute the contribution from the preceding and ensuing macroblocks, respectively.

lized, the fine-tuning of rate is accomplished via a single parameter λ with the desirable outcome that—no matter what bit rate results—the distortion of the GOB will be minimum for that rate. This is in striking contrast to other encoder strategies that typically scale a single parameter such as the quantizer step size to control the instantaneous rate, but cannot guarantee any type of optimal rate-distortion performance.

APPLICATION TO H.263

We now consider the application of the rate-constrained mode switching algorithm to H.263, the International Telecommunication Union's (ITU) draft recommendation for video coding over narrow telecommunications channels [4]. As is the case with the other standards, in H.263 each frame of the image sequence is first subdivided into unit regions called macroblocks which relate to 16 pixels by 16 lines of the luminance component (Y) and the spatially corresponding 8 pixels by 8 lines of both chrominance components (C_B and C_R). As part of H.263, each macroblock can also be coded using any one of several possible modes, the allowable set of which is determined by the picture coding type.

The recommendation for the standard contains two picture coding types, INTRA and INTER which specify the possible macroblock modes that may be used for the current frame. The INTRA picture type is more limiting in that it only allows intra coding for macroblocks. It is typically used only for special purposes, e.g., coding the first frame of a video sequence. In this paper, we concern ourselves with the INTER picture type because within this picture type, individual macroblocks can be coded using a large variety of macroblocks modes, including intra and inter. Specific to H.263 is an additional capability called Advanced Prediction which enforces overlapped motion compensation and permits the use of four motion vectors per macroblock. This function can be set by a single bit and impacts the macroblock modes for an entire frame. For our simulations we include the following standard and optional macroblocks modes: intra (I-mode), inter with one motion vector (P-mode), inter with four motion vectors ($P4$-mode), and uncoded (U-mode) which we now briefly describe.

In the I-mode, the luminance and chrominance components are quantized using a "JPEG-like" coding scheme. The components are initially segmented into 8×8 blocks which are subsequently transformed by the DCT. All AC transform coefficients are then identically scalar quantized with an even step-size value ranging from 2 to 62. Next, the coefficients are "zig-zag" scanned and losslessly encoded using a look-up table that exploits long runs of zeros. Special attention is paid to the quantization of the DC transform coefficient as it is uniformly scalar quantized using an 8 bit codeword. Typically, the quantizer step size is fixed for all macroblocks in a GOB. However, as part of the H.263 standard, the encoder can set a two-bit option in the macroblock header which permits a change in the quantizer step-size of ±1 or ±2 for all succeeding macroblocks. For our experiments, this type of macroblock-by-macroblock parameter adjustment is not considered for now due to the associated complexity required for its optimization, though in principle it is not a fundamental obstacle.

In the P-mode, the current macroblock is first predicted using a single, half-pixel accurate motion vector. Each motion vector points to a 16×16 luminance region and two 8×8 chrominance regions in the previously decoded frame within a horizontal and vertical range of -16 to $+15.5$ pixels. Once determined, the motion vectors are differentially encoded after each vector is first predicted using the median of three candidate vectors. The candidate vectors correspond to the three surrounding motion vectors located directly above, above and to the right, and directly left of the current

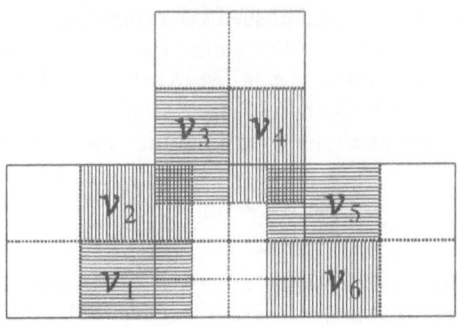

Figure 2. Illustration of H.263 overlapped motion compensation.

motion vector, respectively. Each motion-error term is encoded without loss using a single variable-length codeword from a fixed look-up table. Next, the resulting motion-compensated prediction error is transformed and quantized in the same manner as the I-mode, with the exception that the DC coefficient is not treated separately. The incremental modification of the quantizer step size for individual macroblocks, while allowed by the H.263 standard, is not considered in this paper.

When Advanced Prediction is turned off, both the I and P-modes act very similarly as in past standards such as H.261 and MPEG. In contrast, when the Advanced Prediction bit is set, the P-mode is modified to include overlapped motion compensation [19]. Moreover, by flipping this bit, an additional macroblock mode can be utilized that not only includes overlapped motion compensation, but also specifies four motion vectors per macroblock. In this mode, which we refer to as the $P4$-mode, the macroblock is segmented into four smaller 8×8 blocks, each compensated by one of the four specified motion vectors in the same manner that the larger 16×16 blocks are compensated in the P-mode. An important point is that the $P4$-mode must be used in conjunction with another special functionality of H.263, called the unrestricted motion vector mode, in order to allow the lapping of pixels located outside the frame boundaries. This function is similarly set by a single bit for an entire frame and is defined such that the pixels from the border of the picture are copied to the regions outside. The lapping from the outside into the current macroblock is depicted in Fig. 2. The vectors $\{v_1...v_6\}$ are the motion vectors from the neighboring macroblocks, and the lapping is performed using fixed weighting windows. Within a macroblock, each of the four smaller luminance blocks is similarly predicted by internally applying overlapped motion compensation between the blocks. The exact procedure for differentially encoding the four motion vectors is detailed in the recommendation. Otherwise, the same prediction loop as previously described is applied, and the quantization is performed as explained for the P-mode. Finally, the uncoded mode (U-mode) (which is indicated by just a single bit for a given macroblock) specifies that the current macroblock is to be represented by simply duplicating the contents of the corresponding macroblock in the previous frame.

According to the standard [4], "the criteria for choice of mode and transmitting a block are not subject to recommendation and may be varied dynamically as part of the coding control strategy." In what follows, we consider the application of the mode selection strategy described previously as an encoder control solution for the H.263 standard. Our goal is to determine the optimum mode selection for a given GOB. For all simulations, the GOB is defined as a single, horizontal macroblock stripe across a given frame. For example, a 176×144 QCIF-image consists of 9 macroblock stripes, each containing 11 macroblocks. We restrict ourselves to this scenario so that dependencies only arise between successive macroblocks for the purpose of employing the Viterbi algorithm. This approach also lends itself to wireless scenarios in that the generation of GOB's on a regular interval facilitates the recovery from bit errors which are more likely in the wireless environment.

We note that whereas, in general, the coding of a given macroblock in H.263 is influenced by the selected mode of neighboring blocks, there are two notable exceptions for this type of dependency: the I-mode and the U-mode in which the mode selection can be carried out independently of the surrounding macroblocks. Because there is no transitional cost between modes, the costs for these nodes can be assigned using (4). For the P-mode, the rate term is dependent on three neighboring macroblocks due to the differential encoding of the motion vectors. By restricting the GOB to a horizontal macroblock stripe, we can eliminate the impact on the trellis from above and need only consider those dependencies resulting from the immediately preceding macroblock. Consequently, we can assign a transitional cost from the previous node to the current node using (5).

In the case of Advanced Prediction, for both the P and $P4$-mode, rate and distortion are dependent on the previous choice for the macroblock mode, while the distortion is dependent on the succeeding macroblock mode as well. Using (6), we can compute the cost for the incoming and outgoing transitions of the current node assigned for the P and $P4$-modes as follows. As illustrated in Fig. 2, the distortion of the left half of the macroblock is only influenced by the motion vectors of the macroblock to the left and from the above. The macroblocks modes from above are fixed because they are determined in the previous GOB, and thus, we need only consider the influence from the left when computing the distortion component of $J'(\cdot)$ in (6). Analogously, all distortion influences except those from the right can be eliminated when computing the distortion component in $J''(\cdot)$. Likewise, the distortion for both chrominance components are equally distributed to the in and outcoming transitions. In terms of rate, the cost assignment to the trellis branches is slightly more complicated because the motion vectors on the right half of the $P4$-mode are predicted from the motion vectors to the left. Consequently, a dynamic update for $J'(\cdot)$ and $J''(\cdot)$ based on the decisions for the incoming transitions is required. Finally, the quantizer step size parameter, QUANT, is optimized using the strategy outlined in [9] for each GOB.

CODING RESULTS

Simulation results for the proposed mode switching strategy as applied to H.263 are provided in Figs. 3–4. In the first experiments, the frame rate is held constant at 8.33 frames per second and the Lagrange multiplier λ is varied to generate coded sequences with an overall average rate from 2.9 Kbits per second (Kb/s) to 400 Kb/s. Though fixing λ for the video sequence does not represent a practical implementation since the maximum rate is not constrained, it does provide a means for assessing the relative importance of each mode at different bit rates. For example, Fig. 3 demonstrates the probability of selecting the I, P, $P4$, and U modes after running the algorithm on the well-known color video sequence, "Grandmother." Similar plots for "Mother-Daughter" and "Car Phone" can be found in [9]. Upon close examination, several intuitively appealing results are confirmed by the plots. For instance, the U-mode, as expected, drops to zero at high rates (when $\lambda = 0$) for all of the test sequences. For the "Grandmother" and "Mother-Daughter" sequences, where the motion model is fairly accurate, the I-mode is almost never chosen except at the highest rates. In contrast, for the "Car Phone" sequence which has more complex motion, the algorithm

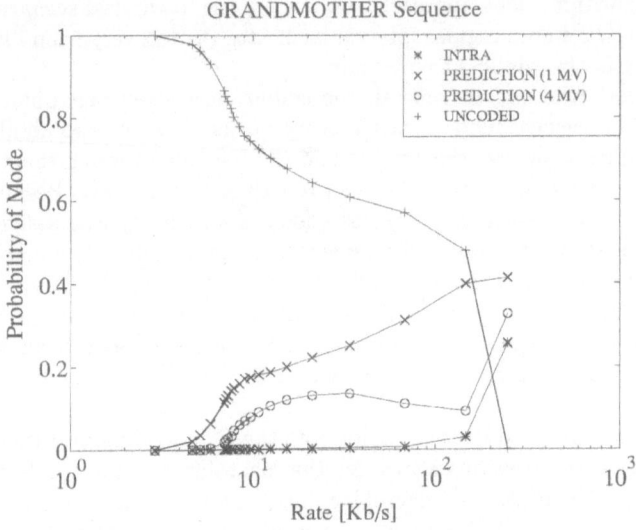

Figure 3. Probability of mode versus rate for the "Grandmother" sequence.

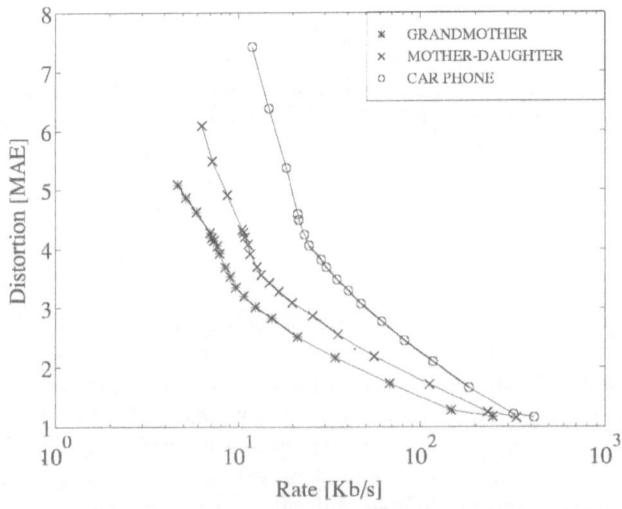

Figure 4. Plot of rate versus distortion for the "Grandmother", "Mother-Daughter", and "Car Phone" video sequences. The Lagrange multiplier λ is varied from 50 to 10 with an equidistant step size of 10, from 10 to 5 with step 1 and from 5 to 0 with step 0.5. Note: the frame skip is held constant at 2 for a frame rate of 8.33 frames per second.

selects the I-mode more frequently as one might anticipate. Note that the seemingly erratic behavior in the mode probabilities at the highest rates can be attributed to the logarithmic scale of the plots, and the fact that only a few data points were computed at these rates since they are outside the recommended usage for H.263.

Next, the overall rate-distortion performance for the three sequences is shown for the mean absolute error (MAE) in Fig. 4. The plots are generated by varying λ from 0 to 50. It is interesting to note that the relationship between λ and distortion (for rates above 10 Kb/s) is rather consistent between the three different sequences, i.e., the same value of λ corresponds roughly to the same value of MAE in all cases. If the primary objective is a constant-distortion coder, then this is good news in that the Lagrange multiplier need not be substantially modified from one frame to the next. Unfortunately, the same desirable relationship does not manifest itself for rate and λ. In fact, depending on the sequence, the same value of λ may correspond to widely varying bit rates. Thus, if the goal is coding for a specified rate, which is more often the case, a method for for controlling the Lagrange multiplier is required. For this reason, we consider a more sophisticated encoder control strategy than before in which the frame-skip is adaptive, and the LMS algorithm is used to update λ on a frame-by-frame basis. These steps are undertaken to generate a more constant rate, and thus, a more practical system. The coding results of the scheme are very encouraging, producing usable video with rates as low as 6.9 Kb/s, 11.0 Kb/s, and 18.3 Kb/s for the "Grandmother", "Mother-Daughter", and "Car Phone" sequences, respectively. The maximum rates for these scenarios correspond to 8 kb/s, 12 Kb/s, and 20 Kb/s. Note that the average rates are very close to the maximum allowable rates for all three of the sequences, confirming the applicability of the LMS algorithm with regards to rate control. Interested readers can also refer to [9] in order to view sample still images taken from these coded sequences.

ACKNOWLEDGMENTS

This work was supported in part by the Ditze Foundation, a National Science Foundation Graduate Fellowship and in part by a University of California MICRO grant with matching supports from Hughes Aircraft, Signal Technology Inc., and Xerox Corporation. The authors are greatful for the helpful comments of Jong Dae Kim.

REFERENCES

[1] ITU-T Recommendation H.261, "Video codec for audiovisual services at $p \times 64$ kbit/s", Dec. 1990, Mar. 1993 (revised).

[2] ISO/IEC 11172-2, "Information technology–coding of moving picture and associated audio for digital storage media at up to about 1.5 mbit/s: Part 2 video", Aug. 1993.

[3] ITU-T Recommendation H.262—ISO/IEC 13818-2, "Information technology–generic coding of moving picture and associated audio for digital storage media at up to about 1.5 mbit/s: Video", (Draft), Mar. 1994.

[4] ITU-T Recommendation H.263, "Video coding for narrow telecommunication channels at less than 64 kbit/s", (Draft), April 1995.

[5] ITU-T, SG15, WP15/1, Expert's group on Very Low Bitrate Video Telephony, LBC-95-193, Delta Information Systems, "Description of mobile networks", June 1995.

[6] ITU-T, SG15 WP15/1, LBC-95-194, Robert Bosch GmbH, "Suggestions for extension of recommendation H.263 towards mobile applications", June 1995.

[7] K. Ramchandran, A. Ortega, and M. Vetterli, "Bit allocation for dependent quantization with applications to multiresolution and MPEG video coders", *IEEE Trans. on Image Processing*, vol. 3, no. 5, pp. 533–545, Sept. 1994.

[8] J. Lee and B.W. Dickinson, "Joint optimization of frame type selection and bit allocation for MPEG video coders", in *Proc. of the Int. Conf. on Image Proc.*, 1994, vol. II, pp. 962–966.

[9] T. Wiegand, M. Lightstone, T.G. Campbell, and S.K. Mitra, "Efficient mode selection for block-based motion compensated video coding", in *Proceedings of the 1995 IEEE International Conference on Image Processing (ICIP '95)*, Washington, D.C., Oct. 1995, (To appear).

[10] M. Lightstone, D. Miller, and S.K. Mitra, "Entropy-constrained product code vector quantization with application to image coding", in *Proceedings of the First IEEE International Conference on Image Processing*, Austin, Texas, Nov. 1994, vol. I, pp. 623–627.

[11] G.D. Forney, "The Viterbi algorithm", *Proceedings of the IEEE*, vol. 61, pp. 268–278, Mar. 1973.

[12] H. Everett III, "Generalized lagrange multiplier method for solving problems of optimum allocation of resources", *Operations Research*, vol. 11, pp. 399–417, 1963.

[13] Y. Shoham and A. Gersho, "Efficient bit allocation for an arbitrary set of quantizers", *IEEE Trans. Acoust., Speech, and Signal Processing*, vol. 36, no. 9, pp. 1445–1453, Sept. 1988.

[14] S.W. Wu and A. Gersho, "Rate-constrained optimal block-adaptive coding for digital tape recording of HDTV", *IEEE Trans. on Circuits and Systems for Video Technology*, vol. 1, no. 1, pp. 100–112, March 1991.

[15] J.E. Dennis and R.B. Schnabel, *Numerical methods for unconstrained optimization and nonlinear equations*, Prentice-Hall,, Englewood Cliffs, NJ, 1983.

[16] K. Ramchandran and M. Vetterli, "Best wavelet packet bases in a rate-distortion sense", *IEEE Trans. on Image Processing*, vol. 2, no. 2, pp. 160–175, Apr. 1993.

[17] S. Haykin, *Adaptive Filter Theory*, Prentice Hall, Englewood Cliffs, NJ, 1991.

[18] M. Rupp, "Bursting in the LMS algorithm", 1995, Submitted for publication.

[19] H. Watanabe and S. Singhal, "Windowed motion compensation", in *Proc. of the SPIE Symposium on Visual Comm. and Image Proc.*, 1991, vol. 1605, pp. 582–589.

STEREOSCOPIC VIDEO CODING

Tihao Chiang and Ya-Qin Zhang

David Sarnoff Research Center
201 Washington Road, Princeton
New Jersey 08543

ABSTRACT

Coding of the stereoscopic video source has received significant interest recently. The MPEG committee decided to form an ad hoc group to define a new profile which is referred to as Multiview Profile (MVP) [4]. The importance of multiview video representation is also recognized by the MPEG 4 committee as one of the eight functionalities to be addressed in the near future. In this paper, we will first review the technical results using temporal scalability (disparity analysis) in MPEG-2 as pioneering by [9] and [10]. Based on temporal scalability, the concept is further generalized to affine transformation to consider the deformation and foreshortening due to the change of view point. Estimation of the affine parameters is crucial for the performance of the estimator. In this paper we propose a novel technique to find a convergent solution which results in the least mean square errors. Our result shows that about 40 percent of the macroblocks in a picture has benefited by using the affine transformation. In our approach, the additional computational complexity is minimal since a pyramidal scheme is used. In one of our experiments, only four interations are necessary to find a convergent solution. The improvement in prediction gain is found to be around 0.77 dB.

1. INTRODUCTION

Coding for stereoscopic television has become an active and important research topic, with the recent advances in video compression algorithms, standardization efforts, and VLSI implementations. There are several applications that have been associated with stereoscopic video processing. For example, it is common in computer vision to estimate 3-D motion and shape information based on stereoscopic source. Stereoscopic image representation is also advantageous for the robot remote guidance. For medical imaging it is desirable to have a stereoscopic presentation of the depth information using CT (Computed Tomography) and MRI (Magnetic Resonance Imaging). In Europe the DISTIMA (Digital Stereoscopic Imaging & Applications) project is currently investigating a compatible transmission of stereoscopic video based on satellite broadcast [3]. Stereoscopic Radar imaging has been found to be an important application [7]. In view of the importance of the stereoscopic video applications, the MPEG committee decided to form an ad hoc group to define a new profile which is referred to as Multiview Profile (MVP) [4] [9] and [10]. The goal is to provide a compression scheme with multiple viewpoint capability while maintaining a backward compatibility with the current monoview MPEG compression scheme. The importance of multiview video representation is also recognized by the MPEG 4 committee as one of the eight functionalities to be addressed in the near future. In this paper we investigate several options and propose new coding tools for efficient representation of video sequences with stereoscopic and multiview formats.

There are several approaches for stereoscopic video coding. The mostly used scheme is based on the disparity analysis and motion analysis. In this approach, the change of viewing angle is considered as a zero time camera movement which is compensated using the disparity analysis. The occlusion is augmented using motion compensation. Such a representation is sufficient for two channels, however, generalization of this approach to multiple channels and viewpoints becomes awkward and inefficient. Since a stereoscopic video sequence is available, an estimation of the 3-D motion based on both channels is possible in achieving better prediction models. In the disparity analysis approach, the deformation and foreshortening are not considered. Using digital image warping technique, the movement of perspective points can be included in the estimator. There are approaches using either affine transformation or quadratic transformation. The main challenge in this approach is the difficulty in interpolating the intermediate viewpoints. Accurate estimation of the model parameters is a difficult problem. One of the disadvantages is the computational complexity in estimating these model parameters. Another disadvantage is that the improvement of prediction may not justify the amount of side information and the increase of complexity.

In this paper, several coding methods are examined. First, we present the experimental results using different forms of temporal scalability to achieve stereoscopic coding. Then we describe a new method based on both block based disparity analysis and affine transformation to interpolate the right channel signals, with a new algorithm in efficiently estimating the affine parameters. Finally, experimental results are shown to demonstrate the advantages of this new technique.

2. TEMPORAL SCALABILITY

2.1 Technical Description

MPEG-2 with main profile at main level (MP@ML) is the most popular format for standard TV broadcasting. In addition to the main profile, there are other profiles that include several scalability tools as defined in MPEG-2 [1]. There are four different scalable modes in MPEG-2 including spatial scalability, temporal scalability, SNR scalability and data partitioning.

A stereoscopic video sequence is composed of two video sequences that represent the left channel and the right channel. Video sequences of both channels are captured at the same time with different viewing angles. Therefore, there is an inherent redundancy between the two channels since they represent snapshots of the same object at the same instant of time with different perspectives. Thus, we use a scalable approach to represent a stereoscopic video sequence in a layered structure. The base layer encoder compresses only the left channel. Then, the enhancement layer encoder compresses the right channel based on the decoded left channel video sequence. Such a layered structure is very desirable in providing services to various users with different display capability and bandwidth. For example, scalability can provide compatibility among different spatial resolutions and scanning formats[8].

2.2. Experimental Results Using Temporal Scalability

The following set of experiments have been performed using a total bandwidth of 6 Mbps. The source material is an interlaced stereoscopic sequence "tunnel" with a resolution of 720x576. We have extensively experimented different possibilities including all combinations in forming the prediction using the permissible modes of temporal scalability. The coding results shown in Table 1 have a significant effect on the final selection of predictor used in the next sections.

The simulcast performs better than the P-picture coding (as shown in the experiments 1, 2, and 3) because the simulcast approach allows the interpolated pictures in the right channels. This is prohibited in the syntax of temporal scalability to prevent from additional delay for frame reordering at the enhancement layer. If we adaptively select the P-picture for prediction (permissible by the temporal scalable syntax), experiment 8 shows slight improvement over simulcast. However, the cost in complexity is unfavorable since the improvement is limited. The simulcast approach also performs better than temporal scalable approach using sources B and C (as shown in the experiment 6). This can be explained by the fact that the perspective deformation has caused the predictions which come from the left channel less efficient. From experiment 4 and 5, we can deduce that source A and B will give the best performance as compared to the simulcast approach. If we further adaptively select the best two frames for prediction, the improvement shown in experiment 7 over experiment 5 is limited.

The conclusions we can deduce from these experiments are the following. The best compression performance is obtained by selecting the best two frames using temporal scalable syntax as described in MPEG [1]. However, the next best approach which uses sources A and B performs comparably with much less complexity. Another observation is that the prediction from the left channel (source B and C) are not as accurate as the prediction from source A. This is due to the perspective deformation from the stereoscopic video source. In the next section, we shall explore new approaches using affine transformation which can better take advantage of the perspective correlations between the two channels.

3. AFFINE TRANSFORMATION

3.1 Formulation of the Approach

In this section, we will first describe the mathematical formulation of three-dimensional motion parameters. We start from the most general case where 12 parameters are required. However, we investigate only the affine transformation since it requires only 6 parameters. Figure 3 shows the block diagram where the affine transformation is used in our frame work.

Different modes to encode the stereoscopic video sequences are investigated in this section. The main idea is to encode a set of parameters on a macroblock basis (16x16 pixels). The encoder can

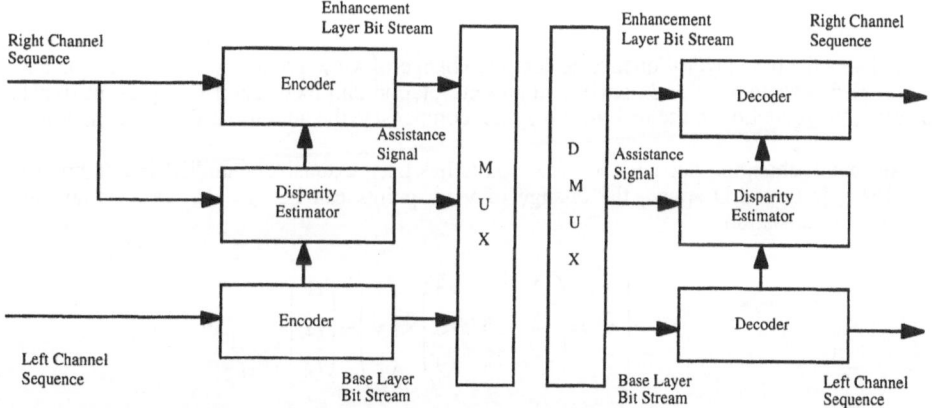

Figure 1. Block diagram for stereoscopic video coding using disparity analysis. The syntax for encoding is based on the temporal scalable profile syntax in MPEG-2.

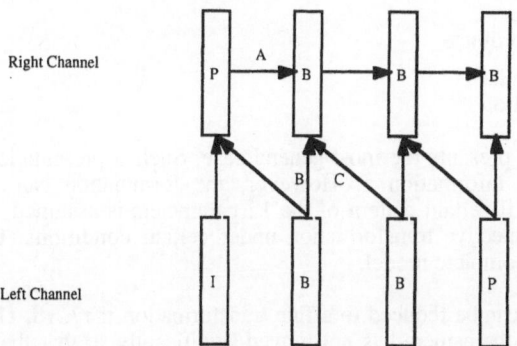

Figure 2. Different prediction modes for estimating the right channel signals. Source A: Most recent decoded right channel picture. Source B: Most recent decoded left channel picture in display order. Source C: Next decoded left channel picture in display order.

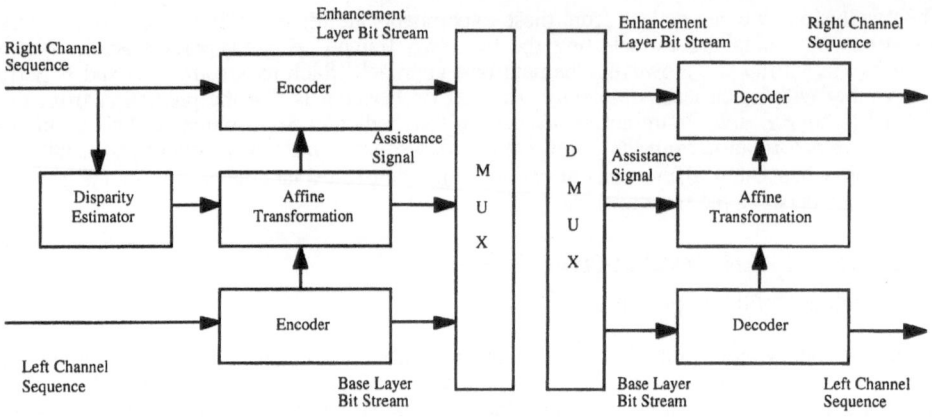

Figure 3. A block diagram for stereoscopic video coding using disparity analysis and affine transformation. The estimation of the affine parameters is based on both block matching algorithm and Gauss-Newton interative method.

manipulate a the base layer sequence so that the change of view point is considered. For the most general case, we can use a 3x3 matrix to represent rotation and 3x1 vector to represent translation. However, the reduction in the residuals may not compensate the additional side information .

The concept of changing perspective view points has been extensively studied in computer vision (CV) [5] [6]. In a 3-D space, the change of view points can be modeled as a combination of rotation and translation.

$$
\begin{pmatrix} \overline{x} \\ \overline{y} \\ \overline{z} \end{pmatrix} = \begin{pmatrix} r1 & r2 & r3 \\ r4 & r5 & r6 \\ r7 & r8 & r9 \end{pmatrix} \times \begin{pmatrix} x \\ y \\ z \end{pmatrix} + \begin{pmatrix} t1 \\ t2 \\ t3 \end{pmatrix}
$$

The correspondence points are found as

$$
\overline{X} = \frac{(r1 \times X + r2 \times Y + r3) \times z + t1}{(r7 \times X + r8 \times Y + r9) \times z + t3}, \quad \overline{Y} = \frac{(r4 \times X + r5 \times Y + r6) \times z + t2}{(r7 \times X + r8 \times Y + r9) \times z + t3}
$$

X, Y: is the original coordinate
$\overline{X}, \overline{Y}$: is the new coordinate
z: is the depth information

The above formulation presents the most general case. Such a presentation requires the accurate estimate of the depth information z. However, the formulation can be reduced to a less sophisticated approach if certain pattern of the 12 parameters is assumed. Specifically, it can be reduced to affine, perspective transformation under certain conditions. If depth information is found, we can use the complete model.

The above formulation can be reduced to affine transformation if r7, r8, t1, t2, t3 are set to zero. The assumption is that the camera has not moved significantly so that the translation is ignored. This is true for the stereoscopic application since the distance of the two cameras are small. The advantage of using affine transformation is that the additional complexity is minimal. Thus, the formulation can be reduced to affine as described in the following equation.

$$
\begin{pmatrix} \bar{x} \\ \bar{y} \\ \bar{z} \end{pmatrix} = \begin{pmatrix} r1 & r2 & r3 \\ r4 & r5 & r6 \\ 0 & 0 & 1 \end{pmatrix} \times \begin{pmatrix} x \\ y \\ z \end{pmatrix} + \begin{pmatrix} 0 \\ 0 \\ 0 \end{pmatrix}
$$

The correspondence relation

$$\bar{X} = r1 \times X + r2 \times Y + r3, \quad \bar{Y} = r4 \times X + r5 \times Y + r6$$

The formulation can be generalized into perspective transformation if t1, t2 and t3 are set to zero. The same formulation is shown with proper setting of parameters.

$$
\begin{pmatrix} \bar{x} \\ \bar{y} \\ \bar{z} \end{pmatrix} = \begin{pmatrix} r1 & r2 & r3 \\ r4 & r5 & r6 \\ r7 & r8 & r9 \end{pmatrix} \times \begin{pmatrix} x \\ y \\ z \end{pmatrix} + \begin{pmatrix} 0 \\ 0 \\ 0 \end{pmatrix}
$$

The correspondence relation has reduced to

$$\bar{X} = \frac{r1 \times X + r2 \times Y + r3}{r7 \times X + r8 \times Y + r9}, \quad \bar{Y} = \frac{r4 \times X + r5 \times Y + r6}{r7 \times X + r8 \times Y + r9}$$

However, the computational load and complexity for the perspective transformation is much higher than that for the affine transformation. Therefore, we will first limit the transformation to be affine assuming that such approximation is acceptable. The main technical challenge is the estimation of the affine parameters for a very small image block (for example, 16x16). An iterative gradient based approach can easily diverge if the initial point is not selected correctly. In the next section, we will explore a pyramidal scheme to find a good initial point in estimating the affine parameters. If a convergent solution is not found, the estimator will fall back to disparity analysis. Thus, we can assure that the performance can be at least lower bounded by the disparity analysis.

3.2. Estimation Algorithm

An accurate and efficient algorithm to estimate the six affine parameters is described in this section. It should be pointed out that the algorithm is not limited to affine transformation. In order to reduce the computational complexity, a spatial pyramid is used for the estimation. The pyramidal approach also can prevent the estimation process from trapping into a local minimum or

Table 1. Coding results of stereoscopic video sequence. The experiments use different prediction modes permissible in the temporal scalable syntax

		Tunnel		
Experiments	Coding Parameters	Y	U	V
Left Channel @ 4 Mbps	N=12 M=3	35.37	39.06	40.32
Right Channel Simulcast @ 2 Mbps	N=12 M=3	32.17	37.84	38.90

	Experiment	Right Channel @2Mbps using the temporal scalable syntax		
	Tunnel	Y	U	V
1	P-picture using source A	32.08	37.48	38.49
2	P-picture using source B	30.41	36.03	35.85
3	P-picture using source C	30.22	35.89	35.55
4	B-picture using source A and B	33.25	37.77	38.57
5	B-picture using source A and C	33.17	37.74	38.52
6	B-picture using source B and C	30.24	35.82	34.92
7	B-picture using A, B and C adaptively	33.35	37.71	38.68
8	P-picture using A, B and C adaptively	32.49	37.40	38.41

divergence. The novelty of our approach is that we combine the concept of block matching algorithm (BMA) and gradient-based Gauss-Newton optimization search. The advantage of such a hybrid approach is that we can achieve global minimum without much computational complexity.

Both the left image and the right video sequence are first subsampled into four layers which are reduced in resolution by half in both the horizontal and the vertical directions. For each layer the estimator follows these steps described below. The initial point is determined by searching a prescribed area using a block matching algorithm. This will yield a stable initial point to estimate the affine parameters. The affine parameters is estimated using the Gauss-Newton optimization technique. Based on the previous initial point, we can find a differential increment to yield a new affine parameters. Using this new affine parameters, we can warp the left channel signal toward the right channel signal. An evaluation of the sum of the square errors(SSE) is performed to determine if the iteration should continue. If a decrease of SSE is observed, another iteration of the Gauss-Newton method is applied. The iteration stops at the point either the SSE is not decreasing significantly or the SSE starts to diverge. In the case of divergence, the new parameters are reset to the initial parameter values which are found using the BMA. Thus, the estimation of a level of the pyramid is completed. All the information acquired at each layer is passed onto the next layer as initial point for optimization.

Our approach has multiple advantages. The computational load is reduced significantly due to the fact that a pyramidal approach is used. The possibility of finding a useful convergent solution is high. In the case of divergent solution, the estimator can always fall back to the solution obtained using the BMA approach. This is of particular importance when the region of interest is reduced to a much smaller size. For example, the Gauss-Newton iteration needs a good initial point when the area size is reduced to the range of 16x16 to 32x32 areas.

3.3. Experimental Results

Experiments have been carried out to compare the estimator based on either disparity analysis or a hybrid of both disparity analysis and affine transformation. The method to estimate the affine parameters is described in the previous section. It is observed that the prediction which comes from the left channel is improved by 0.66 dB and 0.77 dB for "manege" and "tunnel" sequences respectively. The maximum number of iterations for the Newton-Gauss method is only 4 for each 16x16 macroblock. Thus, the computational load is nominal as compared to the additional motion estimation for the affine transformation. If the estimation of the affine transformation diverges, the estimator will observe that using the mean square error measure and fall back to the disparity analysis mode. Thus, the prediction performance is lower bounded by the performance of the disparity analysis approach.

	Disparity Analysis			Disparity Analysis & Affine Transformation		
	Y	U	V	Y	U	V
Manege	22.80	33.07	33.95	23.46(+0.66)	33.17(+0.10)	33.95(+0.00)
Tunnel	27.41	35.43	32.78	28.18(+0.77)	35.65(+0.22)	32.88(0.10)

Table 2: Comparison of the predictor using disparity analysis or a hybrid of both disparity analysis and affine transformation. The iteration number for the affine transformation is 4.

4. CONCLUSIONS

In this paper, we have studied two approaches to encode a stereoscopic video sequence. First, we have reviewed the technical results by using the MPEG-2 temporal scalability syntax. After an exhaustive search of all the possibilities permissible by the current syntax, the conclusions are that the best performance is achieved by adaptively selecting the best two sources of prediction from the three possible sources of prediction. It is found that the performance is degraded by only 0.1 dB if a fixed set of sources of predictions is used. Theses two sources are the past decoded right channel signals and the current decoded left channel signals.

We generalize the concept of disparity analysis to affine transformation. The estimation of the six affine parameters is a technical challenging problem for small image block size. It is crucial to find a convergent and accurate solutions in a few interations. We have proposed an estimation technique which is a hybrid of BMA and Gauss-Newton method. The advantages of our proposal is that the additional computational complexity is nominal. We find convergent and improved solutions for about 40 percent of the image blocks in a picture. The improvement of the prediction is observed to be around 0.77 dB with a maximum of 4 iterations. Therefore, the hardware implementation of the technique is feasible and we expect more improvement in the future investigation.

ACKNOWLEDGMENTS

The authors wish to thank Dr. Huifang Sun, Dr. Joel Zdepski and Dr. Michal Irani for their helpful discussions and assistance. The authors also wish to thank Mr. Bruno Chroquet for providing us with the source stereoscopic test sequences.

REFERENCES

1. ISO, Information Technology-Generic Coding of Moving Pictures and Associated Audio Information: Video, Recommendation H.262, (Paris), May 1994.
2. Chassaing F., Choquet B., Pele D., "A stereoscopic television system (3D-TV) and compatible transmission on a MAC channel (3D-MAC)", Image Communication Nov, 1991
3. International Organisation for Standardisation, "Report of the ad hoc group on MPEG-2 applications for multi-viewpoint pictures", ISO/IEC JTC/SC29/WG11 No. 861 March, 1995
4. International Organisation for Standardisation, "Status Report on the study of Multi-viewpoint pictures", ISO/IEC JTC/SC29/WG11 No. 906 March, 1995.
5. A. Zakhor and F. Lari, "Edge-Based 3-D Camera Motion Estimation with Application to Video Coding," IEEE Trans. on Image Processing, Vol. 2, No. 4, E 1993.
6. Roger Y. Tsai and Thomas S. Huang, "Uniqueness and Estimation of Three-Dimensional Motion Parameters of Rigid Objects with Curved Surfaces," IEEE Trans. on PAMI, Vol. 6, No. 1 Jan. 1984.
7. Randall B. Perlow, Ph.D. Dissertation University of Pennsylvania, "The Application of Stereoscopic Techniques to High Resolution Radar Images for Improved Detection of Targets in Clutter", 1994.
8. Tihao Chiang, Ph.D. Dissertation Columbia University, "Hierarchical Coding of Digital Television", 1995.
9. A. Puri, V. Kollarits and B.G. Haskell, " Stereoscopic Video Compression using temporal scalability", SPIE Visual Communications and Image Processing, Taipei, Taiwan, May 1995.
10. B. L. Tseng and D. Anastassiou, "Compatible video coding of stereoscopic sequences using MPEG-2's scalability and interlaced structure," Workshop on HDTV'94, Torino, Oct. 1994.

COMPRESSED-DOMAIN CONTENT-BASED IMAGE AND VIDEO RETRIEVAL

Shih-Fu Chang

Department of Electrical Engineering &
Center for Telecommunications Research
Columbia University, New York, NY 10027
sfchang@ctr.columbia.edu, http://www.ctr.columbia.edu/~sfchang

INTRODUCTION

With more and more visual material produced and stored in visual information systems (VIS) (*i.e.*, image databases or video servers), the need for efficient, effective methods for indexing, searching, and retrieving images and videos from large collections has become critical. Users of large VIS will desire a more powerful method for searching images than just traditional text-based query (*e.g.*, keywords). Manual creation of keywords for a huge collection of visual materials is too time-consuming for many practical applications. Subjective descriptions based on users' input will be neither consistent nor complete. Also, the vocabulary used in describing visual contents is usually domain specific.

To overcome these shortcomings, there is a recent approach called "content-based" visual query, which allows users to specify query and execute search based on the visual content of the visual material [1, 23]. The term "content" refers to the semantic structure of images and videos at various levels, ranging from pixel patterns, physical objects, to spatial/temporal structures of the visual material. The content-based approach is not intended as a replacement for the keyword approach. Instead, it is considered as a complementary tool, particularly for applications which have large data collection and require fast search response. Provision of the content-based visual retrieval techniques also brings in new synergy between the text-based information and the visual information of the same material. Fusion of different information channels (text and visual in this case) has been used to achieve performance improvement in multimedia databases such as news archive [4].

A content-based visual query system requires several key components, including visual feature extraction, feature indexing data structure, distance (or similarity) measurement, fast search methods, integration of different visual features, and integration of visual features with text-based indices. We focus on issues directly related to image/video processing. Our goal is to investigate automated methods using various useful visual features for content-based visual query. This is a relatively new area, in which some promising work has been reported in the literature for different applications. The QBIC system of IBM [1] provides semi-automatic mechanisms for extracting color, texture, shape, and structure for image query. Pentland *et al* [13] have demonstrated image query systems based on shape, texture, and face. Video indexing based on temporal characteristics has been described in [14, 17, 20, 21, 28]. Other researchers [24, 22] have also proposed techniques for image search based on shape, color, or texture.

In this paper, we focus on automated image retrieval using various types of visual features (including image templates). Specifically, we explore a unique approach which emphasizes the compressed-domain solutions. Given most images and videos stored in the compressed form, it is highly desirable to perform content-based visual retrieval directly in the compressed domain. In other words, extraction of visual features, matching of images and videos, or even manipulations of search results can

Feature Specification Tools

texture, color, shape,
object layout, motion,
camera operations, temporal structure,
face, local, audio, speech

Semantic Structure

"A red car in front of a house",
"A person running on the beach"

Query by Image Example

feature-based
image template matching

—> **(user assistance needed)**   —> **(automatic process possible)**

FIGURE 1. Different Modalities of Image/Video Search and Retrieval

be implemented using the compressed visual data without decompressing (or sometimes with minimal decoding). The advantage of performing these tasks in the compressed domain is obvious. The computational cost can be greatly reduced compared to the alternative which fully decodes the compressed material and executes the above tasks in the uncompressed domain. Not only that the decoding process is avoided, the amount of data in the compressed domain is also lower than in the uncompressed domain. We have applied the compressed-domain approach to image/video manipulation and compositing [2,11], texture/color-based image query [9], image matching [10], video indexing and editing [6, 25].

We first describe different modalities for image/video retrieval in a VIS. We then present an overview of our current work in compressed-domain image/video retrieval, followed by a brief description of an integrated VIS prototype system with Video-on-Demand applications [5]. Finally, we discuss conclusions and future work.

DIFFERENT MODALITIES OF IMAGE SEARCH AND RETRIEVAL

There are several possible ways of indexing and retrieving visual material. From users' point of view, the more methods available, the higher the flexibility they can exercise and adapt to the specific information they are seeking. However, from the system designer's point of view, different methods imply different cost and efficiency. It's important to achieve good overall system performance. Figure 1 shows several possible methods for image query. The traditional keyword-based retrieval methods can be extended to more general semantic-level descriptions such as "a red car in front of a house" or "a person running on the beach". This type of retrieval requires semantic information given by users in the indexing stage. The second type of query allows users to specify a complete image or an image region as the query key. Specific images can be retrieved based on the similarity with the input image itself or the image features derived from the input image. This is usually called *query by image example*. The last type of query shown in the figure is *feature-based image retrieval*. Visual signal features are extracted in the indexing stage and compared in the search stage to find the "similar" images/videos. Typical features include texture, color, shape, object layout, motion, camera operations, face, logo, associated audio and speech, *etc.* Some features are for still images and others are for videos, although in general all still image features are applicable to videos as well. The formulation of input features to the search engine can be provided by user's raw data (*e.g.*, drawing and sketch) or user's selection from system templates. The population of feature sets from each image or video in the database can be automatic or semi-automatic (*i.e.*, with user's assistance). Knowledge of application domain also helps significantly in developing reliable, automatic retrieval techniques. In this paper, we focus on the automatic techniques in the compressed domain, with necessary domain knowledge whenever available.

TEXTURE AND COLOR FEATURE EXTRACTION IN THE SPATIAL-FREQUENCY TRANSFORM DOMAIN

Among various image features, texture, color, and shape are the most typical features used in characterizing the image content for image retrieval. Texture is useful for discriminating different natural objects, such as trees, walls, and textiles. Color has been used as a powerful technique in discriminating colored objects and different video scenes. Shape gives the most direct geometric information about the object of interest. However, it is also more difficult to extract under unconstrained conditions. Spatial and temporal relations of these features can provide higher-level information about the images and usually can help achieve better performance.

We have made progress in handling the first two features, texture and color. Psychological studies have suggested that human eyes recognize different textures based on their frequency and orientation properties. Based on this assumption, spatial-frequency methods have been used to perform texture discrimination and segmentation. Extending this assumption, we define the texture features from several popular transforms used in image compression. Examples include the Discrete Cosine Transform (DCT) and the Subband Transform (wavelet, uniform, and tree structure). Texture features are defined as the energy distributions over different subbands. For DCT, different subbands are formed by re-grouping transform coefficients using a polyphase transform. The above method of feature extraction is considered as a compressed-domain solution, since many practical systems use these transforms as the underlying compression technique. Given images represented by the transform coefficients, texture features can be directly computed without decoding of compressed images.

The above texture feature can be applied to texture classification and segmentation. For classification, we tested the entire Brodatz texture set of 112 texture classes [7]. Using the Fisher Discriminant Analysis technique, we derived a linear transform which transforms the above feature vectors to a new space in which the average distance between different texture classes is maximized and the average distance within each texture class is minimized. The Mahalanobis distance measure in the transformed feature space was used to compute the dissimilarity between any two images. This texture feature has been tested on a texture database containing more than 2000 random cuts from the Broadtz set. It showed satisfactory accuracy at about 89% - 93%, depending on the specific transform used.

In texture segmentation, we adopted a modified quad-tree data structure to assist the segmentation process. Thresholds for merging and splitting neighboring nodes in the quad-tree were established based on heuristics optimized in the training process using the entire Broadtz texture set. Using this quad-tree based texture segmentation method, we extracted prominent texture regions from a general image. Note that we emphasize the term "prominent" here because our goal is not to perform accurate object segmentation in unconstrained images (which is still considered hard). Instead, our primary interest is finding significant regions which have consistent visual properties such as texture.

In another effort to improve the indexing and search efficiency, we further simplified the process of extracting texture features. We still focused on the energy distributions in the transform domain, particularly the wavelet subband domain. We took advantage of the spatial resolution provided by the wavelet subband decomposition and defined a pixel-based texture signature, which is a binary vector indicating the energy level of each pixel in each subband. Neighboring pixels with the same texture signatures were grouped to the same texture region, using some non-linear smoothing operations. Each texture region was indexed by a single texture signature, which then can be used in the searching stage. Note that since the texture feature was defined by a binary vector, distance measurement between different textures can be easily computed (e.g., the hamming distance).

Figure 2 shows a diagram illustrating the above texture extraction algorithm in the subband domain. It also shows the concept of extracting color features from the approximation subband (i.e., the lowest subband). We define the color feature in a way so that prominent image regions with a consistent "single" color or a consistent "color mixture" can be extracted. By quantizing a carefully chosen color space, we first reduce the color resolution of the images. For each quantized color, we calculate the distance map, apply the non-linear smoothing operation, and extract prominent image regions with this color. Prominent image regions with multiple colors (e.g., mixture of red, white, and blue in the U.S. flag) can also be extracted using extension of this method. Each resulting color region is indexed by a binary feature vector, indicating the presence of individual colors in this region.

With the combination of multiple features (e.g., texture and color in this case), users can search for desired images with greater flexibility. The relationship between the color regions and texture regions also provides useful information for image retrieval. In our feature definitions, both color and texture use binary feature vectors, which then can be concatenated easily for joint descriptions.

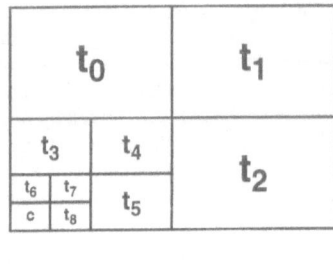

(a)

(b)

FIGURE 2. Extracting texture and color in the wavelet subband domain (a) color from the lowest subband and texture from other subbands (b) an example illustrating the extracted texture regions.

QUERY BY IMAGE MATCHING

As mentioned earlier, one way of image query is for users to provide an input image or image region, which is then matched against all image candidates in the database. The matching procedure can be based on derived features, as described above, or directly on the raw image data. Some papers categorized the former as "feature-descriptor based", the latter simply as "image matching". We focus on image matching in this section.

Image matching provides a brute-force method for finding specific images from the database. Compared to feature-based image retrieval, it does not require the operations for feature extraction. Also, if consistent, prominent features are difficult to extract, image matching can be considered as a valuable alternative. However, image matching does not provide straightforward correspondence to subjective perception. If the input image is not generated from the existing images in the database, it may be hard to evaluate the relevance of the returned images simply based on the relative ranking of the matching scores.

Image matching can be implemented by several possible methods, including correlation, MSE (mean square error), and normalized correlation (*e.g.*, correlation coefficient). Applying the compressed-domain approach, we have investigated a fast algorithm for image correlation directly in the subband domain. It can be shown that other matching criteria, such as MSE and normalized correlation, can be computed based on the correlation operation. Given the subband components of two images, the complete image correlation can be expressed as the summation of the filtered correlations of different subbands, including intra-band and cross-band correlations. For ideal lowpass and highpass filter banks, the cross-band correlations are cancelled. In a related research, Vaidyanathan [27] has used the subband convolvor (time reversed version of correlation) to improve the coding gain and accuracy of convolution.

To take advantage of energy compaction of subband decomposition, we further proposed an adaptive image matching method to discard correlation components with insignificant energy levels. By choosing energy-significant subbands, we can approximate the original correlation with minimal loss of accuracy. Also, the computation cost in the subband domain is much less than that in the pixel domain (more than 10 times faster), since many insignificant cross-band correlation components can be discarded.

Our experiment in satellite image database search showed that the normalized correlation provided the most satisfactory search results, compared to MSE and correlation. One obvious advantage of using normalized correlation was illumination invariance. However, it requires a little bit higher implementation cost than others. We also evaluated different filter banks based on their energy compaction capability and the resulting accuracy in terms of adaptive approximation of image correlation.

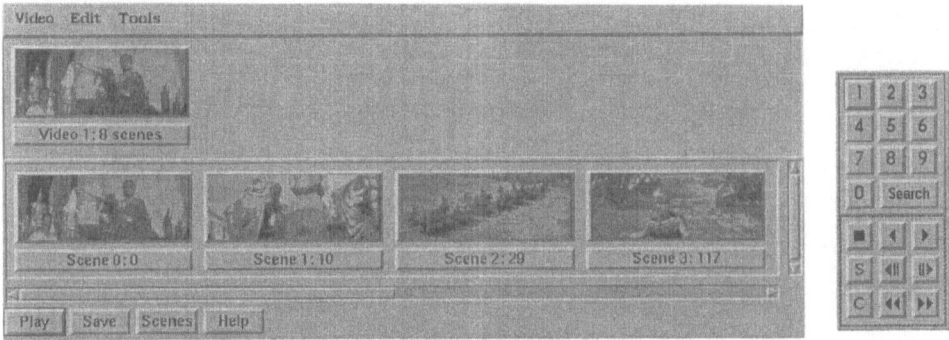

FIGURE 3. Graphic User Interface of Columbia's Video Indexing and Editing System. The VCR control panel is displayed on the right side.

ADDING THE TEMPORAL DIMENSION — VIDEO SEARCH

Video provides a more dynamic dimension in presenting visual content — time. Image objects, their associated attributes, and their spatial relationship may change as time proceeds. There have been two different approaches in deciphering the information carried by the video signal. One approach is to identify and recognize different channels of information and then index their continuous temporal flow structure in the entire video sequence. For example, people, background, weather, and actions can each be an individual information channel and indexed separately. Different channels of information do not need to occupy the same temporal intervals (*e.g.*, scene cuts). Because of the rich semantic information involved, this approach needs intensive input from users. The other approach is based on the hierarchical temporal structure of videos. Each video sequence can be segmented to different scenes and each scene may have multiple shots. Shots of different scenes may be arbitrarily interleaved on the time line based on the preference of the director. If we assume the visual content within each shot evolves in a "trackable" way, we may index each shot with few key frames or an extended view (*e.g.*, image mosaics). One benefit about the scene-based indexing is that automatic detection of scene change and extraction of key frames are feasible. (Note that most literatures used the terms "scene" and "shot" interchangeably, though they differ in film production.) Once the scene unit and key frames are extracted, various retrieval techniques for still images can be applied to the key frames or extended representation of the scenes.

We investigated techniques for scene change detection in the compressed domain (*e.g.*, MPEG). A complete scene change detection should also handle camera operations (*e.g.*, pan, zoom), special editing effects (*e.g.*, dissolve, fade in/out). For the general scene analysis task, the MPEG compression standard provides several clues, including the motion vector distribution (temporal direction, spatial direction, and magnitude) and the transform coefficients. By looking at the motion vector distribution of the B and P frames, our techniques can detect abrupt scene changes at any arbitrary frame (including I frames). For the dissolve special effect, we assumed a quadratic evolution model of the image intensity variance. Using the DC transform coefficients to approximate the image intensity variance, we can detect dissolve in the compressed domain as well. Ideal panning and zooming produce special patterns (*e.g.*, parallel or concentric motion vectors) in the extracted motion field, which can be used to detect such camera operations [25]. Motion features can sometimes be used to segment simple foreground objects and background objects as well. Once the foreground and background objects can be separated, they can be stored and manipulated separately. This has very interesting applications in video content production, such as video games.

Figure 3 shows the user interface of the current prototype of our video indexing and browsing system. Software parsing and decoding of the compressed stream has been developed. Interactive controls (*e.g.*, VCR functions) were also provided. From the individual detected scenes, users may further manipulate (e.g, cut and paste) these scene units to create new visual content directly in the compressed domain, as discussed in the next section.

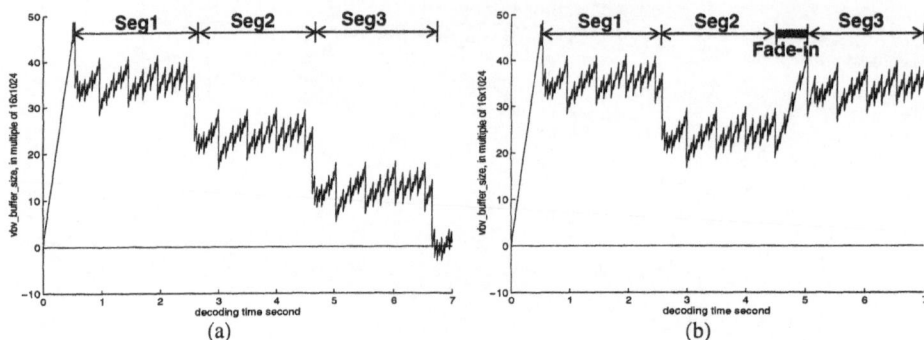

FIGURE 4. (a) A hypothetical example illustrating the buffer overflow problem in cutting and pasting compressed video. (b) The overflow problem is solved by inserting a synthetic fade-in interval.

MANIPULATION OF SEARCH RESULTS IN THE COMPRESSED-DOMAIN

Indexing and editing are two functions that come hand-in-hand in many practical applications. Users of visual information systems need editing functions to formulate the query keys or, more importantly, to manipulate the query results for final presentation or production. Our journey in the compressed-domain world does not stop with the indexing part. Continuing our prior work on compressed-domain image/video compositing, we have developed techniques for editing compressed video in the compressed domain. One example from our prior work is compositing multiple video windows into a single displayable video window. This function was originally intended for multi-point video conferencing, though it can actually be applied to other general applications. A video bridge in the network will take multiple compressed videos, perform size scaling and spatial translation, and then overlay them to the same displayable video window. The final composited video is then distributed to all participants. Our work shows that the compressed-domain approach provides great potential for reducing the computational cost, particularly if the video compression ratio is high and the manipulation operations are simple. Complicated operations (*e.g.*, rotation and shearing) also have compressed-domain equivalents [19]. But their compressed-domain implementations may not pay off due to the large overhead involved.

One simple but useful video editing operation is cut and paste. After retrieving several video segments from the database, users may want to "splice" some interesting segments from different sources and create a new video sequence, hopefully in the compressed domain. Unfortunately, there are some constraints imposed by the practical design of the video decoders. In order to have non-interrupted playback, the hardware decoders need to maintain a certain level of fullness in the buffer all the time. This constraint is essential whenever the input and output of the buffer have different rate patterns (constant bit rate vs. variable bit rate). In this case, the decoder buffer has CBR in and VBR out. Usually, the buffer constraint is enforced in the encoder by some rate control mechanisms (*e.g.*, VBV control in MPEG). Each encoded video stream maintains its rate integrity through this rate control mechanism. This integrity will be broken if we arbitrarily cut and paste random video segments from different streams.

Figure 4 shows a hypothetical case in which a new stream created by multiple cuts and pastes causes the decoder underflow problem due to the large picture size at the beginning of connected video segments. These large pictures are possible particularly when users cut at the scene change points. There are cases in which buffer overflow may occur as well. Examples are scenes starting with dark lighting (*e.g.*, fade in). In order to maintain valid compressed streams without causing buffer violations in the decoder, we have investigated several solutions. Possible solutions for buffer underflow include simple frame repetition at the decoder (so that the buffer has time to fill up) or insertion of a synthetic special effect (*e.g.*, fade-in) between two video segments. The former relies on the specific design of the decoder while the latter can completely remove the buffer violation and thus relieve the responsibility from the decoder. The synthetic fade-in period produces low bit rate and therefore it can help bring the decoder buffer back to the normal status. The fade-in effect minimizes the visual discontinuity of the displayed video at the cost of a minor increase of delay.

Besides the above buffer issue, boundary frames of the cut segments need to be modified so that the new stream will conform to the required format (*e.g.*, the GOP and frame structure in MPEG). But this process involves only few frames and requires relatively little computations. We have developed fast methods to handle required format modification as well [26].

AN INTEGRATED VIS/VOD TESTBED

We have developed a Video on Demand (VOD) testbed [5] in which the visual information system plays a critical component. The VOD system is equipped with real-time parallel graphic processors and high-speed ATM network resources. Real-time video pump, network adaptation, high-level application controls (DSMCC), and real-time client decoder (MPEG-1 and MPEG-2) have been developed. Envisioned applications include education, information, and instruction on demand on the campus. The same VOD testbed is being used for interoperability experiments.

Video indexing and editing tools described above are being incorporated into the VOD testbed. Key frames and/or classified scene categories are used as the basis for video browsing. From the video browsing interface, users can select any arbitrary segments from different sources and "paste" them to form a new compressed video stream, using the compressed-domain technique. The compressed-domain approach also allows us to develop fast software implementation.

CONCLUSIONS AND FUTURE WORK

We presented an overview for the compressed-domain content-based visual retrieval technologies. For still images, we have developed techniques for texture and color feature extraction in the transform domain (particularly the wavelet subband domain). We have also developed adaptive image matching technique in the subband domain. A speedup greater than 10 times can be achieved using the compressed-domain image matching. For videos, we used the compressed data (such as motion vectors and transform coefficients) to detect the low-level visual features, such as scene change, camera operation, and special effects. We have also developed preliminary techniques for editing compressed video streams in the compressed domain. A fast software-based compressed video editor has been implemented. Most of the visual indexing work described in this paper is being incorporated to a Video-on-Demand testbed in the Image and Advanced TV Lab at Columbia.

One future goal is to investigate fusion of multiple visual features and their integration with domain knowledge in specific applications. All techniques described above can be considered as different tools which should be fine-tuned and combined in an optimal way to satisfy the specific requirement of each application. Another direction of future work is to add content accessibility to new image/video compression techniques. Many constraints are imposed by today's compression standards, resulting in limited content access functionality. The task is not easy. But with many exciting applications driving the need, it has captured great interest from researchers in various disciplines, including the new video compression standardization effort, MPEG-4.

ACKNOWLEDGEMENT

The author is thankful for Prof. David Messerschmitt at UC Berkeley for his invaluable guidance and early stimulation of the author's interest in this field. This paper includes joint work with several people, including John Smith, Hualu Wang, and Jianhao Meng. This research is supported in part by the National Science Foundation (under a CAREER award, IRI-9501266), IBM (under a 1995 UPP award), NEC, and Columbia's ADVENT project.

REFERENCES

1. W. Niblack, R. Barber, W. Equitz, M. Flickner, E. Glasman, D. Petkovic, P. Yanker, C. Faloutsos, and G. Taubin, "The QBIC Project: Querying Images by Content Using Color, Texture and Shape", SPIE 1993 Intl. Symposium on Electronic Imaging: Science and Technology, Conf. 1908, Storage and Retrieval for Image and Video Databases, Feb. 1993. (also in IBM Research Report RJ 9203 (81511), Feb. 1, 1993, Computer Science.)

2. S.-F. Chang and D.G. Messerschmitt, "Manipulation and Compositing of MC-DCT Compressed Video," *IEEE Journal of Selected Areas in Communications*, Special Issue on Intelligent Signal Processing, pp. 1-11, Jan. 1995.

3. S.-F. Chang and J. R. Smith, "Extracting Multi-Dimensional Signal Features for Content-Based Visual Query," *SPIE Visual Communications and Image Processing*, Taipei, May 1995. (Best Paper Award)

4. R. K. Srihari, "Automatic Indexing and Content-Based Retrieval of Captioned Images," *IEEE Computer Magazine*, Vol. 28, No. 9, pp. 49-56, Sept. 1995.

5. S.-F. Chang, A. Eleftheriadis, and D. Anastassiou, "Development of Columbia's Video on Demand Testbed," *Journal of Image Communication*, Special Issue on Video on Demand and Interactive TV, 1995, in press.

6. J. Meng, Y. Juan and S.-F. Chang, "Scene Change Detection in a MPEG Compressed Video Sequence," *SPIE Symposium on Electronic Imaging— Digital Video Compression: Algorithms and Technologies*, San Jose, Feb. 1995.

7. P. Brodatz, Textures: a Photographic Album for Artists and Designers, Dover, New York, 1965.

8. J.R. Smith and S.-F. Chang, "Quad-Tree Segmentation for Texture-Based Image Query" *Proceedings*, *ACM 2nd Multimedia Conference*, San Francisco, Oct. 1994.

9. J. Smith and S.-F. Chang, "Automated Image Retrieval Using Color and Texture," submitted to *IEEE Transactions on Pattern Analysis and Machine Intelligence*, Special Issue on Digital Libraries - Representation and Retrieval, Nov. 1996. (Also CU-CTR Technical Report # 414-95-20.)

10. H. Wang and S.-F. Chang, "Adaptive Image Matching in the Subband Domain," submitted to *SPIE/IEEE Visual Communications and Image Processing '96*, Orlando, Florida, March, 1996. (also CU/CTR Technical Report #422-95-28).

11. S.-F. Chang, *Compositing and Manipulation of Video Signals for Multimedia Network Video Services*, Ph.D. Dissertation, U.C. Berkeley, Aug., 1993.

12. R.W. Picard, "Light-years from Lena: Video and Image Libraries of the Future," *IEEE International Conference on Image Processing*, Washington DC, Oct. 1995.

13. A. Pentland, R. Picard, and S. Sclaroff, "Photobook: Tools for Content-Based Manipulation of Image Databases," *Int'l Journal of Computer Vision*, 1995, in press.

14. H. Zhang A. Kankanhalli, S.W. Smoliar," Automatic Parsing of Full-Motion Video," *ACM-Springer Multimedia Systems*, 1(1), pp. 10-28, 1993.

15. F. Arman, A. Hsu, and M.-Y. Chiu, "Image Processing on Compressed Data for Large Video Databases," *Proceedings of ACM Multimedia Conference*, June 1993.

16. Y.Y. Lee and J. Woods, "Video Post Production with Compressed Images," *SMPTE. J. Vol. 103,* pp. 76-84, Feb. 1994.

17. N. Dimitrova and F. Golshani, "Rx for Semantic Video Database Retrieval," *ACM Multimedia Conference*, 1994, Oct., San Francisco.

18. B.C. Smith and L. Rowe, "A New Family of Algorithms for Manipulating Compressed Images," *IEEE Computer Graphics and Applications*, pp. 34-42, Sept., 1993.

19. S.-F. Chang, "New Algorithms for Processing Images in the Transform-Compressed Domain, " *SPIE Symposium on Visual Communications and Image Processing*, Taipei, May, 1995.

20. B.-L. Yeo and B. Liu, "A Unified Approach to Temporal Segmentation of Motion JPEG and MPEG Compressed Video," *IEEE Int. Conf. on Multimedia Computing and Systems*, May 1995.

21. A. Hampapur, R. Jain, and T.E. Weymouth, "Production Model Based Digital Video Segmentation," *Journal of Multimedia Tools and Applications*, Kluwer Academic Publishers, Vol. 1, No. 1, March, 1995.

22. R. Mehrotra and J. E. Gary, "Similar-Shape Retrieval in Shape Data Management," *IEEE Computer Magazine*, Vol. 28, No. 9, pp. 57-62, Sept. 1995.

23. R. Jain, NSF workshop on Visual Information Management Systems, Redwood, CA, Feb. 1992.

24. H.S. Stone and C.-S. Li, "Image Matching by Means of Intensity and Texture Matching in the Fourier Domain," to appear in *Proceedings of SPIE Conference on Image and Video Databases*, San Jose, CA, Jan. 1996.

25. J. Meng and S.-F. Chang, "Tools for Compressed-Domain Video Indexing and Editing," to appear in *SPIE Conference on Storage and Retrieval for Image and Video Database*, San Jose, Feb. 1996.

26. J. Meng and S.-F. Chang, "Buffer Control Techniques for Compressed Video Editing," Submitted to *IEEE Int'l Conference on Circuits and Systems*, ISCAS '96.

27. P. P Vaidyananthan, "Orthonormal and Biorthonormal Filter Banks as Convolvors, and Convolution Coding Gain," *IEEE Transactions on Signal Processing*, Vol. 41, June, 1993.

28. H.S. Sawhney, S. Ayer, and M. Gorkani, "Model-Based 2D and 3D Dominant Motion Estimation for Mosaicking and Video Representation," *Proc. Fifth Int'l Conf. Computer Vision, 1995.*

STATISTICAL ANALYSIS OF MPEG2-CODED VBR MOVIE SOURCE

Daniel P. Heyman[1], T. V. Lakshman[2], and Ali Tabatabai[3]

[1]Bellcore
Red Bank NJ 07701

[2] AT&T Bell Laboratories
Holmdel NJ 07733

[3]Tektronix, Inc.
Beaverton OR 97077

INTRODUCTION

Traffic from video sources are envisaged to be a substantial portion of the traffic in broadband networks. Statistical source models of different types of video traffic are needed to design networks that achieve acceptable picture quality at at minimum cost, and to control and shape the output rate[1]. Source models are needed, for instance, to do admission control and bandwidth allocation[2]. Since a significant fraction of video traffic is expected to be generated by MPEG sources[3], it is imporortant to understand the statistics of traffic from MPEG sources. As noted in [3], this work is "especially critical" because video sources require high bandwidth and are delay sensitive. Moreover, the MPEG coding standard is close to completion, and MPEG-type algorithms are being proposed for HDTV and multimedia services.

This paper summarizes our statistical analysis of a 14,000 frame MPEG2 coded movie sequence. Statistical summaries of the data are given in the next section, and they are followed by statistical models. The models were tested by simulation experiments, and these are described in the penultimate section. Our conclusions are given in the final section.

DATA ANALYSIS

The data (number of bits/frame) were generated for a 10-minute long movie sequence coded in VBR mode using MPEG2 Main Profile/Main Level (MP&ML) and a fixed quantizer step size. I,B, and P pictures with M=3 and N=15 were used. The pictures were coded in progressive mode, and a 3:2 pull down detector was used at the input as a preprocessor.

Multimedia Communications and Video Coding
Edited by Y. Wang *et al.*, Plenum Press, New York, 1996

The fundamentals of MPEG coding are described in [3]; we will summarize the salient features for source modeling. The I,B, and P frames repeat with the following 15 frame pattern:

$$\text{I B B P B B P B B P B B P B B}$$

which is called a *group of pictures*, abbreviated *GOP*. The P frames carry predictive information, and the B frames carry interpolative information. The I frames are coded in the intraframe mode, so they have less compression than the B and P frames. The payload of ATM cells is 48 bytes, and 4 bytes will be used by the ATM adaptation layer (AAL) header, so we divided the data by 352 ($=8 \times 44$) to get the number of cells/frame.

A segment of the data is shown in Fig. 1. The sample path shows the dominant effect of the I frames occuring periodically. From the sample path it is evident that the mean bit-rate is insufficient to describe the traffic requirements of the bit stream. For this segment, the mean rate is 1.042 Mb/s and the peak rate is 3.256 Mb/s; for the entire sequence the corresponding values are 1.060 Mb/s and 6.543 Mb/s. There is a potential for multiplexing gain.

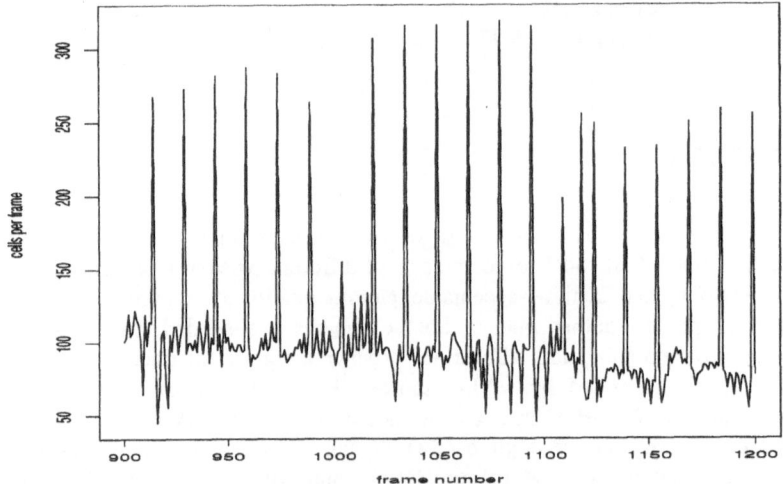

Figure 1. Cells/frame for frames 900 through 1,200

We now look at each type of frame separately. Let B_i be the ith B frame (i=1 through 10) and P_i be the ith P frame (i=1, 2, 3, 4) in the sequence. A boxplot shows that the B and P frames have similar distributions, and confirms that the I frames are typically much larger than the B and P frames. A Q-Q plot showed that even though the B and P frames have a similar distribution, the distributions are distinct.

I Frames

The histogram of sizes of the I frames is unimodal with a long right tail. The histogram of the logarithm of the data looked like a Normal curve, and that was confirmed with a Q-Q plot (shown in Fig. 2(a)). We conclude that the I frames have a log-normal distribution.

The mean and standard deviation of the logarithm (base e) of the number of bits in the I frames are given by

$$\mu[\log(I)] = 11.37 \text{ bits} \quad \text{and} \quad \sigma[\log(I)] = 0.2866 \text{ bits}.$$

The autocorrelation function is adequately matched by the geometric function 0.823^k, where k is the lag (shown in Fig. 2(b)).

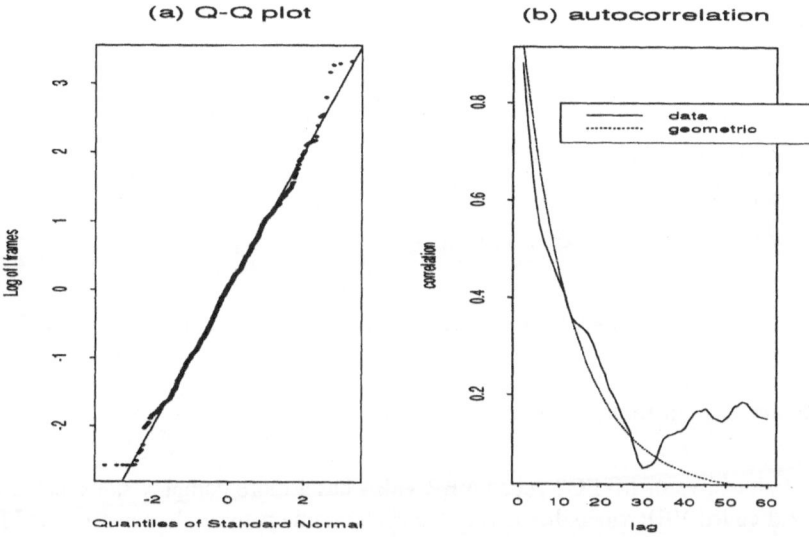

Figure 2. Q-Q plot and correlogram for I frames

B Frames

The histogram and autocorrelation function for the B frames are shown in Fig. 3. From this figure we see that sizes of the B frames have a long-tailed distribution, and that there is a strong and complicated dependence among them. The bump at the right of the histogram indicates that the marginal distribution of a B frame is a mixture of distributions, which makes it difficult to fit it to standard distributions. The jagged nature of the autocorrelation function suggests that there is some structure to the dependency. In Fig. 4 we see that B_{i+1} is strongly dependent on B_i, $i=1$ through 9. These properties of the B frames will be exploited in the next section.

P Frames

Box plots were used to see that the P_1, P_2, P_3 and P_4 frames all had the same distribution. The histogram and autocorrelation function of the P frames are shown in Fig. 5. The scatter plot in Fig. 6 shows that the size of a P frame is primarily determined by the size of the preceding B frame. The nonlinear curves in Fig. 6 are fitted by loess[4] and they are nearly straight. The deviations from the fitted curve occur at the small values. Since it is the large values that make the major contribution to cell losses (which is the ultimate object of interest), the deviations from the fitted curves do not seem to be important. The correlation between the size of a P frame and the size of the previous P frame is 0.770.

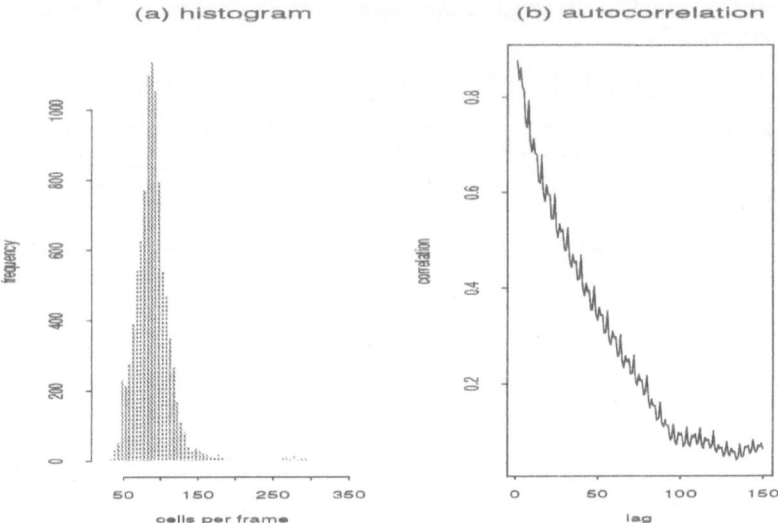

(a) histogram · (b) autocorrelation

Figure 3. Histogram and autocorrelation function for B frames

Preliminary Conclusions

The frame sizes of MPEG2 coded VBR video has a more complex statistical structure than H.261 coded VBR video, because of the cyclic frame pattern. In our data, the I frames possess a straightforward model, but the B and P frames do not. There are two reasons. The first is the bump in the right-hand tail of the histograms; this makes the distribution of the number of cells/frame complicated and hard to estimate. The second is the dependency among the sizes of the B and P frames. This seems to be a result of the coupling among the B and P frames induced by the coding algorithm; the I and P frames are used to construct the B frames and the I frames refresh the image. We have postulated several models for the B and P frames, and tested their accuracy for predicting the cell-loss rate by comparing emulations using the data with simulations using the model, as we have done for video conferences[5]. These experiments showed that our models either consistently under or over estimated the cell-loss rate by more than an order of magnitude, so finding an adequate model is still an open problem.

There are two inadequate models that are worth mentioning. Since the I frames are typically much larger than the B and P frames, one might suspect that they control the cell-loss rate. Our experiments show that ignoring the B and P frames leads to severe under estimates of the cell-loss rate. The sizes of the B and P frames have similar histograms, so a simple model is to treat them as coming from a common distribution (we fitted a weibull) and using a DAR(1) model[5]. This overestimated the cell-loss rate by more than an order of magnitude.

STATISTICAL MODELS

The results in section 2.1 imply that sizes of the I frames can be simulated with a Markov chain with transition matrix P say, given by

$$P = \rho I + (1 - \rho)Q, \tag{1}$$

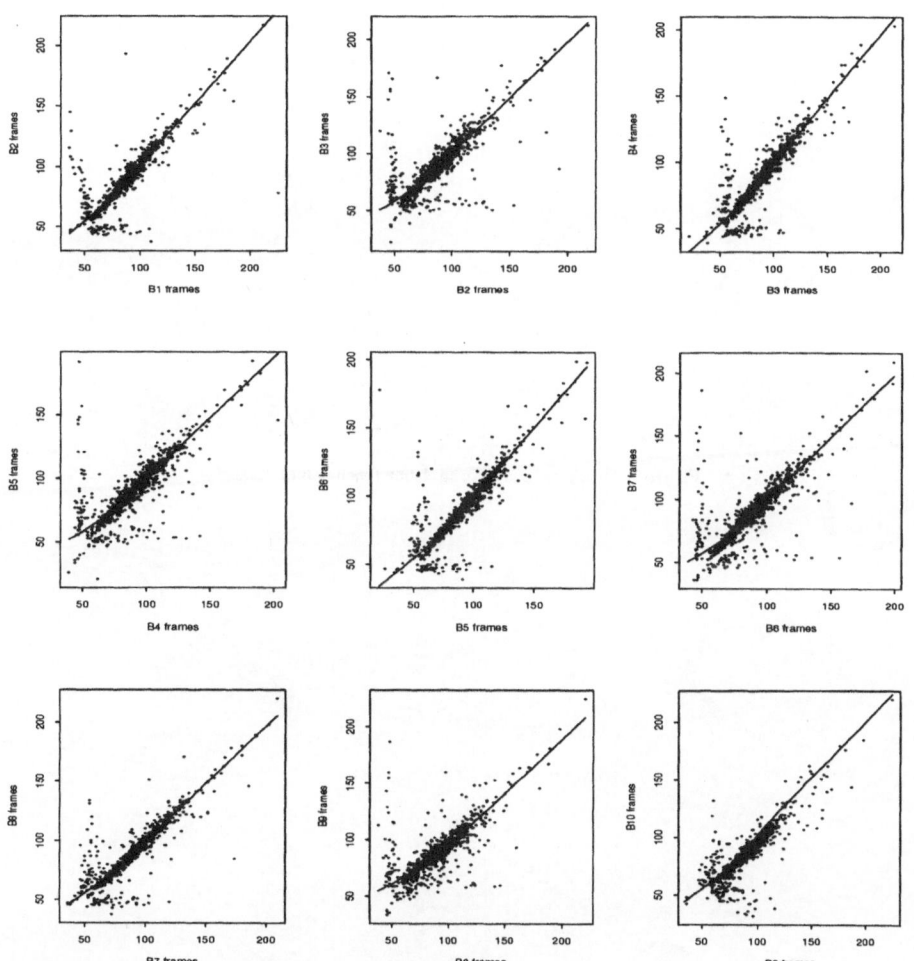

Figure 4. Current B frame sizes vs. lagged B frame sizes

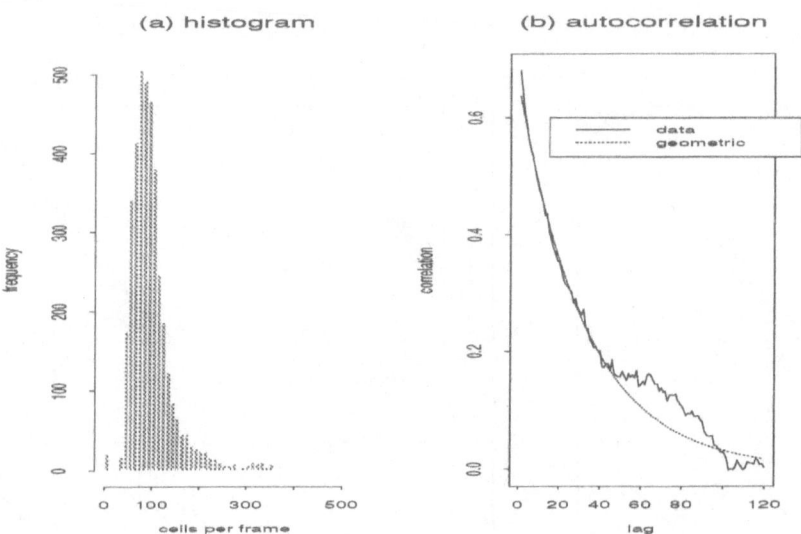

Figure 5. Histogram and autocorrelation function for P frames

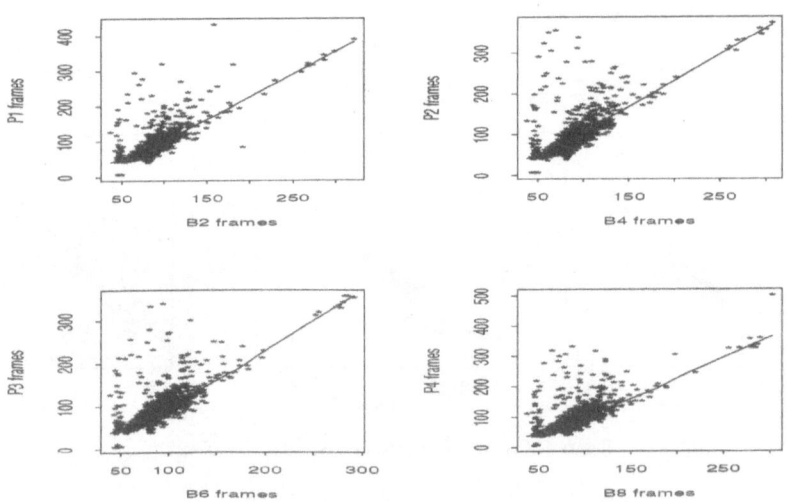

Figure 6. P frame sizes vs. B frame sizes

where ρ is the correlation between successive I frames (ρ=0.902 for the I frames), Q is a square matrix all of whose rows have a discretization of the log-normal distribution (with as many columns as is convenient), and I is the commensurate identity matrix[6]. A Markov chain of this sort (the log-normal distribution is just an example - any discrete non-negative distribution will do) is called a *discrete autoregressive process of order 1*, or DAR(1) for short.

The material in above shows that models for the B and P frames are hard to construct.

Discarding Outliers

An examination of the largest B and P frames showed that they all occured between frames 8234 and 8382. Frames 8234 through 8382 represent 6.2s of high video activity, and can be regarded as a statistical outliers. When these frames are removed, the distributions of the B and P frames become log-normal; this was justified by Q-Q plots. There is a minimal effect on the I frames. The parameters for these distributions are

$$\mu[log\,(B)] = 4.3698 \text{ bits}, \quad \sigma[log\,(B)] = 0.2525 \text{ bits}^2,$$

and

$$\mu[log\,(P)] = 10.31 \text{ bits}, \quad \sigma[log\,(B)] = 0.4264 \text{ bits}^2,$$

In a GOP, the correlation between successive B frames (ignoring intervening P frames, if any) is 0.90. The autocorrelation function for successive B_1 frames is shown in Fig. 7; the geometric function 0.8^k is a usable fit. Consequently, we model the B_1 frames with a DAR(1) processes, using (1) with ρ=0.80 and the rows of Q a discretization of the log-normal distribution with the parameters given above for B frames. Based on figures 5 and 7, the succeeding B and P frames are generated with appropriate pairwise correlations (0.770 between the B and P frames, and 0.900 between successive B frames) in each GOP. This is done by generating pairs of pairwise correlated uniform pseudo-random numbers (this is a simple application of the TES method[7])

Modeling the Outliers

The model above represents our video sequence with the "extreme" scene removed. A complete model would describe the extreme scenes and give a rule for inserting them in the sequence. We cannot do the latter because we observed only one extreme scene; describing the spacings between extreme scene will have to wait until more data are available. We can give a statistical description of the B and P frames in the extreme scene of our data. Q-Q plots showed that the largest B and P frames also follow a log-normal distribution. There is not enough data to give good estimates of the autocorrelation function, but a geometric approximation appears to be reasonable.

SIMULATION EXPERIMENTS

The statistical source model, with the outliers excluded, was tested us ing the procedure introduced in [5]. The data trace (with the outliers excluded) was used to represent 36 independent sources by treating the data as a circular list and picking random starting points for each source. The cells arrive on 8.5 Mb/s access lines. The cell arrival times are chosen so that the cells arrive at a constant rate during each frame interval (this mitigates the source-effect observed in [5]). The simulation used 48 octets for the cell payload and the

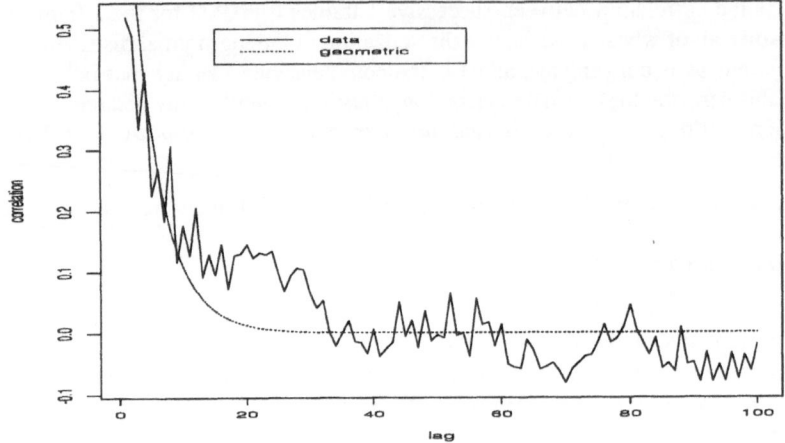

Figure 7. Autocorrelation function of B_1 frames

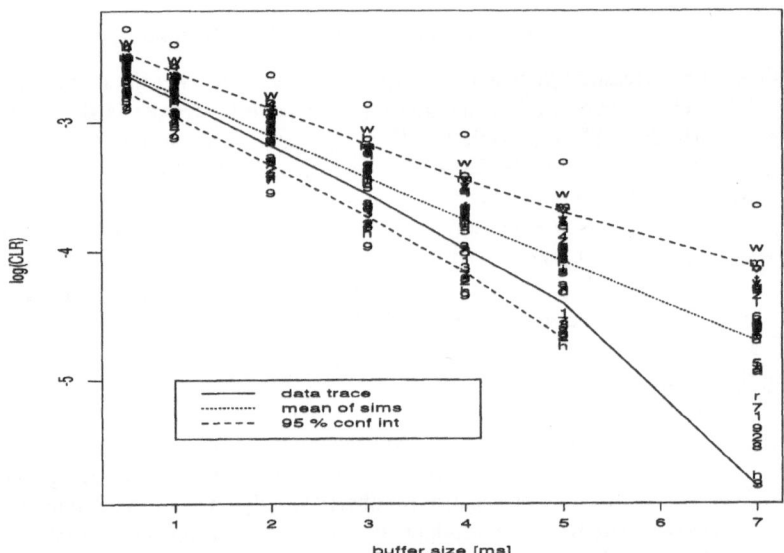

Figure 8. Simulated cell-loss rates for trace and source models, log base 10 scale

cell size, ignoring the 4 octets for AAL and the 5 octets for the header; this should not materially affect the results. The cells are placed in a buffer on a first-come first-served basis. We use buffer sizes of 0.5, 1, 2, 3, 4, 5, and 7 msec, and the buffer is drained at rate 39Mb/s, so each msec of buffer size corresponds to 102 cells. Monte Carlo methods were used to construct 36 sample paths from the statistical models in section 3; they were used in the same way as the trace was used.

The results of the simulations are shown in Fig. 8. Each path is represented by one of the characters 0, 1, ...,9, a, b, ..., z. All but one of the paths lies in the 95 per cent confidence interval (one exception is good, because one out of forty points should fall outside the confidence interval), and are spread fairly uniformly around the trace value.

The cell-loss rate for the largest buffer is based on fewer cell losses than the other points, so it has much more sampling variance, and the discrepancy of this point can be attributed to the sampling variability. The model estimates the order of magnitude of the cell-loss rate correctly.

The peak rate of the (modified) trace is 5.66 Mb/s, and the mean rate is 0.859 Mb/s; the peak to mean ratio is 6.3. The cell-loss rate for the 7 msec buffer is 1.4×10^{-6}. (Since there are only about 11.4×10^6 cells in the trace, there are not enough cell losses for this number to be statistically reliable.) Even so, it seems safe to say that this buffer size and processing speed can accomomdate about 40 sources at a cell-loss rate of about 10^{-6}. If peak rate allocation were used, only 13 sources could be served, with spare capacity that is equivalent to 0.9 of a source. We'll round this to 14 sources with peak rate allocation, giving a multiplexing gain of 40/14 which is about 3. This is the same multiplexing gain we found for VBR video teleconference traffic in [5], but that traffic had a smaller peak-to-mean ratio (five), it is more efficient to transport. The difference may be caused by the periodic peaks of the I frames, rather than the smoother transitions among bit rates in the video conference coding.

A key property exploited in admission control computations is that the log of the cell-loss rate should be linear in the buffer size. This property holds for these data, as shown in Fig. 8. We are adapting the admission control algorithm in [2] to the model described in this paper.

REFERENCES

1. T. Ott, A. Tabatabai and T. V. Lakshman, "A scheme for for smoothing delay sensitive traffic offered to ATM networks", *Proc. IEEE INFOCOM '92*, (1992).

2. A. Elwalid, D. P. Heyman, T. V. Lakshman, D. Mitra and A. Weiss, "Fundamental bounds and approximations for ATM multiplexers with applications to video teleconferencing", *J. Select. Areas Commun. 13*,1004-1016 (1995)

3. P. Pancha and M. El Zarki, "MPEG coding for variable bit rate video transmission", *IEEE Communications Magazine*, 54-66 (1994).

4. W. S. Cleveland, *Visualizing Data*, Hobart Press, Summit NJ (1993).

5. D. Heyman, A. Tabatabai, and T. V. Lakshman, "Statistical analysis and simulation study of video teleconference traffic in ATM networks", *IEEE Trans. Ckts. and Systems for Video Technology*, 2, 49-59 (1992).

6. P. A. Jacobs and P. A. W. Lewis, "Time series generated by mixtures", *J. Time Series Analysis*, 4, 19-36 (1983).

7. B. Melemed, "TES: a class of methods for generating autocorrelated uniform variates", *ORSA J. on Computing*, 3, 317-316 (1991).

SOURCE TRAFFIC DESCRIPTOR FOR VBR MPEG IN ATM NETWORKS

Jorge Mata and Sebastià Sallent

Department of Applied Mathematics and Telematics
Polytechnic University of Catalonia
C/ Gran Capitán, S/N. Módulo C-3, Campus Nord, 08034 Barcelona SPAIN
Tel/Fax +34-[3] 401 6014 / 401 5981, E-Mail : jmata@mat.upc.es

INTRODUCTION

Broadband Networks based on the Asynchronous Transfer Mode (ATM) will support, among others, traffic coming from variable bit rate video (VBR) codecs[1], which are capable of maintaining a constant picture quality of the decoded image. The characterization of such VBR video sources becomes important in the analysis and design of Broadband Integrated Services Digital Networks (B-ISDN). The network architecture and its characteristics, such as cell-loss probabilities, transmission delay, high-speed statistical multiplexing gain, buffering, etc., are strongly related to the statistical properties of the sources and the coding schemes involved.

On the other hand, a characterization of the traffic generated by a VBR source is necessary in order to allocate resources in ATM networks, as well as, to keep a satisfactory quality of service (QoS). In the call establishment phase, service requirements are negotiated between the user and the network to establish a Traffic Contract. The source traffic parameters used to specify the statistical properties are the Peak Cell Rate (PCR), Sustainable Cell Rate (SCR) and Burst Tolerance. The Generic Cell Rate Algorithm (GCRA) is used[2] to provide a formal definition of the traffic conformance. This algorithm depends only on the increment parameter (I) and the limit parameter (L).

Several algorithms have been specified to compress video. MPEG-1 is specially suitable for video-on-demand in ATM networks since it allows VBR mode and a VCR resolution[3]. The analysis of MPEG traffic can be extended to other video coding schemes which use Discrete Cosine Transform (DCT) and motion compensation techniques. MPEG-1 has mainly two coding modes: intraframe and interframe. The frames coded in the intraframe mode are called I and the frames coded in the interframe mode must be distinguished by P or B according to the motion estimation applied (forward o bidirectionally). A video sequence of pictures (SOP) is divided into groups of N pictures (GOP). A GOP consists of subgroups of M pictures where the first is a reference picture, intra or predicted, and the rest are bidirectionally-predicted. The image quality depends on M, N values and the selected quantizer step size (q).

MPEG VIDEO CODING ANALYSIS

To study the VBR MPEG transmission over ATM Networks it is necessary to perform an analysis of parameters q, M and N. Those parameters have to be selected to minimize the traffic rate for a constant signal to noise ratio (SNR). In this paper, a PAL sequence of "Live in Central Park" of the America music band is used. The processed sequence has 4200

Multimedia Communications and Video Coding
Edited by Y. Wang *et al.*, Plenum Press, New York, 1996

frames in CIF format with a 352x288 pels per frame resolution and 25 frames per second picture rate. Eighty sets of parameters q, M and N are chosen to code the video sequence. The compression ratio and the average SNR are analyzed for these triplets (q, M, N).

The analysis of the coded sequences has been carried out for several values of the tripplet (q, M, N). The studied parameters have been the SNR average, the compression ratio (R) and the rate of transmission. The compression ratio has been defined as the ratio between the mean rate per frame of the coded bitstream and the necessary rate to transmit one frame in CIF format.

The results of the codification are presented in the table 1. Four levels of image quality have been studied. The chosen values of q are associated to a level of high quality (q= 4), medium high quality (q= 6), medium low quality (q= 9) and low quality (q= 20). The values of M and N have been selected in order to observe the variations of the SNR and of the compression ratios when more or fewer P or B frames are used in the MPEG-1 algorithm. The results show that the values of SNR depend on q, and are very insensitive to the values of M and N. However, the compression ratio increases quickly when P and B frames are introduced. For levels of high quality, there are not considerable increments of R for values of M higher than 2 and for values of N higher than 4 or 6. For levels of medium and low quality, M and N could reach value 3 and 9 respectively.

Table 1. VBR MPEG-1 parameters analysis.

	SNRav	R	Kbps		SNRav	R	Kbps
N=1, M=1, Q=4	40.1582	10.91	2788	N=1, M=1, Q=9	37.5302	21.20	1435
N=2, M=1, Q=4	40.0306	12.45	2441	N=2, M=1, Q=9	37.0801	30.14	1009
N=2, M=2, Q=4	39.5401	14.08	2160	N=2, M=2, Q=9	36.9667	32.64	932
N=3, M=3, Q=4	39.3161	15.06	2018	N=3, M=1, Q=9	36.8601	34.62	878
N=4, M=1, Q=4	39.8758	13.17	2310	N=3, M=3, Q=9	36.6993	38.34	793
N=4, M=2, Q=4	39.4719	14.80	2054	N=4, M=1, Q=9	36.7048	37.24	817
N=4, M=4, Q=4	39.2052	15.31	1985	N=4, M=2, Q=9	36.6613	40.06	759
N=6, M=1, Q=4	39.7767	13.26	2294	N=4, M=4, Q=9	36.5214	41.06	740
N=6, M=2, Q=4	39.4217	14.95	2034	N=5, M=1, Q=9	36.5912	38.93	781
N=6, M=3, Q=4	39.274	15.36	1979	N=6, M=1, Q=9	36.4986	40.01	760
N=8, M=2, Q=4	39.3832	14.92	2038	N=6, M=2, Q=9	36.5056	42.88	709
N=9, M=3, Q=4	39.2392	15.31	1982	N=6, M=3, Q=9	36.4556	43.54	698
N=12, M=4, Q=4	39.1517	15.29	1987	N=8, M=2, Q=9	36.3969	44.15	689
N=16, M=8, Q=4	39.0463	14.84	2042	N=8, M=4, Q=9	36.3198	44.58	682
				N=9, M=3, Q=9	36.3273	45.00	675
				N=12, M=2, Q=9	36.2435	45.31	671
	SNRav	R	Kbps	N=12, M=4, Q=9	36.2107	45.42	669
N=1, M=1, Q=6	38.8541	15.70	1936	N=16, M=8, Q=9	36.051	43.58	695
N=2, M=1, Q=6	38.5208	20.41	1486		SNRav	R	Kbps
N=2, M=2, Q=6	38.2019	22.79	1334	N=1, M=1, Q=20	34.6725	34.64	878
N=3, M=3, Q=6	37.9339	25.80	1178	N=2, M=1, Q=20	34.1974	53.64	567
N=4, M=1, Q=6	38.2373	23.68	1284	N=2, M=2, Q=20	34.4354	54.34	560
N=4, M=2, Q=6	38.0042	26.16	1162	N=3, M=1, Q=20	33.9536	64.77	470
N=4, M=4, Q=6	37.7772	27.00	1125	N=3, M=3, Q=20	34.2617	65.10	467
N=6, M=1, Q=6	38.0779	24.70	1231	N=4, M=1, Q=20	33.7766	72.10	422
N=6, M=2, Q=6	37.9	27.28	1114	N=4, M=2, Q=20	34.0226	71.49	425
N=6, M=3, Q=6	37.7924	27.89	1090	N=4, M=4, Q=20	34.1118	70.92	429
N=8, M=2, Q=6	37.827	27.68	1098	N=6, M=1, Q=20	33.5468	80.83	376
N=8, M=4, Q=6	37.6679	28.27	1075	N=6, M=2, Q=20	33.8068	79.02	385
N=9, M=3, Q=6	37.7145	28.34	1071	N=6, M=3, Q=20	33.8813	78.29	388
N=12, M=4, Q=6	37.6052	28.46	1069	N=9, M=3, Q=20	33.6808	82.90	366
N=16, M=8, Q=6	37.4546	27.35	1108	N=12, M=4, Q=20	33.5806	83.75	363

In figure 1 are shown the compression ratios for the most representative values of N and M as a function of the quantizer step size. In figure 2, it can be seen the relationship between the SNR and the quantizer step size. The relationship between the factor of compression and the quantizer step are known as the empirical scale factor. The empirical scale factor is practically equal in the cases (N= 4, M= 2), (N= 6, M= 2) and (N= 9, M= 3). In figure 3 the empirical scale factor normalized to the value q = 9 is shown. The empirical scale factor follows the approximated curve: y = 5.686 x$^{-0.774}$. The relationship between the SNR and the quantizer step size can also be observed in figure 4. This relationship could be expressed through the curve: y = 45.879 x$^{-0.1}$.

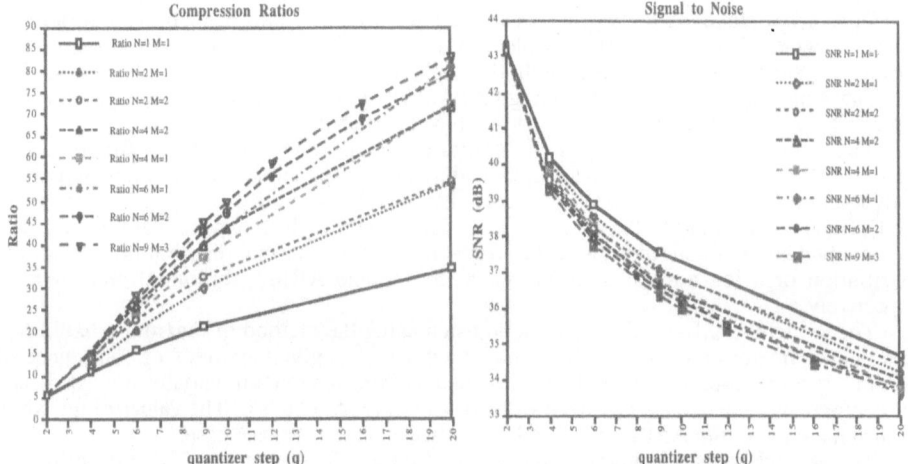

Figure 1. Compression Ratios as a function of MPEG-1 parameters q, N and M

Figure 2. Averaged SNR as a function of MPEG-1 parameters q, N and M

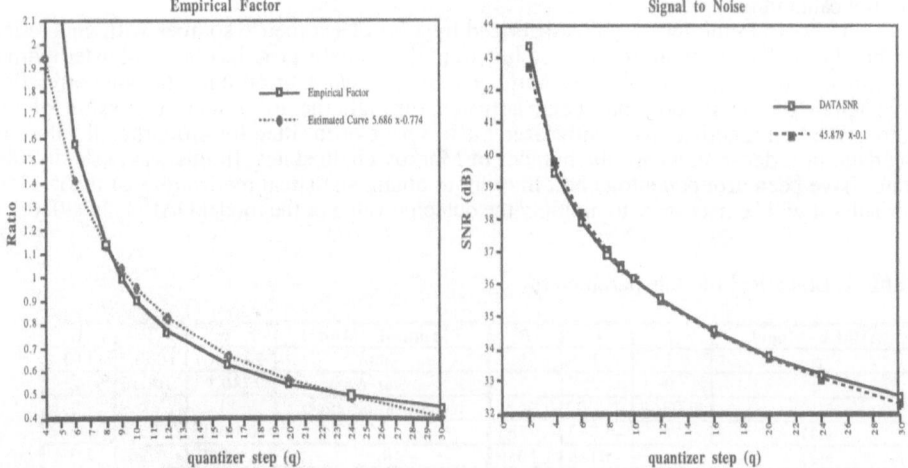

Figure 3. Empirical scale factor fitted curve normalized to the value q=9

Figure 4. Signal to noise ratio fitted curve

According to the results obtained in this analysis, the values of (N, M) should be (4, 2) or (6, 2) for high or medium image quality. For the ATM services, such as video-on-demand, it is more suitable to choose N= 4 and M= 2.

The statistical multiplexing in the ATM networks causes losses of cells for VBR services. These losses forces that the video compression has some error recovery mechanism. The election of N= 4, instead of N= 6, decreases the time between reference I frames. In this way, the error recovery mechanism in the decoder will be employed during less time.

STATISTICAL ANALYSIS OF VBR MPEG TRAFFIC

The analysis of VBR MPEG traffic is achieved on the basis of three random variables associated to the bits for I, P and B frames. The VBR traffic model can be implemented integrating the individual I, P and B models derived. The individual models must be visited according to the physical operation of the VBR MPEG-1 coder, fixed by the parameters N and M.

The mean, standard deviation, autocorrelation function and probability distribution function, obtained for the different triplets (q, M, N), have been evaluated and compared. Various statistical models are fitted with the triplet (q=9, M=2, N=4). The adjusted models are: The Autoregresive Moving Average(ARMA)[4,5], Discrete Autoregresive (DAR)[6] and Discrete-State Continuous Time Markov Model (M/M/∞//S)[7,8,9].

These models have been tuned regarding the probability density function and autocorrelation function. The ARMA model adjusts better the autocorrelation function under study than the binomial and DAR(1) models, therefore it is the most accurate. The binomial and DAR(1) models only fit the autocorrelation function through an exponential function. The probability density function of the real traffic is fitted with a normal distribution, a binomial distribution or a log-normal distribution in the models ARMA, binomial and DAR(1) respectively.

The models ARMA(p,q) have been adjusted using the method of maximum likelihood with the Marquardt algorithm. The residual diagnostic has given an index of confidence of 70% in the worst case. In table 2, the values obtained for each random variable are listed. The AR coefficients have been symbolized with a(i) and the MA with b(j). The values of the white noise at the input of the ARMA(p,q) model are also indicated for each case.

The second adjusted model is the binomial model. The parameters that describe the model could be expressed through aggregated minisources[9]. Each minisource has a state of low generation of A_{min} rate and a high state of $A_{min}+A$ rate. The transition rates between states are a and b respectively. In order to discretize the transition times between states, a time factor of 0.04, corresponding to the length of an image, has been considered. The aggregation of minisources could be expressed with a chain of Markov M/M/∞//S where S is the number of implicated minisources.

The DAR(1) models[6] are characterized by a set of geometric sources with generation probability p. The probability density function of the real traffic has been adjusted with a maximum value K_{max} in each case, through a log-normal distribution. The autocorrelation coefficient of the models has been adjusted through the log-linear regression of the autocorrelation function. The synthesized models have been fitted for proportional values of the data, in order to decrease the number of Markov chain states. In this way, sets of 2000 states have been grouped altogether. In order to obtain statistical realizations of the random variables it will be necessary to multiply the outcome value of the model DAR(1) by 2000.

Table 2. Statistical models parameters

ARMA Model	I	P	B	Binomial Model	I	P	B
p	1	1	1	Amin	2307.07	1090.3	1110.4
q	0	2	3	Step A	7780.93	10036	6742.3
b(0)	0.9368	0.9400	0.9634	Number Minisources	11	7	8
a(1)	x	-0.392	-.2785	a	0.6019	1.0518	0.4816
a(2)	x	-.0155	-.1059	b	1.0069	2.56	2.37
a(3)	x	x	-.0905	Time Scale	0.04	0.04	0.04
gaussian mean	3606.0	2798.8	1248.8				
gaussian std	4430.8	6586.2	3400.4	DAR 1 Model	I	P	B
				Probability (p)	0.73618	0.386	0.705
IPB Source DATA	Peak	Mean	Ratio	N.Geometric Sources	80	9	21
ARMA	120911	30187	4.00	Max Distance (Kmax)	58	38	30
Binomial	110968	30359	3.65	Autocorrelation	0.93767	0.8655	0.8922
Coder	110968	30370	3.65	Proportionality	2000.0	2000.0	2000.0
DAR 1	118000	30280	3.89				

In table 2, it can be checked that the peak rate and the mean rate reached by the models are close to the coder values. The ratio between the peak rate and the mean rate is approximately 4 for the tripplet (q= 9, N= 4, M= 2). The autocorrelation function presented in figure 5 and the probability distribution function, in figure 6, show that the behavior of the models is very similar and they fit well with the autocorrelación function of the real traffic.

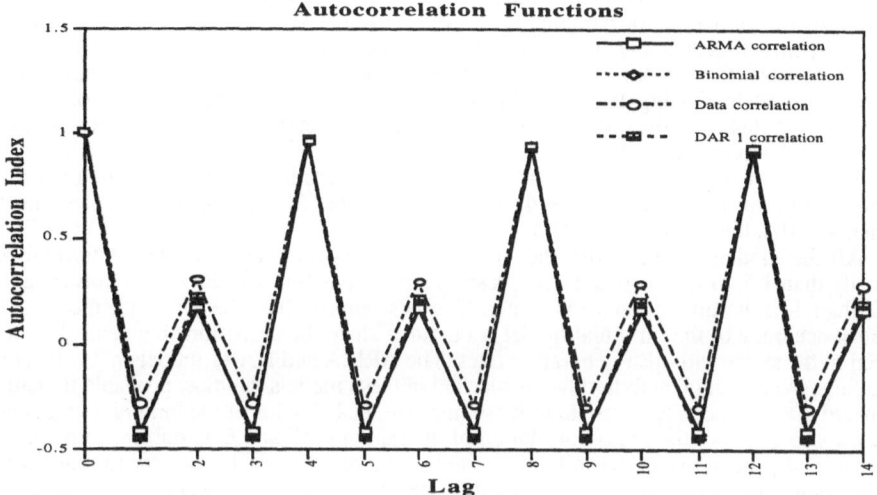

Figure 5. Autocorrelation function for the statistical models and the real traffic

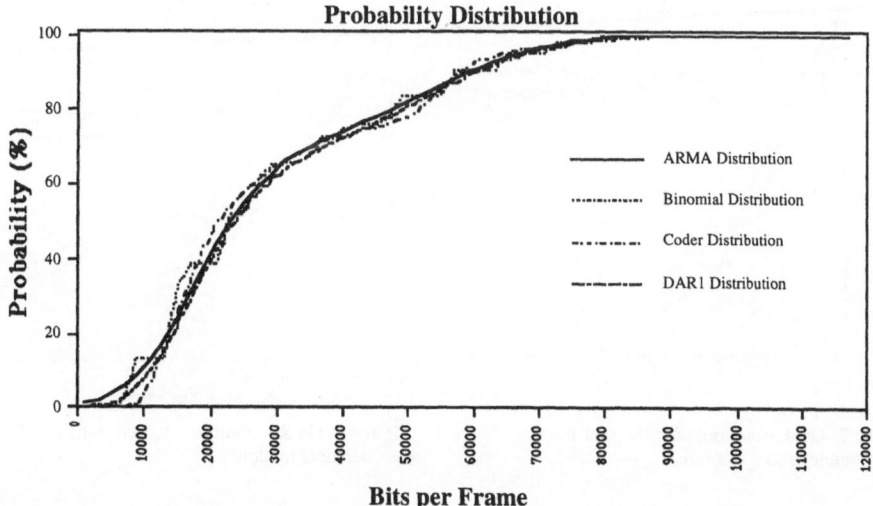

Figure 6. Probability distribution function for the statistical models and the real traffic

VBR MPEG-1 SOURCE TRAFFIC DESCRIPTOR

The concept of GCRA contours [9] is used to characterize the video traffic and to obtain the source traffic parameters[10].The GCRA contours are obtained considering a flow of cells delivered equidistantly for one, two or four image times. The bits per frame are packetized in cells. A cell consists of a 5 byte header, 4 bytes for the Adaptation Layer and 44 bytes for the data segmentation. The evaluated output link rate is equal to 149.76 Mbps (353207.547 cells per second). The curves have been dimensioned according to the bucket length and the normalized capacity (C) of the virtual circuit. The normalized capacity is the ratio between the maximum allocated rate to the virtual circuit and the mean rate of the source (m). The value of the limit parameter L of the GCRA is the value of the queue length represented. The increment parameter I is related to the inverse of the product of the normalized capacity and the mean rate of the source.

The smoothing effects in the source could be observed in figure 7. Smoothing the source, the traffic descriptor parameters decrease strongly. Therefore, the smooth effect is suitable in order to reduce the allocated resources of the virtual circuit. Likewise, there are no advantages when the smooth are achieved for times greater than two frames. To smooth the source, without increasing the delay in the service, the coder should utilize forecasting techniques. In this case, the coder could incorporate a predictor derived of the ARMA filters described in the previous section.

The GCRA contours has been used in order to determine the goodness-of-fit for the models. In figure 8 can be seen the descriptor parameters for the coded sequence and the models ARMA, binomial and DAR(1).

All the models fit well with the traffic descriptors for greater values of normalized capacity than 1.5. Since the real traffic presents a more burstiness behavior, for lower values of C than 1.5, the queue length is larger. The best approach is the binomial model. The greatest accuracy of the binomial model is obtained since the transitions between states are forced to the states with near generation rates. The ARMA and DAR1 models have different behaviors. When a high activity level is reached in these models the most probable transition is toward the intermediate states. In order to improve the behavior of the binomial model it is necessary to consider the cross-correlation of the random I, P and B variables. Considering the marginal probability between these variables, the results could be closer to real traffic. The adjustment of a model allows to extent the results. In figure 9, it is presented the traffic descriptors and the smoothing effects for the binomial model.

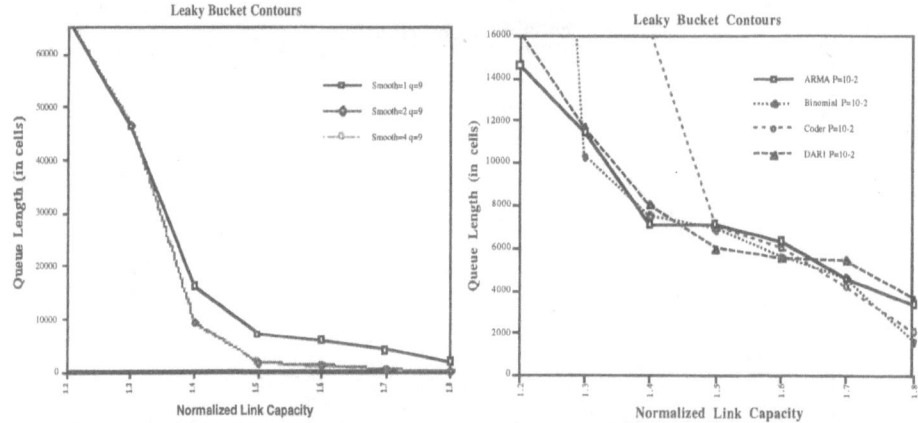

Figure 7. GCRA contours for the real traffic with three levels of smooth

Figure 8. GCRA contours goodness-fit-test for the statistical models

CONCLUSIONS

This work permits to establish a relationship between the coding parameters (q, M and N) of a MPEG coded sequence and the statistical behaviour of its VBR traffic. Analyzing the SNR and the compression ratio of the coded sequence, a suitable selection of the triplets is achieved. Likewise, an empirically determined scale factor for VBR MPEG-1 algorithm and the relationship between the SNR and the quantizer step size are obtained.

Several statistical models are fitted for the activity levels of I, P and B frames when q=9, M=2 and N=4. The AR, DAR and M/M/∞//S models are compared with real traffic in terms of the autocorrelation function, the probability distribution function and the GCRA contours goodness-fit-test. By using these tests, it is determined that the binomial model is closer to the real traffic case.

Various GCRA contours of VBR MPEG traffic are obtained when the source is smoothed using the binomial model. On the basis of these results, it can be concluded that the GCRA parameters are strongly related to the transmission rates of the I and P frames. In addition, it can be noted that to maximize the statistical multiplexing gain is necessary to smooth the VBR MPEG traffic using a traffic shaper. The ARMA model is proposed to forecast the traffic and reduce the smoothing delay. Several extended GCRA contours are presented using the binomial model for loss probabilities of 10^{-2}, 10^{-3} and 10^{-4} regarding the smoothing effects.

These results have been obtained via simulation, by means of an object oriented software for the MPEG coder and an Equivalent Terminal Source Environment simulator, written in C++, and completely developed by the authors.

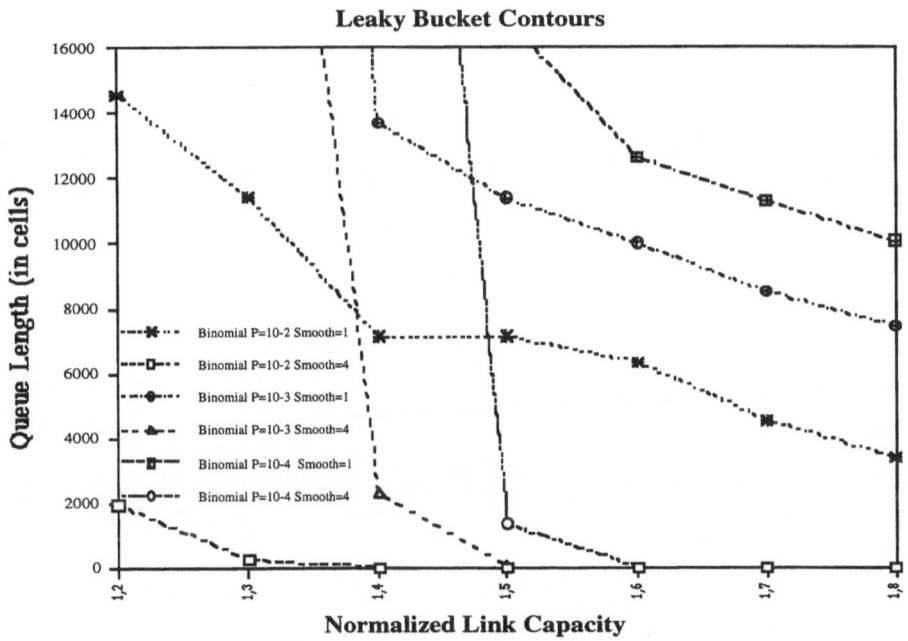

Figure 7. GCRA contours for the binomial model for different smoothing degrees and loss probabilities

REFERENCES

1. W.Verbiest, L.Pinnoo, B.Vosten. *The Impact of the ATM Concept on Video Coding*. IEEE J. on Selected Areas in Commun., SAC- 6, N° 9, pp. 1623-1632, Dec. 1988.
2. ATM User-Nerwork Interface Specification, Version 3.0. Spetember 10,1993.
3. D. Le Gall. *MPEG: A Video Compression Standard for Multimedia Applications*. Communications of the ACM,34(4),pp. 305-313, April 1991.
4. M. Nomura, T. Fujii and N. Ohta. *Basic Characteristics of Variable Rate Video Coding in ATM Environment*. IEEE Journal on Selected Areas in Communications, vol. 7, no. 5, pp. 752-760, June 1989.
5. F. Yegenoglu, B. Jabbari, Ya-Quin Zhang, *Motion-Classified Autoregressive Modeling of Variable Bit Rate Video*. IEEE Trans. on Circuits and Video Techn., Vol. 3, N° 1, pp. 42-53, February 1993.
6. D.P. Heyman, A. Tabatabai, T.V. Lakshman. *Statistical Analysis and Simulation Study of Video Teleconference Traffic in ATM Networks*. IEEE Trans. on Circuits and Video Techn., Vol. 2, N° 1, pp. 49-59, March 1992.
7. B.Maglaris, D.Anastassiou, P.Sen, G.Karlsson, J.D.Robbins. *Performance Models of Statistical Multiplexing in Packet Video Communications*.IEEE Trans. On Commun.,

COM-36, n°7, pp. 834-843, July 1988.

8. P.Sen, B.Maglaris, N.Rikli, D.Anastassiou. *Models for Packet Switching of Variable Bit Rate Video Sources*. IEEE J. on Selected Areas in Commun., SAC. 7, pp. 865-869, June 1989.

9. J. Mata, S. Sallent, J. Balsells, J. Zamora, and A. van der Kolk. *Statistical Models for MPEG Video Standard*. Proceedings of IEE EUSIPCO'94, 1994.

10. D.M. Lucantoni, M. F. Neuts and A. R. Reibman. *Methods for Performance Evaluation of VBR Video Traffic Models*. IEEE/ACM Transactions on Networking, vol 2, N. 2, pp.176-180, April 1994.

GOP-SCALE RATE SHAPING FOR MPEG TRANSMISSION IN THE B-ISDN

Maher Hamdi[1], James W. Roberts[2] and Pierre Rolin[1]

[1] Department of Networks and Multimedia Services
Télécom bretagne, BP 78, 35512 Cesson Sévigné, France
[2] France Télécom - CNET
38 rue du Général Leclerc, 92131 Issy les Moulineaux, France

Abstract

The efficiency of VBR video multiplexing in ATM network is largely dependent upon the rate control algorithm used by the coder since it has detrimental effects on both visual quality and traffic characteristics. In order to build an effective rate control, we establish a relationship between the quantization parameter and the GoP bit rate. We propose to use it to maintain the VBR MPEG output conformity to ATM traffic contract.

Keywords: MPEG, Rate Shaping, Traffic Constraints, ATM Networks.

1 INTRODUCTION

We consider preventive traffic control in the B-ISDN. Concerning VBR video connections, there are two basic approaches to the traffic control problem: either find a suitable characterization of coder output enabling QoS prediction or oblige the coder to make its output conform to predefined characteristics.

The output of certain applications such as video-conferencing may be characterized by a small number of parameters which can be used in Markovian models to evaluate the performance of a network multiplexer [1]. However, the characterization of less stereotyped video is particularly difficult [2]. Indeed, long video sequences seem to systematically exhibit long range dependence whose significant detrimental impact on performance is beginning to be well understood [3].

The alternative approach consisting in modifying the coder output in order to make it more compliant (e.g, to a leaky bucket [4] or a Markov chain [5]) and predictable, has been considered by fewer authors[6,7,8]. Our contribution belongs to this category. A simple rate prediction technique is presented. It allows the quantizer value to be determined according to the target GoP bit allocation. We then give a control algorithm operating at the GoP scale that ensures coder output conformity to the standardized Sustainable Cell Rate and Intrinsic Burst Tolerance parameters[9]. Scene activity is taken into account in the algorithm resulting in graceful degradation of image quality.

2 RATE CONTROL IN MPEG

2.1 Traffic Variability

The MPEG standard [10,11,12,13] offers two coding options: CBR and VBR. CBR control algorithms are essentially based on the quantization parameter Q determining the resolution of the currently coded macroblock. A fixed quantity of bits is allocated to each GoP and apportioned progressively to successive pictures and, within pictures, to successive macroblocks [13]. The drawback of CBR coding is that the same bit rate is generated independently of the scene contents thereby resulting in variable quality.

VBR traffic can be generated using open loop coding where the same quantization parameter is used for all macroblocks, naturally resulting in variable output. Rate variations depend on image complexity and activity. Image quality is said to be constant since the quality reduction is assumed to be similar for all scenes.

2.2 Image Quality

It has been shown that the human ability to perceive image degradations decreases with increasing image spatial frequency and motion. Therefore, full variability of open loop coding is not necessary to maintain a constant subjective quality. We propose to seek a compromise between open loop coding and CBR coding which maintains adequate subjective quality while satisfying traffic constraints.

3 GOP SCALE RATE PREDICTION

It is well known[14] that the generated bit rate R decreases with increasing Q (the quantization parameter). The relationship between Q and R varies in time and depends on instantaneous activity and motion. If known, the function $Q(R)$ gives the right value of Q corresponding to the desired output rate. When performing CBR coding, the control loop acts on the macroblock scale. In this case the precise $Q(R)$ function is too complex and needs to be approximated using for instance predefined codebooks[8]. In VBR coding, the control loop acts on a time scale larger than that of the macroblock.

Let R_i be expressed in bits/GoP (i.e. R_i is the number of bits generated by GoP$_i$) and Q_i the average quantization parameter used for coding GoP$_i$. We then try to derive a relationship beween R_i and Q_i, which we expect to be simpler than that existing at the macroblock scale.

We worked on a 500 frame sequence from the TV program "Spitting Images" in "384x288" format coded with GoP structure IBBPBBPBBPBB. The following approximate relationship between Q and R has been shown experimentally:

$$Q_i = K_i / R_i \tag{1}$$

where K_i is a constant that depends only on the scene complexity (i.e. depends on i). In Figure 1 we plotted R_i^{-1} as a function of Q_i (ranging from 2 to 61) for six randomly chosen,

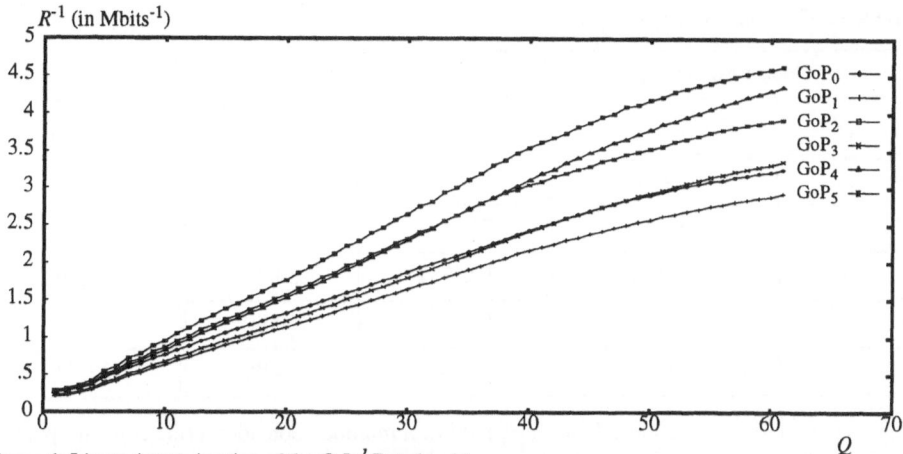

Figure 1: Linear Approximation of the Q-R^{-1} Relationship

open loop coded GoPs (GoP$_i$ starts at image 100*i). Curves of Figure 1 can be approximated by linear functions hence correspond to expression 1. Recall that the approximate principle stated above is significant at the GoP scale where detailed rate-distortion properties are cancelled when averaging R and Q over a GoP.

Expression 1 implies that the product RQ for a given GoP, is independent of the rate control algorithm used by the coder, and for a given algorithm, independent of the quantization values used for coding the GoP. To verify this property, we compressed the video sequence described above in both CBR (using 5 different bit rates) and open loop modes (using 5 values for the quantization parameter). Each time, the RQ product is plotted versus the frame number (see Figure 2). If expression 1 is exact, curves of Figure 2 must be identi-

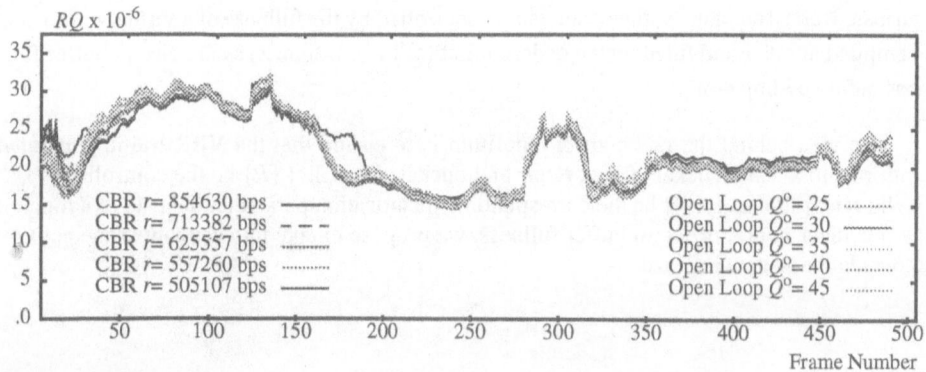

Figure 2: The RQ product

cal. The tiny differences between the curves are caused by the non-linear parts of the curves of Figure 1. Table 1 shows the matching of the averaged values of Q and R for the two coding modes (CBR and Open Loop). Q^o denotes the constant quantizer used in open loop coding and R^o the corresponding variable rate.

Also, as a function of the GoP number i, K_i is a highly correlated process (as shown in Figure 2). In fact it can be considered as a global measure of the scene complexity because it depends only on GoP$_i$ spatial and temporal activity.

Table 1 : The constant RQ product

Open Loop mode		CBR mode		RQ
Q^o	$E[R^o{}_i]$ (in bps)	R (in bps)	$E[Q]$	RxQ^ox10^{-6}
25	854630	854630	25.62	21.36
30	713382	713382	30.44	21.40
35	625557	625557	34.79	21.89
40	557260	557260	39.80	22.29
45	505107	505107	44.86	22.72

This relation is used as a GoP rate prediction method. Consider a rate control operating at the GoP scale to satisfy some traffic constraints, i.e., before coding GoP$_{i+1}$, an algorithm gives the target bit allocation R_{i+1} (in bits) of that GoP. Using expression (1) and approximating K_{i+1} by K_i, we obtain:

$$Q_{i+1} = Q_i R_i / R_{i+1} \qquad (2)$$

This expression gives the quantization parameter value to be used to obtain the desired bit allocation R_{i+1}.

4 THE SHAPING ALGORITHM

If we consider that video traffic has to be compliant to a Sustainable Cell Rate r and Intrinsic Burst Tolerance b, the coder can be controlled by the fullness of a virtual buffer that is emptied at rate r and filled by the coder output (as in [4]). Define X_i as the virtual buffer fullness before coding GoP$_i$.

The idea behind the rate control algorithm is to ensure that the VBR traffic generated conforms to a leaky bucket of leak rate r and bucket size b. Let $\{R\}$ be the controlled variable bit rate traffic and $\{Q\}$ be the corresponding quantization parameter. Instead of acting on the quantizer in proportion to buffer fullness, we propose to take into account scene activity to decide the target GoP size.

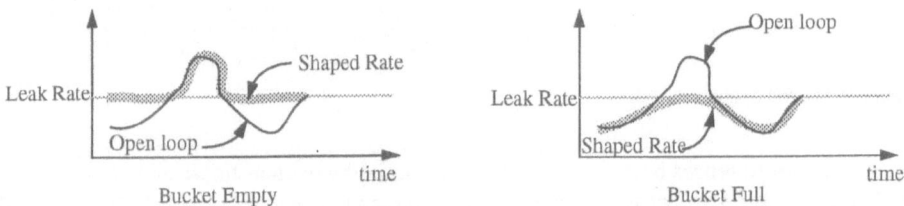

Figure 3: Principle of the Shaping Algorithm

The control principle is shown in Figure 3. Scene activity is measured using a prediction of the equivalent open loop bit rate R^o. A scene with reasonable activity and duration is coded at the rate R^o while excessively long and/or active scenes are "truncated" and their bit rate is reduced to r. This means that for periods where R^o conforms to the traffic contract, the shaping algorithm behaves like open loop control. On the other hand, during overload periods (those where R^o does not conform to the traffic contract) the algorithm switches to a CBR mode of rate r. During these periods, image quality may be reduced to the quality of CBR coding. However, because network resources are dimensioned based on the leaky bucket conformance, this shaping avoids cell loss which could otherwise occur at rate up to R^o-r. Thus, only harmful scenes are shaped. In addition, when the virtual buffer is empty, the minimum allowed rate is r (ensuring that the overall mean rate will be r). A simplified algorithm description follows:

- For high activity scenes ($R_i^o \geq r$):

    when $X_i \approx 0$, the algorithm behaves as in open loop, i.e. R_i is set to R_i^o.

    when $X_i \approx b$, the algorithm behaves like CBR, i.e. R_i is set to r.

- For low activity scenes ($R_i^o \leq r$):

    when $X_i \approx 0$, the algorithm behaves like CBR, i.e. R_i is set to r.

    when $X_i \approx b$, the algorithm behaves as in open loop, i.e. R_i is set to R_i^o.

When the bucket is partially filled, R_i is set to a linear combination of the two extreme cases stated above. The quantization parameter value is then given by expression (2). The algorithm operates at the beginning of each GoP with R^o_{i+1} estimated using expression (1) as:

$$R^o_{i+1} = R_i \, Q_i / Q^o$$

The virtual buffer is filled during complex scenes and emptied during low activity scenes. The target bit rate is set to r when the buffer would otherwise overflow or underflow. More details about the algorithm and its implementation can be found in [15].

Figure 4 shows the bit rate generated by the open loop and shaping algorithms. To remove high frequency variations, the plotted rates are the moving average of 7 consecutive GoPs. It may be noted that the shaping algorithm generates less traffic than open loop in active scenes (frames 1 to 150 and 900 to 1000) and compensates by providing higher rates in calmer periods (200 to 300, 1400 to 1500). Rate variability is maintained but with smaller amplitude. Corresponding variations in the virtual buffer fullness are illustrated in Figure 6. This curve confirms that the algorithm indeed exploits the full range of variability provided by burstiness parameter b. Note that the virtual buffer is filled only during active scenes (those corresponding to $R^o > r$) and emptied during low activity scenes. This would not be the case if Q were set in proportion to the buffer fullness. The shaping algorithm decides whether the scene is active or not using a prediction of R^o based on expression 1. The curve of Figure 6 confirms the validity of the RQ approximated relationship.

To compare quantization parameter variations with those of the CBR case, we have plotted both of them in Figure 5. Quantizer variations are much more stable with the shaping algorithm than in CBR coding.

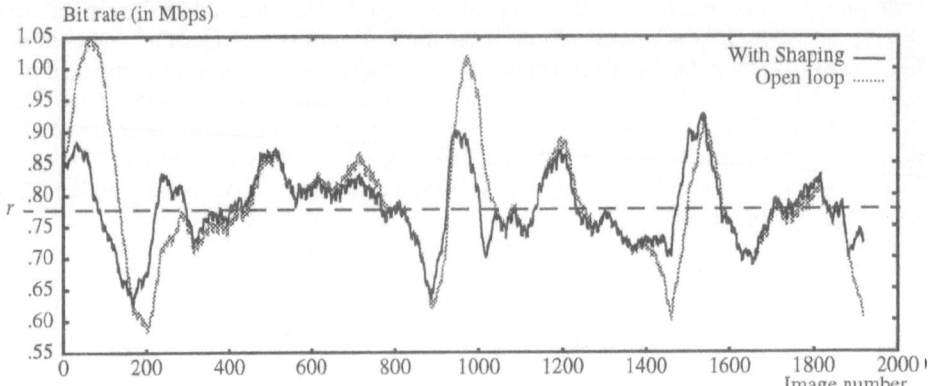

Figure 4: Instantaneous Bit Rate.

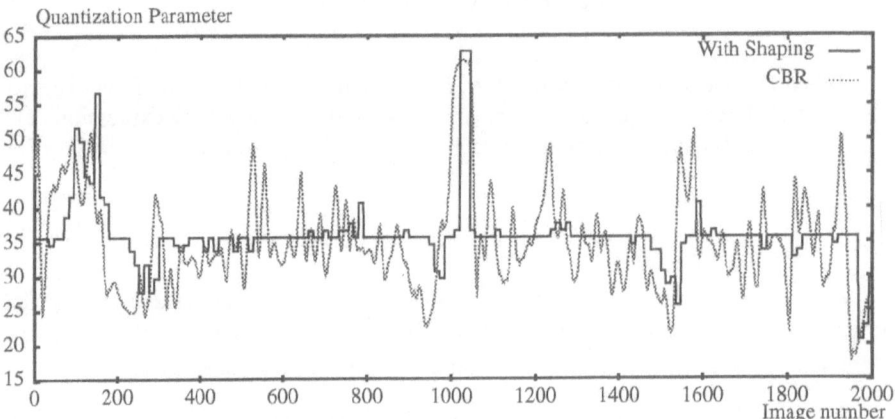

Figure 5:   Quantization Parameter Variation.

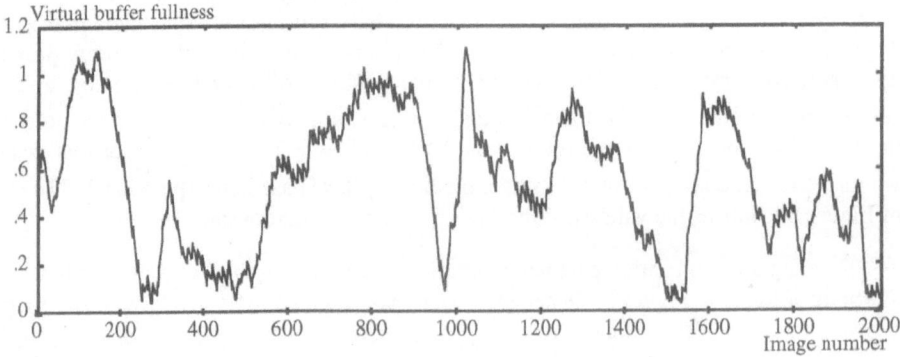

Figure 6: The Bucket Fullness

406

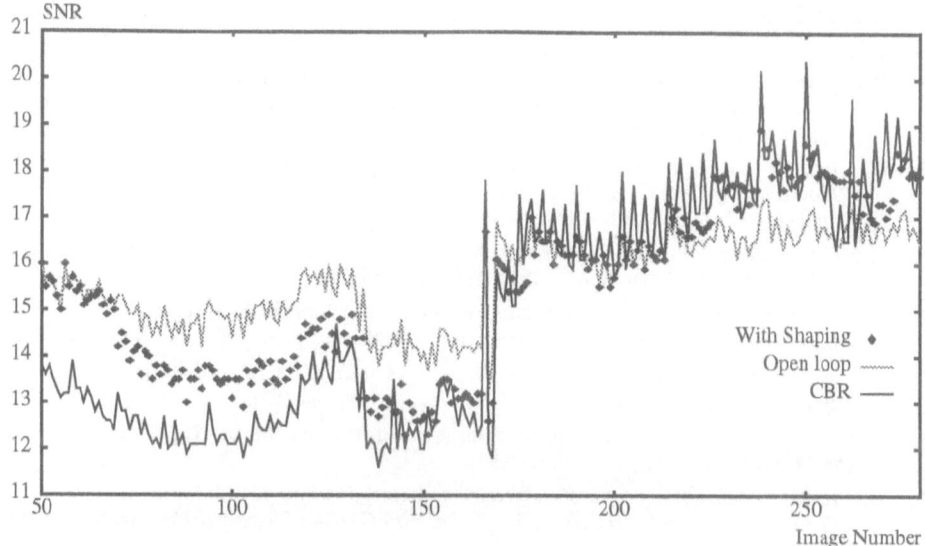

Figure 7: SNR Comparison.

Although only psycho-visual tests can decide about the visual quality, we have plotted the Signal-to-Noise ratio of the three algorithms in Figure 7. For reasons of clarity, only results for frames 50 to 300 have been plotted. First we note that the SNR of the shaped output is always higher than the minimum of the open loop and CBR SNR. It is equal to the maximum of them when the virtual bucket is empty and equal to the minimum of them when the virtual buffer is full. If the Sustainable bit Rate and Intrinsic Burst Tolerance (i.e. r and b) are adequately dimensioned, the SNR is equal to the maximum of open loop and CBR SNR in most of the time since only scenes that cause congestion (buffer full) will be shaped.

5 CONCLUSION

Rate control algorithms have to ensure codec output compliance to traffic constraints while optimizing visual quality. We have proposed a GoP scale rate prediction technique that allows effective rate control guaranteeing conformance to Sustainable Bit Rate and Intrinsic Burst Tolerance traffic parameters. We have suggested that the coder can optimize its use of the defined traffic contract by performing intelligent quality degradation. We argue that this is better for the network than to accomodate uncontrolled burstiness making it difficult to guarantee the required QoS.

We proposed a shaping algorithm for MPEG coders based on GoP by GoP rate adjustment guaranteeing lossless conformance to a leaky bucket controller. It was designed to control traffic burstiness while maintaining the advantage of VBR coding with respect to perceived quality. For a given long term average, visual quality of shaped video is expected to be better than CBR quality since the bit rate allocated to GoPs depends on the current scene activity. It is also significant for interactive video communications that the output bit stream of SBR coders does not need to be buffered in the coder and is delivered to the network at the connection peak rate. Conformance to leaky bucket parameters is controlled using the fullness of a virtual buffer that can be arbitrarily large. The advantage is that, unlike the physical buffer of CBR coding, the virtual leaky bucket queue does not introduce any delay.

REFERENCES

1. D. P. Heyman, A. Tabatabai, and T. V. Lakshman. Statistical analysis and simulation study of video teleconference traffic in ATM networks. *IEEE Transactions on Circuits and Systems for Video Technology*, 2(1):49–59, March (1992).

2. M. Hamdi and P. Rolin, Resources requirements for VBR MPEG traffic in interactive applications, in *Proceedings of Cost237 Conference On Multimedia Transport and Teleservices, Vienna, Nov. 1994*. Springer-Verlag, LNCS 882 (1994).

3. M.W.Garrett and W. Willinger. Analysis, modeling and generation of self-Similar VBR video traffic. In *Proceedings of SigComm*. ACM Press, September (1994).

4. A.R. Reibman and B.G. Haskell. Constraints on variable bit rate video for ATM networks. *IEEE Transactions On Circuits and Systems for Video Technology*, 2(4):361–372, December (1992).

5. H. Heeke. A traffic control algorithm for ATM networks. *IEEE Transactions On Circuits and Systems for Video Technology*, Vol. 3 N. 3, pages 182-189, June (1993).

6. R. Coelho and S.Tohme. Video coding mechanism topredict video traffic in ATM network. In *GLOBECOM'93*, pages 447–451, (1993).

7. M. Hamdi, D. Curet, J. Roberts andG. Madec, Statistical multiplexing of VBR Mpeg2 streams, *Contribution to ISO/IEC JTC1/SC29/WG11, Ref. MPEG94/349*, Singaphore Meeting, Nov (1994).

8. M.R. Pickering and J.F. Arnold. A perceptually efficient VBR rate control algorithm. *IEEE Transactions On Image Processing*, 3(5):527–532, September (1994).

9. I.371 *Recommendation*. Traffic control and resource management in B-ISDN. ITU-T, (1992).

10. ISO Coding of Moving Pictures and Associated Audio for Digital Storage Media at up to 1.5 Mbits/s. ISO-IEC/JTC1/SC29/WG11, *DIS11172-1*, March (1992).

11. ISO Generic Coding of Moving Pictures and Associated Audio: Video. *Recommendation ITU-T H.262, ISO/IEC 13818-2*, November (1994).

12. ISO Generic Coding of Moving Pictures and Associated Audio: Sytems. *Recommendation ITU-T H.222.0, ISO/IEC 13818-1*, November (1994).

13. ISO Coded Representation of Picture and Audio Information. *MPEG Test Model 2*, ISO-IEC/JTC1/SC29/WG11 July (1992).

14. T. Berger, *Rate Distortion Theory, a Mathematical Basis for Data Compression*, Englewood Cliffs, NJ: Prentice Hall, (1971).

15. M. Hamdi, J.W. Roberts, QoS guarantees for shaped bit rate video connections in broadband networks, in *Proceedings of the International Conference on Multimedia Networking, MmNet'95, Sep. 1995, Aizu-Wakamatsu, Japan*. Published by IEEE Computer Society Press (1995).

JOINT ENCODER AND VBR CHANNEL OPTIMIZATION WITH BUFFER AND LEAKY BUCKET CONSTRAINTS *

Chi-Yuan Hsu and Antonio Ortega

Signal and Image Processing Institute
Dept. of EE-Systems
University of Southern California
Los Angeles, CA 90089

INTRODUCTION

Video sequences present varying degrees of complexity and thus, when encoded with most practical compression algorithms at a fixed quality level, generate a variable bit rate (VBR) sequence. In order to transmit such a VBR sequence over a constant bit rate (CBR) channel, a buffer is used to absorb the variations in source bit rate, and a rate control algorithm is needed to prevent buffer overflow. The size of the buffer depends on the total end-to-end delay allowed between encoder and decoder (excluding channel transmission delay). A larger buffer size can allow the encoder to perform better in terms of distortion with the same available channel rate, but such improvement is obtained at the price of larger end-to-end delay.

To improve the video quality without increasing the end-to-end delay one can resort to VBR transmission. For example, VBR video transmission has been proposed and implemented for datagram networks, e.g. those forming the Internet (see [1] [2] for examples of rate control in this scenario.) In these networks, the channel rate depends on the network conditions and there is no guarantee on the maximum delay and channel bandwidth available to the user. This could be a critical shortcoming when, as is the case with video, real time data is being transmitted. We thus concentrate on video transmission over Asynchronous Transfer Mode (ATM) networks where we assume that the quality of service can be guaranteed by the network if the encoder complies with a traffic "contract", established at the connection set up stage. In the ATM environment the video encoder can now select the number of bits to be sent to the network. Our goal is to study the potential benefits of *rate control algorithms which can adjust both the encoder rate and the channel rate*. The choice of channel rate cannot be arbitrary since it will be constrained by the agreed upon traffic contract, and monitored through

---

*This work was supported in part by the National Science foundation under award MIP-9502227 (CAREER)

some network policing function [3]. Therefore, both the network policing function and the physical buffer sizes will restrict both the encoding rate and transmission rate [4].

While previous work, [4], has considered the buffering constraints in the VBR case, in this paper we introduce the concept of *effective buffer size* which establishes the link between the channel rate and the buffer size required to prevent data loss in VBR transmission. Using the effective buffer size makes it simple to use our knowledge of the CBR case to better understand the VBR case. Given the constraints imposed by the effective buffer size and the network policing function, we introduce new techniques that allow joint optimization of the choice of encoder and channel rates. We focus on optimal techniques that will provide bounds on achievable performance and can also be used in off-line compression environments. Based on our results we can derive conclusions that are applicable to simpler coding scenarios.

We start by discussing the bit rate constraints imposed by the end-to-end delay and channel rate in the CBR and VBR channel environments. The encoder optimization in CBR channel by dynamic programming [5] is briefly reviewed. We then formalize the joint optimization problem in the VBR case and extend the optimization techniques used in the CBR to find the optimal solution. Finally, we provide some experimental results.

BUFFER CONSTRAINTS IN CBR AND VBR CHANNEL

When a VBR source is transmitted through a CBR channel, buffers are used at both encoder and decoder to absorb the variations on the bit rate produced by the encoder. For a fixed total bit budget available to code a video sequence, the encoder can achieve less total distortion if a larger buffer size is used, but the size of buffer cannot be arbitrarily large due, as will be seen, to end-to-end delay considerations.

In a video communications system, the end-to-end delay, ΔT, is the time it will take for one frame to be transmitted. It can be written as

$$\Delta T = \delta t_e + \delta t_c + \delta t_d, \tag{1}$$

where δt_e and δt_d are, respectively, the delays in encoder and decoder buffer, and δt_c is the channel transmission delay. The total end-to-end delay ΔT has to be constant. Although the channel delay can be variable (this corresponds to the delay jitter in networking environments) we assume that the delay variations are relatively small compared to the other delay components and thus assume that the sum of delay in encoder buffer and decoder buffer $\delta t_e + \delta t_d = \Delta T - \delta t_c$ is also constant. We will use this combined buffer delay as the end-to-end delay parameter in system design. Here we consider the frame interval as our basic time unit, but similar analyses could be derived for smaller time intervals (e.g. down to block level). Assume the length of the time interval for one video frame is $\mathcal{T}$. Given ΔT, the end-to-end delay due to buffering (expressed in number of frames rather than in seconds) is

$$\Delta N = \frac{\Delta T - \delta t_c}{\mathcal{T}}. \tag{2}$$

The encoder has to keep both encoder and decoder buffers from overflowing to avoid the ensuing loss of data. As for underflow, while encoder buffer underflow can be easily avoided by bit-stuffing, decoder buffer underflow is a much more serious problem since it means that the coded data in the encoder buffer is not arriving to the decoder within the required end-to-end delay. If decoder underflow occurs the corresponding video frame is considered lost. Note that data loss due to decoder buffer underflow may occur *even if the physical size of the encoder buffer is large enough to hold all the coded data.*

We now describe how decoder buffer underflow can be prevented by the encoder. To do so we introduce the concept of *effective buffer size*, B_{eff}, which we define as *the maximum level of buffer occupancy that the encoder can reach without violating of end-to-end delay constraint* (i.e. without producing decoder underflow.) Note that the effective buffer size may be smaller than the physical buffer size that is available at the encoder.

Effective Buffer Size in CBR Channels

Before tackling the VBR case, we consider the effective buffer size in the CBR channel case. Suppose $B^e(i)$ and $B^d(i)$ are, respectively, the encoder buffer and decoder buffer occupancies at time i (i.e. during the i-th frame interval.) C is the channel bit rate in one frame interval and $R(i)$ is the number of bits used by the encoder to code the ith frame of video. If neither encoder or decoder buffers ever overflow or underflow, then the encoder and decoder fullness at time i will be:

$$B^e(i) = \sum_{j=1}^{i} R(j) - i \cdot C \qquad (3)$$

$$B^d(i) = \begin{cases} i \cdot C - \sum_{j=1}^{i-\Delta N} R(j), & \text{when } i \geq \Delta N, \\ i \cdot C, & \text{when } i < \Delta N, \end{cases} \qquad (4)$$

where ΔN, the end-to-end delay, is also the time that the decoder waits after receiving the first bits of information before starting to decode. (As can be seen, data is removed from the decoder buffer only after $i = \Delta N$.) By combining the encoder buffer occupancy (3) at time i and decoder buffer occupancy (4) at time $i + \Delta N$, we can get

$$B^d(i + \Delta N) = \Delta N \cdot C - B^e(i) \quad \text{when} \quad i \geq \Delta N \qquad (5)$$

If we can keep $\Delta N \cdot C - B^e(i) \geq 0$ all the time, then $B^e(i + \Delta N)$ will always be greater than 0 and therefore the decoder will not underflow. We can thus define the effective buffer size in the CBR case as

$$B_{eff} = \Delta N \cdot C \qquad (6)$$

and we can guarantee that if the encoder buffer fullness $B^e(i)$ is always smaller than B_{eff}, then the decoder buffer will not underflow. From (5) we also know that the maximum value of encoder buffer occupancy, $B^e(i)$, will be $\Delta N \cdot C$ when $B^d(i) = 0$. Therefore for the encoder, a buffer size as large as $\Delta N \cdot C = B_{eff}$ will be enough to hold all possible coded data within the end-to-end constraint. A similar argument can also apply to the decoder buffer.

If the physical buffer size is smaller than B_{eff}, then there will be more constraints on the buffer occupancy besides B_{eff}. The buffer occupancy will then be upper bounded by the physical buffer size as $B^e(i) \leq B^e \quad B^d(i) \leq B^d$.

Effective Buffer Size in VBR Channels

There is a detailed discussion on the buffer dynamics in the VBR channel environment in Reibman and Haskell's paper [4]. Essentially the same constraints as in the CBR case apply except that now the channel rate is time varying. Denote $C(i)$ the channel rate at time i, then (3) and (4), the encoder and decoder buffer occupancies can be rewritten as

$$B^e(i) = \sum_{j=1}^{i} R(j) - \sum_{j=1}^{i} C(j) \qquad (7)$$

$$B^d(i) = \begin{cases} \sum_{j=1}^{i} C(j) - \sum_{j=1}^{i-\Delta N} R(j), & \text{when } i \geq \Delta N \\ \sum_{j=1}^{i} C(j), & \text{when } i < \Delta N \end{cases} \tag{8}$$

Combining the encoder buffer occupancy (7) at time i and decoder buffer occupancy (8) at time $i + \Delta N$, we can get the following equation:

$$\begin{aligned} B^d(i + \Delta N) &= \sum_{j=1}^{i+\Delta N} C(j) - \sum_{j=1}^{i} R(j) = \sum_{j=i+1}^{i+\Delta N} C(j) - (\sum_{j=1}^{i} R(j) - \sum_{j=1}^{i} C(j)) \\ &= \sum_{j=i+1}^{i+\Delta} C(j) - B^e(i) \end{aligned} \tag{9}$$

In order to prevent the decoder buffer from underflowing, we have to keep the right side of (9) always greater than 0. Therefore in the VBR case we can again define the effective buffer size as the maximum buffer occupancy at the encoder buffer that will not cause decoder buffer underflow. The main difference is that from (9) we have

$$B_{eff}(i) = \sum_{j=i+1}^{i+\Delta N} C(j) \tag{10}$$

and the effective buffer size depends on the frame interval and is equal to the sum of future ΔN channel rates as. This is an intuitively obvious result since it just states that in order to arrive within ΔN frames, the amount of data to be transmitted cannot exceed the total channel rate available during that period.

This formulation can also be used to design rate control algorithms in other VBR scenarios such as time-varying wireless channels [6], where future rates are not known, but stochastic models can be used instead. Here we concentrate on the case where the future ΔN channel rates, and thus the effective buffer size, can be planned or controlled in advance, but where the choice of the channel rates is subject to a constraint, as will be the case in ATM networks.

OPTIMAL RATE CONTROLS IN CBR AND VBR CHANNEL

Encoder Optimization in CBR Channel

The optimal encoder bit-allocation for a discrete set of quantizers with buffer constraints in the CBR channel can be solved by dynamic programming [5]. In this formulation, assuming a sufficiently long end-to-end delay, a trellis can be formed where each branch represents a choice of quantization for the frame and has associated a distortion. We can find out the trellis path with minimum distortion using the Viterbi algorithm (VA) [7]. In each stage, each node represents the buffer occupancy of the encoder buffer and accumulated distortion of the video. Each branch represents a possible quantizer choice with its associated bit rate and distortion. Therefore, the process of connecting the node in current stage to the nodes in the next stage by all the possible branch represents all the possible transition of buffer occupancy and accumulated distortion when the quantizer corresponding too that branch is chosen by the encoder to code the current video frame. Since the buffer occupancy is constrained by the maximum applicable buffer size, this transition of buffer occupancy cannot exceed the maximum buffer size in every stage. The branches which can cause the buffer to overflow are then eliminated. For the trellis paths arriving at the same node, the one with smallest accumulated distortion is chosen, and the rest are pruned out Refer to Fig. 1 and to [5] for the details.

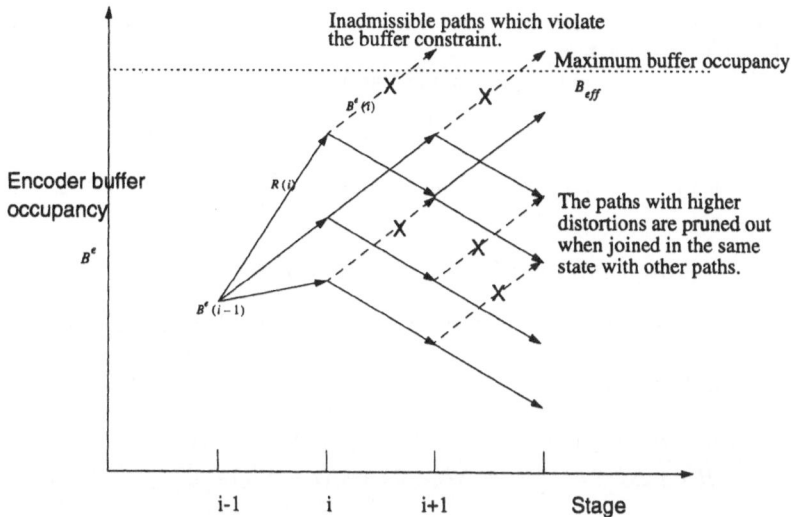

Figure 1: Buffer constrained optimization in the CBR channel case.

Joint optimization of encoder and VBR channel

Our goal is to jointly optimize the encoder and VBR channel bit-allocation with buffer and leaky bucket constraints. The encoder can choose not only the coding bit rate $R(i)$, which is constrained by the effective buffer size, but also the channel rate $C(i)$, which is constrained by leaky bucket policing function.

We consider the VBR channel case with leaky bucket constraint as an example of VBR rate constrained scenarios. Let $\tilde{C}$ be the leaky bucket drain rate and let LB_{max} be the leaky bucket size in bits (i.e. if the bucket can store L tokens and each token represents r bits to be transmitted, then $LB_{max} = r \cdot L$). Refer to [3] for a more detailed description of the leaky bucket mechanism. In order to apply dynamic programming to solve this joint optimization problem, we assume that the number of possible channel rates available to the encoder is finite. The encoder will choose one rate from this finite set of possible channel rates to transmit the data in each time interval. The allowable bit rates will be constrained by the leaky bucket mechanism.

Assume the buffer and leaky bucket never underflow or overflow, and the physical buffer size is large enough to hold all the coded data. Under this assumption we do not have to take into account the physical buffer size in optimization. We define this joint optimization problem as follows.

Given the encoder buffer occupancy, $B^e(i) = B^e(i - 1) + R(i) - C(i)$, the leaky bucket fullness, $LB(i) = LB(i - 1) + C(i) - \tilde{C}$ and the effective buffer size $B_{eff}(i) = \sum_{j=i+1}^{i+\Delta N} C(j)$, find $R(i)$ and $C(i)$ for all i, to minimize the total distortion of the coded video sequence:

$$\min \sum_{j=1}^{N} D(i) \tag{11}$$

subject to $\quad LB(i) \le LB_{max} \quad \text{and} \quad B^e(i) \le B_{eff}(i) = \sum_{j=i+1}^{i+\Delta N} C(j), \quad \forall i \tag{12}$

Two-Variables Viterbi Algorithm

The effective buffer size constraint is needed to prevent decoder underflow and it depends on the choice of channel rates ΔN frames ahead. At time i, the variables to be chosen by the encoder are thus the encoding bit rate $R(i)$ at time i and channel bit rate $C(i + \Delta N)$ at time $i + \Delta N$.

From (9) the decoder buffer occupancy can be expressed as

$$B^d(i + \Delta N) = B^d(i + \Delta N - 1) + C(i + \Delta N) - R(i) \tag{13}$$

except for the first ΔN frames. The leaky bucket fullness at time $i + \Delta N$ can also expressed as

$$LB(i + \Delta N) = LB(i + \Delta N - 1) + C(i + \Delta N) - \bar{C} \tag{14}$$

The goal of this constrained optimization is to choose $R(i)$ and $C(i + \Delta N)$ which will not cause decoder buffer underflow nor exceed the leaky bucket constraint. Such constraints can be expressed as

$$B^d(i + \Delta N) = B^d(i + \Delta N - 1) + C(i + \Delta N) - R(i) \geq 0 \tag{15}$$
$$LB(i + \Delta N) = LB(i + \Delta N - 1) + C(i + \Delta N) - \bar{C} \leq LB_{max} \tag{16}$$

From these applicable $R(i)$ and $C(i + \Delta N)$ for all i, we want to find out the ones with minimum accumulated distortion. Such choice of $R(i)$ and $C(i + \Delta N)$ with minimum cost can be found by using the Viterbi algorithm (VA). The method is similar to that used in the CBR case except that there are two state variables: $B^d(i + \Delta N)$ and $LB(i + \Delta N)$. In the VA, if there are two or more trellis paths which end in the same state at some intermediate stage, all the paths can be pruned out except for the one with minimum cost. The buffer and leaky bucket constraints are used within the VA to prevent the trellis paths from going into the inadmissible states, i.e. the states where $B^d(i + \Delta N) < 0$ or $LB(+\Delta N) > LB_{max}$. Conceptually the trellis paths at each stage can extend into a two dimensional space defined by all possible values of the state variables $B^d(i + \Delta N)$ and $LB(i + \Delta N)$. The whole "state grid" propagates along the direction of the third, or time, axis, where each stage represents a frame interval. Refer to Fig. 2.

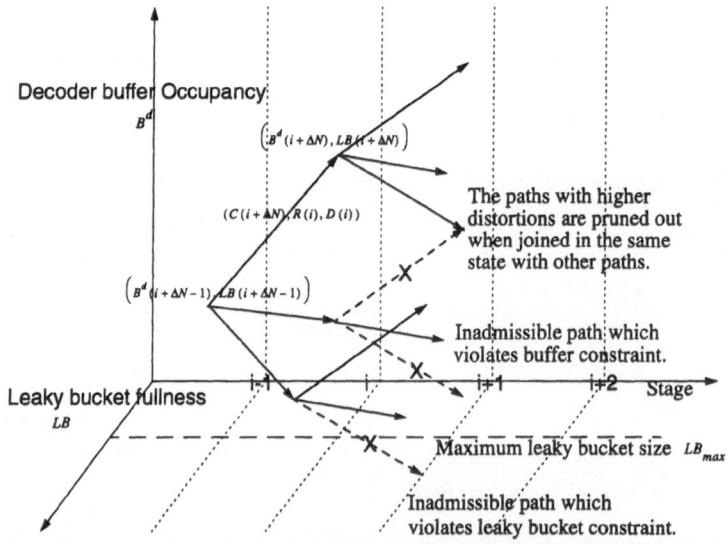

Figure 2: Buffer and leaky bucket constrained optimization in the VBR channel case.

In this two-variable VA, each node in stage i represents a possible state of $B^d(i + \Delta N)$, $LB(i + \Delta N)$ and the accumulated distortion of the system. Each branch represents a possible choice of $R(i)$, $C(i + \Delta N)$, and the associated distortion $D(i)$ when the video frame at time i is coded by bit rate $R(i)$.

A node in stage $i - 1$ is connected to the nodes in next stage i by branches with associated choice of $R(i)$ and $C(i + \Delta N)$ when such choice of $R(i)$ and $C(i + \Delta N)$ meet the $LB(i + \Delta N) \leq LB_{max}$ and $B^d < 0$ constraints. The update of the two state variables and the accumulated distortion then are:

$$LB(i + \Delta N) = LB(i + \Delta N - 1) + C(i + \Delta N) - \bar{R} \qquad (17)$$
$$B^d(i + \Delta N) = B^d(i + \Delta N - 1) + C(i + \Delta N) - R(i) \qquad (18)$$

For the trellis paths arriving at the same node, the one with smallest accumulated distortion is chosen, and the rest are pruned out. At the final stage N, we can find the node with smallest accumulated cost: $\min \sum_{j=1}^{N} D(j)$. Tracing back the path from that node. Then the $R(i)$ and the $C(i)$ on the branches along this path will be the optimum choice of bit rate assignments for encoder and channel.

EXPERIMENTAL RESULTS AND CONCLUSIONS

We implement this joint optimization algorithm in the "Star Wars" sequence from frame 1 to frame 2000 in the VBR channel case defined in above. The buffer constrained optimization in CBR channel case is also implemented as a comparison to show how much improvement of video quality can be achieved by controlling the channel rate and encoding rate jointly according the complexities of the video sequence.

Fig. 3 is the PSNR of coded video sequences in VBR and CBR channel cases. The end-to-end delay varies from 5 frames to 25 frames, and the bit rate budget R equals to 60,000 bits per frames in both VBR an CBR case. In the VBR channel cases, leaky bucket sizes equal to $10 \times C$, $15 \times C$ and $20 \times C$ are implemented.

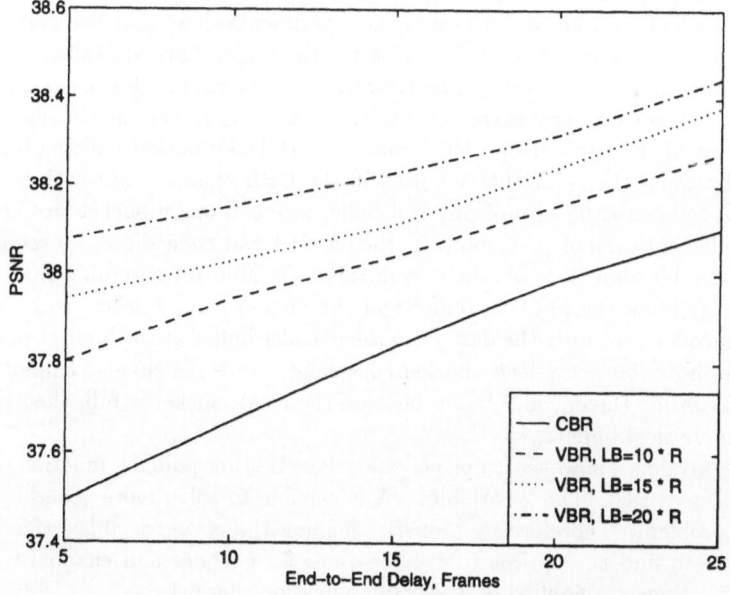

Figure 3: PSNR of coded video sequence in VBR and CBR channel.

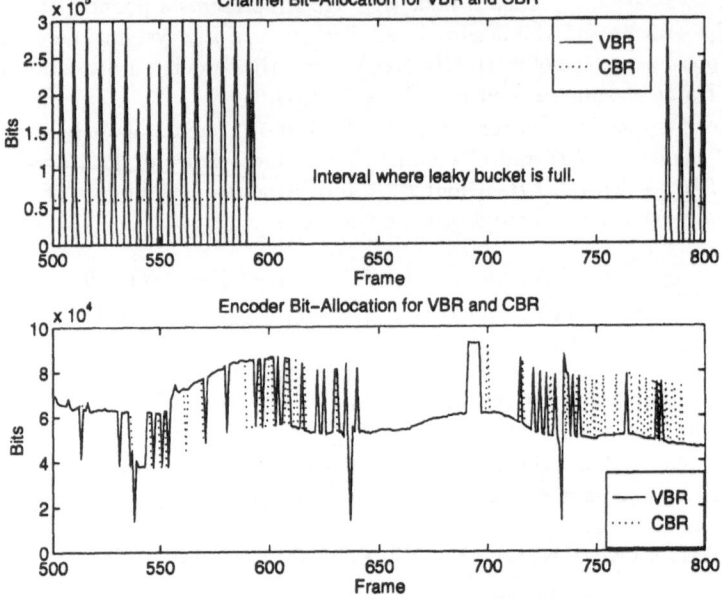

Figure 4: Bit rates used by encoder and channel in VBR and CBR channel.

Fig. 4 is the evolution of the bit rates used by encoder and channel in CBR and VBR channel case. The end-to-end delay equals to 15 frames in both VBR and CBR channel cases. The leaky bucket size of the VBR channel equals to $20 \times C$.

Reibman and Haskell [4] have shown that in the VBR channel with leaky bucket policing function, the constraint for the encoder to choose the bit rate for each frame is the same as in the CBR channel case with virtual encoder buffer size which equals to the sum of physical buffer size and leaky bucket size. The equivalent channel rate is equal to the leaky bucket drain rate. Our experiments show that the performance of these two cases in terms of total distortion are the same. This equivalence comes from the fact that the leaky bucket can be treated as an internal buffer inside the network, but which will not cost any extra delay to the data flow. In this specific case, the joint optimization of the encoder and VBR channel with leaky bucket policing function can be done by using the 1-variable VA used in the CBR channel case with larger buffer size which equals to the sum of physical buffer size and leaky bucket size. In the real world implementation of such solution, the encoder can code the video sequence with the optimum bit-allocation obtained from this 1-variable VA assuming that the larger buffer size (adding the physical buffer and the virtual leaky bucket size) is available. The the encoder transmits the data from the encoder buffer at the highest possible rate which does not violate the leaky bucket constraint. Once the channel cannot carry the coded data inside the encoder buffer because the leaky bucket is full, then the data is left in the physical buffer.

This equivalence however is not necessary true if other policing function other than leaky bucket is used. The 2-variables VA is capable to solve more general joint optimization problem. Therefore by properly defining the state variables, this algorithm can applied to find the optimum bit-allocations for encoder and channel when other policing functions are applied on the communication channel.

416

References

[1] H. Kanakia, P. P. Mishra, and A. Reibman, "An adaptive congestion control scheme for real-time packet video transport," in *Proc. ACM SIGCOMM'93*, (San Francisco, CA), Sept. 1993.

[2] J. C. Bolot and T. Turletti, "A rate control mechanism for packet video in the internet," in *Proc. Infocom '94*, 1994.

[3] E. P. Rathgeb, "Modeling and performance comparison of policing mechanisms for ATM networks," *IEEE J. on sel. area in comm.*, vol. 9, pp. 325–334, April 1991.

[4] A. R. Reibman and B. G. Haskell, "Constraints on variable bit-rate video for atm networks," *IEEE Trans. on CAS for video tech.*, vol. 2, pp. 361–372, Dec. 1992.

[5] A. Ortega, K. Ramchandran, and M. Vetterli, "Optimal trellis-based buffered compression and fast approximations," *IEEE Trans. on Image Proc.*, vol. 3, pp. 26–40, Jan. 1994.

[6] A. Ortega and M. Khansari, "Rate control for video coding over variable bit rate channels with applications to wireless transmission," in *ICIP'95*, (Washington, D.C.), Oct. 1995.

[7] G. D. Forney, "The Viterbi algorithm," *Proc. of the IEEE*, vol. 61, pp. 268–278, Mar. 1973.

MODELING TWO-LAYER MPEG-2 VIDEO TRAFFIC

Kavitha Chandra and Amy R. Reibman
AT&T Bell Laboratories
Holmdel, NJ 07733

1. INTRODUCTION

Recently, there has been much interest in sending video over Broadband Integrated Services Digital Networks (B-ISDN) using Asynchronous Transfer Mode (ATM). ATM affords a flexible multiplexing and switching capability for integrated delivery of bursty traffic. As such, video can be transported either with a constant bit-rate (CBR) or with a variable bit-rate (VBR). VBR video has several potential advantages over traditional CBR video: improved image quality and shorter delay. In addition, through statistical multiplexing, improved channel allocation may be obtained compared to CBR transport.

However, statistical multiplexing invariably results in buffering delays and losses, which can significantly degrade video quality. To minimize the *amount* of delay and loss, the networking community has focused on the development of effective and implementable congestion control schemes, including connection admission control (CAC) and usage parameter control (UPC). To minimize the *impact* of delay and loss, the video community has focused on developing good error concealment algorithms and designing efficient two-layer coding algorithms[2,3] for use in combination with the dual-priority transport provided by ATM networks. For example, while one-layer MPEG-2[1] produces generally unacceptable video quality with a cell loss ratio of 10^{-3}, losses at this rate with SNR scalability (one of the four standardized layered coding algorithms of MPEG-2) are generally invisible, even to experienced viewers[11].

To determine a set of techniques appropriate for CAC and UPC, traffic models that accurately represent the statistical nature of very high-speed bursty services are necessary. Models for low to medium activity videophone and teleconferencing video are compared in [4], while Garrett and Willinger[5] propose a self-similar process to model long-range dependence features found in the compressed "Star Wars" movie. However, much less is known about the statistical characteristics of two-layer video. In two-layer coding algorithms, the base layer can be decoded independently to produce a lower quality picture, and should be transported at high priority with negligible loss. The enhancement layer, which contains the remaining information, can be transported at low priority.

Pancha and El Zarki explore the statistical characteristics of a non-standard approach for splitting MPEG bitstreams similar to data partitioning[6], while Ismail et. al.[7] examine modeling MPEG-2 data partitioning using TES. However, both of these assume that the base layer is VBR, which is a problem for CAC and UPC on the important base layer. Earlier work[8,9,10] suggests that the overhead incurred by two-layer coding is large enough to offset any gains in improved cell loss resilience. However, these studies are based on less efficient compression algorithms than are currently standardized in MPEG-2.

In this work, we focus on modeling SNR scalability, in which the base layer consists of a coarsely quantized version of the video, and the enhancement layer contains the refinement information. We choose SNR scalability because it provides the best trade-off between error resilience and complexity among the standardized layered coding algorithms[11]. Our goal is to characterize the statistical process of the VBR enhancement-layer traffic, given that the base layer is CBR. Our video source is 15 minutes from "The Blues Brothers".

Section 2 describes the SNR scalability coding algorithm. Section 3 describes the traffic model used for VBR video, both for the one-layer video and for the enhancement layer of the two-layer video. In section 4, we present our two-layer model for SNR scalable video where the base layer is transported with constant bit-rate (CBR) and the enhancement layer has variable bit-rate (VBR). Section 5 presents the statistical multiplexing gains that can be achieved for one- and two-layer VBR video. Section 6 concludes the paper.

2. SNR SCALABILITY CODING ALGORITHM

SNR scalability provides a way of transmitting two layers at the same spatio-temporal resolution but with different qualities. The base layer encoding process is identical to that for a non-layered encoder. The quantized DCT coefficients from the base layer (after being dequantized) are subtracted from the input DCT block. The resulting quantization error from each block is next re-quantized and encoded to form the enhancement-layer bitstream.

In the standardized decoder for SNR scalability, the dequantized base- and enhancement-layer DCT-coefficients are first obtained by independently processing the respective bitstreams. At this point the two sets of coefficients are summed blockwise, and the IDCT is applied to this sum. To this result is added the temporal prediction signal to produce the output pels, which are also fed back into the motion compensation loop. More details on the coding algorithm for SNR scalability can be found in[1,11].

3. SOURCE MODEL DESCRIPTION

Characterizing the compressed video source is necessary for defining traffic parameters for resource reservation and billing. Here, we use the single-source model developed in[12] for a one-layer coder. The source model for the MPEG-2 encoded video is derived by considering two basic features that characterize the encoding process. First, in the encoder considered here, Intra-frames (I-frames) are generated only at a scene change. Therefore their frequency depends on the video source. Second, between scene changes, frames are coded predictively (P-frames). Since there are no significant changes in the information in successive P frames of a scene, the bit rates of these frames tend to cluster around an average value. In addition, there is significant correlation between the number of bits per frame in consecutive frames of a scene. The correlation between adjacent P frames can be seen in the two-dimensional state space of frame bit rates $r[n]$ vs. $r[n+1]$, where $r[n]$ represents the number of bits in the n-th frame[12]. A mapping of the sequence of one layer VBR data shown in Figure 1 to the two-dimensional state space is shown in Figure 2. The central diagonal region labelled "PP" corresponds to the P frames. This region, we hypothesize, is formed by a composition of clusters with individual average values extending across the diagonal of the state space. The I frames, which do not exhibit any dependence on the preceding frame, occupy the region outside of the diagonal cluster. In particular, the I frames can be identified by the frame $r[n+1]$ along the y axis for the values of n that occupy the region labelled "PI". This feature renders a mechanism for extracting the I frames from the encoded VBR video sequence.

Based on the aforementioned hypotheses, the source model is derived from the mea-

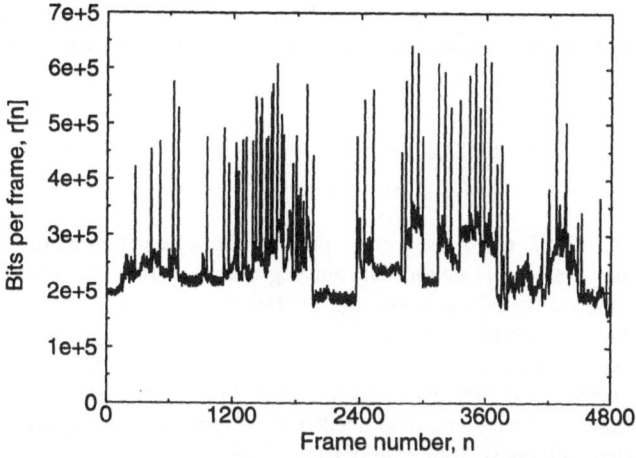

Figure 1. One-layer video data, The Blues Brothers

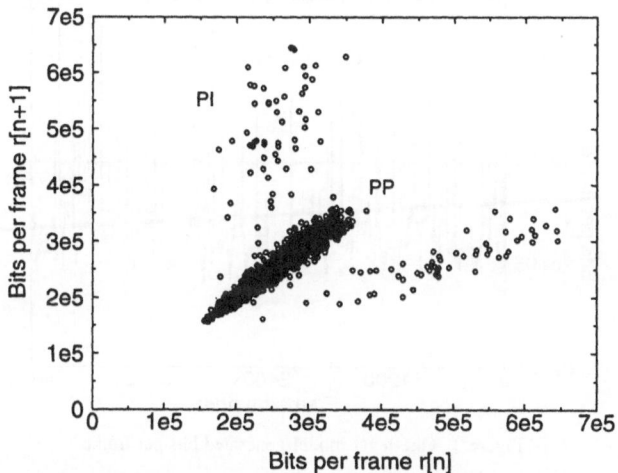

Figure 2. Two dimensional state-space of Figure 1

surements by performing four basic operations. (1) Segment the I-frames from the data using the state space representation and hypothesize the distribution of the I-frame bit rates; (2) Segment the P-frame region (PP) into a finite set of K clusters and map the video bit rate sequence to the K+1 state indices; (3) estimate the probabilities of transition between states; (4) model the correlation between successive frames in each of the K states characterizing the P-frames. The state segmentation is carried out using the *K-means* clustering algorithm [14].

The optimal number of clusters chosen to represent the "PP" region is governed by application dependent criteria. For the VBR video model, the value of K is successively increased until the source model matches selected statistical features found in the measured data. These features are the histogram of the frame bit rates, a visual agreement of the QQ plots and a matching of the bit loss characteristics in a leaky bucket policing function.

We use K clusters to characterize the P frames. The generation of bits when in a particular state is governed by a first order autoregressive process. This model is based on matching the exponentially correlated structure in the bit rate sequence of individual states. The transitions between the states is modeled as a Markov chain, with the transition probabilities obtained empirically from the measurements. These transition probabilities are found to be typically diagonally dominant, providing one validation of our hypothesis that successive P-frames cluster in a state during a scene. In summary, the traffic model for VBR video can be described by the following set of equations, given the number of states K and the transition probabilities.

$$r_i[n] = m_i + \alpha_i(r_j[n-1] - m_j) + g_i[n] \quad when(i = j)$$
$$r_i[n] = m_i + g_i[n] \qquad\qquad when(i \neq j)$$

(1)

where $i, j = 1,.....K+1$, and i and j correspond to the state occupied in frame n and $n+1$ respectively. The $-1 \leq \alpha_i \leq 1$ are the state autoregressive coefficients with $\alpha_{K+1} = 0$, m_i corresponds to the estimated mean bit rate in state i., and $g_i[n]$ represents a zero mean Gaussian process with estimated variance σ_i.

The one layer VBR data was modeled by $K=15$ states. One ensemble of the simulated VBR video sequence is depicted in Figure 3. The horizontal line in Figure 3 at 390000 depicts the bit rate value that would be needed for comparable video quality if one were to transmit the video with constant bit rate (CBR).

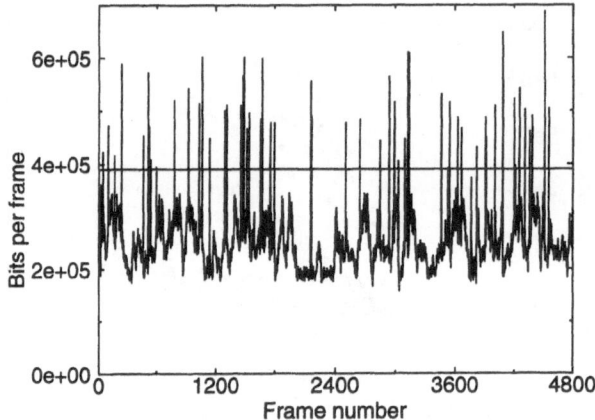

Figure 3. One-layer model-generated bits per frame

One statistic for validating the model is the bit losses that occur in a leaky bucket counter. The leaky bucket drain rate R and the buffer size M expressed in terms of the time to drain (seconds) were varied from $r_{avg} \leq R \leq 5r_{avg}$ and $0.001 \leq M \leq 1$ respectively, where r_{avg} is the actual average number of bits per frame over the sequence. The variation of the bit loss ratio (BLR) as a function of the drain rate is plotted in Figure 4 for three values of M. The results for the model output are seen to be in good agreement with the bit loss ratios for the data. Matching the trends in the leaky bucket loss ratios is indicative of how well the traffic generator models the burstiness inherent in the measurements. This feature can not be validated by simply matching aggregate measures such as the distribution function and the moments of the bit rates.

4. TWO-LAYER MODEL

In [13] we present a model for one-layer video that incorporates feedback for buffer control, as shown in Figure 5. This model enables characterization of a video source compressed for a constant-rate channel. The source model generates the output bit-rate of a constant quality VBR encoder with quantization parameter fixed at Q=4, while a multiplicative scaling factor accounts for variations in bit-rate caused by using a different quantization parameter to ensure no buffer overflow or underflow. The model in Figure 5 was shown to accurately capture the video quality when CBR transport is used [13].

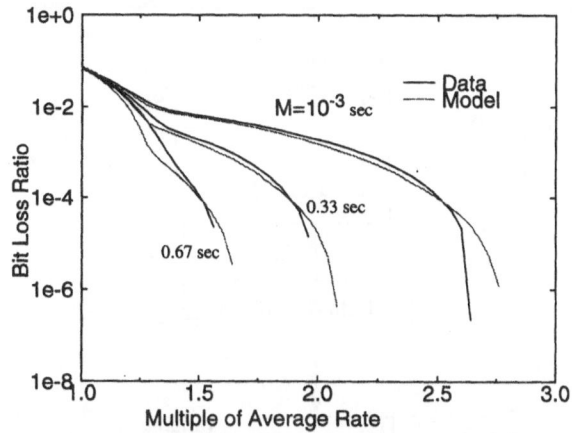

Figure 4. Bit-loss rate for one-layer data and model

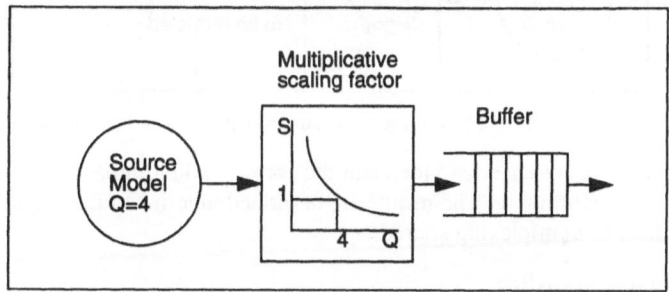

Figure 5. Rate control for one-layer video coder

We extend this concept to obtain a model for two-layer video when the base layer is transported with CBR, shown in Figure 6. The base and enhancement layer data were measured using a constant quantizer step-size of $Q_b=8$ in the base layer and $Q_e=3$ in the enhancement layer. Since the base layer is to be transported at a constant bit-rate, the quantization parameter actually used when encoding the base layer must be chosen such that the base-layer buffer neither overflows or underflows. By changing the actual Q_b to be different than 8, the base-layer bit-rate varies, as does the enhancement-layer bit-rate. These variations are well characterized by a multiplicative scaling factor plus an offset, $R_Q = A_Q R_8 + B_Q$ The parameters A_Q and B_Q for these scaling functions are derived from the measured data using least squares. Separate scaling functions are used for the I frames and the P frames. As depicted in Figure 6, when Q_b increases, less data is coded in the base layer and more is left to be coded in the enhancement layer.

The rate control algorithm used for buffer control is as follows. The quantization step (q-step) is adjusted based solely on the buffer fullness. When the buffer is empty, the minimum q-step of 2 is used, while when the buffer is full, the maximum q-step of 31 is used. Between these end-points, the chosen q-step is an exponential function of the buffer fullness.

For each value of the base layer CBR rate chosen, the rate control algorithm generates a different sequence of VBR enhancement layer. This data is then fit using the source model described in the previous section. The model generated data is then used to compare the tradeoffs in one and two layer encoding.

5. SOURCE STATISTICS and STATISTICAL MULTIPLEXING

The goal of this study has been to model and evaluate the relative statistical characteris-

423

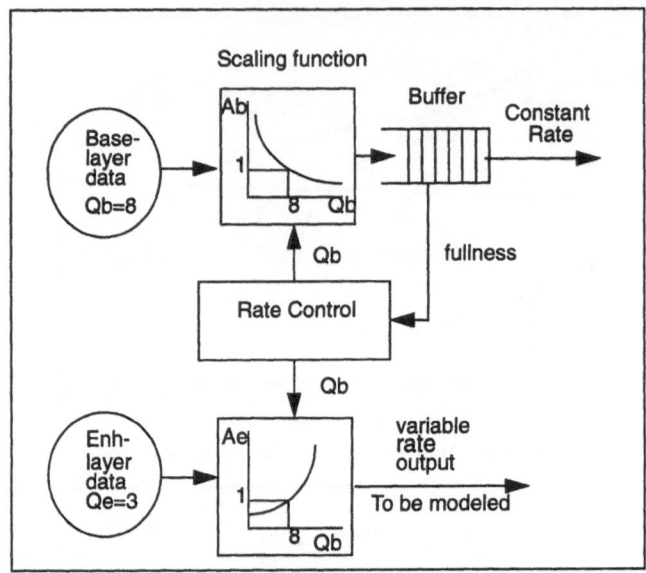

Figure 6. Effect of base-layer rate control on enhancement-layer bit-rate

tics of one and two layer encoded video and the corresponding impact these characteristics have on network performance. The results are described in terms of the single source statistics and the effect of multiplexing sources.

5.1 Single Source Statistics

Two layer video was modeled for base layer CBR values ranging from 1.5 to 3.5 Mbps. This range translates to a factor of 16-37% of the CBR characterizing one-layer video (9.6 Mbps). In the model, the encoder buffer size corresponded to a six-frame duration. The single source statistical characteristics are tabulated in Table 1, in terms of the overall mean rate of the video signal (CBR base rate + mean rate of VBR enhancement layer), the variance of the video, and the peak to mean ratio of the enhancement layer.

Table 1. Single Source Characteristics

	One-layer	Two-layer base bit-rate				
		1.5	2	2.5	3	3.5
Mean bit-rate (Mbits/frame)	0.2438	0.2355	0.2421	0.2494	0.2574	0.2673
Variance (enhancement)	$2.48*10^9$	$1.17*10^9$	$1.05*10^9$	$9.4*10^8$	$8.69*10^8$	$7.24*10^8$
Peak/mean (enhancement)	2.64	1.85	1.77	1.86	1.95	1.97

The impact of the rate control algorithm on the VBR enhancement layer can seen in two ways. First, as the base layer CBR rate increases, the average base-layer q-step decreases, decreasing the average bit rate of the enhancement layer. The rate of decrease is determined by the parameters of the scaling functions that characterize the video source. For the Blues Brothers example considered here, for values of base rate up to around 2 Mbps, the overall average rate of the two layer signal remains below that of the one-layer video. Beyond this value, the decrease in the enhancement-layer average rate is too slow to offset the chosen increase in the base rate. Second, the variability of the enhancement layer data, as captured by the estimated value of the variance, decreases monotonically with increasing base rate. This feature is to be expected, since by increasing the base rate the I-frame bit rates are absorbed by the base layer, effectively smoothing the enhancement-layer bit rates. This reduction in the variability of the video can also be seen in the peak-to-mean ratios of the video. Again, the decrease in the average rate of the two layer data is not large enough for base rates greater than 2 Mbps to warrant a monotone decrease in this statistic.

The impact of reduced variance is important in that traffic parameters for the UPC function can be more robust relative to the one layer video. This feature is illustrated in Figure 7 by comparing the bit loss ratios for the two-layer and the one-layer VBR video signal. The two-layer result is for a base CBR rate of 1.5 Mbps. For a fixed bucket size of 1 ms, the rate of decay of the bit losses with increasing drain rate is significantly faster than that exhibited by the one-layer data.

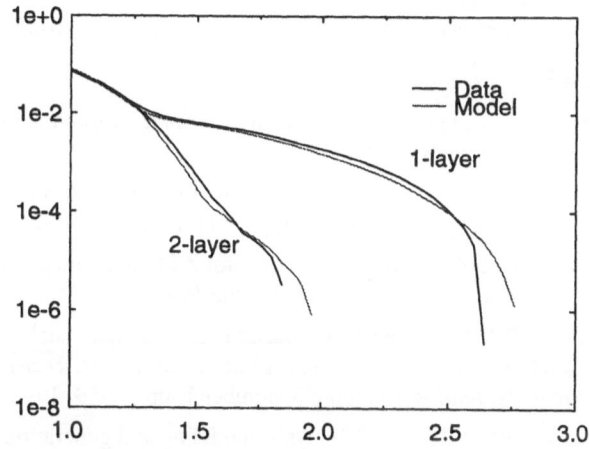

Figure 7. Bit loss rate for one- and two-layer video

5.2 Statistical Multiplexing

The expected advantage of VBR video coding is the bandwidth savings a network provider could achieve by statistically multiplexing the video sources. In Table 2, we compare the statistical multiplexing efficiency of one- and two-layer video. The buffer size chosen constrains the maximum queueing delay to be 20 milliseconds. The channel capacities are 155 and 600 Mbps. The first observation is that statistical multiplexing efficiency increases with increasing channel capacity for the one-layer video. The required per-source bandwidth decreases by 0.4 Mbits as the capacity increases from 155 to 600 Mbps. The corresponding difference for two layers is about half that. This feature results because the overall variability of the two-layer VBR signals is reduced, thus requiring fewer sources to be multiplexed for convergence of the statistical gain. The second observation is that for base rates below 2.5 Mbps, two-layer coding has a lower per-source bandwidth than one-layer coding. Again, because the statistical gain converges faster for two layers, the bandwidth savings is higher for the lower channel capacity. The multiplexing performance becomes comparable to the one-layer case for base rates larger than 2.5 Mbps, for reasons outlined in section 5.1.

6. CONCLUSIONS

A technique for modeling one- and two-layer VBR video has been presented. The traffic model has been used to statistically characterize and evaluate the relative network performance of one and two layer video. The two-layer encoding is found to be advantageous in having statistically smoother variations in its enhancement layer relative to one layer VBR video. This feature allows a tighter characterization of UPC parameters. It also results in a

Table 2. Statistical Multiplexing Gains

Capacity		one-layer BLR=10^{-6}	Two-layer (BLR=10^{-3})				
			1.5	2.0	2.5	3	3.5
155	# sources	24	26	26	25	24	23
	per source BW	6.46	5.96	5.96	6.2	6.46	6.73
600	# sources	99	105	102	99	96	92
	per source BW	6.06	5.71	5.88	6.06	6.25	6.52

faster convergence of statistical multiplexing gains with increasing channel capacity. Two-layer encoding results in a per-source bandwidth savings if the average rate of the aggregate base and enhancement layer is less than the average rate of the one-layer VBR video.

7. REFERENCES

1 ISO/IEC 13818-2 | Rec. ITU-T H.262, "Generic coding of moving pictures and associated audio," November 1994.

2 M. Ghanbari, "Two-layer coding of video signals for VBR networks," *IEEE J. on Selected Areas in Communications*, vol. 7, pp. 771-781, June 1989.

3 S. Tubaro, "A two layers video coding scheme for ATM networks," Signal Processing: *Image Communication*, vol. 3, pp. 129-141, June 1991.

4 D. P. Heyman, A. Tabatabai, and T. V. Lakshman, "Statistical analysis and simulation study of video teleconference traffic in ATM networks", *IEEE Trans. on Circuits and Systems for Video Technology*, volume 2, number 1, pp. 49-59, 1992.

5 M. W. Garrett and W. Willinger, "Analysis, modeling and generation of self-similar VBR video traffic", Proc. ACM SigComm, London, September 1994.

6 P. Pancha and M. El Zarki, "Prioritized transmission of VBR MPEG video", GLOBECOM '92, pp. 1135-1139, 1992.

7 M. R. Ismail, I. E. Lambadaris, M. Devetsikiotis, and A. R. Raye, "Modelling Prioritized MPEG video using TES and a frame spreading strategy for transmission in ATM networks",. INFOCOM '95, pp. 762-769, April 1995.

8 G. Morrison and D. Beaumont, "Two-Level Video Coding for ATM Networks," *Signal Processing: Image Communication*, volume 3, pp. 179-195, June 1991.

9 J. R. Louvion, "2-Layer Versus 1-Layer Video Codecs: A Network Performance Approach," 4th International Workshop on Packet Video, Kyoto, Japan, August 1991.

10 J. W. Roberts, editor, "Performance Evaluation and Design of Multiservice Networks," COST 224 Final Report, Commission of the European Communities, Brussels, 1992.

11 R. Aravind, M. R. Civanlar, and A. R. Reibman, "Packet loss resilience of MPEG-2 scalable coding algorithms", to appear in *IEEE Trans. on CSVT*, 1996.

12 K. Chandra and A. R. Reibman, "Modeling traffic and statistical gains for multimedia applications", 2nd Workshop on Community Networking, Princeton NJ, June 1995.

13 D. M. Lucantoni, M. F. Neuts, and A. R. Reibman, "Methods for performance evalution of VBR video traffic models", *IEEE/ACM Trans. on Net.*, vol. 2, no. 2, pp. 176-180, May 1994.

14 J. A. Hartigan and M. A. Wong, "A K-means clustering algorithm", *J. Royal Statistical Society, Ser. C, Applied Statistics*, vol. 28, pp. 100-108, 1979.

TRANSPORT OF SCALABLE MPEG-2 VIDEO
OVER ATM BASED NETWORKS

Wenjun Luo and Magda El Zarki

University of Pennsylvania
Department of Electrical Engineering
200 South 33rd Street
Philadelphia, PA 19104

1 Introduction

Networks based on Asynchronous Transfer Mode (ATM)[1] offer the possibility to support source coding at variable bit rate (VBR), which has the advantages of consistent picture quality, bandwidth savings and delay reduction. However, in times of simultaneous peak rate from different connections, network congestion can occur due to statistical multiplexing. The resultant cell losses and delay jitter will cause picture quality degradation. Hence for a high bandwidth-delay product ATM based network, preventive approaches, such as traffic control, need be taken to alleviate or solve this problem. Traffic control for real-time video services over an ATM based network is mainly implemented at the user network interface (UNI).

On the network side, traffic control for a high bandwidth-delay product network incorporates two main functions: the Call Acceptance Control (CAC) and the Usage Parameter Control (UPC).

- CAC is implemented during the call setup to ensure the admission of a call will not jeopardize the existing connections and also enough network resources are available for this call. It is also referred to as *call admission control*. The result of CAC is a service contract.

- UPC is performed during the whole connection period. It is performed to check if the source traffic characteristics respect the service contract specification. If excessive traffic is detected, it can be either immediately discarded or tagged for selective discarding if congestion is encountered in the network. The UPC is also referred to as *traffic monitoring, traffic shaping, bandwidth enforcement* or *cell admission control*. Leaky Bucket (LB) control is a widely accepted implementation of the UPC function.

On the user side, the traffic control mainly takes the form of controlling the coding procedure so that the generated traffic characteristics conform to the service contract specification. It is also known as *source rate control*. Generally, the rate control on the encoding procedure will cause picture quality degradation. Layered Source Coding (LSC) can be adopted to overcome this drawback and maintain the advantage of VBR coding. LSC generates a high priority (HP) layer (or base layer) which contains the most important video information and whose transport should be guaranteed to ensure the delivery of an acceptable video quality stream. The remaining information can then be transmitted as a low priority (LP) layer (or enhancement layer). The LP layer can exploit the characteristics of statistical multiplexing as loss of video information at this level is not so critical. LSC also has the advantages of providing scalable video quality, error resilience, etc.

Previous studies on CAC/UPC and LSC are often disjointed. Studies on LSC naturally assume that the HP data will be guaranteed to be delivered by the underlying network. However, in an ATM network, CAC/UPC functions generally do not consider the cell content/importance, i.e., they enforce bandwidth usage at the UNI regardless of the video information distribution over the cell stream. This can be reflected by the fact that only one priority bit is provided in the cell header. Hence, even if an HP layer carries high video quality, it still can not be guaranteed if its traffic characteristics do not comply with the UPC algorithm. On the other hand, if the HP data stream could meet the traffic characteristics requirement, it is still not acceptable unless the HP layer carries enough video data thereby compromising quality.

Hence for high-end real-time video services, the performance of a LSC scheme can be measured by the following two criteria: 1. The HP data can be ensured to be delivered in a simple and efficient manner by a network UPC function such as the LB control; 2. The HP layer should carry enough high visual quality. The paper first studied two MPEG-2[2,3,4] layered source coding schemes for this purpose, which are respectively Data Partitioning (DP) and SNR Scalability (SNRS). In DP, the first β DCT coefficients are put into the HP layer together with all the header information. The remaining 64-β coefficients are put into the LP layer with certain header information. In SNRS, the HP layer are obtained by quantizing the DCT coefficients with a coarse quantization scale Q_{HP}. The error signal can then be quantized with a fine quantization scale Q_{LP} and put into the LP layer. Experimental results revealed that both the static LSC (layering parameter is fixed during the coding period) and the constant-bit-rate (CBR) LSC (layering parameter is dynamically adjusted to generate a CBR HP bitstream) cannot meet the two criteria simultaneously. Based upon the characteristics of the human visual system and the MPEG traffic characteristics, an adaptive LSC algorithm is implemented that achieves an acceptable compromise between high visual quality and low bit rate requirement during periods of long bursts. It outperforms both the static and CBR LSC in terms of video quality carried by HP layer and the efficiency of network resource utilization.

This paper is organized as follows: In section 2, the lossless transmission of HP data under the LB control is addressed. An overview of the key LB control parameters is briefly described in section 2.1. Section 2.2 reveals the reasons why LB is not effective/efficient enough to control the layered MPEG-2 video. The problem can be alleviated by utilizing the human perception properties, which is derived in section 3.1. Section 3.2 presents the performance results of different LSC schemes. The summary is given in section 4.

2 Leaky Bucket Control Over Layered MPEG-2 Video

An essential part in any CAC/UPC scheme is the function that separates conforming cell arrivals from nonconforming arrivals (or excessive traffic). This function is referred to as the Police Criterion Calculator (PCC). Many algorithms have been proposed, such as Moving Window, Fixed (Jumping) Window, Leaky Bucket (LB) and Exponentially Weighted Moving Average[9,10,11]. Judging from the performance of these algorithms, most research has revealed that the LB scheme is an acceptable alternative. For example, when comparing the ability to deal with *cell delay variation*, LB simply scores the best[12]. In this study, we focused on the LB algorithm, which has gained widespread acceptance.

An overview of the key LB control parameters is briefly described in section 2.1. Section 2.2 reveals the reasons why LB is not effective/efficient enough for the lossless transmission of HP data.

2.1 Leaky Bucket Parameters

In the LB scheme (see Fig. 1), a certain transmission rate (bandwidth) is allocated for the connection in terms of generating cell admission tokens to the sender at the assigned rate, which is defined as the token generation rate γ. There is also a token pool (also referred to as *bucket*) to hold the residual tokens when the encoder has less cells to send. For each cell generated by the encoder, if there is no token in the token pool, the cell is tagged and in times of congestion in the network, the tagged cells will be discarded. Note that we assume no input buffer. If any buffering is to occur, it will be done at the source for smoothing out the traffic. If on the other hand, there are tokens in the

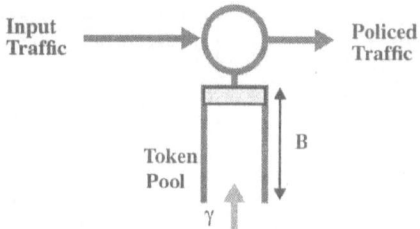

Figure 1. Leaky Bucket Control Diagram.

token pool, the cell is not tagged and will be guaranteed to be delivered. Let's define the maximum token pool size as B. It is apparent that during any time interval T, the maximum allowable burst cell rate will be:

$$(\gamma \times T + B)/T = \gamma + B/T \qquad (1)$$

We can define a BW efficiency measurement variable θ as:

$$\theta = \gamma/\alpha \qquad (2)$$

where α is the average video source rate. Hence a large θ indicates that the network bandwidth is under utilized since the token arrival rate is larger than the average source rate.

2.2 Leaky Bucket Control

LB was not initially designed for video services. In this section, we study the feasibility of LB for complex layered video streams, especially, we will look into the problem of how to dimension the LB parameters to guarantee the lossless (non-tagged) transmission of HP data.

2.2.1 Lossless Transmission of statically layered Video

Lossless transmission can be achieved by choosing an appropriate combination of γ and B. For a given γ, B can always be increased to exceed certain value such that the token pool has enough tokens to hold the largest burst period (assume cells from a frame are evenly spaced). Given a B, it is always possible to increase γ up to the peak rate to transmit all the cells without being tagged.

We present in Fig.2 the relationship between θ and B to guarantee the lossless transmission of the HP cells obtained for different β (note β remains fixed during the coding period). We define this as static DP. We can observe that first, no matter what value β is, if we restrict the token pool size to be small, the required γ should be much higher than the average source rate. This will result in inefficient BW utilization. On the other hand, if we want a high BW utilization (i.e., θ close to 1), then using a β of 1 can reduce the buffer requirement greatly however, the visual quality is very poor. Generally in ATM switches the buffers are not very large to control the variance in the cell transfer delay. In particular for real-time services, if the delay exceeds a certain threshold, a decoder will regard the delayed cells as lost. So it is desirable to restrict the token pool size to be small.

In Fig. 3, we present the result for static SNRS with different values of Q_{HP} A similar observation can be made: Q_{HP} of 64 can greatly reduce the buffer requirement but yield an unacceptable video quality with BW allocation close to the average source rate. If the token buffer is limited to be small, BW utilization will be inefficient.

2.2.2 Cell Tag Rate and Cell Loss Pattern

If the HP layer carries enough high visual quality data (small Q_{HP} or large β) and network resource allocation is not sufficient, e.g., the buffer size is restricted to be small, cell loss may occur in the HP layer. Even if a very small amount of HP cells are lost, the resulted video quality may be

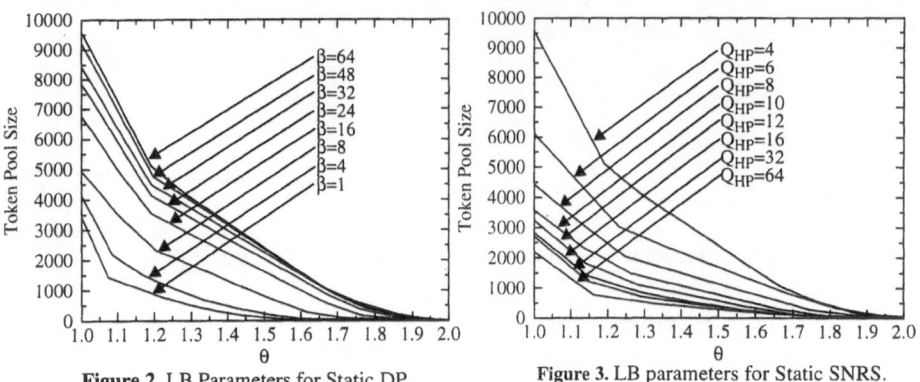

Figure 2. LB Parameters for Static DP.   **Figure 3.** LB parameters for Static SNRS.

greatly impacted. This is because video quality is affected not only by the number of cell losses, but also the cell loss pattern. We know that cell losses due to congestion mainly happen during the peak rate period. By drawing a time diagram, it can be shown that LB tags the cells in a distributed way in times of peak cell arrival rate. However, tagging cells in such a way actually is detrimental for MPEG-2 video transmission. This is because MPEG-2 video is very sensitive to this type of cell loss pattern.

An example is given here. Let's assume during a certain picture period, (an Intra-coded frame for instance), the bitrate is 1.5 times higher than the reserved BW and the token pool size is limited to be very small. According to the LB algorithm, the approximate distribution pattern of tagged cells is shown in Fig. 4. If all the tagged cells are discarded because of congestion, the decoder will only receive the untagged cells. However, are all these received cells useful? Can they be used to reconstruct the picture? Actually it turns out that they can be useless. The reason is that, in an I frame, the information of a macroblock (MB) is often spread out over several consecutive cells. For example, let's say, cells "i", "i+1" and "i+2" contain a MB's information. If cell "i" is lost, the decoder will have no idea what information cells "i+1" & "i+2" contain because variable length coding is used. Furthermore, differential coding of the DC coefficients of consecutive MBs will also make subsequent cells useless as well because they are based on the information of the previous cells. So "i+4", "i+5"... are of no use to the decoder until a new slice header is encountered and synchronization is regained. Hence we can conclude that, for MPEG video, if tagged cells are lost due to congestion, untagged HP cells are also possibly useless for reconstructing the image and transmitting them will actually waste bandwidth. Or, we can say, the actual HP cell loss rate will higher than the tag rate under the congestion conditions.

We can conclude therefore that static LSC is not a good solution. A natural next step is to implement dynamic LSC, i.e., to generate a dynamically controlled HP bit stream that fits the LB parameters such that the BW efficiency is close to 1 in conjunction with a small token pool size.

3   Adaptive Layered Source Coding

3.1 Adaptive Rate Control of the HP Layer

The design of a dynamic LSC scheme needs to be carefully studied and implemented to retain a high visual quality. For example, in MPEG, the I frames generally yield a bit rate higher than the average rate. If the dynamic LSC scheme is to generate a CBR HP bit stream (there are some widely studied CBR rate control schemes available for this purpose), then it will decrease β or increase Q_{HP} to produce a lower rate I frame and thus cause picture quality degradation. However, between the I frames, the full picture quality is generally maintained in contrast due to the general low rate of P pictures. Since I pictures occur every N pictures, the degradation happens periodically. Hence during certain periods, the contrast between the nearly full quality and the degraded quality and its frequent/periodic occurrence can easily catch the attention of the viewers and appears to be annoying as the

i   i+1 i+2   i+4

Figure 4. The cell tag pattern of Leaky Bucket.

subjective evaluation results reveal. Thereby, in order to design a LSC with better performance, we may need to examine both the psychophysical property of the human visual system and the MPEG video traffic characteristics.

Picture quality degradation is inevitable if we have to reduce the bit rate of a certain picture. However, the amount of picture quality degradation perceived by a viewer varies for different segments and/or regions of the video sequence. In other words, the visual quality may still be maintained to a certain degree even if actual degradation exists, as studies in the early eighties revealed[15,16]. The phenomenon can be explained in terms of *visual masking*. In this context, it means that the visibility of a degradation can be affected or masked by other visual stimuli such as the complex spatial and temporal features in the video sequences. For example, picture degradation is less noticeable when it happens in a scene change. An interesting test we conducted revealed that if we substitute the first few pictures after a scene change with lower quality pictures, the viewers hardly sense the substitution, as the previous scene is persistent in the human's memory for a certain period. Other similar tests include the degradation of the original area of certain fast-moving objects in a series of high-motion-packed pictures. The motion catches the attention and hence the viewer pays less attention to the original area. The degradation of certain regions in a very complex picture can also be less perceptible, the complex luminance details can disguise the degradation very well.

One of the major explanation of visual masking is that not all the pels have the same visual importance. Besides the pixel values of the picture, redundancies exist in the representation of the visual information, this is not fully exploited in MPEG as MPEG mainly explores the redundancy in the mathematical representation of pixel values. For the same example of a fast moving object, it will have the same visual effect no matter how much detail its original area contains in each single picture, as it is the object and the motion that catches one's attention. This indicates that we can compress more on the representation of the object's original position without losing much in visual quality. It is our belief that for certain pictures, a dynamic LSC scheme could further exploit such visual redundancy to achieve an acceptable compromise between the high visual quality and the low bit rate requirement.

On the other hand, when we examined the MPEG video traffic and the corresponding token buffer occupancy, we could make the following observations:

1. Bursts can be categorized into two classes: short bursts (lasting for one or a few frames) and long bursts (lasting longer). Short bursts are mainly caused by the intra coding mode or abrupt scene changes. Long bursts are mainly caused by a period of high motion outburst and/or the background keeps shifting rapidly;

2. Short bursts generally do not deplete the token buffer, they normally consume less than the peak number of cells generated by a picture. This is because the P pictures before these pictures generally yield lower rates and hence enough tokens accumulate to accommodate the short bursts. It is therefore not necessary to smooth these pictures, especially the periodic I pictures to avoid the periodic quality degradation we discussed above. It is the long bursts that quickly deplete the token pool and hence require an extremely large token pool when the BW allocation is comparable to the average source rate. Hence it is necessary to smooth them without much loss in the visual quality.

Because the long bursts correspond to a period of high motion with or without rapid scene change in the video sequence, and tests have shown that visual masking on degradation can be very effective in these scenarios, we believe that the visual quality may be maintained to a certain degree (i.e., degradation is imperceptible) while reducing the burst rate of long bursts to a lower rate close to the allocated BW. Based on this analysis, we designed an adaptive LSC scheme within the MPEG framework which has the capability of detecting long bursts and react accordingly, such that the visual impairments are minimized while the burst rate is reduced.

Because burst is a relative term to the average rate and not all the high motion/scene changes

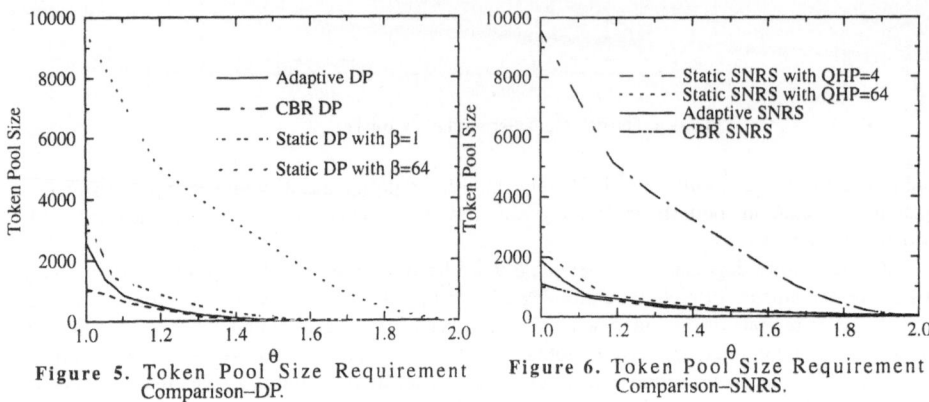

Figure 5. Token Pool Size Requirement Comparison–DP.

Figure 6. Token Pool Size Requirement Comparison–SNRS.

will cause a higher rate than the average rate and quickly deplete the token pool, the rate control module is only activated when the current frame is P frame and the token pool fullness of an internal logical LB is less than a certain threshold B_{Th}. After being activated, the control module will need to identify the masking regions for further bit allocation and compression.

We know that long bursts are caused mainly by fast motion and/or continuously rapid scene changes, which in turn can be reflected by the coding statistics. Object with high motion will generally leave an intra-coded region in its original position as there is no matching area for it in the previous frame, provided that the object is identified correctly by the motion vectors during the motion search. If the motion search fails to identify the fast moving object or its components because of the large magnitude or complexity of the motion, the object itself or some parts of it such as the edges will also be coded in intra mode. Scene changes can always be reflected by the large number of intra-coded MBs as well. These intra-coded MBs generally cause burst rates for P pictures.

Hence in the control module, we first check the distribution of intra-coded macroblocks. If the number of intra-coded MBs exceeds a certain threshold I_{Th}, further compression should be implemented on these intra-coded regions to reduce the rate while maintaining a high visual quality. Bit allocation is then implemented to allocate bits between intra and non-intra coded regions in the current frame according to the target bit rate, which can be set to be the allocated BW. After the bit allocation, rate control is then implemented by varying the layering parameters according to the bit rate allocated to each MB. For DP, the rate control is achieved by adjusting the break point until the rate is reduced low enough. For SNRS, Q_{HP} is computed according to an algorithm slightly modified from the one in [17] to achieve an acceptable bit rate. In the next section, we will present the experimental results for the performance of this algorithm.

3.2 Experimental Results

In Fig. 5, we present the token pool size requirement for the sequence *Indiana Jones* using four schemes: 1) the full image (i.e., static DP with $\beta=64$), 2) static DP with $\beta=1$, 3) CBR DP with the target HP data rate equal to the average source rate and 4) adaptive DP also with the target rate to be the average source rate. It shows that, the token pool size drops dramatically for the adaptive scheme when θ is increased to 1.1. At that point the buffer performance is comparable to CBR DP which has the lowest token pool size requirement for all θ. Both the two dynamic schemes (CBR and adaptive LSC) outperform the static scheme even for $\beta=1$. Similar results can be observed for SNRS as in Fig. 6. Question we now ask is how is the visual quality impacted? If the adaptive scheme can yield a much better quality than the CBR scheme, its relative complexity can be then well compensated for.

In Table 1, we present the visual quality results based upon the carried HP data only for several video sequences. Three schemes are compared here: the static LSC with $\beta=1$ or $Q_{HP}=64$, CBR DP or SNRS with $\theta=1.1$ and a maximum token pool size equal to 1000, and the adaptive DP or SNRS scheme with $\theta=1.1$ and the maximum token pool size equal to 1000. The quality rating system is: 5-Excellent, 4-Good, 3-Medium, 2-Poor, 1-Unacceptable, and is obtained through subjective viewing and an objective video quality assessment system based on human perception[14]. The results are normalized by the full quality image (i.e., static LSC with β=64 or Q_{HP}=4). We observe that the adaptive scheme outperforms the other two schemes. The difference between CBR scheme and the adaptive scheme is small for the sequence *Last Emperor* for which we have observed very long periods of either low or high bit rates. Similarly for *Star Trek* the difference is not large and that is due to the fact that there are very few long bursts and so the adaptive scheme has no much advantage over the CBR scheme. Furthermore, judging from the simulation results, the SNRS scheme generally outperforms the DP given that it delivers higher visual quality under the same network resources. The may be interpreted by the flexibility in the SNRS rate control algorithm, as the QHP can be varied for each MB while for MB, the rate control parameter β can be changed only on the basis of slice.

Video Sequence	Normalized Quality		
	β=1/Q_{HP}=64	CBR DP/SNRS	ADAPTIVE DP/SNRS
Last Emperor	0.2012/0.2003	0.8954/0.8896	0.9017/0.8936
Indiana Jones	0.2045/0.2010	0.8834/0.8928	0.9372/0.9443
Star Wars	0.2103/0.2021	0.8971/0.9088	0.9283/0.9293
League	0.2001/0.2015	0.9101/0.9072	0.9377/0.9325
Taxi Driver	0.2024/0.2009	0.8824/0.8911	0.9356/0.9414
Star Trek	0.2003/0.2013	0.9238/0.9311	0.9245/0.9337

Table 1: Normalized quality recovered from HP cells

4 Summary

ATM-based networks offer the possibility to support source coding at VBR. However, because of the highly bursty nature of VBR video, implementing only the Leaky Bucket at the network interface turns out not to be effective if high utilization of network resources is desired. By making use of LSC at the source, this problem is expected to be alleviated. For high-end real-time video services, a LSC scheme should generate a HP layer which meets the following two criteria: 1. The HP data can be ensured to be delivered in a simple and efficient manner by a network UPC function such as the LB control; 2. The HP layer should carry enough high visual quality. This paper studied the DP and SNRS for this purpose. The adaptive LSC scheme we designed is certainly not optimal. However, it provides a baseline performance, showing the improvement one can achieve by utilizing the characteristics of human perception in its compression. More sophisticated schemes can be developed but at the expense of higher complexity. One needs to consider the complexity/performance trade-off.

References:

[1] Martin de Prycker, "Asynchronous Transfer Mode, solution for broadband ISDN", Second Edition, Ellis Horwood, 1993, TK 5103.5.P79 1993.

[2] P. Pancha and M. El Zarki, "A look at the MPEG video coding standard for variable bit rate video transmission", Proceedings of IEEE INFOCOM'92, May 1992.

[3] B. DeCleene, P. Pancha, M. El Zarki, H. Sorensen, "Comparison of Priority Partition Methods for VBR MPEG", Proceedings of IEEE INFOCOM'94, June. 1994, Volume 2, pp 689-696.

[4] ISO/IEC 13818-2, "Generic Coding of Moving Picture and Associated Audio", MPEG-2 Draft International Standard, May, 1994.

[5] Y-Q Zhang, W. W. Wu, K.S. Kim, R.L. Pickholtz, and J. Ramasastry, "Variable Bit Rate Video Transmission in the Broadband ISDN Environment", Proceedings of the IEEE, 79(2):214–221, February 1991.

[6]   P. Pancha and M. El Zarki. "Bandwidth Allocation Schemes for Variable Bit Rate MPEG Sources in ATM Networks," IEEE Transactions on Circuits and Systems for Video Technology, 3(3):190-198, June 1993.

[7]   S. Chong, S. Li, J. Ghosh, "Dynamic Bandwidth Allocation for Efficient Transport of Real-Time VBR Video over ATM", Proceedings of IEEE INFOCOM '94, June. 1994, Volume 1, pp. 81-91.

[8]   H. Kanakia, P. P. Mishra, A. Reibman, "An Adaptive Congestion Control Scheme for Real-Time Packet Video Transport", Proceeding of SIGCOMM'93.

[9]   RACE-1022, TG6, "Policing, Connection Acceptance Control and related Functions", Present Position Paper of R-1022 Taskgroup 6; issue F; July 1989.

[10] E. Rathgeb, "Modeling and performance Comparison of Policing Mechanisms for ATM Networks", IEEE JSAC, Vol. 9, No. 3, April 1991, pp. 325-334.

[11] L. Dittman, S.B. Jacobsen, K. Moth, "Flow Enforcement Algorithms for ATM networks", IEEE JSAC, Vol. 9, No. 3, April 1991, pp. 343-350.

[12] M. Dirksen, K. an der Wal, "Safety Margins in ATM Policing Function Algorithms", IEEE Trans. on Comm., 1992.

[13] P. Pancha and M. El Zarki, "Leaky Bucket Access Control for VBR MPEG Video", Proceedings of IEEE INFOCOM'95, Boston, MA, April 1995.

[14] A. A. Webster, C. T. Jones, M. H. Pinson, S. D. Voran, S. Wolf, "An objective video quality assessment system based on human perception", Proceedings of Human Vision, Visual Processing, and Digital Display TV, SPIE'93.

[15] W. Glenn, K. Glenn and C. Bastian, "Imaging System Design Based On Psychophysical Data", Proceedings of SID, Vol. 26, No. 1, 1985, pp. 70-77.

[16] J. Westerink, C, Teunissen, "Perceived Sharpness in Moving Images", Proceedings of SPIE, Vol. 1249, 1990, pp. 77-86.

[17] M. Pickering, J. Arnold and M. Cavenor, "A VBR Rate Control Algorithm with Perceptual Adaptive Quantization and Traffic Shaping", International Worshop on Packet Video'94.

DISTORTION POLICY OF BUFFER-CONSTRAINED RATE CONTROL FOR REAL-TIME VBR CODER

Jinho Choi

Multimedia Lab., DACOM R&D Center
34 Gajeong-Dong, Yuseung-Gu
305-350, Daejeon, Korea

INTRODUCTION

For fixed capacity (or bandwidth) networks, e.g., circuit switching networks, constant bit rate (CBR) coders are quite common for speech or video signals transmission. By increasing interest in packet switching networks, e.g., asynchronous transfer mode (ATM) networks, variable bit rate (VBR) coders are now highly desirable[7]. A major advantage of the VBR coder is constant quality (or distortion). Since speech or video/image signals are nonstationary, bit rate cannot be fixed to keep constant quality. In this case, the VBR coder can provide nearly constant quality by allowing VBR. For CBR links, the buffer has been used at the output side of the VBR coder to absorb fluctuation of bit rate. It is well known that overflow or underflow of buffer cannot be prevented without buffer control methods.

In Choi and Park[1] and Ortega et al.[6], rate control algorithms have been proposed not only to prevent the buffer overflow or underflow, but also to minimize distortion as possible. The relationship between rate and distortion has been used to develop those algorithms. The main idea of the rate control algorithm is quite simple. To ensure that the buffer is not overflowing or underflowing, a feedback signal is required to the VBR coder. If the buffer is reaching an "almost full" threshold, the VBR coder is informed and urged to produce less information (bit rate). This results in decreasing quality. If the buffer is reaching an "almost empty" threshold, the VBR coder is urged to generate more information. To implement the above idea, the VBR coder should be controllable. The controllable VBR (CVBR) coder is a VBR coder that can control the quality of reconstructed signal as well as bit rate. The CVBR coder and its output buffer are considered in Figure 1. According to change of the quality, the rate is changed. For example, if the quality of the reconstructed signal becomes poor, the rate would be small, and vice versa. To construct the CVBR coder, a set of quantizers (or adaptive quantizer) can be used. For example, the quality control in the JPEG coder can be implemented using the quality factor Q_f of the (adaptive) quantizer. By increasing the value of Q_f, bit rate becomes small, and vice versa. In Ding[4], a set of data to interpret the relationship between bit rate and Q_f can be found. With distribution assumption, the relationship between bit rate and Q_f (R-Q) and the relationship between distortion and Q_f (D-Q) are obtained in Gormish[5].

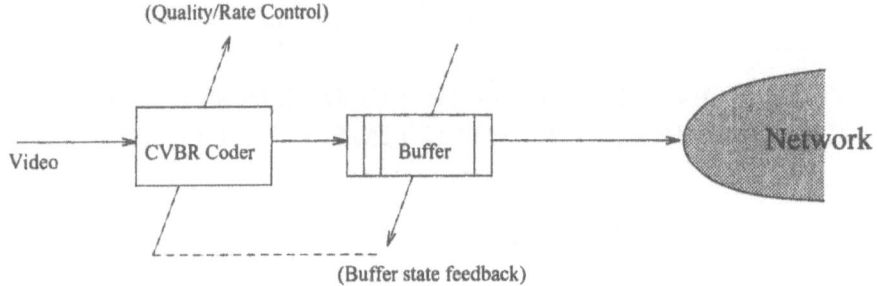

Figure 1. Rate control using the CVBR coder

In this paper, we consider the CVBR coder and its output buffer to provide nearly constant quality for CBR or VBR links without the buffer overflow or underflow. Some policies are considered not only to stabilize the state of the buffer occupancy, but also to keep the fluctuation of quality as small as possible (it is noteworthy that we do *not* intend to minimize distortion, but to minimize fluctuation of quality).

The paper is organized as follows. In Section 2, the distortion policy is considered, and adaptive methods for the distortion policy are proposed in Section 3. Application of the policies to compressed image is considered in Section 4, and summary and further work are given in Section 5.

DISTORTION POLICY

In Figure 2, the relationship between various rate and distortion sequences are shown. Typical rate and distortion sequences that are produced by a CBR coder are shown in Figure 2 (a). It is seen that although the rate is constant, the distortion has large fluctuation. For the VBR coder case, typical sequences are shown in Figure 2 (b). In this case, the distortion is constant, but the rate has large fluctuation. It is desirable, however, not only to stabilize the state of the buffer occupancy, but also to keep fluctuation of the distortion small as shown in Figure 2 (c). The advantage of the CVBR coder is the capability that can provide nearly constant quality (distortion) over time varying characteristics of signal.

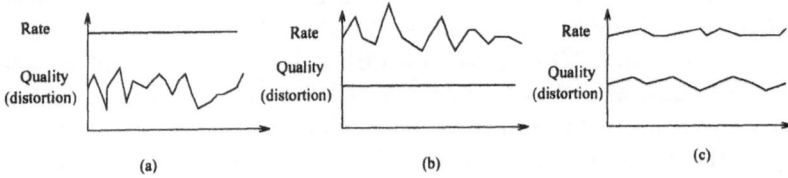

Figure 2. The relationship between rate and distortion sequences: (a) from the CBR coder, (b) from the VBR coder, and (c) from the CVBR coder.

In general, the following relationship between the distortion and the rate of any coders can be obtained

$$d = W(\sigma^2)e^{-\alpha r}, \tag{1}$$

where σ^2 is the variance of signal and d and r are the distortion (that is measured by the mean square error (MSE) between the original and reconstructed signals) and the rate, respectively. Here $W(\cdot)$ is a function and α is a constant. For a well designed noisy source coder, we can observe that the relationship between the distortion and the rate follows (1). For instance, in the case of the high resolution approximation, it is expected that

$$d = \sigma^2 2^{-2r}.$$

Therefore, $W(\sigma^2) = \sigma^2$ and $\alpha = 2\log_e 2$ (hereafter, we use log for $\log_e$). It is noteworthy that the relationship between distortion and rate is obtained using statistical average for stationary signals. Thus the above relationship between distortion and rate should be modified for practical cases. Although the above relationship can be satisfied in the statistical average sense, a pair of distortion and rate from a practical coder does not strictly follow the relationship. Suppose that a sequence of distortion-rate pair from a practical coder is denoted by $\{(d_1, r_1), (d_2, r_2), \ldots\}$, where d_n are r_n are the square error (not MSE) and the rate, respectively, for the nth block of signal. From the high resolution approximation, we can consider a simple model for the pair of distortion and rate[1] from a practical coder that is given by

$$d_n = \epsilon_n \sigma_n^2 e^{-\alpha r_n}, \tag{2}$$

where σ_n^2 is the variance of the nth block and $\{\epsilon_n\}$ is the (positive) perturbation sequence. By taking log in (2), we also have

$$\phi_n = s_n - \alpha r_n + \beta_n, \tag{3}$$

where $\log \sigma_n^2 = s_n$, $\phi_n = \log d_n$, and $\beta_n = \log \epsilon_n$.

Let us consider the recursion of the state of the buffer occupancy that is given by

$$q_{n+1} = q_n + r_n - \gamma_n, \tag{4}$$

where q_n and γ_n are the state of the buffer occupancy and the outputing rate of the buffer, respectively. For CBR links (e.g., circuit switching networks), γ_n is a constant. On the other hand, in the VBR link case (e.g., packet switching networks), γ_n is varied for each n. It is noteworthy that networks supporting guaranteed quality of service (QoS) provide bounded γ_n and the bounds can be prescribed. In this paper, we assume that γ_n can be time varying. By replacing r_n in (4) with ϕ_n in (3), we can obtain the following equation

$$
\begin{aligned}
q_{n+1} &= q_n + \frac{1}{\alpha}(s_n - \phi_n + \beta_n) - \gamma_n \\
&= q_n + \check{s}_n - \check{\phi}_n + \check{\beta}_n - \gamma_n,
\end{aligned} \tag{5}
$$

where $\check{s}_n = s_n/\alpha$, $\check{\phi}_n = \phi_n/\alpha$, and $\check{\beta}_n = \beta_n/\alpha$.

We can observe that the recursion (4) contains the rate term r_n, while the recursion (5) contains the distortion term $\check{\phi}_n$. To develop the rate control with distortion policy, the recursion (5) is of importance. We now consider the cost function approach. Using the cost function in terms of the state of the buffer and distortion, a rate control can be obtained to make the state of the buffer stable as well as keep small fluctuation of distortion.

At the nth stage, with known q_n, we need to determine r_n or $\tilde{\phi}_n$ in (4) or (5), respectively. Using the following cost function

$$g_d(q_{n+1}, \tilde{\phi}_n) = (q_{n+1} - q^*)^2 + \lambda(\tilde{\phi}_n - \tilde{\phi}^*)^2, \tag{6}$$

where $\tilde{\phi}^*$ is the desired distortion and λ is the weighting factor, we can decide the distortion quantity $\tilde{\phi}_n$. For a large weighting factor λ, it is expected that fluctuation of distortion becomes small. When the stability of the state of the buffer is highly desired, the weighting factor λ should be small to emphasize the term of the state of the buffer in (6). The distortion policy can be obtained by finding $\phi(q_n)$ which minimizes (6).

Since s_n or $\tilde{s}_n$ is available at the nth stage, the (normalized) distortion policy $\tilde{\phi}(q_n) = \phi(q_n)/\alpha$ can be given by

$$\phi(q_n) = \arg \min_{\tilde{\phi}} \{(q_{n+1} - q^*)^2 + \lambda(\tilde{\phi} - \tilde{\phi}^*)^2\} \tag{7}$$

with given q_n and $\tilde{s}_n$. From (5) and (7), we have the distortion policy

$$\tilde{\phi}(q_n) = \frac{q_n - q^* + \tilde{s}_n - \hat{\gamma}_n + \lambda\tilde{\phi}^*}{1 + \lambda}, \tag{8}$$

with the estimate $\hat{\gamma}_n$ of γ_n. From a set of quantizers, we choose the quantizer that produces close distortion to the distortion in (8). Since the number of quantizers is finite, there is the perturbation term w_n^ϕ and we have

$$\tilde{\phi}_n = \tilde{\phi}(q_n) + w_n^\phi. \tag{9}$$

As is easily observed, the term w_n^ϕ becomes small when the number of quantizers is increased.

ADAPTIVE DISTORTION POLICIES

Since the outputing rate of the buffer and the variance of (block of) signal have time varying characteristics, adaptive approaches are required. In this section we consider the following two adaptive approaches that can be used alone or simultaneously.

A1 Switching the value of the weighting factor λ depends on the state of the buffer and the variance of signal

A2 Adjusting the desired distortion quantity $\tilde{\phi}^*$ depends on the outputing rate of the buffer and the variance of signal

Switching the value of λ

The value of λ affects the stability of the state of the buffer[2]. If λ approaches 0, the state of the buffer becomes more stable, and vice versa. From this observation, an algorithm can be derived: if the state of the buffer is close to the desired state of the buffer q^*, the value of λ would be large for small fluctuation of distortion, and when the buffer is reaching an "almost empty" threshold or an "almost full" threshold, the value of λ should be small to emphasize the state of the buffer term in the cost function (6). As a result, the state of the buffer can approach the desired state of the buffer q^* with small λ.

For a large value of λ, the distortion quantity $\tilde{\phi}_n$ can be close to $\tilde{\phi}^*$. For a small value of λ, $\tilde{\phi}_n$ cannot be close to $\tilde{\phi}^*$ and varied as $\tilde{s}_n$ varying. Thus it is reasonable that if fluctuation of $\tilde{s}_n$ is large, the value of λ would be large, and vice versa, to maintain small fluctuation of distortion.

We show an approach to make q_n as close to q^* and/or to keep $\tilde{\phi}_n$ nearly constant through switching of the value of λ. For example, we can use the following algorithm, called switching λ method: 1) determine a set of the four values of λ, $\lambda_1 > \lambda_2$, $\lambda_3 > \lambda_4$, the buffer threshold B, and the variance threshold V, 2) switch the value of λ according to the following conditions:

$$
\begin{aligned}
|q_n - q^*| < B \text{ and } v_n^s \geq V &\Rightarrow \lambda = \lambda_1 \\
|q_n - q^*| \geq B \text{ and } v_n^s \geq V &\Rightarrow \lambda = \lambda_2 \\
|q_n - q^*| < B \text{ and } v_n^s < V &\Rightarrow \lambda = \lambda_3 \\
|q_n - q^*| \geq B \text{ and } v_n^s < V &\Rightarrow \lambda = \lambda_4,
\end{aligned}
\tag{10}
$$

where v_n^s is an on-line measurement of the variance of $\tilde{s}_n$.

Adjusting the desired distortion quantity

Denoting $\bar{\tilde{s}}$ and $\bar{\gamma}$ as the mean values of $\tilde{s}_n$ and γ_n, respectively. To ensure that the equilibrium state of the buffer becomes the desired state of the buffer, we should have $\tilde{\phi}^* = \bar{\tilde{s}} - \bar{\gamma}$ from the equilibrium analysis[2]. Thus $\tilde{\phi}^*$ cannot be arbitrary determined. The mismatching between $\tilde{\phi}^*$ and $\bar{\tilde{s}} - \bar{\gamma}$ causes the bias of the average state of the buffer from the desired state of the buffer when $\lambda > 0$. In general, $\tilde{s}_n$ often does not have its mean value, and only the average value over the short duration is available. From image sequences, it is observed that characteristics of $\tilde{s}_n$ can be significantly changed. For example, a smooth region of image yields slowly varying $\tilde{s}_n$. On the other hand, the region which has edges can yield rapidly varying $\tilde{s}_n$. In addition to time varying characteristics of $\tilde{s}_n$, the buffer outputing rate can be widely changed. For VBR links, γ_n is not a constant, and the network congestion makes the value of γ_n small. From the above observation, the desired distortion quantity $\tilde{\phi}^*$ cannot be a constant and it should be adjustable. To maintain nearly constant distortion, however, $\tilde{\phi}^*$ should be a constant.

From this point of view, it is obvious to consider a tradeoff. The quantity $\tilde{\phi}^*$ is slowly varying and it is adjustable to close to the *current* average value of $\tilde{s}_n - \gamma_n$. The quantity $\tilde{\phi}^*$ is no longer a constant and we use $\tilde{\phi}_n^*$ instead of $\tilde{\phi}^*$. We can consider slowly varying $\tilde{\phi}_n^*$ that is close to the current average value of $\tilde{s}_n - \gamma_n$ with the following recursion

$$
\tilde{\phi}_{n+1}^* = \omega_3 \tilde{\phi}_n^* + (1 - \omega_3)(\tilde{s}_n - \gamma_n),
\tag{11}
$$

where $0 < \omega_3 < 1$. That is, the current average of $\tilde{s}_n - \gamma_n$ is obtained using the recursion (11) with the forgetting factor ω_3, and this quantity is used for the slowly varying desired distortion quantity. As ω_3 approaches 1, the variation of the quantity $\tilde{q}_n^*$ becomes slower. Moreover, it is shown that the value of $\tilde{\phi}_n^*$ in (11) is adaptively changed in accordance with the change of $\tilde{s}_n$ and γ_n. Using (11), the difference between the equilibrium state of the buffer and the desired state of the buffer can be reduced.

APPLICATION TO COMPRESSED IMAGE TRANSMISSION

For image transmission over VBR links, a JPEG-like CVBR coder has been used. A set of 16 quantizers of the quality factor $Q_f = 10, 20, \ldots, 160$, has been used to provide various rate-distortion pairs. From this we can choose a proper quantizer from the set of quantizers to yield the distortion according to the distortion policy.

Let us consider the rate control using the distortion policy. From the distortion policy we can obtain small fluctuation of the distortion sequence with increasing λ. As a result, we can have large fluctuation of the state of the buffer. Simulation has been performed with $q^* = 20$ and $\breve{\phi}^* = 1.5$. The mean value of the buffer outputing rate $\bar{\gamma}$ is 1.5. The degrees of fluctuation of the state of the buffer and the distortion can be indicated by their variances. In Table 1, the sample variances of q_n and $\breve{\phi}_n$ are obtained for various λ. It is shown that the variance of q_n becomes small as λ approaches zero. On the other hand, when λ becomes large, the variance of $\breve{\phi}_n$ becomes small.

Table 1. The variances of the state of the buffer and the distortion for various value of λ

λ	0.1	0.2	0.5	1	2	5	10
$var(q_n)$ from simulation	0.886	0.929	0.967	1.441	2.887	8.667	20.295
$var(\phi_n)$ from simulation	1.030	1.004	0.924	0.771	0.586	0.342	0.245

In Figure 3, the sequences of the state of the buffer and the distortion are shown. For comparison, we consider a strict rate control for the state of the buffer with $\lambda = 0$. The strict rate control chooses the quantizer only for the stability of the state of the buffer (i.e., for the minimization of the cost $(q_{n+1} - q^*)^2$, where q^* is given by 20). On the other hand, the rate control using the distortion policy is considered with $\lambda = 1$. In Figure 3 (a), the sate of buffer is shown. It is seen that the strict rate control provides smaller fluctuation than the rate control using the distortion policy. While, in Figure 3 (b), we observe that the rate control using the distortion policy has smaller fluctuation of distortion than the strict rate control. It is noteworthy that the average value of the state of the buffer (= 21.394) is greater than the desired state of the buffer $q^* = 20$. The bias is due to the mismatching between $\breve{\phi}^*$ and $\bar{s} - \bar{\gamma}$. If the bias of the state of the buffer is significant, the buffer overflowing or underflowing can be frequently occurred.

Based on the state of the buffer and the variance of $\breve{s}_n$, the value of λ is switched by $\lambda_1 = 5$, $\lambda_2 = \lambda_3 = 1$, or $\lambda_4 = 0$ in Figure 4. The desired state of the buffer q^* is 20, $B = 15$, and $V = 4$. This adaptive method has some disadvantage because of the difficulty of the parameters setting. In Figure 4, large fluctuation of the state of

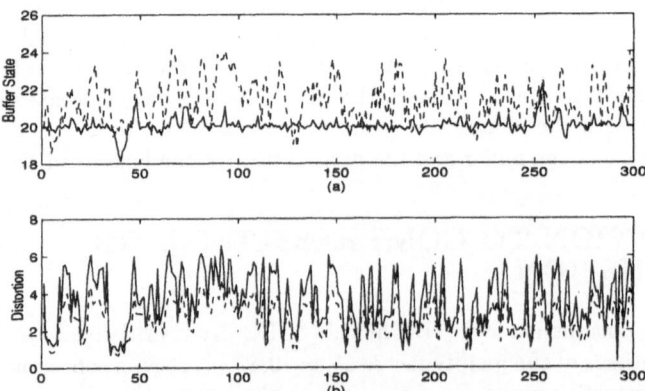

Figure 3. Rate control using distortion policy: (a) buffer state and (b) distortion (solid line: strict rate control, dashed line: rate control using distortion policy)

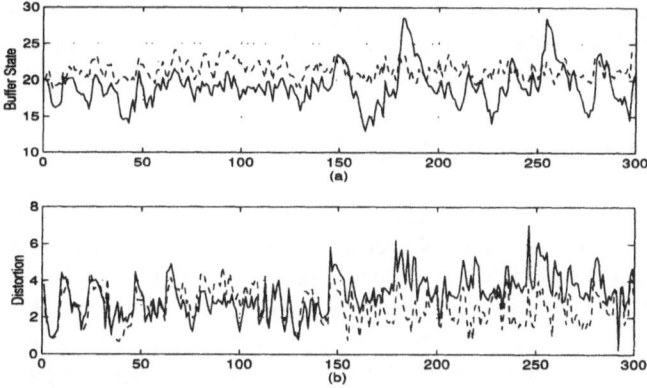

Figure 4. Adaptive distortion policy using λ: (a) buffer state and (b) distortion (solid line: adaptive distortion policy, dashed line: distortion policy)

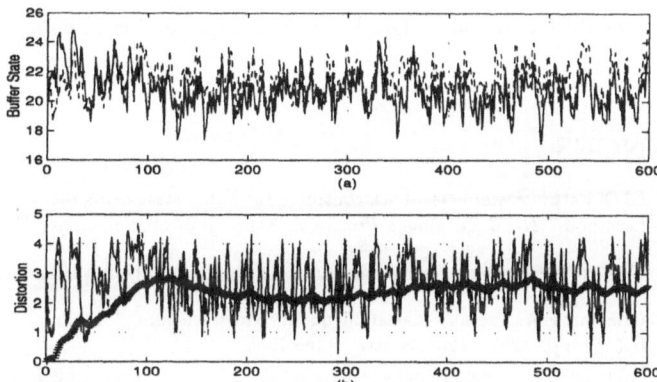

Figure 5. Adaptive distortion policy by adjusting desired distortion: (a) buffer state and (b) distortion (solid line: adaptive distortion policy, dashed line: distortion policy, hard solid line: adjusted desired distortion)

the buffer and large distortion are observed for the switching λ method. We observe, however, that the average value of the state of the buffer ($= 19.143$) is close to the desired state of the buffer ($= 20$), i.e., the bias of the state of the buffer is improved.

The approach of adjusting the desired distortion quantity $\tilde{\phi}^*$ is applied and the results are shown in Figure 5. In this simulation, $\lambda = 1$ and $\omega_3 = 0.98$. In Figure 5, it is shown that the desired distortion quantity $\tilde{\phi}_n^*$ is slowly varied in accordance with time varying characteristics of signal and the outputing rate of the buffer. The state of the buffer seems to be bounded and has its average value around $q^* = 20$, and the fluctuation of $\tilde{\phi}_n$ is not significant. This approach provides both small fluctuation of the state of the buffer and distortion.

It is observed that the average value of the state of the buffer is greater than $q^* = 20$ when the fixed desired state of the buffer is used (the dashed line in Figure 3 (a) or Figure 5 (a)). It comes from the mismatching between the actual average distortion and the desired distortion. By adjusting the desired distortion, this mismatching can be resolved. As a result, the bias of the state of the buffer is improved. In Figure 5 (b), it is seen that the average value of the state of the buffer is close to $q^* = 20$ when we adjust the desired distortion (the average values of q_n with and without adjusting the desired state of the buffer are 20.625 and 21.394, respectively). While the two distortion sequences (with and without adjusting the desired distortion) are close to each other.

SUMMARY AND FURTHER WORK

The distortion policy of the buffer-constrained rate control for VBR coders based on the adaptive quantization has been considered. From the distortion policy, we can obtain not only the stability of the state of the buffer, but also small fluctuation of distortion. To reflect the time varying characteristics of signal and the outputing rate of the buffer, the two adaptive methods in the distortion policy are proposed.

For further work, we will modify the cost function to incorporate perceptual quality[3] and apply the distortion policy to the MPEG coder.

ACKNOWLEDGMENT

The author would like to thank Dr. K.-H. Yoo for careful reading and constructive comments.

REFERENCES

1. J. Choi and D. Park, "A stable feedback control of the buffer state using the controlled Lagrange multiplier method," *IEEE Tr. Image Proc.*, vol. 3, pp. 546-558, September 1994.

2. J. Choi, "Some policies of buffer-constrained rate control for real-time VBR coder," *manuscript.*

3. Ismail Dalgic and F.A. Tobagi, "Constant quality video encoding," in *Proc. IEEE Int. Conf. Comm., ICC'95*, pp. 1255 - 1261, Seattle, June 1995.

4. W. Ding and B. Liu, "Rate-quantization modeling for rate control of MPEG video coding and recording," in *Proc. SPIE Electronic Imaging - Digital Video Compression, San Jose*, vol-2419, February 1995.

5. M.J. Gormish, *Source Coding with Channel, Distortion, and Complexity Constraints, Ph.D. Dissertation*, Dept. E.E., Stanford, March 1994.

6. A. Ortega, K. Ramchandran, and M. Vetterli, "Optimal buffer-constrained source quantization and fast approximations," in *Proc. IEEE Int. Sym. on Circ. and Syst., ISCAS'92*, San Diego, May 1992.

7. A. Ortega, M.W. Garrett, and M. Vetterli, "Toward joint optimization of VBR video coding and packet network traffic control," in *Proc. Fifth Packet Video Workshop*, Berlin, March 1993.

MPEG-2 BASED DIGITAL VIDEO TRANSMISSION OVER SATELLITE CHANNELS

Kou-Hu Tzou

COMSAT Laboratories
22300 COMSAT Drive
Clarksburg, MD 20871

INTRODUCTION

Satellite video broadcasting provides an effective way for point-to-multipoint video distribution. It has been widely used in video distribution to cable headends and to satellite TVRO (TV Receive Only) users for years. Due to recent developments in high-powered Ku-band satellite transponders, satellite video broadcasting to small home antennas has become feasible. The cost of a consumer satellite receive system, including receive dish antenna/LNB and Integrated Receiver/Decoder (IRD) is very affordable. Furthermore, due to advances in digital video compression technology, the capacity of the satellite transponder has been increased substantially. Today, digital TV with 100 or more channels per satellite is being broadcast in the North America.

The MPEG-2 video and audio coding standards are capable of achieving high video/audio quality with great bandwidth efficiency. The MPEG-2 system standard offers an effective way of multiplexing video, audio, user data, and various service information using a packet format. The adoption of MPEG-2 based standards (transport, video, and audio) for satellite video distribution not only creates an environment for equipment interoperability, but also opens the door for potential multimedia services to the mass population.

In this paper, we will overview characteristics of digital satellite channels and examine enabling technologies for direct-to-home digital video broadcasting. We will discuss multimedia services suitable for satellite broadcasting.

OVERVIEW OF SATELLITE VIDEO TRANSMISSION SYSTEM

In an analog satellite transmission system, the baseband video signal is FM modulated and transmitted from an uplink site to a geo-stationary satellite. The signal is received by the satellite and retransmitted downward at a different frequency. At a receive site, the

Multimedia Communications and Video Coding
Edited by Y. Wang *et al.*, Plenum Press, New York, 1996

signal is received by the receive antenna, block frequency converted to a lower frequency band, and carried through a coax cable to an indoor IRD unit. A simplified system is shown in Figure 1.

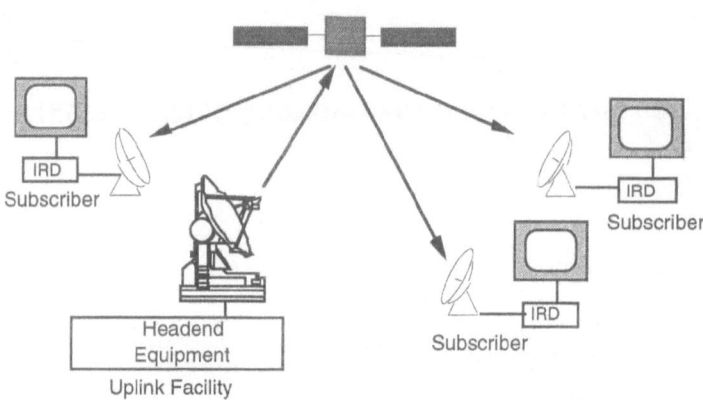

Figure 1. A satellite video transmission system.

The satellite is powered by solar cells with backup batteries. Due to the constraint on the power limit, the satellite transmitters are normally operated in the saturated mode, which introduces system nonlinearity and causes waveform distortion. The available signal-to-noise ratio for satellite channels is usually much lower than that for cable channels. In order to overcome the nonlinearity as well as to improve the signal-to-noise ratio, the FM technique is always used for analog TV transmission over satellites. For Ku-band applications, a 27-MHz bandwidth is normally allocated to carry one analog TV signal.

For digital transmission over satellite, the QPSK (Quadrature Phase Shift Keying) modulation is the most popular technique. The QAM (Quadrature Amplitude Modulation) technique, which requires a linear system response, is not suitable for satellite applications. In the North America region, the Ku-band DBS uses the 12/14 GHz frequency band (14 GHz for uplink and 12 GHz for downlink), which allows the subscribers to use a smaller dish antenna. However, the Ku-band link is more susceptible to rain fading than the C-band link and therefore, more margin for rain fading is required for the Ku-band link. Due to the typical low signal-to-noise ratio available for satellite links, powerful coding techniques are required to order to achieve a high-quality link. For an MPEG-2 video streams at 10 Mbit/s, an average of one-hour error-free transmission will require a channel Bit Error Rate (BER) of 2.778×10^{-11}.

Channel Modulation and Coding

The satellite link has been notorious for its nonlinearity and relatively low carrier-to-noise ratio. Due to the nonlinearity, amplitude modulation techniques are discouraged in satellite environment. Without forward error correction coding, typical satellite links can only achieve a channel BER between 10^{-2} and 10^{-5}. The above BER is far from the targeted quality of service for compressed digital video, which requires an error-free transmission interval to be in the order of one hour. For typical MPEG-2 compressed bit streams, this error-free interval translates into a channel BER between 10^{-10} and 10^{-11}. Concatenated inner and outer codes developed for satellite communications are capable of reducing the channel BER to below 10^{-12} from 10^{-4}.

Recently, European Broadcasting Union (EBU) launched a project intended to set a standard for digital video transmission over satellite, cable and Satellite Master Antenna TV (SMATV) channels. A draft standard [1] was published by EBU/European Telecommunications Standards Institute (ETSI). This draft specified a powerful error correction scheme based on the concatenation of convolutional and Reed-Solomon (RS) codes. The convolutional code can be configured to operate at the following different rates, 1/2, 2/3, 3/4, 5/6, 7/8, and 1, to optimize the transponder power and bandwidth performance. At the receive end, Viterbi decoding with soft-decision is often used to decode the convolutional code. By using the convolutional code alone, a BER between 10^{-3} and 10^{-8} may be achieved for typical satellite links. However, this is still not adequate for real-time digital video applications.

In order to further improve the BER performance, an outer code using the Reed-Solomon (RS) code is applied to correct errors remaining uncorrected by the convolutional code. Channel errors generated at the output of the Viterbi decoder tend to occur in bursts. The Reed-Solomon code operates on byte-oriented data and is effective in correcting burst errors. To improve the effectiveness of the RS code, an interleaver is usually used between the convolutional code and the RS code. The RS code appends a number of redundant bytes to a block of data to achieve error correction. Usually, $2n$ redundant bytes can correct up to n byte errors. In order to maintain the structure of the MPEG-2 transport packets, the (204,188) RS code has been chosen which append 16 redundant bytes to each MPEG-2 transport packet. By using the (204,188) RS code and a convolutional interleaver of depth 12, the BER of $2 \cdot 10^{-4}$ for the convolutional code can be improved to approximately 10^{-11}. A recent report [2] showed that a channel BER of approximately 10^{-11} may be achieved for typical high-powered DBS with bit rates ranging from 23 to 41 Mbit/s by using concatenated convolutional and RS codes.

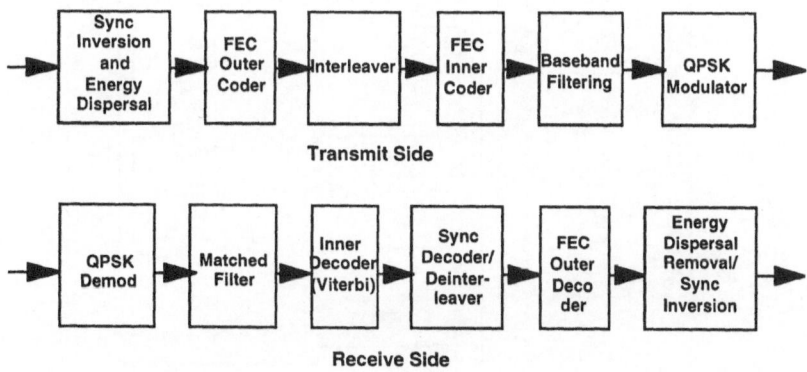

Figure 2. Block diagram of coding and modulation for satellite channels

Service Information

In order to facilitate versatile video services over broadcast channels, some service associated information must be transmitted so that the transport and distribution networks, and subscriber set-top boxes can utilize the information and react accordingly. For example, a subscriber might wish to view the program guide and set a Timer Record function on the screen by pointing the cursor to the selected program. To accommodate this, program information and system time/date information are needed. Some program specific information (PSI) has been defined in the MPEG-2 System standard to facilitate multiplexing of multiple programs into a single bit stream and to accommodate transport of user data and other service related information such as conditional access messages. However, some PSI specifications in the MPEG-2 System standard, particularly the Network Information Table (NIT), are very preliminary, and need to be extended or augmented to maximize end-to-end compatibility. For example, more information about the physical network and its parameters are needed, which may include frequency, orbit position, polarization, FEC inner code and outer code, symbol rate and modulation.

The int Technical Committee (JTC) of the European Broadcasting Union and the European elecommunications Standard Institute (EBU/ETSI) has produced a draft standard to fill the missing portion of the service information [3]. The EBU/ETSI Draft specifies additional data which complements the MPEG-2 PSI by providing data to aid automatic tuning of IRDs, and additional information intended for user to manipulate. The additional service tables defined in the EBU/ETSI Draft include Bouquet Association Table (BAT), Service Description Table (SDT), Event Information Table (EIT), Running Status Table (RST), Time and Date Table (TDT), and Stuffing Table (ST). Figure 3 shows all the service information accommodated in the EBU/ETSI DVB standard. The service information tables are mapped into MPEG-2 transport stream packets by using the Section syntactic structure, as defined in the MPEG-2 System standard. With the service information implemented, a subscriber set-top box can be enhanced with many new features such as Electronic Program Guide (EPG), parental lock, and addressable receiver.

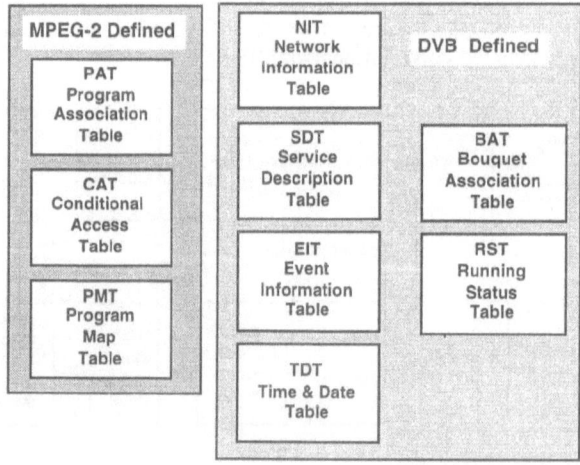

Figure 3. Service information specified in the EBU/ETSI DVB standard.

Conditional Access

In satellite TV broadcasting, premium programming and value-added services are broadcast to all receivers regardless of their subscription status. In order to protect the service from un-authorized reception, conditional access must be used to control the viewing rights of individual receivers. Conditional access systems usually rely on addressable decoders and scrambling schemes. Most manufacturers prefer to keep the details of their conditional access subsystem wrapped in order to deter any effects to crack the system. However, a proprietary approach may cause equipment compatibility problem. In order to maintain compatibility while allowing different conditional access system to co-exist, the Conditional Access Table (CAT) in MPEG-2 and DVB only specifies the syntactic structure of the CA sections. It still leaves room for manufacturers to implement a proprietary CA systems.

A draft recommendation [4] on conditional access broadcasting systems is proposed by CCIR (now ITU-R), which specifies a system architecture for a conditional access subsystem. The system uses a three-level key hierarchy and is fairly secure and flexible to accommodate conditional access control. A system block diagram of the CCIR recommendation is shown in Figure 4. The only permanent secret at the subscriber site is the distribution key, which is specific to each set-top box or each security card. This secret information is usually programmed into a secure device which prevents the information from being accessed externally.

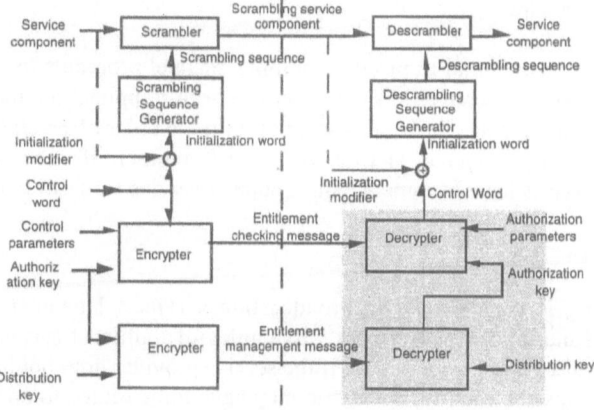

Figure 4. The three-level key hierarchy used in the CCIR draft recommendation.

In the CCIR proposed system, no CA information needs to be transmitted through a secure channel. The first-level key is the Control Word (CW) which is used to initialize the pseudo random number generator. The generated pseudo random binary bit stream is then used to scramble the digital data. At the receive end, if the receiver can recover the CW and synchronize it with the transmitter, the scrambled data stream can be decrypted. In order to enhance the security, the CW is usually changed several times per second. The second-level key is the authorization key which is used to encrypt the CW. The authorization key is usually associated with the program. The CW, encrypted by using the authorization key, is transmitted over the broadcast channel as the Entitlement Control Message (ECM). If the subscriber is authorized to receive a program, a valid authorization key should be able to be recovered from the ECM. The third-level key is the distribution key which is used to encrypt the authorization key and the encrypted authorization key is transmitted as the Entitlement Management Message (EMM). Since the scrambled authorization key is encrypted by user-specific distribution key, only the intended receiver will be able to recover the authorization key. Since the recovered authorization key will stay valid for a relatively long period (hours, days, or weeks), it has to be protected from being revealed. The shaded area in Figure 4 indicates that a secure implementation is required for the decrypters to permit the authorization key and CW to be recovered. The implementation of the scrambler and encrypters, key size, and methods of key generation are left open in the CCIR draft recommendation.

POTENTIAL SERVICES OF MPEG-2 BASED DBS

Digital satellite broadcasting provides an economical way for distributing high-speed data to households in wide areas. With the MPEG-2 packet-based transport standard adopted, digital DBS may become a vehicle to offer a variety of multimedia services to customers. However, due to the unique characteristics of the DBS channel, certain services may not be suited. One of the most notorious characteristics of satellite transmission is the long delay, which is approximately 0.25 second for a one-way transmission. For broadcasting services, the delay may not be noticed at all by the users. However, for interactive services, the long delay may have some impacts on the quality of service.

Broadcasting Service

A DBS has the potential to provide multiple-channel programs to tens or hundreds of millions of subscribers scattered in wide areas. For this application, the economy of DBS can be easily justified since the transmission cost (in the order of several million dollars per year and per transponder) is shared by a huge number of subscribers. Therefore, broadcasting services have become the most popular service for DBS today.

Multicasting Service

Multicasting is very similar to broadcasting service. However the signal is only intended for a number of subscribers. Examples of multicast service include distance learning and multi-point business TV. If the service provider does not have direct access to a satellite transmission facility, the traffic may have to be routed to an uplink site through terrestrial links such as ATM/SONET. On the other hand, the signal received at a satellite receiving site may be relayed locally to multiple subscribers through ATM networks if the subscribers do not have satellite receivers. When the number of subscribes is large or the distances between the service provider and subscribers are large, the DBS-based multicasting will likely be more attractive than other alternatives.

On-Demand Service

On-Demand is a type of interactive service where the down stream information (video, audio, images, or data) is under the user's control. Such applications include video-on-demand (VOD), near video-on-demand (NVOD), image database retrieval, Web site browsing, and software down loading. The transmission cost for DBS is so high that many personalized services are impractical. A typical cost for a transponder lease is 2 million dollars per year, which is translated to approximately $3.8 per minute, or about $76 for a 2-hour movie (assuming 6 digital video signals in each transponder). Therefore, full VOD cannot be afforded over DBS due to the high transmission cost. Nevertheless, NVOD is still possible if more than a thousand of subscribers are expected to watch each staggered showing.

Although the $3.8 per minute transmission cost is prohibitively high for VOD, it is still very attractive for data services. For a transponder capacity of 20 Mbit/s, the cost for data services is less than 3 cents per Mega Byte. Compared with data services using modems over telephone lines, the transmission cost is much cheaper and the data downloading time is much quicker. Since DBS is a one-way link, the return channel for sending user commands has to rely on other media such as telephone lines.

OTHER TECHNICAL ISSUES

Video Multiplexing

For DBS transmission, all of the signals are combined at the uplink site by using a packet-based multiplexing technique. The signals may come from various sources with different time bases and in various formats. For example, the video may come from an ATM network in the MPEG-2 transport system format, may be in a compressed MPEG-2 program stream format from a digital storage device, or may be generated by a real-time MPEG-2 encoder in the transport stream format. The MPEG-2 system standard utilizes Program Clock Reference (PCR) and Presentation Time Stamp (PTS) to recover the time base and to synchronize time-sensitive service data. It is a great challenge to receive compressed digital video data from various sources with different time bases and remultiplex them together in a seamless fashion.

Return Channel

The DBS link can only provide downward broadcasting. It has to rely on other media to provide a return link. Today, a simple way to provide a return link is to use the existing public-switched telephone networks (PSTN). However, wireless radio in combination with the TV remote control may offer a user friendly solution since it does not require an additional phone line.

SUMMARY

In this paper, a brief tutorial on satellite video broadcasting has been presented by addressing key enabling technologies. Today, the DBS systems have been designed, by using powerful coding techniques, to offer a channel BER performance as good as the cable system. DBS is ideal for broadcasting services intended for a large number of subscribers. For selected multicast services, DBS may also appear as a favorable alternative. By using the MPEG-2 packet-based multiplexing, DBS can offer a high-speed data pipe to households in wide areas, which makes DBS very attractive for economical high-speed data services.

References

1. EBU/ETSI JTC, Draft digital broadcasting system for television, sound and data services; Framing structure, channel coding and modulation for 11/12 GHz satellite services, Draft prETS 300 421, June 1994.
2. M. Cominetti and A. Morello, Direct-to-home digital multi-programme television by satellite, Int. Broadcasting Convention, pp. 358-365, June 1994.
3. EBU/ETSI JTC, Draft service information (SI) in digital video broadcasting (DVB), Draft prETS 300 468, Nov. 1994.
4. CCIR, Draft new recommendation: Conditional-access broadcasting system, Doc. 11/BL/33-E, 26 May, 1992.

DATA PARTITIONING AND UNBALANCED ERROR PROTECTION FOR H.263 VIDEO TRANSMISSION OVER WIRELESS CHANNELS

Hang Liu and Magda El Zarki

Video Processing and Telecommunications Laboratory
Department of Electrical Engineering
University of Pennsylvania
Philadelphia, PA 19104

I. INTRODUCTION

Recently, the growing demand and potential market for new services and applications has spurred the research on video transmission and multimedia services over wireless networks. In terms of channel characteristics, the radio channels in wireless networks are error prone and their bandwidth is limited due to the availability of accessible frequency spectrum. The video bit stream, after efficient source coding, is very vulnerable to channel errors, it therefore presents a challenging design problem to obtain acceptable quality video over lossy radio channels. Classified (layered) source coding and unbalanced (unequal) error protection provide a possible mechanism to enable the reliable transmission of compressed video bit streams over wireless links.

There are two main issues related to classified source coding and unbalanced error protection. One is how to partition the data into classes. The other is how to choose the appropriate error protection schemes for the different classes. In order to obtain optimal transmission quality, these two issues have to be jointly considered in conjunction with the channel characteristics.

Transmission of still images and MPEG video sequences via radio channels has been studied by several researchers.[1,2] ITU-T SG 15 recently finalized a very low bit rate video coding standard, H.263, for video telephony service over the General Switched Telephone Network (GSTN) at data rates less than 64 kbit/s.[3] The expert's group is starting to adapt H.263 for wireless applications because the low bit rate makes it best suitable for the narrowband wireless networks. A couple of proposals are currently being evaluated.[4]

The transmission of H.263 coded video sequences over wireless channels is studied in this paper. Powerful forward error correction (FEC) codes are required in order to obtain acceptable video quality. However this results in excessive overhead. Based on the analysis of the H.263 coding structure, the channel error effect and the phychophysical properties of the human visual system, we present a two class data partitioning and unbalanced error protection scheme in order to increase bandwidth utilization. The performance of the scheme is investigated. The capabilities and trade-offs of several channel coding and interleaving strategies are also analyzed.

This paper is organized as follows: Section II briefly describes the system model. In section III, the H.263 coding structure and syntax are outlined and an analysis of the error effect on the reconstructed pictures is addressed. Section IV discusses the capabilities and trade-offs of different FEC codes and interleaving strategies. The simulation results and performance comparison of classified and non-classified transmission schemes are presented in section V. Section VI concludes our work.

II. COMMUNICATION SYSTEM MODEL

Fig.1 shows the block diagram of the mobile video transmission system under investigation. Classified source coding is performed first. Then the two priority classes from the video source coder employ different error protection schemes, i.e. FEC and interleaving strategies, depending on the importance of the information. A sequential multiplexing mechanism is used to multiplex the encoded data from the two priority classes and input the data to the buffer for transmission.[4] The two classes of data are serially interleaved with each other in the multiplexed stream. Every class is preceded by a class header in order to allow demultiplexing at the receiver. For each frame, the data of the high priority class with the prefixed class header is transmitted first, followed by the low priority class header and data.

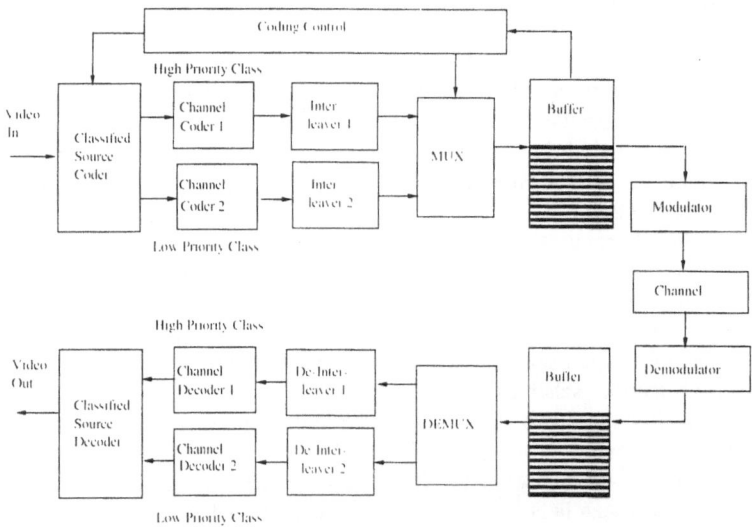

Figure 1. Diagram of two class video transmission system.

If block FEC codes are employed and the class length from a coded frame is not exactly a multiple of the FEC block code length (i.e., when the coded data from one class of a frame is divided into blocks, the last block does not have enough bits), the last block includes some bits from the same class of the next transmitted frame. In our simulation, block codes with 127 bits long are employed. The class header is of fixed length and includes three fields. The first 1 bit field indicates the class type. The second field is 7 bits long and indicates how many bits in the last block come from the next frame. The third field is 8 bits long and indicates the length of the data following the header in the units of 127 bits. The length is variable depending on the coded frames. Generally, 8 bit field is enough to represent the length of encoded QCIF format video sequences. In case the length is longer than what the length field can represent, the field is extended by another 8 bits. Then the value of the first 8 bit length field is 2^8-1 and the value of the last 8 bit length field is the difference of the length and 2^8-1. If convolution codes are used, the class header only needs two fields, one for class type and the other for the length in bits because the data comes from exactly only one frame.[4]

Note that the picture start codes are not necessary because the class header information indicates where a picture starts. Therefore, the picture start code will be removed before channel coding at the transmitter and added after channel decoding at the receiver. The class headers are well protected for transmission so that the errors in the class headers are negligible. A buffer control scheme will adaptively adjust the source coding rate to the channel rate to guarantee that no underflow and/or overflow occurs in the encoder and decoder buffers as described in the H.263 standard.[3] At the receiver, the two classes of the bit streams are first reconstituted by the demultiplexer, then channel decoding and source decoding are performed.

III. H.263 VIDEO CODER AND EFFECT OF TRANSMISSION ERRORS

H.263 coded data is arranged in a hierarchical structure with four syntax layers as shown in figure 2.[3] From top to bottom they are: Picture, Group of Blocks (GOB), Macroblock (MB) and Block. Variable length coding (VLC) is employed at the macroblock and block layers. The impact of a single bit error will depend very much on which bit is hit. A single bit error in the DCT coefficients will at least corrupt one block as the DCT coefficients of each block are coded using run-length and variable length coding. If the loss of synchronization occurs, all the subsequent blocks in the GOB may be destroyed. As motion vectors are differentially encoded for the intercoded macroblocks of the same GOB, then a bit error in a motion vector may result in the corruption of this macroblock and the following predicted macroblocks. In addition, because VLC is also employed in the macroblock headers, synchronization will probably be lost when a bit error occurs. Therefore, a bit error in the macroblock headers may at worst damage a complete GOB. The start codes in the picture and GOB headers provide the synchronization in the spatial domain and stop error propagation. One or more bit errors in a GOB does not affect other GOBs. Only if a bit error occurs in the control information symbols of the picture headers, may it seriously impair the total frame.

In the temporal domain, predictive coded pictures (P-pictures) are coded using motion compensated prediction from a past intra or predictive coded picture and are generally used as a reference for further prediction. The error will propagate among the P-pictures until the next intra coded picture (I-picture) which is coded using information only from itself and provides error resilience in the temporal domain. Furthermore, as the higher frequencies are visually less important than the lower ones, a bit error occurring in the high frequency DCT coefficients will have very little impact on video quality.

Wireless channels are much less reliable than wirelinks; both random and bursty errors generated by noise and fading exist when signals are transmitted. Powerful FEC is required in order to combat the errors. However, FEC adds much overhead to the system which at worst could render the delivery of acceptable quality video impossible because the required data rate exceeds the available channel capacity. A natural approach is to rearrange the coded video information such that the most important information is most likely to be received correctly.

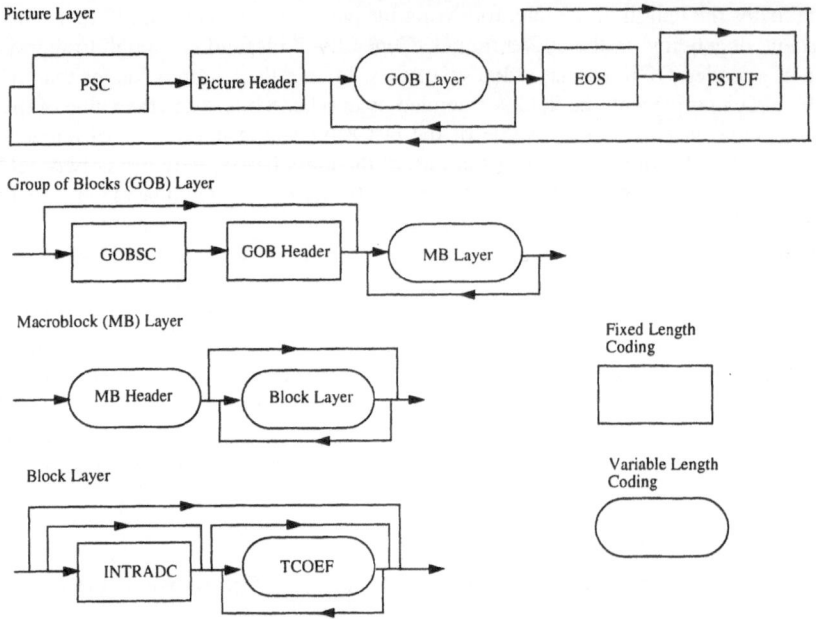

Picture Layer

Group of Blocks (GOB) Layer

Macroblock (MB) Layer

Fixed Length
Coding

Block Layer

Variable Length
Coding

Figure 2. Simplified syntax diagram for the H.263 video bitstream.

Based upon the above observations, a simple approach is to partition the H.263 coded data into two priority classes. The high priority class (or base layer) consists of the most important data, which contains the control information, motion vectors and maybe lower order DCT coefficients, while the low priority class (or enhancement layer) contains the higher frequency DCT coefficients. The GOB start code and GOB number are redundantly copied in the low priority class to facilitate synchronization and error recovery. A priority breakpoint in the picture headers and GOB headers indicates what elements are to be included in the high priority class, the remainder of the bitstream is to be placed in the low priority class. This is similar to the MPEG-2 data partitioning syntax.[5] The controller can change the priority breakpoint so that the I-pictures have more of the data in the high priority class than the P-pictures. The high priority data, including a class header are transmitted first followed by the low priority class header and data for each frame. This data partitioning scheme has the advantage of minimum complexity, good compatibility with single class H.263 coders and a bit rate efficiency very close to the single class coder. Furthermore, the high priority class could carry enough information to produce an acceptable visual quality image even if the low priority data is in error.

IV. CHANNEL ERROR PROTECTION

The selection of FEC codes depends on several factors: (1) the capability of the error correction code which depends very much on the channel error patterns, (2) the overhead or code rate, (3) the block size for each code and (4) the cost and complexity. Interleaving spreads the bursty errors due to Rayleigh fading into random errors required by most FEC codes. It should be noted that a single bit error could have the same impact on the reconstructed pictures as if all the bits of the GOB are in error. This is because of error propaga-

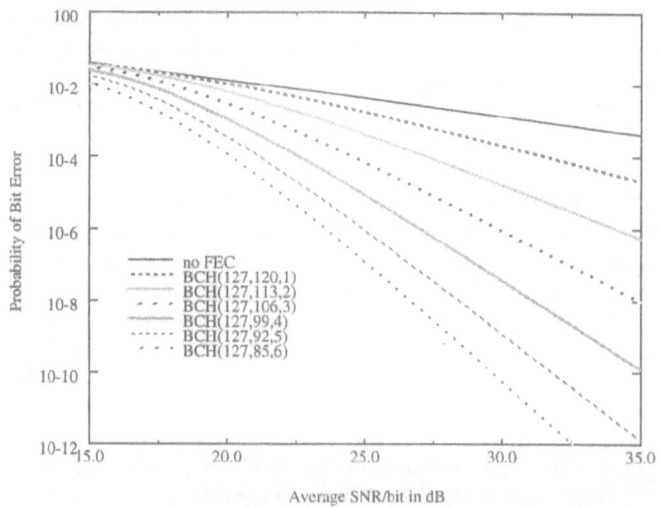

Figure 3. BER performance of BCH codes on interleaved Rayleigh fading channels.

tion in a GOB in the compressed video bit stream. The spreading of errors, via interleaving, in a compressed video bit stream, may damage more GOBs thereby having a negative effect on the overall performance unless the used channel coding is strong enough to correct those erroneous bits. It is the same as MPEG coded video[2] and our simulation also proves it. Therefore, interleaving has to be carefully deployed in applications with compressed video bit streams.

BCH codes provide a good trade-off in terms of error correction capability versus complexity. They can effectively correct random errors and have also been successfully applied in bursty error environments when combined with interleaving. For our simulation, we use BCH codes. Figure 3 depicts the BER performance versus channel SNR over a flat Rayleigh fading channel for BCH(127, 120, 1), BCH(127, 113, 2), BCH(127, 106, 3), BCH(127, 99, 4), BCH(127, 92, 5) and BCH(127, 85, 6) codes, where we assume the modulation scheme is $\pi/4$-DQPSK and interleaving is used to randomize the bursts of errors. The associated overheads are shown in table 1.

Table 1. List of overhead for BCH codes.

BCH code	(127, 120,1)	(127, 113, 2)	(127, 106, 3)	(127, 99, 4)	(127, 92, 5)	(127, 85, 6)
Overhead	5.5%	11%	16.5%	22%	27.5%	33%

V. SIMULATIONS AND RESULTS

Simulations are performed on a QCIF format (176x144 pixels) "CNN" video sequence, which contains typical video telephony-like images. We consider a few video coding and channel coding rates with a fixed overall rate of 22 kbit/s. The original video sequence with 30 frames per second is encoded by the source coder at 5 frames per second. A fading simulator is used to generate the error patterns of the radio channel. We assume that the channel coherence bandwidth is much larger than the signal bandwidth, which is a reasonable assumption for microcell environments, otherwise, equalizers must be used to alleviate intersymbol interference. The average SNR is 22 dB, the mobile speed is 6 mi/h and the modulation scheme is π/4-DQPSK with a carrier frequency of 1.9 GHz which is in the frequency band of the emerging personal communication system (PCS).

Table 2 shows the average peak SNR of seventy decoded frames using BCH codes with block length n=127 bits and different error correction capabilities "t". The standard single class H.263 coded bit streams are used with/without 127-bit-depth interleaving (INV). In order to reduce the sensitivity of the average PSNR to the error location, each simulation was run five times using the different starting time of the transmission or the different random number seeds in the fading simulator. The average value over all runs for each simulation is presented in the table 2. For no FEC and BCH(127, 120, 1), interleaving results in a lower average PSNR of the decoded pictures as was discussed in previous section.

Table 2. Average PSNR of the 70 frames of decoded "CNN" video sequence under different n = 127 BCH codes with/without interleaving (INV).

PSNR (dB)	no error	no FEC	BCH t=1	BCH t=2	BCH t=3	BCH t=4	BCH t=5	BCH t=6
With INV	33.11	14.22	15.02	16.27	20.04	25.88	30.21	32.26
No INV	33.11	15.33	16.11	16.19	16.85	17.23	17.85	18.11

For the data partitioned source coder, all picture, GOB, macroblock syntax elements are included in the high priority class. If a macroblock is intracoded, the first six (last, run, level) DCT coefficient events are placed in the high priority class, and the remainder after the six is in the low priority class. If a macroblock is intercoded, all the DCT coefficients are placed in the low priority class. Different combinations of FEC codes are deployed for the high and low priority classes with/without interleaving in the simulations.

In figure 4, the typical PSNR of a decoded video sequence is depicted for comparison under the four scenarios: (1) single class H.263 coded bit stream without FEC over an ideal error free channel. (2) single class H.263 coded bit stream over the above wireless channel but with no FEC and no interleaving. (3) single class H.263 coded bit stream over the same wireless channel with BCH(127, 99, 4) and 127-bit-depth interleaving. (4) data partitioned H.263 coded bit stream over the same channel, BCH(127, 85, 6) with interleaving for high priority class and BCH(127, 120, 1) with no interleaving for low priority class.The total overhead of scenario (4) due to channel coding and data partitioning is the same as the channel coding overhead of scenario (3), i.e. scenario (3) and scenario (4) have the same source data rate. For scenario (4), however, the PSNR of the decoded pictures is in the acceptable range.

Subjective evaluation tests also indicate that there is no visible damage on the decoded pictures of the video sequence under the unbalanced error protection. The overall visual quality is greatly improved over that of a single class coder. Figure 5a-5d display the fourteenth frame of the decoded sequence under the four scenarios, respectively.

VI. CONCLUSION

In this paper, we have studied the transmission of H.263 video sequence over wireless channels under different BCH codes and interleaving strategies. Powerful FEC codes are required to deliver acceptable video quality. This increases the data rate a lot. A two class

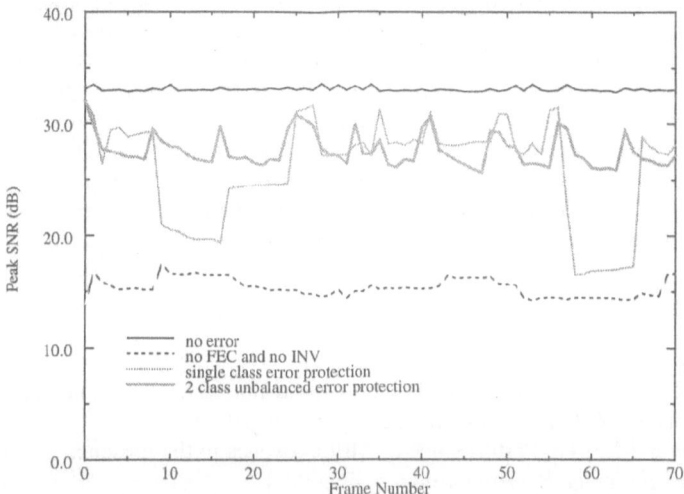

Figure 4. PSNR of decoded frames of "CNN" video sequence transmitted over a wireless channel at 22 kbit/s under single class and 2 class unbalanced error protection.

Figure 5. The 14th frame of decoded "CNN" sequence after transmission over the wireless channel. (a) no error, (b) no FEC and no INV, (c) single class error protection, (d) 2 class unbalanced error protection.

data partitioning scheme based on a H.263 video coder is described in order to make the best use of the limited channel bandwidth. Simulation results show that the errors in the low priority class have much less impact on the visual quality of the reconstructed pictures when the important information in the high priority class is well protected. This simple two class source coding scheme combined with adequate unbalanced error protection can greatly improve the video transmission quality compared to the non-classified transmission scheme. Error concealment has not been used. As a matter of fact, concealment techniques are expected to be more effective in the case of unbalanced error protection because most of the errors come from less important information and are more easily concealed.

REFERENCES

1. Y. Zhang, Y. Liu, and R.L. Pickholtz, Layered image transmission over cellular radio channels, *IEEE Trans. Vehicular Tech.,* 43:786 (1994).
2. Y. Zhang and X. Lee, Performance of MPEG codecs in the presence of errors, *J. Visual Comm. and Image Representation,* 5:379 (1994).
3. ITU-T/SG15, *Draft Recommendation H.263, Video Coding for Low Bitrate Communication* (1995).
4. Robert Bosch GmbH and IENT, RWTH Aachen, *ITU-T/SG15 Proposal, LBC-95-003, Proposal for Extension of Recommendation H.26P for Mobile Application: H.26P/M* (1995).
5. ISO/IEC JTC1/SC29/WG11 (MPEG-2), *Generic Coding of Moving Pictures and Associated Audio-Part 2: Video* (1994).

A REAL-TIME SOFTWARE BASED END-TO-END WIRELESS VISUAL COMMUNICATIONS SIMULATION PLATFORM

Andria H. Wong, Li Fung Chang, and Ming-Ting Sun

Bellcore, 331 Newman Springs Road, Red Bank, NJ 07701

Ting-Chung Chen

CLI, 350 East Plumeria, San Jose, CA 95134

T. Russell Hsing

Bellcore, 445 South Street, Morristown, NJ 07960-6438

ABSTRACT

For visual communication services to be truly ubiquitous, transmission of video signals over the wireless medium should be feasible with acceptable quality. Wireless transmission characteristics are dominated by channel effects like multipath propagation and Doppler frequency which result in severe error bursts and time-varying signal strengths. Such channel effects pose serious challenges to real-time visual communications systems. In this paper, we describe a software based wireless visual communications simulation platform which can be used to evaluate wireless video performance in real-time. The platform consists of two personal computers serving as hosts. Major components of each PC host include a real-time programmable video codec, a wireless channel simulator, and a network interface for data transport between the two hosts. These three components are interfaced in real-time to show the interaction of various wireless channels and video coding algorithms. The programmable features in the above components give us the flexibility to study various wireless channel effects without physically carrying out these experiments. Physically building such systems can be very time-consuming, costly, and the experiments are only limited in scope.

Using this simulation platform as a testbed, we have experimented with several wireless channel effects including Rayleigh fading, various multipath delays, Doppler frequency, various transmitted signal power, antenna diversity, and packet loss.

1 INTRODUCTION

With the new advancements in signal processing and circuit technologies, real-time visual communications is becoming an important service in multimedia communications. Conventionally, real-time visual communications requires isochronous and low delay transport and is achieved using digital networks like ISDN or T1/E1. In the communication and networking area, however, wireless access is becoming the fastest growing business. In planning visual communications for the future, it is thus necessary to consider wireless transport because of its growing potential and flexibility.

Wireless communications conventionally carry only voice and data. Noisy channels, power and complexity limitations have limited the transport bandwidth to mainly low bit-rate operations. To extend wireless communications to carry real-time video, bandwidth needs to be increased and performance needs to be improved. The video source coding algorithm also needs to be robustly designed for error mitigation. These requirements pose technical challenges on both wireless communications and video source coding.

To verify the performance of a wireless visual communications system, it is necessary to investigate the overall combined operation. Since hardware implementation takes considerable time and money, it is desirable to simulate as much of the overall operation in software as possible. Currently, video signal processors are becoming popular and it is practical to have real-time programmable video codecs. However, to simulate various wireless access scenarios, a programmable modem used over various transport environments is required. This part of the process is time consuming and requires much effort. To speed up the turn-around time, it is desirable to simulate the modem and transport environments to the largest extent such that both emulate the effects of real operations.

In this paper, we describe the experimental prototyping of a real-time software-based end-to-end wireless visual communications simulation platform. This paper is organized as follows. Section 1 gives an overview of the simulation platform which is built upon several important components. The programmable video codec and the related software are described in section 2. The packet network interface over Ethernet is described in section 3. Section 4 presents the third component, the wireless channel simulator. Some low bit-rate simulation results are presented in section 5. A summary of the investigation is given in section 6.

2 PLATFORM

A high level system diagram of the simulation platform is shown in Fig. 1.

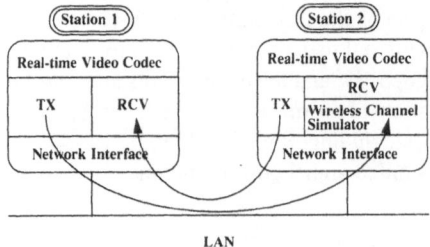

LAN

Figure 1: A real-time software-based wireless video simulator

The platform consists of 2 stations (PCs) continuously transmitting video signal to each other over the Ethernet. The LAN is used for convenient transport purpose and is transparent to the system performance. There are two major components within each station. The first component is a real-time video codec which is a programmable signal processor residing in the PC. The second component is a packet network interface which is used to connect the two PCs to simulate the backbone of the future wireless-ATM network. To observe the wireless channel effects on real-time video, a third component, a wireless channel simulator, is inserted between the video codec and the packet network interface of station 2 to simulate various wireless modems and access environments. This channel simulator generates error patterns based on the programmed wireless parameters and imposes the errors on to the signal transmitted by station 1. After decoding, the impaired video can be viewed on station 2. Station 1 is always displaying a clean signal so that video qualities from both the clean and the impaired channels can be compared side by side.

3 VIDEO CODEC

A programmable real-time video codec was selected for use in this prototype. The codec allows programmable bit rate, frame rate, picture size, and picture format. Various coding algorithms such as H.261, JPEG, and MPEG can be downloaded for operation when the corresponding microcodes are available. In our applications for low bit-rate visual communications, we select the H.261 algorithm commonly used in video conferencing as a reference standard.[1]

The software used in the video codec consists of three parts. The first part consists of computationally intensive functions such as motion compensation, DCT, zig-zag scanning, quantization, and run-length coding. These functions are microcoded and supplied with the codec. Algorithm parameters such as quantization values are programmable by the user. The second part controls the passage of data through the codec. It performs pre-processing, Huffman coding, and buffer interface. These functions can be coded in a higher level language and are then cross-compiled to generate microcodes to be downloaded for execution. The third part contains the PC driver functions which are coded in a higher level language and reside in the PC processor for interfacing between the PC and the video codec.

4 ETHERNET INTERFACE

The Ethernet software uses UNIX socket functions, which are built on top of a TCP/IP kernal and are integrated into the PC video driver functions. This allows the data generated by the video codec to be packetized into the TCP/UDP/IP protocol and be communicated between the two PC's. Packetizing parameters such as packet size, transmitter buffer size, receiver buffer size, acknowledge window size, etc. which affect data transport and delay can be programmed by the user to get a viable performance. To assure reliable communications between the two PC's, TCP/IP protocol and a private Ethernet are used where the Ethernet is transparent and does not affect the system performance.

To integrate the concurrent operation of these two video codecs, control software is required. Basically, this program implements the following four steps to permit full duplex operation, where the first two steps handle coding and transmission, and the last two steps handle receiving and decoding. Fig. 2 illustrates the flow chart of the control software with the following operations.

1. The video coder writes coded data to the transmit buffers in the PC memory,
2. The PC sends these data to the Ethernet transmitter,
3. The Ethernet receiver reads data to the receive buffer in the PC memory,
4. The PC sends data to the video decoder.

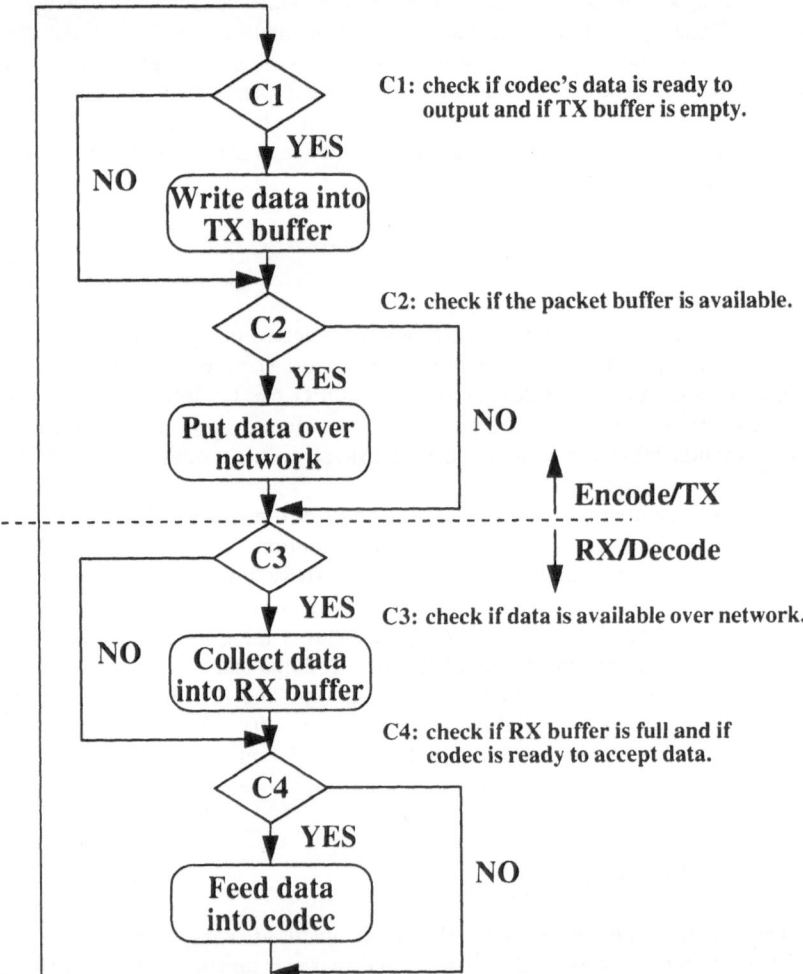

C1: check if codec's data is ready to output and if TX buffer is empty.

C2: check if the packet buffer is available.

C3: check if data is available over network.

C4: check if RX buffer is full and if codec is ready to accept data.

Figure 2: The client video terminal's control procedure

In each step, there is a condition check to decide if the action should be executed. If the condition is not satisfied, it will advance to the next step. The control software continuously cycles through these four steps without waiting. Hence, the possibility of network blocking and video codec hang-up can be minimized.

When a wireless channel simulator is incorporated, introduction of wireless impairments happens between steps 3 and 4. The modified video data is then sent to the video decoder for output evaluation.

5 WIRELESS CHANNEL SIMULATOR

To investigate the performance of wireless visual communications for a fixed modem configuration, the easiest way would be to integrate a wireless LAN modem to the described PC and video codec platform and carry out the measurements through different wireless access environments. However, this approach only allows the testing of a small set of uncontrollable and unrepeatable wireless environments.

An alternative to the above is to simulate both the modem and the channel in software and then integrate them into the PC and video codec platform. This approach has the advantage of improved flexibility and transmission environment programmability, but its fidelity must be carefully calibrated. Another major advantage is that unlimited number of channel characteristics can be tested in a short time and at a very low cost.

An implementation issue here is the processing capability limitation of PCs. The PC driver used to interface the video codec and Ethernet packetizer has already used up most of the PC processor power. Additional wireless channel processing will effect extra delay, and the video codec will not be able to perform real-time operations. To alleviate this problem, we choose to run the wireless channel simulator off-line in a non-real-time fashion. A typical simulator is shown in Fig. 3.

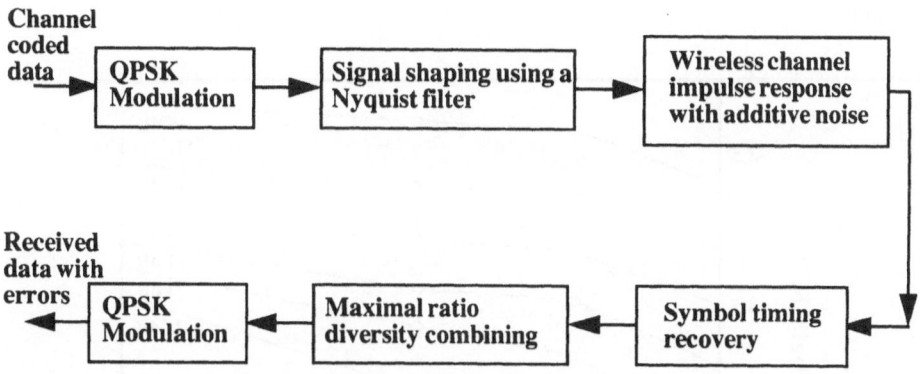

Figure 3: A typical wireless channel simulator

The operation of the wireless channel model is as follows. Channel coded data are converted to symbols using QPSK modulation with Gray code mapping. A matched Nyquist filter with roll-off factor of 0.5 is used. The channel impulse response with time correlated Rayleigh fading of multipath delay spread are calculated.[3] The radio channel transmission rate is 192k symbols per second which corresponds to an available channel bandwidth of 384kbps. A coherent receiver, with optimal symbol timing recovery and perfect carrier recovery is assumed. An ideal maximal ratio combiner for antenna diversity combining is used. The time-correlated Rayleigh fading was simulated using the technique described by Jakes.[2] The autocorrelation function of the simulated fading waveform approximates the Bessel function $J_0(2\pi f_d \Delta t)$, where f_d is the maximum Doppler frequency and Δt is the time separation between fading samples.

A long stream of random bits are input to the simulator. The received data with errors are compared with the original random bits to generate bit error patterns which are used to corrupt the real-time data by masking it. The effects of channel errors on a compressed video bit stream can now be observed on the video platform. The generation of error bit patterns is done off-line, while the masking operation is done in real-time. This bit error pattern file must also be large enough such that repeated usage of it will not generate correlated error pattern.

In our current implementation, the parameters that can be changed in the wireless model include the maximum Doppler frequency f_d, the propagation power file (modeled as n-rays with different inter-path delays and power), signal power, and antenna diversity.

6 PERFORMANCE EVALUATION AND COMPARISON

Experiments were performed to simulate the effects of various wireless channel and modem characteristics. Of these initial experiments, the Doppler Shift Frequency of 0.5 Hz is assumed which corresponds to the Doppler effect caused by an almost stationary receiver within an indoor environment. Different Doppler Frequency values affect the error bursts characteristics and not the bit error rate (BER). The antenna diversity is set to 2. We simulate the wireless channel characteristics of two-ray fading with an inter-path delay of 0 (one-ray or flat fading), 1/16, 1/8, and 1/4 of the symbol period respectively. The last setting corresponds to the worst case in a typical indoor environment. We observe the BERs due to the environment under the range of operating SNRs from 15 dB to 25 dB with every increment step of 2 dB. Fig. 4 summarizes the result.

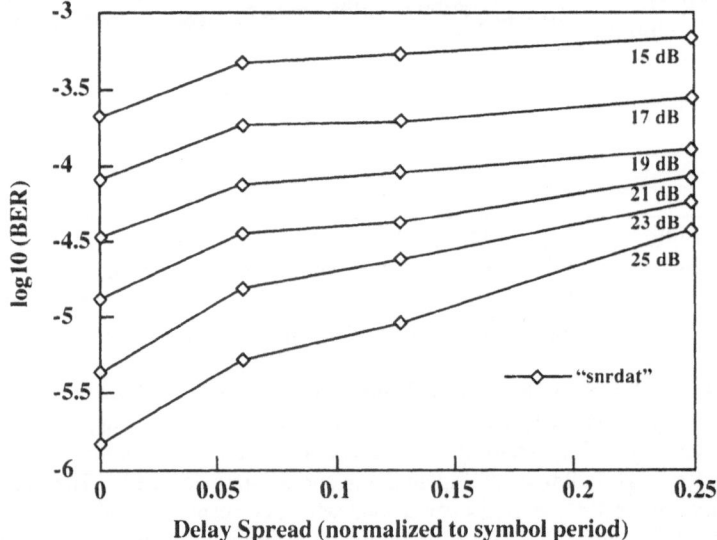

Figure 4: Effect of delay spread on BER at SNR = 15, 17, 19, 21, 23 and 25 dB

We observed the impaired picture quality of real-time coded video based on H.261[1] on CIF images at a bit rate of 64kbps and a frame rate of 15 fps. The subjective quality at 21 dB gives an acceptable quality for video-telephony applications. At the typical operating SNRs of 17 dB to 19 dB, viable services are possible with simple error correction mechanisms. Selection of error mitigation methods and evaluation of their impacts will be the subject of further study.

7 CONCLUSION

This paper discusses the implementation of a software based wireless visual communications simulation platform which allows experimentation of real-time end-to-end visual communication over a wireless channel. This is a flexible platform that allows evaluation of video quality based on different video coding algorithms under transmission of a variety of wireless environments. Different channel characteristics can be emulated using a wireless channel simulator software integrated into a PC which serves as a communication host. Different video coding algorithms can be simulated by incorporating the appropriate microcodes into the programmable video codec. Low bit-rate visual communications based on H.261 on CIF images at 64 kbits and 15 frames per second under a specific environment setting were evaluated using this system. The subjective quality at 21 dB gives an acceptable quality for video-telephony applications. At the typical operating SNRs of 17 dB to 19 dB, viable services are possible with simple error correction mechanisms. Selection of error mitigation methods and evaluation of their impacts will be the subject of further study.

8 REFERENCES

[1] CCITT Study Group XV - Report R 95, Recommendation H.261, *Video Codec for Audiovisual Services at p x 64 kbit/s*, May 1992.

[2] William C. Jakes, Jr., Editor, *Microwave Mobile Communications*, John Wiley and Sons, 1974.

[3] Donald C. Cox, IEEE, *Universal Digital Portable Radio Communications*, Proceedings of the IEEE, vol.75, no.4, April 1987.

THE INTERACTION OF SOURCE AND CHANNEL CODING IN WIRELESS AND PACKET VIDEO

Martin Vetterli, Masoud Khansari, and Steve McCanne

Dept. of Electrical Engineering and Computer Sciences
University of California at Berkeley
Berkeley, CA 94720

Abstract

Interaction between source coding algorithms and channel coding and transmission methods are discussed, in particular for time-varying channels as found in wireless and packet transmission, as well as in broadcast/multicast scenarios. The traditional separation between source and channel coding is not valid, and instead, a tighter coupling of the source coder with the transmission is used. Typically, this requires a layered source coder which interacts with a layered transmission system.

1. INTRODUCTION

A guiding principle in communication system design is the separation principle first established by Shannon [16]. It allows the design of a channel coder with transmission rate close to channel capacity, as well as a source coder realizing compression down to the transmission rate, and this independently of each other.

In this paper, we discuss cases where such a separation does not hold, and instead, a close interaction between source and channel coding is necessary. Two cases of particular interest are investigated, namely the transmission of video over time-varying channels like the wireless and packet channel, and the case of broadcast and multicast channels.

In the case of time-varying channels with a delay constraint, it is advantageous to operate a multimode coder that can adapt to channel conditions, since buffering will not be able to smooth long periods of bad channel conditions. Such a scheme is practical when information on the channel state is available at the encoder, like for example in two way wireless communications.

Several cases of informed transmitters and receivers can be considered, namely [7]:
- informed transmitter and receiver
- informed transmitter or informed receiver
while the classic scenario involves no information to either transmitter or receiver.

Broadcast channels do not necessarily allow for a separation of source and channel coding. In the case of broadcast or multicast channels, we discuss the use of layered coding to distribute the best possible quality to most users, as opposed to a single quality to a subset of users.

A particularly interesting application is multicast of real-time video over the internet. We will discuss a layered coder developed for real-time encoding and decoding of multiresolution video, as well as an associated layered multicast tree scheme for the internet [11].

2. SUCCESSIVE APPROXIMATION SOURCE CODING

Successive approximation coding, also called embedded, multiresolution or scalable coding, denotes a source compression scheme were a substream of bits leads to a coarse approximation of the signal [17, 18]. From a theoretical point of view, only a subset of sources allows such successive approximation coding without a penalty in the distortion-rate sense [3]. In practice, it was a subject of debate if efficient successive approximation source coding was achievable until some convincing examples were shown. In particular, in image compression, the most successful compression scheme happens to be also a successive approximation scheme [15].

In the following, we assume that an efficient successive approximation source coder is used. That is, we assume that a source has a distortion-rate function $D(R)$ and that we split the total rate R_T into $R_T = R_C + \Delta$ where R_C is the rate for the first approximation and Δ is the added rate for the full approximation. Then, we say that the successive approximation source coder is efficient if its operational distortion-rate function at R_C and after adding rate Δ are both "close" to $D(R_C)$ and $D(R_T)$, respectively. In practice, R_C and Δ might not be totally arbitrary, but sufficiently general to permit adequate successive approximation.

3. REACTIVE CODING, UEP AND RESOURCE ALLOCATION

In general, for time-varying channels where the statistics of the channel change with time, it is advantageous to adapt the source encoder parameters (e.g. encoding rate) to the current channel conditions. This is a departure from the classical open-loop design approach where the source encoder is designed based on the long-term statistics of the channel and is to perform the best over all the foreseeable situations. Unequal error protection (UEP) is an example of such open-loop approach where different segment of the transmitted information are protected differently by anticipating possible channel states and hence preventing from catastrophic loss of information.

If the source encoder is to react to the channel changes, it is necessary to provide it with the current status of the channel. In the interactive applications such as tele-conferencing this information can be piggy-backed on the reverse link. In this situation it is up to the receiver to estimate the current channel conditions. There are many different methods to achieve this goal. For example, the error rate of the received bits or frames can be used. Also, *soft* measures such as received SNR can be employed. Note that many of these channel estimation methods have already been employed, e.g. in the closed and open-loop power control algorithms for the IS-95 standard. Also retransmission based error recovery schemes such as Automatic Repeat Request (ARQ) can provide such channel information through monitoring the status of the transmitted packets. In [9], a video transmission system for micro-cellular environment is proposed where by monitoring the occupancy level of the ARQ buffer, the source rate is lowered during the fades. This can be done since during the fades the occupancy of the buffer increases due to the backlog of unacknowledged frames. It is shown that this method can significantly reduce the overall end-to-end distortion.

In reactive schemes the delay at which the channel side information is provided at the transmitter is of importance. Ideally, one would want for this delay to be as small as possible. This is however not the case in most practical situations. The tolerable delay is a function of both the rate at which the channel changes as well as the speed at which the source encoder can utilize and react to this information. The faster the rate of the change in the channel, the more stringent the delay requirement. For channels with long propagation delay such as satellite transmission, it is usually not possible to provide the channel side information at the transmitter in time and as a result closed-loop methods are of limited or no utility. In summary, when it is possible to effectively use channel side information (i.e. delay requirement is not violated) then significant improvement in the overall performance of the system can be expected. For example, in [13], it is shown that by incorporating channel side information in the buffer control algorithm the end-to-end distortion can be reduced.

An alternative to reactive methods are open-loop schemes such as UEP. Typically, in these methods the information stream is split into many streams based on the importance of these information in the reconstruction procedure at the receiver. One way to quantify the "importance" of each segment of information stream is to measure the amount of distortion caused by not or erroneously receiving that segment. One method of UEP is to use different channel coding for different streams and to protect more important information with more powerful codes. Alternatively, each stream can be transmitted with different power and hence resulting in different BER for different streams. Also, embedded modulation constellation can be used to embed the least significant into more important information [14].

In general the end-to-end distortion can be decomposed into two (not necessarily independent) components. First, distortion caused by compression at the source encoding stage, d_s, and secondly the distortion due to undetectable and uncorrectable channel errors, d_c. This latter component is in fact due to the constraint on the delay that can be tolerated at the channel encoding stage and $d_c = 0$ had we assumed no constraint on this delay.

Let us for the time being assume that there is no constraint on the channel delay and find the optimal power allocation for each stream. Let D_i be the end-to-end distortion of the ith stream and let P_i be the power allocated to that stream. We assume that the channel is an additive white Gaussian noise channel with noise power being N, moreover, that the streams are independent and therefore the overall distortion is the sum of each stream distortion or $d = \sum_i D_i$. We assume that the distortion-rate function for the ith stream is $D_i(R) = \epsilon_i \sigma_i^2 2^{-2R}$ where σ_i^2 is the energy of the ith stream. This distortion-rate function can closely approximate the performance of high-rate quantizers. We would like to find P_i's such that d is minimized subject to the constraint $1/M \sum_i P_i = \bar{P}$ where M is the number of streams and $\bar{P}$ is the average transmitted power. The solution to this convex optimization problem is:

$$P_i = \gamma_i \bar{P} + (\gamma_i - 1)N \tag{1}$$

where γ_i is defined as

$$\gamma_i = \frac{\sqrt{\epsilon_i \sigma_i^2}}{1/M \sum_{j=1}^{M} \sqrt{\epsilon_j \sigma_j^2}}. \tag{2}$$

Note that streams with $\gamma_i < 1$ (low energy streams) are penalized in favor of the high energy streams ($\gamma_i > 1$).

We define the coding gain G (in a similar fashion as is done in the rate allocation problem) as the ratio of the overall end-to-end distortion by setting $P_i = \bar{P}$ $\forall i$ to the distortion resulted using the optimal power allocation policy (1). Then

$$G = \frac{\dfrac{1}{M} \sum_{j=1}^{M} \epsilon_j \sigma_j^2}{\left(\dfrac{1}{M} \sum_{j=1}^{M} \sqrt{\epsilon_j \sigma_j^2} \right)^2}, \tag{3}$$

which is the ratio of the arithmetic of $\{\epsilon_j \sigma_j^2, j = 1, \cdots, M\}$ to the square of the arithmetic mean of $\{\sqrt{\epsilon_j \sigma_j^2}, j = 1, \cdots, M\}$ and as a result as one would expect $G \geq 1$. The equality holds if and only if $\epsilon_j \sigma_j^2$ is the same for all the bands. One can show that the KLT is still the best unitary transform to maximize this gain [8]. Now since the arithmetic mean of a positive number is lower bounded by their geometric mean,

$$G \leq \left(\frac{1}{M} \sum_{j=1}^{M} \epsilon_j \sigma_j^2 \right) / \left(\prod_{j=1}^{M} \epsilon_j \sigma_j^2 \right)^{1/M}. \tag{4}$$

In other words, fundamentally, if the channel encoding delay is of no concern and very complex channel encoders can be used, it is more advantageous to consider the information of all the layers as one stream and allocate all the power to maximize the transmission rate for this stream than unequal power allocation for different layers.

Let us now consider the general case where the channel encoding delay is finite. We define the distortion-power function by relating the minimum achievable end-to-end distortion (including both d_s and d_c components) to the power used to transmit this information stream. Clearly this is a non-increasing function. Moreover it is a convex function, since if not, one can achieve better performance through time-sharing of two operating points. Let $D_i(P)$ be the distortion-power function associated with the ith stream, then the solution to the above optimization problem $\{P_l^*\}$ is:

$$\begin{cases} \frac{\partial d}{\partial P_l}\big|_{P_l=P_l^*} = \theta & \text{if } P_l^* > 0 \\[2mm] \frac{\partial d}{\partial P_l}\big|_{P_l=P_l^*} < \theta & \text{if } P_l^* = 0 \end{cases} \tag{5}$$

for some $\theta < 0$. The above policy is also known as *constant slope policy*. Therefore, the main problem becomes that of calculating $D(P)$ for each stream.

We assume that the distortion introduced by channel errors is independent of the distortion due to the source compression and these two components are additive, or $d = d_s + d_c$. We can now define distortion-rate function $\bar{D}(R)$ for each stream as: $D(\bar{R}) = p_e D(0) + (1 - p_e)D(R)$ where R is the rate at which the stream is being compressed and p_e is the probability of the bit error. The problems with this approach are two-fold; first it is necessary to define bit-level distortion and secondly it is not clear what R should be chosen.

Effective rate (R_e), after capacity, is probably the most important parameter in characterizing a communication channel. This is the rate that sets an upper bound on the acceptable performance of the convolution coders. Also, the exponent of the bound on the probability of bit error is linear while operating at or below this rate. Therefore it seems to be a reasonable assumption to use R_e as the transmission rate for each stream. We now briefly show how this parameter is defined [12]. Let X and Y be the input and the output channel alphabets and $p(y|x)$ be the probability of receiving $y \in Y$ if $x \in X$ is being transmitted. Let us define $j(x_1, x_2)$ as

$$j(x_1, x_2) = \sum_{y \in Y} \sqrt{p(y|x_1)p(y|x_2)}. \tag{6}$$

The effective rate can then be defined as:

$$R_e = \max_{p(x)} \{ - \log_2 \mathbf{E}(j(x_1, x_2)) \} \tag{7}$$

where the expectation operation is taken over x_1 and x_2 which are independent copies of the random variable X having probability density $p(x)$. For example for the binary symmetric channel with cross-over probability ϵ, we have

$$R_e = - \log_2 \left(\frac{1}{2} + \sqrt{\epsilon(1 - \epsilon)} \right). \tag{8}$$

There are other definitions for the effective rate that can be shown to be equivalent to the above [4].

Since most practical channel encoders decode the received bit streams in a frame-based fashion, the primary figure of merit is the probability of frame error (p_E) instead of probability of bit error. In other words, we propose to consider the information stream as a frame-based rather than a bit-based stream. Therefore we define the distortion-rate function of each stream as:

$$\bar{D}(R_e) = p_E D(0) + (1 - p_E)D(R_e). \tag{9}$$

One has to note that both p_E and R_e are a function of the allocated power used to transmit that information stream. Also, it is reasonable to define frame-level distortion and assume that all the frames carry an equivalent amount of information. The same is not necessarily true for bits. The optimization problem is then to allocate power to each stream such that the overall distortion is minimized given the constraint on the average transmission power. The distortion of each stream is found using the above relation where the assumption is made that each stream is encoded at the effective rate set by the power allocated to that stream.

4. BROADCAST CHANNELS

We consider the broadcast channel and a source compressed with a successive approximation source coder. Such a scenario has been considered for digital broadcast of television in [14]. We show below that the choice of multiplexing rate is dependent on the distortion rate function when minimizing the combined distortion of two receivers. In the following, we assume a two receiver scenario with capacities C_1 and C_2 where $C1 > C_2$. We also consider multiplexing scenarios only, rather than superposition codes for simplicity.

First, note that if one wants simply to maximize the sum of rates $R_T = R_1 + R_2$, then the operating point is either $[C_1, C_2]$ when $C_2 < C_1 < 2C_2$ or $[C_1, 0]$ when $C_1 > 2C_2$ (when $C_1 = 2C_2$, then all multiplexings are optimal as well).

As to be expected, if one considers the sum of distortions (or a weigthed sum in general), then the optimal operating point changes depending on the distortion-rate function, showing a dependence between source and channel coding.

Proposition: Assume a source with a successive approximation distortion-rate function $D(R) = \alpha 2^{-\beta R}$. Consider a two receiver broadcast channel with capacities C_1 and C_2 where $C_2 = C$ and $C_1 = \gamma C$, $\gamma > 1$. The optimal multiplexing rate which minimizes the combined distortion of the two receivers, $D = D_1 + \lambda D_2$, $\lambda > 0$, is in general dependent on β, that is, dependent on the distortion-rate function.

Proof: The joint rates $[R_1, R_2]$ which are possible with multiplexing cover a quadrilateral surface with corners $\{[0,0], [\gamma C, 0], [0, C], [C, C]\}$ [2]. Only the multiplexing along the line $[C, C] - [\gamma C, 0]$ is of interest, since the optimal point must lie on it (other points on the surface lead to higher distortion). Thus we consider the following rates

$$R_2(R_1) = \frac{1}{\gamma - 1}(\gamma C - R_1), \quad R_1 \in [C, \gamma C]. \tag{10}$$

The distortions are

$$D_1(R_1) = \alpha 2^{-\beta R_1} \tag{11}$$

$$D_2(R_1) = \alpha 2^{-\frac{\beta}{\gamma-1}(\gamma C - R_1)} \tag{12}$$

and the combined distortion $D_T(R_1) = D_1(R_1) + \lambda D_2(R_1)$ has derivative

$$\frac{\partial D_T(R_1)}{\partial R_1} = \alpha\beta \ln 2(-2^{-\beta R_1} + \frac{\lambda}{\gamma - 1}2^{-\frac{\beta}{\gamma-1}(\gamma C - R_1)}). \tag{13}$$

Setting it to zero leads to

$$2^{-\beta\gamma R_1/(\gamma-1)} = \frac{\lambda}{\gamma-1} 2^{-\beta\gamma C/(\gamma-1)}. \tag{14}$$

This leads to

$$\hat{R}_1 = C + \frac{\gamma-1}{\beta\gamma} \log_2 \frac{\gamma-1}{\lambda}. \tag{15}$$

Then, the best choice for R_1, R_{opt} is $\hat{R}_1$ if $\hat{R}_1 \in [C, \gamma C]$. If $\hat{R}_1 < C$ then $R_{opt} = C$ (this is the case for example if $\gamma - 1 < \lambda$). Finally, if $\hat{R}_1 \geq \gamma C$ then $R_{opt} = \gamma C$. In general, (15) shows that R_{opt} depends on β, that is, the distortion-rate function.

Note that we did not consider the case of embedded codes for transmission [2], or more general convex distortion rate functions, but we conjecture that the result will still hold.

The gist of the above result is that successive approximation source coding and appropriate transmission allocation is useful in broadcast and multicast situations.

5. LAYERED CODING FOR MULTICAST OVER THE INTERNET

One example of a layered source coder combined with a layered transmission system is the scheme for multicast video transmission over the Internet described in [11]. Current Internet video transmissions are carried out either by transmitting single-layer video "open-loop" in the hope that the best-effort IP network can support the traffic [6], or transmitted using source-based adaptive control where receiver feedback is used to throttle the source transmission rate [1].

Unfortunately, both of these schemes are not well-matched to the heterogeneity inherent in the Internet. By using a single-layer compression algorithm, a uniform quality of video is transmitted to all users even those situated along high bandwidth paths from the source. In the open-loop algorithm, low bandwidth portions of the multicast distribution are flooded resulting in poor quality delivered to a possibly large fraction of the users. On the other hand, the source-based control scheme prevents congestion but forces the system to run at the most constrained rate, resulting in unnecessarily low quality delivered to high bandwidth users.

Our solution to the heterogeneity problem is to utilize a layered source-coder in tandem with a layered transmission scheme. The layered source-coder produces an embedded bit stream that can be decomposed into arbitrary number of hierarchical flows. These flows are distributed across multiple multicast groups, allowing receivers to "tune in" to some subset of the flows. Each layer provides a progressively higher quality signal.

We utilize only mechanisms that presently exist in the Internet and leverage off a mechanism called "pruning" in IP Multicast. IP Multicast works by constructing a simplex distribution tree from each source subnet. A source transmits a packet with an "IP group address" destination and the multicast routers forward a copy of this packet along each link in the distribution tree. When a destination subnet has no active receivers for a certain group address, the last-hop router sends a "prune" message back up the distribution tree, which prevents intermediate routers from forwarding the unnecessary traffic. Thus, each user can locally adapt to network capacity by adjusting the number of multicast groups — i.e., the number of compression layers — that they receive.

Our source-coder is a simple, low-complexity, wavelet-based algorithm that can be run in software on standard workstations and PC's. Conceptually, the algorithm works by conditionally replenishing wavelet transform coefficients of each frame. By using a wavelet decomposition with fairly short basis functions, we can optimize the coder by carrying out conditional replenishment in the pixel domain and then transforming and coding only those blocks that need to be updated. The wavelet coefficients are bit-plane coded using a representation similar to the well-known zero-tree decomposition [15]. All zero-tree sets are computed in parallel with a single bottom up traversal of the coefficient quad-tree, and all layers of the bit-stream are computed in parallel using a table-driven approach.

Our coder has been prototyped in version 2.7 of the UCB/LBNL video conferencing tool, *vic* [10]. The compression and run-time performance of the current prototype are comparable to the "Intra-H.261" scheme that vic presently uses. We are in the process of tuning both aspects of performance and expect to deploy the system in the next release of vic, as a favorable alternative to current methods of Internet video transmission.

6. CONCLUSION

We have seen several scenarios where a close interaction between source and channel coding is useful. Some of these scenarios, like UEP's, are already in use (e.g. CELP coded speech on wireless channels), while others are part of current prototype systems. Yet, a systematic study of the interaction of source and channel coding is only starting to appear. The potential benefits, in particular for highly time-varying channels and broadcast/multicast scenarios, are very interesting.

References

[1] J.-C. Bolot, T. Turletti I. Wakeman, "Scalable Feedback Control for Multicast Video Distribution in the Internet," In Proceedings of SIGCOMM '94, Sep. 1994, University College London, London, U.K.

[2] T.Cover and J.Thomas, Elements of Information Theory, Wiley, New York, 1991.

[3] W. H. Equitz and T. M. Cover, "Successive refinement of information," IEEE Tran. on IT, Vol. 37, No. 2, March 1991, pp 269-275.

[4] R.G.Gallager, Information Theory and Reliable Communications, Wiley, New York, 1968.

[5] M.W.Garrett and M.Vetterli, "Joint source/channel coding of statistically multiplexed real-time services on packet networks," IEEE/ACM Trans. on Networking,Vol. 1, No. 1, Feb. 1993, pp.71-80.

[6] R. Frederick, "Experiences with real-time software video compression," Proceedings of the Sixth International Workshop on Packet Video, Sep. 1994, Portland, OR.

[7] M.Khansari and M.Vetterli, "Transmission of sources with fidelity criteria over time-varying channels with side-information," submitted to IEEE Trans. on Communications, Aug. 1995.

[8] M. Khansari and M.Vetterli, "Layered transmission of signals over power-constrained wireless channels," to appear, IEEE ICIP-95.

[9] M.Khansari, A.Jalali, E.Dubois and P.Mermelstein, "Low bit-rate video transmission over fading channels," Accepted for publication in IEEE Trans. on Circuit and Systems for Video Technology.

[10] S. McCanne and V. Jacobson, "vic: A Flexible Framework for Packet Video," In Proceedings of ACM Multimedia '95, Nov. 1995, San Francisco, CA.

[11] S.McCanne and M.Vetterli, "Joint source/channel coding for multicast packet video," to appear, IEEE ICIP-95.

[12] R.J.McEliece and W.E.Stark, "Channels with block interference," IEEE Trans. on Info. Theorey, Vol. IT-30, No. 1, Jan. 1983, pp.44-53.

[13] A.Ortega and M.Khansari, " Rate control for video coding over variable bit rate channels with applications to wireless transmission," to appear, IEEE ICIP-95.

[14] K.Ramchandran, A.Ortega, K.M.Uz and M.Vetterli, "Multiresolution broadcast for digital HDTV using joint source-channel coding," IEEE JSAC, Special issue on High Definition Television and Digital Video Communications, Vol. 11, No. 1, Jan. 1993, pp.6-23.

[15] J. M. Shapiro, "Embedded Image Coding Using Zerotrees of Wavelet Coefficients", IEEE Tran. on SP, Vol. 41, No. 12, Dec. 1993, pp. 3445-3462.

[16] C. E. Shannon, Communications in the presence of noise, Proc. of the IRE, 37, 10–21, January, 1949.

[17] M.Vetterli and K.M.Uz, "Multiresolution coding techniques for digital video: a review," Special Issue on Multidimensional Processing of Video Signals, Multidimensional Systems and Signal Processing, Kluwer Acad. Pub., No. 3, pp. 161-187, 1992.

[18] M.Vetterli and J.Kovacevic, **Wavelets and Subband Coding**, Prentice-Hall, Englewood Cliffs, NJ, 1995.

SCALABLE VIDEO CODING WITH MULTISCALE MOTION COMPENSATION AND UNEQUAL ERROR PROTECTION

Bernd Girod,[1] Uwe Horn,[1] and Ben Belzer[2]

[1] Telecommunications Institute
University of Erlangen-Nuremberg
Cauerstrasse 7/NT, D-91058 Erlangen, Germany
[2] UCLA Electrical Engineering Department
405 Hilgard Avenue, Los Angeles, CA 90024-1594

1   INTRODUCTION

Scalable video codecs should support variation in image resolution, partial decodability of the bit-stream and computation-limited coding and decoding. Scalability can also be utilized within wireless video systems or for digital broadcasting, where the variable bit error rate of the radio channel is a major obstacle for many conventional, non-scalable video compression schemes.

In this paper, we present some of our recent results obtained for a scalable video codec based on a spatiotemporal resolution pyramid combined with E_8-lattice vector quantization. We first introduce spatiotemporal pyramids and appropriate coding schemes. We discuss the problem of optimum bit-allocation and multiscale motion compensation. In the second part we present simulation results concerning coding performance, software-only decoding, and digital video broadcasting.

2   SPATIOTEMPORAL RESOLUTION PYRAMIDS

Encoding of spatial resolution pyramids can be done by predictive coding schemes as shown in Fig. 1(a) and (b). Without loss of generality we consider initially only two layer pyramids. Fig. 1(a) shows an open-loop coder whereas Fig. 1(b) shows a closed-loop coder which results from Fig. 1(a) by including noise feedback. Common to both approaches are filters for downsampling and interpolation. The decoder for both coder structures is identical (Fig. 1(c)). Critically sampled subband coding schemes[1, 2, 3] need carefully designed filters to achieve perfect reconstruction and aliasing cancellation. In pyramid decompositions, filters can be designed more freely according to aspects like subjective image quality and complexity. Another advantage is that multiscale motion compensation can be easily incorporated as described in section 2.2. Filters we us in our simulations have impulse responses [1/2 1/2] prior to downsampling and [1/4 3/4 3/4 1/4] after upsampling. For two-dimensional filtering they are applied separately in the horizontal and vertical direction. Both filters can be implemented without multiplications. Particularly, interpolation at the decoder is very inexpensive.

For encoding, the input signal is first filtered and then downsampled. The coarse resolution layer is quantized and transmitted. The open-loop coder (Fig. 1(a)) uses the coarse

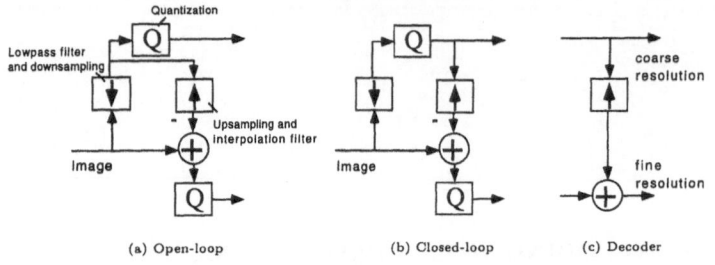

Figure 1: Two different pyramid encoders (a), (b) with corresponding decoder (c)

resolution signal *prior* to quantization to generate a prediction for the fine resolution layer. With noise-feedback (Fig. 1(b)) the signal *after* quantization is interpolated and used as prediction. In both cases the resulting prediction error is quantized and transmitted to the decoder where it it is used to reconstruct the fine resolution layer. The bitrate in each layer can be controlled independently by the corresponding quantizer. In the given example two bit-streams are transmitted. At a low bitrate only the coarse resolution layer is decoded. Decoding of the fine resolution layer needs a higher bitrate for additional transmission of the interpolation error.

2.1 Optimum Bit-Allocation

Given a certain amount of bits to be spent for one frame, how does one allocate these bits among the various layers? By using a Lagrange approach, it can be shown that the allocation is optimal if it meets the equal distortion-rate slope condition.[4, 5]

Bit-allocation found by this approach often leads to an unacceptably low quality in the low resolution layer. Quality in coarse layers can be improved only by leaving the overall optimum. As shown by Ramchandran et al.,[5] and also verified by our own experiments the closed-loop approach is less sensitive to suboptimal solutions. Additional constraints like comparable picture qualities in all layers can be added without noticeable penalty. Therefore, we focus on closed-loop coders for the remaining part of the paper.

2.2 Multiscale motion compensation

The predictive coders described above efficiently exploit redundancy in the spatial domain. Further improvement can be gained by applying motion compensation (MC) which also exploits redundancy in the temporal domain. For partial bit-stream decodability and efficient compression at all bitrates it is essential that MC is used within all spatial resolution layers which is commonly referred to as multiscale motion compensation (MSMC).

Motivated by results from a theoretical model for single layer motion compensation,[6] we investigate two different block-based MSMC approaches. For the first approach (Fig. 2) we add motion compensation loops to the spatial pyramid decomposition resulting from the closed-loop coder shown in Fig. 1(b). In pyramids with more than two layers it can be seen that, except for the lowest resolution layer, MC applies to interpolation error signals. Since those signals have bandpass character we refer to this method as bandpass compensation.

The second approach can be seen in Fig. 3. In contrast to bandpass compensation, MC is applied to the original frame and all of its lowpass filtered and downsampled versions. We refer to this method as lowpass compensation.

Both approaches can switch between an inter- and an intra-mode. In intra-mode where MC is not used both schemes work like the coder of Fig. 1(b). Selection of coding modes and estimation of motion vectors is explained later in this section.

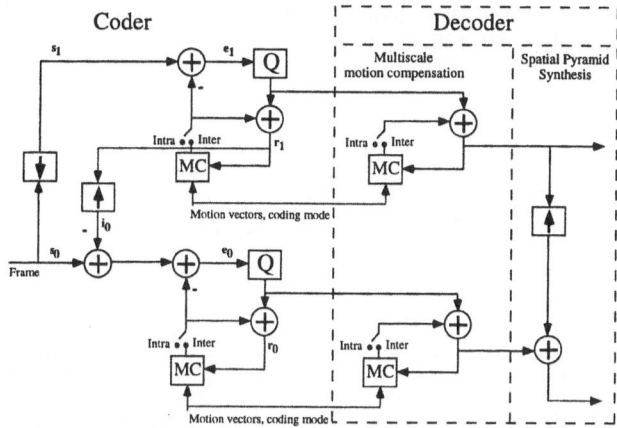

Figure 2: Two-layer pyramid codec with bandpass MC

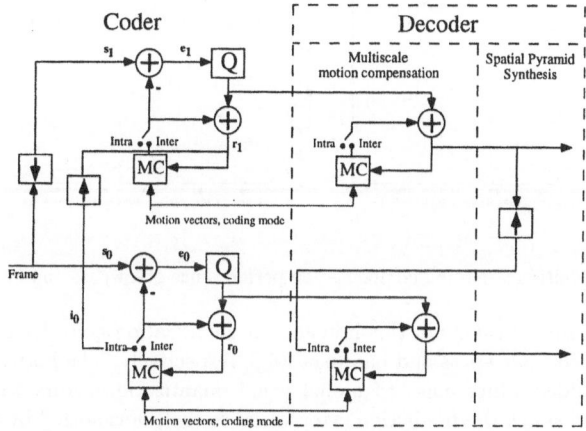

Figure 3: Two-layer pyramid codec with lowpass MC

It can be seen that the decoder of the approach shown in Fig. 3 is less complex. Interpolation from lower resolution is employed only in intra-mode whereas in Fig. 2 it is always needed, independent from the selected coding mode.

2.2.1 Comparison of lowpass and bandpass MC. To get a first insight into the performance of both MC schemes we compare the codecs corresponding to Figs. 2 and 3. We use frames generated by gradient based spatial interpolation[7] from odd fields of an interlaced test sequence. Image sizes are 704×480 for the fine resolution layer and 352×240 for the coarse layer. For block-based motion compensation blocksizes are set to 16×16 and 8×8, respectively. Motion vectors with half-pel accuracy are computed based on a mean-squared error (MSE) criterion independently for each layer. The same set of motion vectors is used for both MC schemes. Pixels at half-pel positions are obtained by bilinear interpolation. Coding mode is selected for each block based on a minimum MSE criterion.

Since for a two-layer pyramid motion compensation in the coarse layer is applied to identical signals in both approaches we can focus on the signal e_0 to be quantized and coded in the fine resolution layer. In intra-mode e_0 depends in both approaches on the signals s_0 and i_0 where i_0 furthermore depends on r_1. Therefore e_0 contains quantization noise introduced by both quantizers. A different situation results from inter-mode. With lowpass MC e_0 only depends on s_0 whereas with bandpass MC it still depends on s_0 and i_0. One would expect that lowpass MC is less sensitive to quantization noise in the lower resolution layer since this noise influences e_0 only in intra-mode.

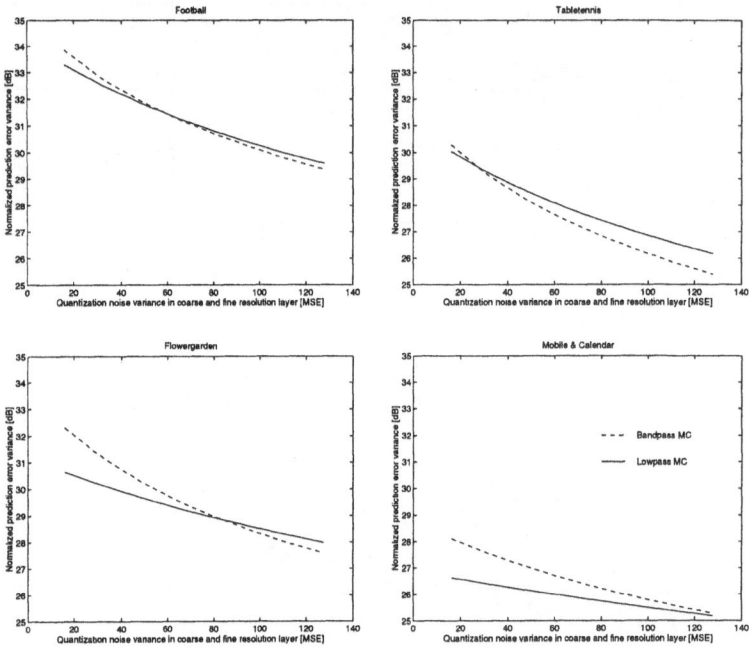

Figure 4: Multiscale motion compensation performance of lowpass and bandpass MC

Fig. 4 yields some insights in this influence of quantization noise for a two-layer closed-loop codec extended by lowpass and bandpass MC, respectively. The horizontal axis denotes the variance of added white noise to model equal quantization errors within both layers. The vertical axis denotes the prediction error variance (e_0), normalized by the square of the amplitude range $0 \ldots 255$ expressed in dB. It can be seen that for higher noise variances lowpass MC results in a lower prediction error variance. That means in terms of bitrate that for low bitrate applications the less complex lowpass compensation scheme is appropriate whereas at higher bitrates it is advantageous to apply bandpass compensation.

2.2.2   Motion estimation techniques.   Performance of motion compensation depends on how motion vectors for each resolution layer are estimated. Three different methods are considered. The first approach estimates motion vectors only for the fine resolution layer. Motion vectors for coarser resolution layers are obtained by appropriate downscaling. The second approach estimates motion vectors for each layer independently. As a third and most natural technique for resolution pyramids hierarchical estimation can be applied.[8, 9] Besides quality of the motion vector field it is important to consider how efficiently motion vectors can be encoded. Simulations show that predictive coding using upscaled motion vectors from the lower resolution layer as a predictor gives better results than using the preceding motion vectors of the same resolution layer.

As expected, best compensation results are achieved with independent estimation which has the highest complexity. Hierarchical estimation performs nearly as well as the latter at a much lower complexity. Furthermore, motion vectors obtained by hierarchical estimation can be encoded with the smallest amount of bits. In terms of compensation performance, motion vector encoding costs, and encoder complexity, hierarchical estimation offers the best compromise.

2.3 E_8-Lattice Vector Quantization

Vector quantizers are known to have a low decoder complexity since decoding is carried out by a simple table-lookup. Discussions of VQ including codebook training by the LBG algorithm and entropy-constrained VQ can be found in various publications.[10, 11, 12, 13] However, for higher bitrates, the computational burden of an unstructured codebook search at the coder becomes prohibitive. This search can be avoided by lattice VQ where a highly structured codebook is used.[11, 14] With lattice VQ, the representative for a given input vector can be computed with much less effort compared to unstructured codebook search.

It is reported that for image subbands entropy-coded E_8-lattice VQ performs as well as unstructured entropy-constrained VQ even at bitrates below 0.5 bits/sample.[15] Another interesting property of the E_8-lattice is its correspondence to the densest sphere packing for eight dimensions.[16] Furthermore fast quantization and decoding algorithms can be used.[17, 18] We improved lattice VQ performance by using a hybrid quantization approach where a codebook of feasible size is used for the 'most popular' code vectors on a small region of the lattice.[15] Instead of using the lattice point as the representative a trained codebook vector is used. Less popular code vectors are quantized to nearest lattice points and coded by a Voronoi code.[18]

3   SIMULATION RESULTS

Fig. 5 shows the spatiotemporal resolution pyramid we used as basis for the applications described in the following sections.[19] As can be seen, layers 2 and 1 roughly correspond to QCIF and CIF resolution, respectively. The highest resolution layer contains an interlaced video signal in its full spatial and temporal resolution according to the ITU-R 601-4 standard.[20] On the right spatial and temporal resolutions of all layers are listed. Typical bit rates with increasing spatiotemporal layer resolution are 64 kbit/s, 330 kbit/s, 2.0 Mbit/s and 4.7 Mbit/s.

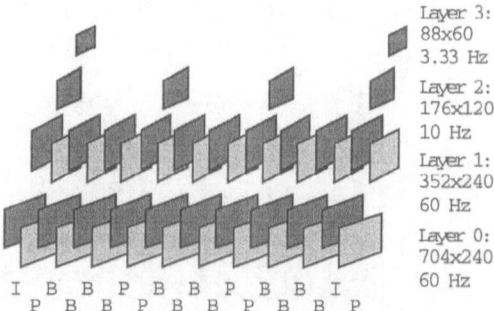

Layer 3:
88x60
3.33 Hz

Layer 2:
176x120
10 Hz

Layer 1:
352x240
60 Hz

Layer 0:
704x240
60 Hz

I   B   B   P   B   B   P   B   B   I
 P   B   B   P   B   B   P   B   B   P

Figure 5: Spatiotemporal pyramid setup

Similar to MPEG,[21, 22, 23] our coding scheme allows three different picture types namely I-, P- and B-pictures. I-pictures are coded without reference to other frames. P-pictures can take reference to frames in the past whereas B-pictures can reference frames in the past as well as in the future. Within B-pictures it is also possible to form a temporal prediction by bi-directional interpolation between a past and a future reference frame. The picture type always remains fixed through all spatial resolution layers. The group of pictures structure used for temporal prediction is shown at the bottom of Fig. 5.

Motion vectors for multiscale motion compensation are obtained by hierarchical motion estimation. Motion compensation is applied according to the closed-loop codec shown in Fig. 3 (extended to four layers). No automatic bitrate control is used. Quantizers are adjusted manually for the different picture types at the start of the sequence to yield comparable picture quality within each resolution layer.

3.1 Coding performance and software-only decoding

Fig. 6 shows the performance of the codec for four test sequences. For each sequence PSNR values for decoding at four different bitrates corresponding to the four resolution layers are shown. PSNR for the original image resolution was measured after interpolation of the decoded layer to the finest resolution layer by the [1/4 3/4 3/4 1/4] filter.

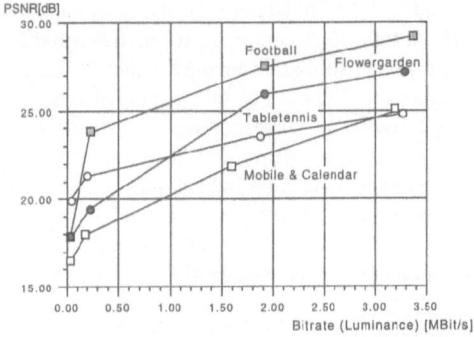

Figure 6: Performance of Codec

The not fully optimized software-only implementation of the decoder is able to decode and display layers 3 and 2 in real-time on a Sun SparcStation5 as well as on an SGI Indigo2 (Fig. 7). Decoding of the full resolution layer would need a ten times faster CPU. It can be seen that required CPU power is roughly proportional to the decoded bitrate.

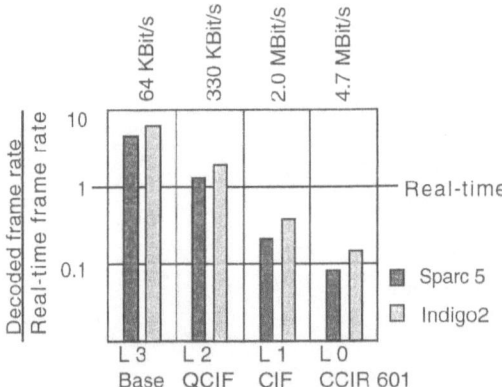

Figure 7: Real-time performance of the decoder

3.2 Digital TV broadcasting

To achieve graceful degradation in high bit error rate environments like digital TV broadcasting we combined the scalable source coder with an unequal error protection scheme (UEP) by using rate-compatible punctured convolutional codes (RCPC codes)[24] of different code rates. We first encoded 5 seconds of the Flowergarden test sequence with our scalable coder. According to the importance of the data, coderates as shown in Table 1 were chosen. The entry K/N means that K input bits are coded with N output bits. As more output bits are generated for one input bit, error protection increases.

We compare this unequal error protection scheme with an equal error protection (EEP) where all packets are protected with a coderate of 4/5. In both cases the increase in bit rate is 26% due to error protection. Simulation results for five different bit error rates on a binary

Picture type	Coderates				
	Header	Layer 3	Layer 2	Layer 1	Layer 0
I-Picture	1/2	1/2	4/7	2/3	1/1
P-Picture	2/3		2/3	4/5	1/1
B-Picture	2/3			2/3	1/1

Table 1: RCPC Coderates used for unequal error protection

symmetric channel (BSC) can be seen in Fig. 8.

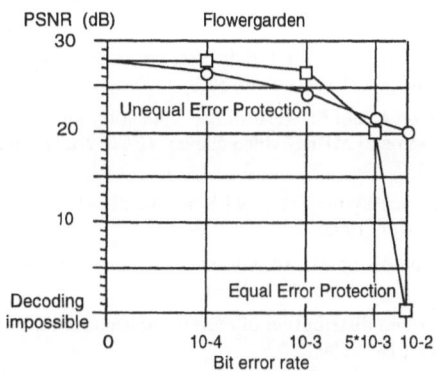

Figure 8: Comparison between UEP and EEP

Average PSNR over the whole sequence is compared to bit error rate. As expected unequal error protection performs better as bit error rate increases. At a bit error rate of 10^{-2} decoding of the equal error protected bit-stream becomes impossible while unequal error protection still results in a 20 dB PSNR.

The bit-stream at the output of the channel decoder may still contain corrupted bits which cannot easily be detected. Residual bit errors within a slice are detected by illegal code words or at synchronization marks. Error concealment replaces the whole slice by decoded information taken either from the lower resolution layer or, if in the base layer, from the corresponding slice of the preceding frame.

4   CONCLUSION

We presented a scalable video codec based on a spatiotemporal pyramid. Low decoder complexity is achieved by simple filters and E_8-lattice vector quantization. It was verified that closed-loop pyramid coders are less sensitive to sub-optimal bit allocations which makes them more suitable for real applications than open-loop coders.

Two different multiscale motion compensation approaches have been investigated. Interestingly, at lower bitrates the less complex lowpass compensation performs better than bandpass compensation. Hierarchical motion estimation offers the best compromise in terms of compensation gain, motion vector encoding costs, and encoder complexity.

Our experiments concerning computation limited decoding showed for the chosen simulation model that CPU power required for decoding is roughly proportional to bitrate. The digital TV broadcast scenario showed how scalability can be efficiently combined with unequal error protection to yield graceful degradation over channels with varying quality.

5 ACKNOWLEDGEMENT

The work of Niko Färber in implementing the RCPC encoding and decoding routines for the digital TV broadcasting simulations is gratefully acknowledged.

REFERENCES

1. A.N. Akansu and R.A. Haddad. *Multiresolution Signal Decomposition*. Academic Press, San Diego, 1992.

2. J.W. Woods (ed.). *Subband Image Coding*. Kluwer Academic Publishers, Boston, 1991.

3. J. Katto and Y. Yasuda. Performance evaluation of subband coding and optimization of its filter coefficients. *Journal of Visual Communication and Image Representation*, 2(4):303–313, December 1991.

4. Y. Shoam and Allen Gersho. Efficient bit allocation for an arbitrary set of quantizers. *IEEE Trans. on Acoustics, Speech, and Signal Processing*, 36(9):1445–1453, Sep. 1988.

5. K. Ramchandran, A. Ortega, and M. Vetterli. Bit allocation for dependent quantization with applications to multiresolution and MPEG video coders. *IEEE Trans. on Signal Processing*, 3(5):533–545, Sep. 1994.

6. B. Girod. Motion-compensating prediction with fractional-pel accuracy. *IEEE Trans. on Communications*, 41(4):604–612, Apr. 1993.

7. K. Jensen and D. Anastassiou. Spatial resolution enhancement of images using nonlinear interpolation. In *Proc. ICASSP 90*, pages 2045–2048. IEEE, 1990.

8. B. Chupeau. Estimation and distribution of motion information. *Signal Processing: Image Communication*, 5(5-6):539–552, Dec. 1993.

9. T. Hanamura, W. Kemeyama, and Tominaga H. Hierarchical coding scheme of video signal with scalability and compatibility. *Signal Processing: Image Communication*, 5(1-2):159–184, Feb. 1993.

10. P.A. Chou, T. Lookabaugh, and R.M. Gray. Entropy-constrained vector quantization. *IEEE Trans. on Acoustics, Speech, and Signal Processing*, 37(1):31–42, Jan. 1989.

11. A. Gersho and R. Gray. *Vector Quantization and Signal Compression*. Kluwer, 1991.

12. Y. Linde, A. Buzo, and R.M. Gray. An algorithm for vector quantizer design. *IEEE Trans. on Communications*, COM-28(1):84–95, Jan. 1980.

13. N.M. Nasrabadi and R.A. King. Image coding using vector quantization: A review. *IEEE Trans. on Communications*, 36(8):957–971, Aug. 1988.

14. D.G. Jeong and J.D. Gibson. Lattice vector quantization for image coding. In *Proc. ICASSP'89*. IEEE, 1989.

15. T. Senoo and B. Girod. Vector quantization for entropy coding of image subbands. *IEEE Trans. on Image Processing*, 1(4):526–532, Oct. 1992.

16. J.H. Conway and N.J.A. Sloane. *Sphere Packings, Lattices and Groups*. Springer, 1988.

17. J.H. Conway and N.J.A. Sloane. Fast quantizing and decoding algorithms for lattice quantizers and codes. *IEEE Trans. on Information Theory*, IT-28(2):227–232, Marv. 1982.

18. J.H. Conway and N.J.A. Sloane. A fast encoding method for lattice codes and quantizers. *IEEE Trans. on Information Theory*, IT-29(6):820–824, Nov. 1983.

19. U. Horn and B. Girod. Pyramid coding using lattice vector quantization for scalable video applications. In *Proc. PCS '94*, Sacramento, CA, 1994.

20. ITU-R. Encoding parameters of digital television for studios. Recommendation 601-4, 1982.

21. ISO/IEC. Document 13818-2, Generic Coding of Moving Pictures and assciated Audio, Part 2: Video, Recommendation H.262, Draft International Standard, März 1994.

22. Didier J. LeGall. MPEG: A video compression standard for multimedia applications. *Comm. ACM*, (34):46–58, 1991.

23. Didier J. LeGall. The MPEG video compression algorithm. *Signal Processing: Image Communication*, 4(2):129–140, Apr. 1992.

24. J. Hagenauer. Rate-compatible punctured convolutional codes (RCPC codes) and their applications. *IEEE Trans. on Communications*, COM-36:389–400, Apr. 1988.

A GENERALIZED FRAMEWORK FOR SCALABLE VIDEO CODING

Susie J. Wee, Michael O. Polley, William F. Schreiber

Department of Electrical Engineering and Computer Science
Research Laboratory of Electronics
Massachusetts Institute of Technology
Cambridge, Massachusetts 02139
swee@image.mit.edu

INTRODUCTION

Conventional scalable video coders are typically based on single-quality video coding techniques. However, additional issues arise when designing a scalable system. For example, the design of a single-quality system is typically motivated by the desire to minimize the perceived distortion. However, in a recursive scalable system, the coded video is not only viewed, but is also used to form a prediction for the higher levels of video. Therefore, the coding artifacts introduced in the lower levels propagate to the higher levels, reducing the coding efficiency of the entire system.

In this paper, we introduce the concept of *conditioning*, which provides a solution to the artifact propagation problem in a scalable video coding system. In each level of a scalable video coder, conditioning controls artifact propagation into the immediate and subsequent levels of the coder, thereby improving the coding efficiency of the system. The video entering and leaving each level of a scalable system is conditioned to remove the artifacts which may affect the overall coding efficiency.

We first present a generalized framework for scalable video coding. The concept of conditioning is then described, and some simple examples are given to illustrate the effectiveness of conditioning in the context of spatially scalable image and video coding systems. Finally, concluding remarks are made and future directions are suggested.

SCALABLE VIDEO CODING

The scalable video coding problem may be stated as follows: What is the best way to encode original video into multiple data streams? In this work, we consider recursive scalable schemes in which the video is encoded into a base data stream and one or more enhancement data streams. At the decoder, the base data stream can be used to recover baseline-quality video; additional enhancement data streams can also be used to recover increased-quality video[1].

Many forms of scalability exist. For example, spatial, temporal, and SNR scalability provide variations in spatial, temporal, and amplitude resolution of the video. Object scalability involves delivering objects in the video in order of importance. While our results apply to many forms of scalability, we specifically address the problem of *spatial scalability*. In a spatially scalable system, the base data stream can be used to reconstruct low-resolution video, and additional enhancement data streams can be used to recover video of increased spatial resolution.

Much work on the topic of spatially scalable image and video coding has appeared in the literature. An overview of spatially scalable coding techniques can be found in [1]. The JPEG and MPEG-2 image and video compression standards incorporate various forms of spatial scalability [2, 3]. Many approaches to spatially scalable video coding have been proposed; a brief sampling of these approaches includes [4, 5, 6, 7, 8].

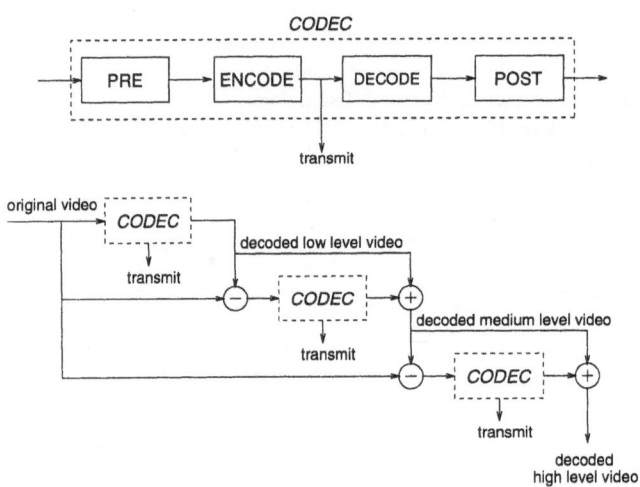

Figure 1: *A Generalized Framework for Scalable Video Coding.* This framework encompasses current approaches and accommodates an improved class of scalable video coders. A novel aspect lies in the use of pre- and post-conditioning within each level of the coder.

A GENERALIZED FRAMEWORK FOR SCALABLE VIDEO CODING

In this section, we introduce a generalized framework for scalable video coding. The framework encompasses conventional approaches and accommodates an improved class of coders. The improvement is due to the incorporation of pre- and post-conditioning within each level of the scalable video coder. The concept of conditioning is further discussed in the next section.

The generalized framework is shown in Figure 1. The overall architecture resembles a pyramid in that the video is coded in multiple levels. In the lowest level of the pyramid, baseline-quality video is coded into an appropriate data stream. In each higher level of the pyramid, an enhancement signal is coded into an appropriate data stream. Notice that the video reconstructed at each level is subtracted from the original to form the input signal of the next higher level. This can be thought of as using the lower-level reconstructed video as a prediction for the next level, which must then encode the resulting residual.

In each level of the scalable video coder, the signal is processed with a codec that consists of a pre-conditioner, encoder, decoder, and post-conditioner. The pre-conditioner extracts

---

[1]It is interesting to note that since the coded data streams are prioritized according to importance, scalable video coding inherently supports a form of joint source/channel coding.

the signal to be coded in the particular level. In addition, it conditions the signal so that it is easier to code. The coder encodes the signal into an appropriate data stream that meets the requirements of the system. The decoder processes the data stream to reconstruct the same signal decoded at the receiver. Finally, the post-conditioner processes the signal to form an appropriate prediction of the original video. In doing so, it should condition the video so that the input signals in the higher levels of the pyramid are easier to code.

In addition, at each level the fully decoded video may also be conditioned. Note that this automatically occurs in the base level, but not in the higher levels where only the enhancement signal is conditioned. The opportunity to condition the completely-decoded video is highly advantageous. For example, the enhancement video reconstructed at each level may have characteristics which seem unusual when viewed alone, but seem normal when viewed in conjunction with the lower-level video. In some instances, it may be advantageous to condition the enhancement video, and in other instances it may be advantageous to condition the fully-decoded video. To simplify the presentation of this material, we will not discuss this topic further. Also note that we have not included this additional conditioning in Figure 1.

CONDITIONING

Scalable systems are often based on well-understood, single-quality video coding techniques. Further consideration should be given to issues unique to scalable video coding, thereby potentially leading to significant improvements in performance. One such issue is the propagation of artifacts among the various scalable levels. In this section, we introduce the concept of conditioning, which can lead to a number of benefits in a scalable system, including control of the propagation of artifacts. We describe the motivation behind conditioning and, in the next section, we illustrate its usefulness by applying it in two conventional scalable systems.

Maximum performance in a scalable system requires the best possible performance in each level of the coder. This is not a trivial task. In a scalable system, the output of the lower levels are both viewed and used to predict the higher-level signals. Intuitively, to achieve maximum efficiency, a good prediction must be formed for each higher level. More specifically, the reconstructed lower-level video will typically be afflicted with coding artifacts that are not true features of the original video. If these artifacts are not specifically addressed, the prediction formed for the next level will also contain these artifacts. The coder in the next level would therefore have to expend capacity to cancel out these artifacts. This effect can occur at each level of the pyramid, leading to a significant reduction in performance. Improved performance can be achieved by removing these artifacts before they propagate to higher levels.

This motivates the concept of conditioning in each level of the scalable video coder. The main idea is to pre-condition the video before coding to make it easier to code, and to post-condition the video after coding to prevent the propagation of artifacts. The system goal is to efficiently allocate the available capacity to the elements that must be coded.

The function of the pre- and post-conditioners in a scalable video coder differs slightly from the function of the pre- and post-processors in a single-quality system. In single-quality systems, the immediate goal of the pre- and post-processors is to enhance the visual quality of the output video. In scalable systems, the immediate goal of the pre- and post-conditioners is to condition the video so that the resulting signals in the immediate and higher levels are easier to code, thereby achieving higher coding efficiency for the system and improved video quality at the receiver. In the higher levels, the pre- and post-conditioning are applied to the enhancement or residual signals at each level, as opposed to conventional pre- and post-processing which are applied to the original and fully reconstructed video.

CONDITIONING IN A SPATIALLY SCALABLE CODER

In this section, we use two examples to illustrate the potential benefits of conditioning. We consider two conventional spatially scalable coders, one for images and one for video. In both cases we employ a simple conditioning technique to improve the performance of the coders.

The two systems share the same basic two-level architecture. In the low level, the pre-conditioner extracts the low-resolution image/video by lowpass filtering and subsampling by a factor of $3/4 \times 3/4$ in the horizontal and vertical spatial dimensions. A coder compresses the image/video into a data stream. A decoder reconstructs the decoded image/video. The post-conditioner upsamples and filters this image/video to form a prediction of the full-resolution image/video. This prediction is subtracted from the original image/video, and the resulting residual is coded in the higher level.

Image Coder

In the first example, an intra-frame, 8×8 block-DCT coder is employed in each level of the spatially scalable system. A major shortcoming of this approach stems from the blocking artifacts that afflict the decoded low-resolution image. Since this image is upsampled to form a prediction for the high level, the low-level blocking artifacts propagate through the system. These artifacts must be cancelled to achieve an acceptable high-resolution output image; this cancellation requires the use of extra channel capacity.

In typical spatially scalable systems, the resolution increments are restricted to octaves, or factors of two. In this case, the DCT block boundaries of the upsampled, low-resolution image line up with the DCT block boundaries in the high-resolution codec. However, factor of two resolution increments are highly restrictive for scalable video coders. Therefore, we investigate the use of non-octave subsampling ratios.

The $3/4 \times 3/4$ (non-octave) subsampling ratio causes additional difficulties because the block boundaries of the two levels do not coincide. Block discontinuities that appear in the upsampled decoded low-resolution image cause similar discontinuities in the high-level residual. Because they do not coincide with the high-level block boundaries, these discontinuities throw excess energy into the high-level DCT coefficients, making the system less efficient. This problem can be alleviated with conditioning. Specifically, by post-conditioning the decoded low-resolution image to reduce the blocking artifacts, a smoother prediction can be made for the high-level coder. This reduces the extraneous discontinuities in the residual which must be encoded in the higher level.

The 512×512 Lena image was processed with the conventional scalable system. The reconstructed high-resolution image had a PSNR of 34.15 dB. Conditioning was applied to the decoded low-resolution image by adaptively lowpass filtering the boundary pixels across the block boundary [9]. A simple 3-tap lowpass filter was used, and only those boundary pixels whose amplitude differences were below a set threshold were filtered. The resulting reconstructed high-resolution image had a PSNR of 34.70 dB, achieving an increase of 0.55 dB. The post-conditioned reconstructed image achieved a considerable improvement in visual quality.

Video Coder

In the second example, we consider a two-level scalable video coder which contains an MPEG coder in each level of the pyramid. Once again, coding artifacts in the coded low-resolution video propagate to the higher level, causing extraneous energy in the residual which must be cancelled in the higher level, thereby decreasing the overall efficiency of the system. Also, because of the $3/4 \times 3/4$ subsampling ratio, the low-level MPEG block boundaries do not coincide with the high-level MPEG block boundaries, causing further difficulty in coding.

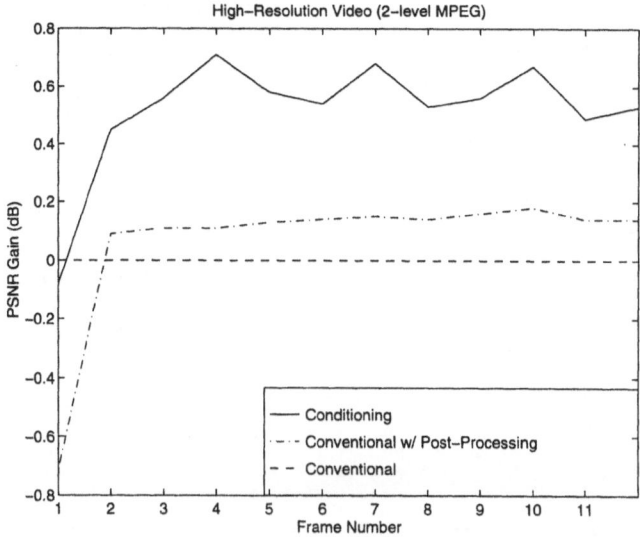

Figure 2: *Conventional approaches vs. conditioning.*

A conventional approach to improving this system would be to post-process the fully decoded video prior to display. Our proposed approach involves post-conditioning the video in each level of the scalable video coder, prior to reconstructing the fully decoded video. In this experiment, we compare the performance of the two approaches to the performance of the conventional system without post-processing and without post-conditioning.

For illustrative purposes, a simple method of reducing blocking artifacts was used for the post-processing and post-conditioning operations. The method involves adaptively filtering the pixels located on the 8×8 block boundary. If the amplitude difference across the block boundary exceeds a predetermined threshold, a simple 3-tap filter (.2,.5,.3) was applied across the block boundary. Neither the filter coefficients nor the threshold was optimized for the particular video sequence.

The HDTV Kodak balloon sequence was processed with three approaches to scalable video coding: the conventional approach, the conventional approach with post-processing, and the proposed approach with post-conditioning in each level of the coder. In all cases, the video was compressed to a total bit rate of .35 bits/pixel, of which 9/16 of the data was allotted to the low-level data stream and 7/16 of the data was allotted to the high-level data stream. The group of pictures contained 12 frames, with 1 I-frame, 3 P-frames, and 8 B-frames.

The improvement in PSNR for the reconstructed high-resolution video is shown for one group of pictures in Figure 2. Notice that with this simple filtering technique, neither approach improves the performance in coding the I-frame. This is partially due to the very high quality achieved on the I-frame by the MPEG coder and the very simple conditioning applied. In the remaining frames, the conventional post-processing technique offers about a .1 *dB* improvement in PSNR, while the proposed post-conditioning technique offers a .5 to .7 *dB* improvement in PSNR. Most importantly, a substantial improvement is made in the visual quality of the reconstructed high-resolution video. A 200×150 pixel portion of the 4th frame of each reconstructed high-resolution video sequence is enlarged and shown in Figure 4. Notice that the blocking artifacts are virtually eliminated with the proposed approach.

A further experiment was performed to better understand the source of improvement achieved by conditioning. In Figure 3, the gain over conventional coding is compared to using post-conditioning only in the low level, only in the high level, and both in the low and high

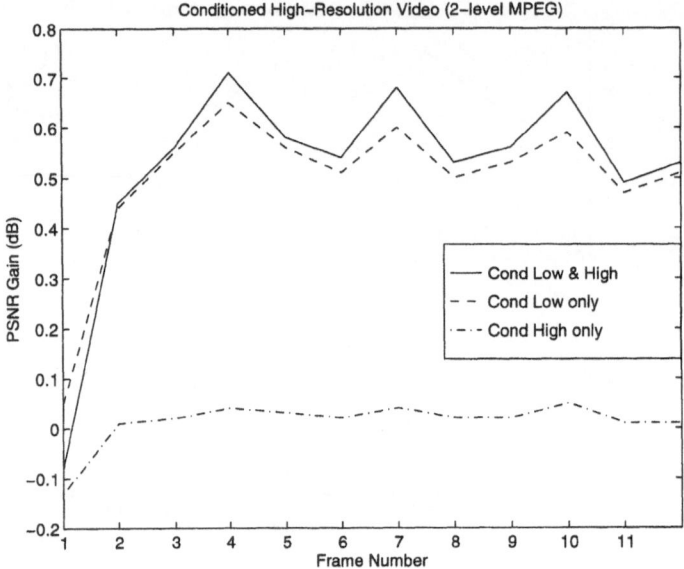

Figure 3: *Effectiveness of conditioning.*

levels. This plot illustrates that the majority of the gain results from post-conditioning the low-level video. This can partially be attributed to the fact that most of the signal energy is in the low-resolution video, and also because the identification and reduction of artifacts is better understood for the low-resolution video. As we improve our understanding of the nature of artifacts in the higher levels and/or apply more sophisticated conditioning techniques, higher gain should be achievable in the higher levels of the system. Furthermore, it is very interesting to note that the improvement in the post-conditioned low-resolution video is negligible, only about .01 *dB*. Despite this negligible quantitative gain in the low resolution, it leads to a gain of about .5 *dB* in the high resolution.

REMARKS and FUTURE WORK

These examples illustrate that improved performance can be achieved with little added complexity by incorporating conditioning in a scalable video coding system. Very simple conditioning examples were purposely chosen to clearly demonstrate the effectiveness of conditioning. Greater improvements may be achieved by employing more sophisticated conditioning techniques, such as [10]. Further research is being performed along these lines.

DCT-based coders were used because of their prevalence in current video compression schemes and because their blocking artifacts are easy to visualize. However, it should be noted that the effectiveness of conditioning is not unique to DCT coders. All lossy coding techniques suffer from coding artifacts, and all recursive scalable coders suffer from artifact propagation. By better understanding the nature of these artifacts, conditioning can be used to increase the efficiency of such scalable systems.

The effectiveness of conditioning in multiple-level systems is also being examined. Preliminary results with a three-level spatially scalable system indicate that conditioning in the upper levels is effective, but the majority of the gain results from proper conditioning of the lowest-level video. It should be noted that the enhancement video encoded in the higher levels of the pyramid are fundamentally different in nature from the video encoded in the lowest level. A better understanding of the nature of the residual video and its coding artifacts

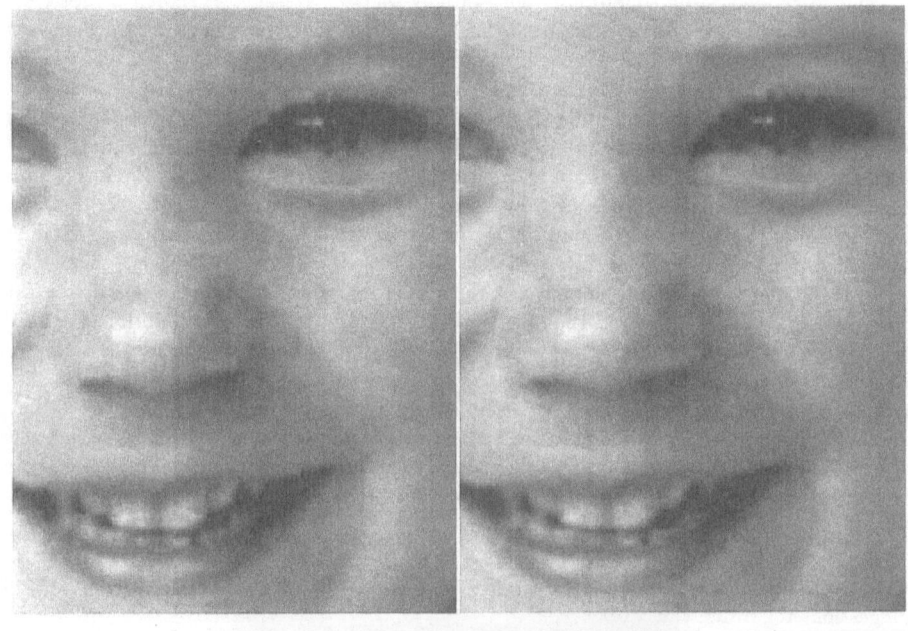

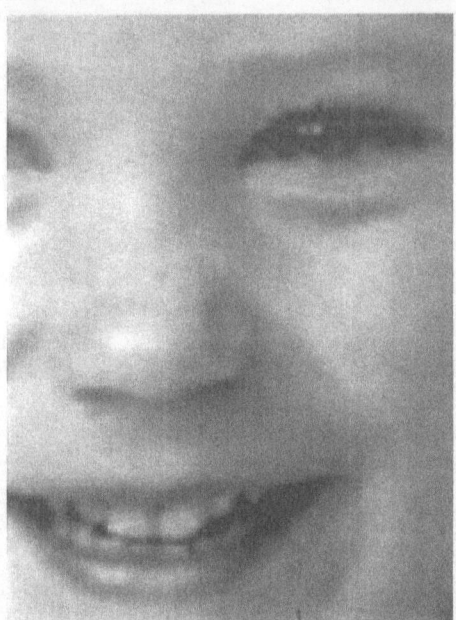

Figure 4: *Performance comparison.* A video sequence was coded in a 2-level spatially scalable video coder. An MPEG coder was used in each level. An enlarged view of the fourth frame of the reconstructed high-resolution video is shown for the conventional approach (37.22 dB, upper left), conventional approach with post-processing (37.33 dB, upper right), and proposed approach featuring conditioning (37.93 dB, bottom).

may lead to improved performance in high-level conditioning and improved overall system performance.

This work focused on conditioning for spatially scalable systems. It should be noted that the same concepts apply to other types of scalability. For example, temporal, frequency, and SNR scalable systems could also benefit from conditioning. One goal in each of these cases is to prevent expending capacity on the cancellation of lower-level artifacts.

Finally, further work should be performed in investigating the benefits of conditioning via image enhancement and restoration techniques to form better predictions in higher levels of the coder. For example, major improvement can be achieved by enhancing the resolution of the decoded low-resolution video, thereby making a better prediction and reducing the energy in the residual that must be coded in the higher levels. One such scheme employing nonlinear interpolation was proposed by Anastassiou [11].

In this paper, we pointed out that additional issues arise when designing a scalable video compression system as opposed to a single-quality system; in particular, we focused on the issue of artifact propagation among the scalable levels. We presented a generalized framework for scalable video coding, which encompasses conventional approaches and accommodates an improved class of coders via the novel feature of conditioning. We showed how very simple post-conditioning techniques can alleviate the artifact propagation problem and lead to sizable improvements both in PSNR and visual quality. We are currently investigating more sophisticated pre- and post- conditioning approaches, which may lead to further improvements in performance.

Acknowledgements The authors wish to thank John Apostolopoulos for his comments and suggestions regarding this work.

REFERENCES

[1] M. Vetterli and K. Uz, "Multiresolution coding techniques for digital television: A review," in *Multidimensional Systems and Signal Processing*, vol. 3, pp. 161–187, Kluwer Academic Publishers, 1992.

[2] G. Wallace, "The JPEG still picture compression standard," in *Communications of the ACM*, April 1991.

[3] "MPEG-2 video compression standard."

[4] P. Burt and E. Adelson, "The Laplacian pyramid as a compact image code," *IEEE Transactions on Communications*, vol. COM-31, pp. 532–540, April 1983.

[5] K. Uz, M. Vetterli, and D. LeGall, "Interpolative multiresolution coding of advanced television with compatible subchannels," *IEEE Transactions on Circuits and Systems for Video Technology*, vol. 1, March 1991.

[6] T. Naveen and J. Woods, "Motion compensated multiresolution transmission of high definition video," *IEEE Transactions on Circuits and Systems for Video Technology*, vol. 4, pp. 29–41, February 1994.

[7] D. Taubman and A. Zakhor, "Multi-rate 3-d subband coding of video," *IEEE Transactions on Image Processing*, vol. 3, pp. 572–588, September 1994.

[8] S. Wee, M. Polley, and W. Schreiber, "A scalable source coder for a hybrid HDTV terrestrial broadcasting system," in *Proceedings of the IEEE International Conference on Image Processing*, vol. 1, pp. 238–242, November 1994.

[9] H. Reeve III and J. Lim, "Reduction of blocking effects in image coding," *Journal of Optical Engineering*, vol. 23, pp. 34–37, January/February 1984.

[10] Y. Yang, N. Galatsanos, and A. Katsaggelos, "Regularized reconstruction to reduce blocking artifacts of block discrete cosine transform compressed images," *IEEE Transactions on Circuits and Systems for Video Technology*, vol. 3, December 1993.

[11] D. Anastassiou, "Generalized three-dimensional pyramid coding for HDTV using nonlinear interpolation," in *Proceedings of the Picture Coding Symposium*, pp. 1.2.1–1.2.2, March 1990.

DIGITAL IMAGE CODING FOR ROBUST
MULTIMEDIA TRANSMISSION

Sheila S. Hemami

School of Electrical Engineering
Cornell University
332 Engineering & Theory Center Bldg.
Ithaca, NY   14853

INTRODUCTION

Reliable reception of still images is imperative in providing robust multimedia transmission over networks offering non-guaranteed transmission, such as ATM and other packet-based networks, satellite channels, and wireless networks. Not only are images used in home-shopping catalogs and on-line information services, but they also form the anchor frames for many video coding algorithms, notably MPEG and H.261. When considering digital image coding, generally the rate-distortion curve is the predominant criteria for measuring the goodness of an algorithm — an algorithm that produces a high PSNR at a high compression ratio is desirable. However, when such a compressed stream is segmented into packets and transmitted over an unreliable network, packet loss can have catastrophic consequences. Depending on the type of source coding employed and the information lost, damaged images may exhibit large segments of missing data, aliasing, or generally poor quality. Robustness is therefore vital for digital image transmission.

Current approaches to provide robustness in non-guaranteed transmission environments include forward error correction (FEC), layered coding and transmission, and automatic retransmission query protocols (ARQ). Packet-level FEC has been proposed, in which error correction codes are applied across packets and additional error correction packets are transmitted.[1,2] Layered coding involves segmenting the data stream into high and low priority streams and transmitting the higher priority information with guaranteed transmission.[3,4] While the Cyclic-UDP protocol allows guaranteed transmission, the more popular internet protocols such as TCP-IP do not provide for it. Finally, ARQ retransmits information as requested by the receiver. In doing so, it lowers the data transmission rate while resending requested data, and can increase the network congestion that initially induced packet loss.[5] This paper presents an alternative to the above techniques for providing robust digital image transmission — reconstruction of lost information from correctly received information. As such, reconstruction for robust transmission requires no special FEC or protocol features.

Image data is highly correlated, and when the properties of the human visual system are considered, lost visual data can be reconstructed from received data such that the reconstructed data is reasonably correct (high PSNR) and visually pleasing. These two criteria are not the same — it is possible to have a higher PSNR with a visually less pleasing result, and a lower PSNR with a better looking result. Providing reconstruction capability as well as compression places requirements on coding of the image data, where coding is defined to include both the actual source coding and the packetization of the compressed data. This paper describes coding of digital images that considers not only compression but also ease and quality of reconstruction to allow for robust transmission. Block-based transform coding techniques are considered so that direct extensions to JPEG-like and MPEG-like coding are possible. First, the trade-offs among decoder computation, transmission bandwidth, and visual quality are described, leading to two approaches to providing robust image transmission. Next, packetization requirements for reconstruction are considered. An approach to block-based reconstruction is described, and three specific solutions are given as examples, requiring different combinations of transmission bandwidth and decoder computation and all providing good visual quality. The reconstruction techniques presented are non-iterative and computationally efficient, and can therefore be easily implemented in software on the personal computers and workstations that will comprise multimedia terminals.

RECONSTRUCTION IN IMAGE TRANSMISSION SYSTEMS

Given that an image transmission system will incorporate reconstruction, several interrelated considerations arise. Most importantly, the visual quality of the reconstructed images must be acceptable. High quality reconstruction algorithms are therefore imperative. Computational requirements at the decoder in order to provide maximum visual quality must therefore be considered. In general, higher quality requires more computation, and the availability of more computation allows higher quality reconstruction. Transmission bandwidth can also be included as a consideration; additional information to facilitate reconstruction and alleviate some computational requirements at the decoder can be used at the expense of transmission bandwidth.

In real systems, computational power and transmission bandwidth are limited. Therefore, trade-offs must be made between quality, computation, and bandwidth. Two dual approaches to providing robust image transmission through reconstruction capabilities exist, *decoder-based adaptive reconstruction* and *reconstruction-optimized source coding*. In *decoder-based adaptive reconstruction*, an adaptive algorithm is designed to exploit the characteristics of the compressed visual information and is implemented at the decoder. When the compressed image is received at the decoder, the received information is used to reconstruct the lost information using the adaptive algorithm. Because the lost data is not known, image characteristics such as spatial and temporal correlation and source distributions must be estimated or at least assumed at the decoder, and used in the reconstruction algorithm. As such, decoder-based adaptive reconstruction generally requires more compute power than reconstruction based on *reconstruction-optimized source coding*.

In *reconstruction-optimized source coding*, the image compression algorithm is designed while considering the reconstruction requirement. This approach provides reconstruction capabilities when the decoder computation is limited, when simpler, non-adaptive algorithms are preferred because they are less computationally intensive. However, such algorithms generally provide poorer visual results. To alleviate this problem, source coding techniques are designed with the non-adaptive reconstruction in mind. The data is compressed with an algorithm designed to provide the best possible visual reconstruction performance using the specified non-adaptive algorithm.

CODING & PACKETIZATION REQUIREMENTS FOR RECONSTRUCTION

To ensure that reconstruction is possible, several coding and packetization requirements must be met. While compression efficiency may suffer slightly, the ability to provide robust image transmission through reconstruction is far greater with the following system requirements. The following discussion assumes an image is transformed with a block-based transform, followed by quantization and entropy coding.

Transform coefficients may be vector or scalar quantized. Any quantizer resolutions are acceptable, providing that blocking artifacts in the decompressed image are minimized, which is primarily accomplished by appropriately quantizing the DC coefficient. No differential encoding of the DC coefficient is performed, thus limiting the effect of loss of a DC coefficient to only one block. Any lossless compression technique for the resulting quantized coefficients is allowable, providing that the stream can be packetized such that loss of a packet does not affect decoding of subsequent packets. Arithmetic coding of quantized coefficients spread over several packets is not acceptable as loss of any packet destroys synchronization at the decoder.

When the source coded data is packetized, image data is transmitted in a known order. When combined with sequence numbers to identify lost packets, this assumption allows identification of the locations of lost data within an image or frame. If bursty errors occur in the network, image data is interleaved before packetization. Interleaving avoids loss of large contiguous areas within an image or a frame, at the expense of latency at the decoder. Reconstruction is thus facilitated by isolating lost portions of data within the image rather than concentrating it in one large area.

THE BLOCK-BASED RECONSTRUCTION PROBLEM

When a digital image is coded as described above and transmitted over a network with a best-effort or non-guaranteed transmission, data loss affects specific blocks in the received image. To provide robust image transmission, the lost data can be reconstructed from known, received data. In one dimension, the block-based reconstruction problem is stated as follows: given the location of a lost coefficient block C_Z and the available adjacent blocks C_{adj}, reconstruct the lost block from the adjacent blocks as $\hat{C}_Z = f(C_{adj})$, such that an error measure $d(\hat{C}_Z, C_Z)$ is minimized. Previous work on decoder-optimized reconstruction has focused on generating each lost coefficient or pixel independently, and possibly iteratively.[6, 7] An alternative approach is to allow the interblock correlation to determine the structure of the block, rather than assuming a structure.

To limit computational complexity at the decoder, the problem is restricted so that f is a linear operator. Specifically, functions of the form $\hat{C}_Z = \sum w_{adj} C_{adj}$ are considered, where C_{adj} refers to adjacent blocks and the weights w_{adj} are scalars. In two dimensions, referring to the adjacent blocks by locations, this yields

$$\hat{C}_Z = w_{Top} C_{Top} + w_{Bottom} C_{Bottom} + w_{Left} C_{Left} + w_{Right} C_{Right} \quad . \tag{1}$$

Linearity of the transform operation allows reconstruction using (1) in either the coefficient domain or in the pixel domain. With (1), the problem of providing up to N^2 unknowns is reduced to finding only four unknowns, and the structure of the reconstructed block is determined by the surrounding blocks rather than by a global assumption. The remainder of this paper describes three specific reconstruction techniques to provide robust digital image transmission, two decoder-based adaptive techniques and one using reconstruction-optimized source coding. Each is suited to a different combination of decoder computation and transmission bandwidth. A technique can therefore be selected based on the image transmission system characteristics to maximize the visual quality.

DECODER-OPTIMIZED ADAPTIVE RECONSTRUCTION

Optimal weight generation minimizes the squared error between a block and its reconstruction and provides the best results, but requires impractically high overhead transmission. Two non-optimal but practical techniques for weight generation are described, both providing good quality reconstruction. Each is suited to a different combination of transmission overhead and decoder computation. The first technique represents *a posteriori* reconstruction, in which all calculations are performed at the decoder without requiring transmission of additional information, and as such is best suited to low bandwidth channels where additional decoder computation can be tolerated. The second technique provides *a priori* reconstruction, in which all computations for weight generation are performed at the encoder where the entire image is known, and extra information for reconstruction must be transmitted. It is thus best suited to channels with excess bandwidth, when minimal decoder computation is desired.

Smoothing Criterion Reconstruction

Smoothing Criterion Reconstruction[9] (SCR) provides reconstruction when no excess transmission bandwidth is available, but decoder computation is available. In this case, in addition to the reconstructed blocks themselves, the weights must also be generated at the decoder, where knowledge of the image is limited to what is received. In SCR, global assumptions about block structure are minimized by imposing a condition that is obviously met in well-coded images — the absence of blocking artifacts. This technique attempts to match the edge pixels of the reconstructed block with the abutting edge pixels of adjacent blocks, thereby assuming that the reconstructed block connects smoothly to its neighbors and implicitly minimizing blockiness. SCR minimizes the squared difference between pixels across block boundaries. The weights w are determined at the decoder by solving the least squares problem

$$\text{Find } w \text{ to minimize} \sum_{\substack{\text{available} \\ \text{adjacent} \\ \text{blocks}}} \sum_{\text{edge pixels}} (p - \hat{p})^2 \tag{2}$$

where $\hat{p}$ represents the reconstructed pixel using (1) in the coefficient domain and then inverse transformed to the pixel domain.

SCR is suitable for low bandwidth links with available decoder power. The computational complexity of SCR for an 8×8 block with 4 available neighbors and reconstruction in the pixel domain is less than 2.2 times the number required for a fast, recursive DCT and less than 1.2 times the number required for a non-recursive DCT. [10]

Vector Quantized Linear Interpolation

If the decoder computation required by SCR is unavailable, Vector Quantized Linear Interpolation[11] (VQLI) can provide reconstruction of equivalent visual quality, with the requirement of less than 10% transmission bandwidth overhead. Vector quantization (VQ) is used at the encoder to select an appropriate codeword for the weights w from a codebook. VQ and reconstruction are combined into a single step by defining a distortion measure to be used when quantizing:

$$d(P_Z, \hat{P}_Z(\hat{w})) = \left\| P_Z - \hat{P}_Z(\hat{w}) \right\|^2 \tag{3}$$

where P_Z represents the original pixel block and $\hat{P}_Z(\hat{w})$ represents the reconstructed pixel block using the quantized weight vector and all four adjacent blocks. The quantization step is then described as

$$\text{Find the index } i \text{ such that } d(P_Z, \hat{P}_Z(\hat{w}_i)) \text{ is minimized.} \tag{4}$$

Use of (2) as the distortion measure attempts to best reconstruct the original block, so if the original block does not exhibit blockiness in the image, then VQLI implicitly minimizes blockiness. Note that this approach differs from traditional vector quantization, in which the weights would be first determined, and then independently vector quantized using the standard Euclidean distortion measure. Though the quantization is performed assuming all of the adjacent blocks are present, the codebook as designed for the given distortion measure provides weights to use when any combination of adjacent blocks are available.[11]

Results

Some general results applying to both techniques. Because the interpolation as given by (1) uses only blocks that are horizontally and vertically adjacent, various structures can be either well or poorly represented by such a solution. Those structures that are reconstructed well include horizontal, vertical, near-horizontal, and near-vertical edges, lines, and patterns. High frequency details in textures are preserved because all coefficients are used in the reconstruction process. Gradual luminance gradients are also maintained without blocking artifacts. Strong diagonal edges are not reconstructed well by this technique. A diagonal edge cannot be generated from a linear sum of shifted diagonal edges, which is conceptually understood in the pixel domain and can be demonstrated mathematically using the shift properties of the DCT. In this case, the quality of the reconstructed block is dependent on the luminance levels of the blocks surrounding the lost block. The difference in luminance levels or chrominance values across an edge dictates how well the reconstructed block blends into its surroundings. For example, an error in an edge between black and white regions is more objectionable than if the regions were both shades of gray. Details or image artifacts whose size is on the order of the block size may or may not be reconstructed well, depending on the technique used and whether the details are replicated in the surrounding blocks. Thus, the reconstructed image quality is content dependent. When the algorithm is applied to

Figure 1 Segment of *couple* (272×272 pixels) with 10% random loss and no reconstruction. PSNR = 15.1 dB.

Figure 2 Segment of *couple* with 10% random loss as shown in Figure 1, reconstructed using SCR. PSNR = 28.3 dB.

chrominance components, color gradients are smoothly reconstructed and slight bleeding is visible only on strong diagonal edges. Figure 1 shows a segment of the *couple* image with 10% random loss, and Figures 2 and 3 show the reconstructed image using SCR and VQLI, respectively.

SCR. Blockiness is explicitly minimized in this technique, and as a result only appears on strong diagonal edges and in areas where the detail size is on the order of the block size. Details in the centers of the blocks are smudged or blurred, because SCR essentially matches only the edge pixels. However, the visual quality of the reconstructed images demonstrates that matching only the edge pixels generates high quality reconstructions.

VQLI. Full-search vector quantizers of sizes 1 to 128 were designed using a splitting algorithm and a training set consisting of 15,508 5-tuples of 8×8 luminance and chrominance blocks taken from 13 database images. A codebook size of 32 produces reconstructed images that are virtually indistinguishable from those reconstructed using non-quantized weights. The most obvious visual difference between images reconstructed using the quantized weight vectors and the non-quantized weight vectors is in detail blocks, for example, around the eyes in *lena*. Since detail blocks are often unique and may not resemble the surrounding blocks, their weight vectors may not be well represented by quantized vectors that represent general trends. However, since VQLI weight generation is performed at the encoder where the original image is known, detail blocks are better matched in VQLI than in SCR, when only edge pixels are matched.

Transmission of non-entropy coded vector indices for a codebook size of 32 requires an overhead of 5 bits/block. For 512×512 monochrome images coded using an 8×8 block size as in the JPEG standard, an extra 2560 bytes must be transmitted with each compressed image. Using JPEG with the standard quantization matrix, this data represents overheads of 8% for *lena* and 6% for *couple*.

RECONSTRUCTION-OPTIMIZED SOURCE CODING

In reconstruction-optimized source coding, a block-based source coding technique is designed to maximize the reconstruction performance when reconstruction is given by (1) with fixed weights for all blocks. To minimize computation, a simple mean is taken: $w_{Top} = w_{Bottom} = w_{Left} = w_{Right} = 1/4$. As such, all reconstruction algorithms will produce the same results for traditional block-based transforms. Therefore, performance can only be modified by using a transform in which adjacent coefficient blocks are not independent. A lapped orthogonal transform (LOT)[12-16] with 50% overlap generates such coefficient blocks. Reconstruction of lost LOT coefficient blocks *a posteriori* at the decoder has been investigated[17], as has use of the

Figure 3 Segment of *couple* with 10% random loss shown in Figure 1, reconstructed using VQLI. PSNR = 28.6 dB.

Figure 4 Segment of *couple* coded using reconstruction-optimized LOT (α = 0.8) with 10% random loss as shown in Figure 1 and reconstructed using mean-reconstruction. PSNR = 29.2 dB.

extended lapped transform with no reconstruction to provide robustness to loss in transmission.[18] The approach presented here differs from previous work in that the transform is explicitly designed to maximize reconstruction performance. The transform is designed one-dimensionally, and is then performed on images independently in both the horizontal and vertical directions.

The input signal x is modelled as a Markov-1 signal, with correlation given by $E\{x_n x_{n+k}\} = \sigma^2 \rho^k$. The reconstruction criterion is formulated as a minimization of the MSE in each reconstructed transform coefficient. It can be shown that this is equivalent to attempting to equally distribute the mean squared reconstruction error among coefficients if the basis functions are designed sequentially. There are N basis functions to be designed, of length $2N$ each. Representing the LOT basis function i by T_i, the MSE for each reconstructed coefficient $\hat{c}_i$ is given by

$$\text{MSE}(\hat{c}_i) = E\{(c_i - \hat{c}_i)^2\} \equiv T_i' R_{err} T_i \tag{5}$$

where $R_{err}(i,j) = (3/2)\rho^{|i-j|} - (\rho^{|i-j+M|} + \rho^{|j-i+M|}) + (1/4)(\rho^{|i-j-2M|} + \rho^{|j-i-2M|})$ and is a $2N \times 2N$ matrix. R_{err} is formed when substituting the inner product of the signal and the basis function for the coefficient. The MSE of the entire reconstruction coefficient block is then given by $\text{MSE}(\hat{c}(n)) = \sum \text{MSE}(\hat{c}_i) = \text{Trace}(T' R_{err} T)$ where T is an $2N \times N$ matrix containing the N LOT basis functions. It can be shown that the particular LOT used does not affect the MSE of the *entire coefficient block*. However, though the sum is the same, the individual reconstructed coefficient errors $\text{MSE}(\hat{c}_i)$ differ. A reconstruction-optimized LOT should distribute the coefficient errors to maximize the visual performance. It can be shown that this condition is equivalent to attempting to equally distribute the errors across the coefficients.

Typically, only coding gain is considered in transform design, given by

$$G_{TC} = \left(\frac{1}{N}\sum_{i=0}^{N-1}\sigma_i^2\right) \Big/ \left(\prod_{i=0}^{N-1}\sigma_i^2\right)^{1/N} \tag{6}$$

where σ_i^2 is the ith diagonal entry of the transformed input autocovariance matrix, $Z' R_1 Z$, and Z represents any transform matrix. Optimality in terms of compression performance is generally argued as $\rho \to 1$ [13], in which case $G_{TC} \to \infty$. However, in the case of reconstruction, as $\rho \to 1$, $\text{MSE}(\hat{c}_i(n)) \to 0$, and this case provides a degenerate reconstruction condition. Therefore, for

Transform	G_{TC}	G_R
DCT-LOT	7.70	0.401
T6	6.07	0.488
T7	5.22	0.545
T8	4.74	0.581
T9	4.47	0.600

Table 1 Coding and reconstruction gains of the designed LOTs and the DCT-LOT.

reconstruction-optimized transforms, "optimality" in the traditional sense of considering performance as $\rho \to 1$ is not appropriate. However, reconstruction performance can be quantified by the "reconstruction gain." Because an equal error distribution is desirable, the reconstruction gain is given as the reciprocal of (6) with R_1 replaced by R_{err}:

$$G_R = \left(\prod_{i=0}^{N-1} T_i' R_{err} T_i \right)^{1/N} \Big/ \left(\frac{1}{N} \sum_{i=0}^{N-1} T_i' R_{err} T_i \right) . \tag{7}$$

The overall performance of a transform can therefore be gauged by considering both the coding gain G_{TC} and the reconstruction gain G_R. The sequential design procedure presented in [12] is used, with a modified objective function incorporating both coding gain and reconstruction gain formed as $\tilde{G}_{TC,R} = T_i' (R_1 - \alpha R_{err}) T_i$, where the reconstruction parameter α controls the emphasis placed on reconstruction capability. Therefore, by selecting appropriate values of the reconstruction parameter α, reconstruction capability can be included in the design of the basis functions. Because the overall reconstruction error is fixed and the basis functions are designed sequentially, each to minimize the reconstruction error, the LOTs designed have a reconstruction error that is more equally distributed across the coefficients.

Results

Four length-16 LOTs ($N = 8$) were designed for $\alpha = \{0.6, 0.7, 0.8, 0.9\}$ with $\rho = 0.95$, and will be denoted T6, T7, T8, and T9, respectively. The compression performance of the designed LOTs was evaluated using the transform coding gain and by compressing images using a modified JPEG algorithm that follows the JPEG syntax but uses the LOT coefficients rather than DCT coefficients, where the quantization matrices are modified for each LOT. The transform coding gains and reconstruction gains for the four designed transforms are given in Table 1, along with the DCT-LOT.[13] The transform coding gains monotonically decrease as the reconstruction parameter α increases, and are reduced by approximately 42%, while the reconstruction gain increases by 50% across the transform range.

The desired more equalized error distribution across coefficients is achieved. As α increases, less error occurs in the first three coefficients, while the cumulative error is approximately equal by the fourth coefficient. Distributing the error in the coefficient domain also produces a "minimax" effect and benefit in the pixel domain — as α increases, the maximum expected error in the pixel domain decreases. With a smaller maximum expected error, the reconstructed results are more visually pleasing.

As expected, the equal-MSE property of the LOTs yields PSNRs that are equal, within 0.3 dB, on the reconstructed unquantized images. However, visual results differ dramatically. T6, the LOT with the least weight on the reconstruction criterion, provides visual results not much better than the DCT-LOT, but T7, T8, and T9 provide dramatic improvement. The better visual performance can be explained by the minimax pixel error. If there is a threshold below which individual pixel errors cannot be perceived, then the designed transforms exhibit smaller, though perhaps more, pixel errors above this threshold. Figure 4 shows the performance of T8 on the *couple* segment.

CONCLUSIONS

This paper has presented two approaches to providing robust image transmission, decoder-optimized adaptive reconstruction and reconstruction-optimized source coding. Three specific solutions have been described, two under the first approach, and one under the second. Smoothing criterion reconstruction provides *a posteriori* reconstruction at the decoder, using only the received image data, while vector quantized linear interpolation requires transmission of reconstruction information at a bandwidth overhead of less than 10% but needs less computation at the decoder. A family of reconstruction-optimized lapped orthogonal transforms allow non-adaptive reconstruction, and as the reconstruction capabilities increase, the compression decreases. Each solution provides reconstruction with a different combination of transmission bandwidth and decoder computation, and hence a technique can be selected based on the specific multimedia transmission system.

REFERENCES

1. N. Shacham and P. McKenney, "Packet recovery in high-speed networks using coding and buffer management," *Proceedings IEEE Infocom '90*, Los Alamitos, CA, 1990, Vol. 1, pp. 124-31.
2. A. Albanese et. al., "Priority encoding transmission," *Proc. 35th IEEE Annual Symp. on Foundations of Computer Science*, 1994, pp. 604-12.
3. M. Ghanbari, "Two layer coding of video signals for VBR networks," *IEEE J. Select. Areas Comm.*, Vol. 7, No. 5, pp. 771-781, June 1989.
4. G. Morrison and D. Beaumont, "Two layer video coding for ATM networks," *Signal Processing: Image Communication*, Vol. 3, No. 2-3, pp. 179-95, June 1991.
5. N. MacDonald, "Transmission of compressed video over radio links," *BT Technology Journal*, Vol. 11, No. 2, April 1993, pp. 182-5.
6. R. N. J. Veldhius, "Adaptive restoration of unknown samples in discrete-time signals and digital images," Ph.D. thesis, Katholieke Universiteit te Nijmengen, The Netherlands, 1988.
7. Y. Wang, Q.-F. Zhu, and L. Shaw, "Maximally smooth image recovery in transform coding," *IEEE Transactions on Communications*, Vol. 41, No. 10, pp. 1544-51, Oct. 1993.
8. W.-M. Lam and A. R. Reibman, "An error concealment algorithm for images subject to channel errors," *IEEE Trans. Image Processing*, Vol. 4, No. 5, May 1995, pp. 533-42.
9. S. S. Hemami and T. H.-Y. Meng, "Transform-coded image reconstruction exploiting interblock correlation," *IEEE Trans. on Image Processing*, Vol. 4, No. 7, July 1995, pp. 1023-27.
10. S. Sheng, A. Chandrakasan, and R. W. Brodersen, "A portable multimedia terminal," *IEEE Communications Magazine*, Vol. 30, No. 12, pp. 64-75, Dec. 1992.
11. S.S. Hemami and R.M. Gray, "Image reconstruction using vector quantized linear interpolation," *Proc. ICASSP '94*, Adelaide, Australia, April 1994, Vol. 5, pp. 629-32.
12. P. Cassereau, "A new class of optimal unitary transforms for image processing," S.M. thesis, Dept. of Elec. Eng. and Comp. Sci., Mass. Inst. of Tech., Cambridge, MA, May 1985.
13. H. S. Malvar and D. H. Staelin, "The LOT: Transform coding without blocking effects," *IEEE Trans. ASSP*, Vol. 38, No. 4, April 1989, pp. 553-559.
14. A. N. Akansu and F. E. Wadas, "On lapped orthogonal transforms," *IEEE Trans. Signal Processing*, Vol. 40, No. 2, February 1992, pp. 439-42.
15. H. S. Malvar, "Lapped transforms for efficient transform/subband coding," *IEEE Trans. ASSP*, Vol. 38, No. 6, June 1990, pp. 969-78.
16. B. W. Suter and M. E. Oxley, "On variable overlapped windows and weighted orthonormal bases," *IEEE Trans. Signal Processing*, Vol. 42, No. 8, August 1994, pp. 1973-82.
17. P. Haskell and D. Messerschmitt, "Reconstructing lost video data in a lapped orthogonal transform based coder," *Proc. ICASSP '90*, Albuquerque, NM, April 1990, pp. 1985-8.
18. R. L. De Queiroz and K. R. Rao, "Extended lapped transform in image coding," *IEEE Trans. Image Processing*, Vol. 4, No. 6, June 1995, pp. 828-32.

ENHANCEMENT OF DCT-JPEG COMPRESSED IMAGES

Kook Yeon Kwak and Richard A. Haddad

Electrical Engineering Department
Polytechnic University
Six Metrotechcenter, Brooklyn, NY 11201

ABSTRACT

We present a method for enhanced decoding of standard DCT encoded images. Our algorithms, based on theory of Projections Onto a Convex Set($POCS$), provide a fast convergence to the decoded-image with blocking artifacts removed, but directional features enhanced. These goals are achieved using a pair of new constructs: the $POCS$ using a slope-difference-and-average(SDA) constraint at the block boundaries and the spatially localizable DCT ($SLDCT$), a new DCT-based linear block transform with spatial localizability. Our method combines these theoretic constructs with a strategy for multiscale classification of the received DCT blocks. Simulations on standard test images demonstrate that this decoder produces images of high fidelity to the original as measured by $PSNR$ metric, and of enhanced perceptual quality, especially at directional edges.

INTRODUCTION

The DCT[1] was adopted as the core algorithm for video or still image compression in international standards such as $H.261$, $JPEG$, $MPEG1$, and $MPEG2$[2]. While the block DCT coding is nearly perfect for highly correlated images at a reasonable bit-rate, it produces visually annoying blocking artifacts in a decoded image at low-bit rates. But even in the standard coding system, the bit rate cost and the quality of the decoded image should be jointly optimized. An enhanced decoder should maximize the quality of the decoded image for the bit-rate given by the encoder. For enhancing the $JPEG$-decoder, a set of image-vectors or codebook[3] was developed for reconstructing block-images from received DCT coefficients; these replace the set of DCT basis vectors. But the quality of the decoded image is sensitive to the training images used in constructing the image vector codebook. Several authors formulated the enhanced decoder problem as that of finding the optimal estimator in the ill-posed conditions resulting from the dequantizer in the decoder. As a departure from filtering methods[4] [5] [6] they take blocking artifacts into account as one item in the apriori information used for estimating the encoded image. Some authors[7] [8] [9] used maximum *a posteriori(MAP)* estimation and *DC* calibration[9] for the most consistent image with the assumed stochastic image model. Though these methods are inherently spatially adaptive, they depend on parameters in the stochastic model determined from an abstract set of general images. Meanwhile, authors[10] [11] [12] [13] investigated iterative techniques based on the projections onto convex sets(*POCS*)[14] [15]. In these techniques, parameters bounding convex sets can be directly obtained from data of the very image to be encoded. These works defined smoothness of an image as differences of adjacent pixels. While this definition can represent smoothness in active regions, it is not good enough for slowly-changing regions. Furthermore, they have not yet considered enhancing directional edges in the image. In this paper, we develop an enhanced decoding method based on the *POCS* theory for *DCT* coded images. This method has a number of features: it is spatially adaptive; it provides fast convergence; it enhances directional features in an image. Our method is based on two background theories, *POCS* and the Time Localizable Linear Transform(*TLLT*)[16].

PROJECTION ONTO CONVEX SETS

For a vector space V, a subset C of V is convex if for every $\mathbf{f}_1, \mathbf{f}_2 \in C$ and any real scalar α such that $0 \leq \alpha \leq 1$, $\alpha\mathbf{f}_1 + (1-\alpha)\mathbf{f}_2$ is included in C. Projecting a vector onto the closed convex set C is equivalent to finding the vector in C closest to the given vector in the sense of a given norm of V. Let $P(\cdot)$ denote the projection operator; then for $\mathbf{f}^0 \in V$,

$$P(\mathbf{f}^0) = \mathrm{argmin}_{\mathbf{f} \in C}\|\mathbf{f}^0 - \mathbf{f}\|. \tag{1}$$

Let $C_1, C_2, \ldots, C_m$ be closed convex sets in V and $P_1, P_2, \ldots, P_m$ be their projection-operators, respectively. The recursion[15]

$$\mathbf{f}^{n+1} = \mathbf{f}^n + \lambda_n(P_{n(mod\ m)+1}(\mathbf{f}^n) - \mathbf{f}^n) \tag{2}$$

converges to a vector in the intersection of $C_1, \ldots, C_m$ with relaxation coefficients λ_n such that $0 < \lambda_n < 2$. Consider a space of $(N^2 \times 1)$ vectors generated by row-wise or column-wise reordering of $(N \times N)$ images. Let B denote the block DCT operator and let Q denote the quantization operator on $B\mathbf{f}$ for a given set of quantization steps. Let $\mathbf{F}_{tr}$ be the transmitted DCT coefficients. In the literature[10] [11] [12] [13], a convex set based on the known DCT quantization table was developed as:

$$C_q = \{\mathbf{f} : QB\mathbf{f} = \mathbf{F}_{tr}\}. \tag{3}$$

Projection on C_q is determined as: $P_q(\mathbf{f}^0) = B^{-1}P_Q(B\mathbf{f}^0)$, where $\mathbf{f}^0$ is an arbitrary vector. The i-th element of $P_Q(B\mathbf{f}^0)$ is obtained from:

$$[P_Q(B\mathbf{f}^0)]_i = \begin{cases} [\mathbf{F}_{tr}]_i + \frac{1}{2}\triangle_i, & \text{if } [B\mathbf{f}^0]_i > [\mathbf{F}_{tr}]_i + \frac{1}{2}\triangle_i, \\ [\mathbf{F}_{tr}]_i - \frac{1}{2}\triangle_i, & \text{if } [B\mathbf{f}^0]_i < [\mathbf{F}_{tr}]_i - \frac{1}{2}\triangle_i, \\ [B\mathbf{f}^0]_i, & \text{otherwise,} \end{cases} \tag{4}$$

for $i = 1, \ldots, N^2$, where $[\mathbf{F}_{tr}]_i$ is the i-th element of $\mathbf{F}_{tr}$ and $\triangle_i$ is the quantization step for the i-th element in the DCT vector. Another convex set, C_p, and projection, P_p, were defined based upon pixel-range information:

$$C_p = \{\mathbf{f} : 0 \le [\mathbf{f}]_i \le U, i = 1, \ldots, N^2\}, \quad [P_p(\mathbf{f}^0)]_i = \begin{cases} U, & \text{if } [\mathbf{f}^0]_i > U, \\ 0, & \text{if } [\mathbf{f}^0]_i < 0, \\ [\mathbf{f}^0]_i, & \text{otherwise,} \end{cases} \quad (5)$$

for $i = 1, \ldots, N^2$, where U is the upper bound of pixel-value.

We can realistically assume that pixels at the boundaries of the (8×8) image are reasonably correlated. This suggests that a smoothness measure along block boundaries in an image is bounded. We introduce an efficient definition of smoothness along block boundaries in terms of differences and average of pixel slopes along the boundaries. Since this definition of smoothness is related to second differences involving four pixels near the block boundaries, this smoothness requirement provides a better way to define a set of smooth images than the pixel-difference approach. Let the i-th pixel column of an image be denoted as $(N \times 1)$ $\mathbf{f}_i$. For $\mathbf{f}_i$ and $\mathbf{f}_{i+1}$, the slope vector, $\mathbf{s}_i$, is defined by $\mathbf{s}_i = \mathbf{f}_i - \mathbf{f}_{i+1}$. Let an $(N \times N)$ image be split into $(M \times M)$ blocks and $L = N/M$. Vertical borders between horizontally adjacent blocks define a vertical block boundary spanning from top to bottom of the image. For the j-th vertical block boundary, vectors $\mathbf{a}_j$ and $\mathbf{d}_j$ are defined as:

$$\mathbf{a}_j = \frac{1}{10}(3\mathbf{s}_{jM-1} + 4\mathbf{s}_{jM} + 3\mathbf{s}_{jM+1}), \quad \mathbf{d}_j = \mathbf{s}_{jM} - \frac{1}{2}(\mathbf{s}_{jM-1} + \mathbf{s}_{jM+1}). \quad (6)$$

Let $\mathbf{a}, \mathbf{d}$ denote the slope-average vector and difference vector determined by vertical concatenation of $\mathbf{a}_j, \mathbf{d}_j, j = 1, \ldots, L - 1$, respectively. We introduce linear operators, A, D, such that $\mathbf{a} = A\mathbf{f}, \mathbf{d} = D\mathbf{f}$. Two convex sets, C_v^a, C_v^d, are defined for given values of s_a, s_d as:

$$C_v^a = \{\mathbf{f} :\| A\mathbf{f} \| \le s_a\}, \quad C_v^d = \{\mathbf{f} :\| D\mathbf{f} \| \le s_d\}, \quad (7)$$

where $\| . \|$ is the ℓ^2 norm. The projectors for these sets, P_v^a, P_v^d, have been found to be[17],

$$P_v^a(\mathbf{f}^0) = \mathbf{f}^0 - 10\alpha_a A^t A\mathbf{f}^0, \quad P_v^d(\mathbf{f}^0) = \mathbf{f}^0 - \frac{2}{5}\alpha_d D^t D\mathbf{f}^0, \quad (8)$$

where $\alpha_a = \frac{1}{2}(1 - \frac{s_a}{\|A\mathbf{f}^0\|})$ and $\alpha_d = \frac{1}{2}(1 - \frac{s_d}{\|D\mathbf{f}^0\|})$. The operators A and D have the nice property that $AD^t = 0$. This property suggests that P_v^a and P_v^d are independent of each other in the sense that projecting on one set has no effect on the parameter used for projecting on the other. An operator can be defined with relaxation coefficients λ_a and λ_d, by:

$$T_v = (I + \lambda_a(P_v^a - I))(I + \lambda_d(P_v^d - I)) = I + \lambda_a(P_v^a - I) + \lambda_d(P_v^d - I), \quad (9)$$

where $0 < \lambda_a, \lambda_d < 2$.

In a similar way, we can define the linear operator mapping an image to the slope-difference and average vector at horizontal block-boundaries spanning from left to right of the image. The corresponding convex sets are denoted by C_h^d and C_h^a, and the projection operators by P_h^d, P_h^a, and T_h.

The optimal values for λ_a and λ_d in Eq.(9) for rapid convergence are not known apriori. We first suppose that an image $\mathbf{f}^0$ satisfies $A\mathbf{f}^0 = cD\mathbf{f}^0$ for some scalar c. Then we define a specific projection operator by Eq.(10) in terms of a single parameter β, which directly control the operator T_v,

$$T_v(\mathbf{f}^0)_i = \begin{cases} \mathbf{f}_i^0 + \beta\mathbf{e}_j^0, & \text{if } i = jM - 1, \\ \mathbf{f}_i^0 + \mathbf{e}_j^0, & \text{if } i = jM, \\ \mathbf{f}_i^0 - \mathbf{e}_j^0, & \text{if } i = jM + 1, \\ \mathbf{f}_i^0 - \beta\mathbf{e}_j^0, & \text{if } i = jM + 2, \\ \mathbf{f}_i^0, & \text{otherwise}, \end{cases} \tag{10}$$

for $i = 1, \ldots, N$ and $j = 1, \ldots, L - 1$. The term $\mathbf{e}_j^0$ is the change in $\mathbf{f}_{jM}^0$ induced by the operator T_v. Let $\mathbf{e}^0$ denote a vector consisting of $\mathbf{e}_j^0$, then we can establish that

$$\mathbf{e}^0 = \frac{2}{\beta - 3}\lambda_d \alpha_d \mathbf{d}^0, \quad \beta = \frac{15\lambda_a \alpha_a c - \lambda_d \alpha_d}{5\lambda_a \alpha_a c + 3\lambda_d \alpha_d}. \tag{11}$$

The projection operator is now specified by parameter β and given bound s_d. The constant s_a is implied by the value of β. We can use Eq.(11) to determine if values of λ_a and λ_d in the acceptable range are compatible with the selection of β at each iteration. Thus, we have shifted our focus to one controlling parameter β, whose value can be empirically set depending on the classification scheme described in the next section. The T_h can be expressed correspondingly in terms of its parameter.

THE SPATIALLY LOCALIZABLE DCT ($SLDCT$)

The $SLDCT$ is a special case of the Time Localizable Linear Transform ($TLLT$) published by the authors[16]. The $TLLT$ basis function is a linear combination of certain basis functions of a given linear block transform which can concentrate its energy in desired time interval. Consider a $\ell^2(Z)$ space S spanned by the set of DCT bases. We partition S into orthogonal subspaces spanned by associated subsets of the DCT bases. And for each subspace, we find a set of new orthonormal bases by a linear combination of DCT bases such that new basis functions concentrate maximally their energies in certain specified time-intervals and these time-intervals evenly cover the entire transform length. Since the subspaces are orthogonal and the sum of the dimensions of each subspace equals the dimension of S, then the union of these new sets constitutes the bases of the $SLDCT$. Let N be the dimension of S and M_k be that of the k-th subspace. Let $\psi_{k,i}(n)$ be the $SLDCT$ basis function in the k-th subspace with energy concentrating on the i-th location; then we choose

$$\psi_{k,i}(n) = \frac{1}{\sqrt{M_k}}\phi_{k_0}(n) + \sqrt{\frac{2}{M_k}}\sum_{l=1}^{M_k-1}\cos(\frac{2i+1}{M_k}l\pi)\phi_{(k_o+l)}(n) \tag{12}$$

for $n = 0, \ldots, N - 1, i = 0, \ldots, M_k - 1$, where, $\phi_l(n)$ is the usual DCT basis function with transform index l and length N: $\phi_l(n) = \alpha(l)\cos(\frac{2n+1}{N}l\pi)$, with $\alpha(0) = 1/\sqrt{N}$ and $\alpha(l) = \sqrt{2/N}$ for $l \neq 0$. It can be easily recognized that the k-th subspace represents roughly the frequency band $\frac{k_0}{N}\pi \leq \omega \leq \frac{k_o+M_k-1}{N}\pi$. The basis functions of $SLDCT$ have two parameters: k for indexing the subspace and i for the location of energy concentration. These are similar to the scaling and shifting parameters in the wavelet transform. Notice that there are many possible $TLLT$'s for a given linear transform arising from different partitions of the space S. Each $SLDCT$ generates its own tiling of

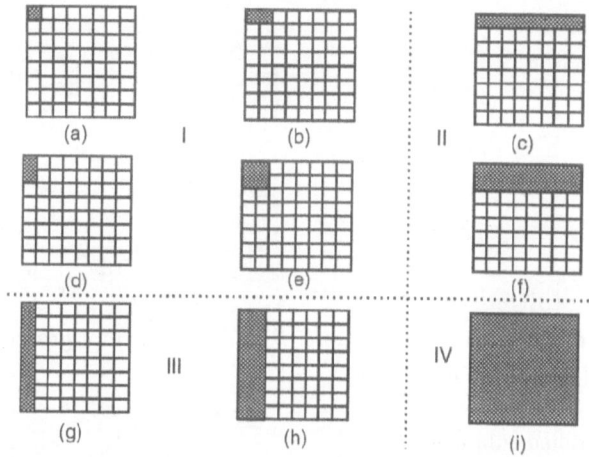

Figure 1: Typical (8×8) DCT coefficients of block for categories (a)-(i). Shady coefficients may have non-zero values.

the time-frequency plane. One of the advantages of $SLDCT$ is that its coefficients can be calculated directly from the DCT coefficients even though the $SLDCT$'s are defined on the spatial signal. Consequently, enhancement of a codec system using the spatial localization feature of $SLDCT$ can be easily embedded into the standard DCT structure. Another advantage of $SLDCT$ is that its basis functions can be changed from block to block in an adaptive manner by simply taking different partitions. Furthermore, $SLDCT$ coefficients for a block preserve characteristic properties of DCT coefficients in the block. The blocking artifacts in the DCT-reconstructed image *remain confined to the block boundaries in the low frequency component of the SLDCT-constructed multiscale decomposition.* These can then be removed by smoothing the $SLDCT$ coefficients using $POCS$ at the block boundaries.

THE ENHANCED DECODER

For $(N \times N)$ encoded image, our decoding operation is as follows:

1) Classify (8×8) blocks into 9 categories depending on *activity* of pixels in the block. Fig.1 shows typical (8×8) DCT coefficients of block for categories (a)-(i). The unshaded DCT coefficients are zero.

2) Perform (8×8) $IDCT$ for all blocks in each category, and obtain $(N \times N)$ subimage for that category. There are 9 subimages where each subimage consists of blocks which are disjoint from blocks in any other subimage.

3) Modify subimages to remove this disjunction of blocks in subimages so that lesser gray block contiguous to denser gray block in Fig.1 is also included in the subimage of he denser block. We now have 9 modified subimages. (Step (2) and (3) are conceptual operations.)

4) Partition the 9 categories into 4 groupings as indicated in Fig.1, and associate an $SLDCT$ operation with each group. The $SLDCT$ operation is done on the DCT coefficients domain for each (8×8) block as shown in Fig.2. There are now 9 $SLDCT$

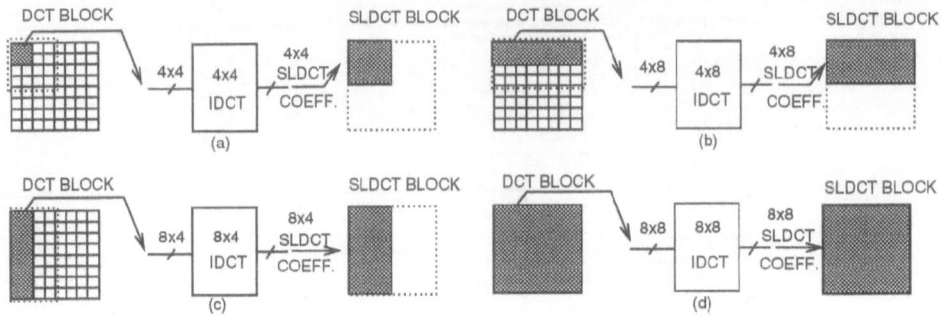

Figure 2: $SLDCT$ operation for each group of categories, (a) for group I, (b) for group II, (c) for group III, and (d) for group IV.

transforms of modified subimages.

5) Define convex sets with slope difference and slope average on $SLDCT$ transform domain. We have 2 operators T_v and T_h for each of the 9 $SLDCT$ transformed subimages.

6) Perform $POCS$ starting with the initial vector associated with the blocky image reconstructed directly from quantized DCT coefficients. T_v and T_h are performed in cascade over all the 9 categories, and the inverse $SLDCT$, P_q, and P_p are performed. An image results in by these operations, which is used for the initial vector for the next iteration of $POCS$. These operations are repeated up to a certain number of iterations, which is empirically obtained. We restrict the number of iterations to prevent divergence of the $POCS$. Thus we can omit the calculations required for verification of relaxation coefficients for convergence.

EXPERIMENTAL RESULTS AND DISCUSSION

For computer experiments, we assume that the (8×8) DCT coefficients are quantized in the encoder using a table previously used[11] yielding around 0.24bpp with *Lena*. The quality of the reconstructed image is measured by $PSNR$. The values for β and s_d are shown Table 1. These were chosen empirically based computer simulations. Our algorithm for enhanced decoding is applied to three popular images, (512×512) *Lena*, (256×256) *Girl*, and (256×256) *MIT* building. We calculate $PSNR$ of restored images after each iteration up to the 5-th of our method. From these experimental results, we find that the proper restriction of the number of iteration is 3. For comparison, $PSNR$ of restored image after 3 iterations of $POCS$ with pixel-difference constraints is calculated. These results are shown in Table 2. Fig.3 shows (256×256) facial part of original *Lena*, *Lena* reconstructed by $IDCT$ of quantized DCT coefficients, by 3 iterations of $POCS$ based on pixel-differences, and by 3 iterations of our method. Our enhanced

Table 1: The values for β and s_d in T_v and T_h for each category. pc is slope difference measured from pixel columns inside blocks in the initial image, and pr from pixel rows.

categories		(a)	(b)	(c)	(d)	(e)	(f)	(g)	(h)	(i)
T_v	β	1/3	0.1	-0.1	1/3	0.1	-0.1	1/3	0.1	-0.1
	s_d	0.0	0.0	pc	0.0	0.0	pc	0.0	0.0	pc
T_h	β	1/3	1/3	1/3	0.1	0.1	0.1	-0.1	-0.1	-0.1
	s_d	0.0	0.0	0.0	0.0	0.0	0.0	pr	pr	pr

Table 2: $PSNR[db]$ of popular images reconstructed without enhancement, by $POCS$ of pixel-differences, and by our method.

	w/o Enhancing	Pixel-$POCS$	Our Method				
Iter.	0	3	1	2	3	4	5
Lena	29.768	30.425	30.641	30.643	30.641	30.640	30.640
Girl	30.442	31.111	31.322	31.326	31.324	31,323	31.323
MIT	25.752	26.472	26.503	26.547	26.555	26.559	26.562

Figure 3: (upper-left) *Lena* original, (upper-right) *Lena* reconstructed without enhancement, (lower-left) *Lena* restored with 3 iterations of $POCS$ based on pixel-differences, and (lower-right) *Lena* restored with 3 iterations of our method.

decoder produces almost a blocking-free image as seen in these images. Furthermore, we emphasize that our method enhances directional edges remarkably, as can be seen clearly in her hair.

While our method works very well on smooth images such as *Lena* and *Girl*, it has little advantage over the existing $POCS$ with pixel-differences constraints for images whose major area is an active region like the *MIT* building. This can be explained by the fact that our method for enhancing decoding is largely based on the assumption of highly correlated pixels in the original image. The slope-based constraint is not so effective for active regions. Though our work deals with *JPEG*-coded still images, our method can also be used to enhance a moving picture coding system by embedding it into the image-reconstruction part for motion estimation.

REFERENCES

1. N. Ahmed, T. Natarajan, and K.R. Rao, *Discrete Cosine Transform*, IEEE Trans. on Computers, Vol.23, pp.90-94, (1974).
2. K.R. Rao, *Digital image/video/audio; international standards*, Material for Short Course at SPIE-94 (1994).
3. S.W. Wu and A. Gersho, *Improved decoder for transform coding with application to the JPEG baseline system*, IEEE Trans. on Comm., Vol.40, No.2, pp.251-254, (1992).
4. B. Reeve, and J. Lim, *Nonlinear space-variant postprossing of block coded images*, Opt. Engin., Vol.34, No.5, pp.1258-1268, (1986).
5. B Ramamurthi, and A. Gersho, *Nonlinear space-variant postprossing of block coded images*, IEEE Trans. on ASSP, Vol.34, No.5, pp.1258-1268, (1986).
6. J.D. Mcdonnell, R.N. Shorten, and A.D. Fagan, *An edge classification based approach to the post-processing of transform encoded images*, Proc. of ICASSP-94, pp.329-332, (1994).
7. R.L. Stevenson, *Reduction of coding artifacts in transform image coding*, Proc. of ICASSP-93, pp.401-404, (1993).
8. J.C. Brailean, T. Ozcelik, and A.K. Katsaggelos, *Restoration of low bit rate compressed images using mean field annealing*, Proc. of ICASSP-94, pp.237-240, (1994).
9. J. Luo, C.W. Chen, K.J. Parker, and T.S. Huang, *A new method for block effect removal in low bit-rate image compression*, Proc. of ICASSP-94, pp.341-344, (1994).
10. A. Zakhor, *Iterative procedures for reduction of blocking effects in transform image coding*, IEEE Trans. on CAS for Video Tech, Vol.2, No.1, pp.91-95, (1992).
11. Y. Yang, N.P. Galatsanos, and A.K. Katsaggelos, *Regularized reconstruction to reduce blocking artifacts of block discrete cosine transform compressed images*, IEEE Trans. on CAS for Video Tech., Vol.3, No.6 pp.421-432, (1993).
12. Y. Yang, N.P. Galatsanos, and A.K. Katsaggelos, *Projection-based spatially-adaptive reconstruction of block-transform compressed images*, SPIE Proc. of VCIP-94, Vol.2308, pp.1477-1488, (1994).
13. K.Y. Kwak and R. A. Haddad, *Projection based eigen vector decomposition for reducing blocking artifacts in DCT coded images*, will be presented in Proc. of ICIP-95 of IEEE, (1995).
14. D.C. Youla, and H. Webb, *Image Restoration by the method of convex projections: part 1-theory*, IEEE Trans. on Medical Imaging, Vol.1, No.2, pp.81-94, (1982).
15. P.L. Combettes and H. Puh, *A fast parallel projection algorithm for set theoretic image recovery*, Proc. of ICASSP-94, pp. 473-476, (1994).
16. K.Y. Kwak and R. A. Haddad, *A new family of orthonormal transforms with time localizability based on DFT*, SPIE Proc. of VCIP-94, Vol.2308, pp. 1158-1169, (1994).
17. K.Y. Kwak, *Time localizable linear transforms and enhancement of DCT coded images*, Ph.D. Dissert. in Polytechnic University, (1995),

INDEX